# MECHANISMS AND CONSEQUENCES OF PROTON TRANSPORT

# MECHANISMS AND CONSEQUENCES OF PROTON TRANSPORT

*edited by*

**Tetsuro Urushidani**
*Graduate School of Pharmaceutical Science*
*The University of Tokyo*

**John G. Forte**
*Dept. of Molecular & Cell Biology*
*University of California at Berkeley*

**George Sachs**
*Center for Ulcer Research & Education*
*University of California, Los Angeles*

**KLUWER ACADEMIC PUBLISHERS**
**Boston / Dordrecht / London**

**Distributors for North, Central and South America:**
Kluwer Academic Publishers
101 Philip Drive
Assinippi Park
Norwell, Massachusetts 02061 USA
Telephone (781) 871-6600
Fax (781) 681-9045
E-Mail <kluwer@wkap.com>

**Distributors for all other countries:**
Kluwer Academic Publishers Group
Post Office Box 322
3300 AH Dordrecht, THE NETHERLANDS
Telephone 31 786 576 000
Fax 31 786 576 474
E-Mail <services@wkap.nl>

Electronic Services <http://www.wkap.nl>

**Library of Congress Cataloging-in-Publication Data**

Mechanisms and consequences of proton transport / edited by Tetsuro Urushidani, John G. Forte, George Sachs.
p. cm.
Includes bibliographical references and index.
ISBN 1-40207-059-4 (alk. paper)
1. Gastric mucosa. 2. Hydrogen ions—Physiological transport. 3. Proton pump inhibitors. 4. Gastric acid. I. Urushidani, Tetsuro. II. Forte, John G. III. Sachs, George, 1935-

QP151 .M415 2002
612.3'2—dc21 2002018650



*Printed on acid-free paper.*
Printed in the United States of America

**The committee acknowledges the following foundation or companies as major sponsors of "The 9th international Proton Transport Conference, Mechanisms And Consequences of Proton Transport" 2001 in Australia.**

American Digestive Health Foundation

National Institute of Diabetes & Digestive & Kidney Diseases

Astra Zeneca

GlaxoSmithKline

Takeda Chemical Industries Ltd.

TAP Pharmaceutical Products Inc.

Wyeth-Ayerst Pharmaceuticals

Yamanouchi Pharmaceutical Co. Ltd.

# Preface

This volume represents the proceedings of the 9th International Proton Transport Conference, "Mechanisms and Consequences of Proton Transport" held in Leura, Australia, August 19 - 21, 2001. This conference has been held since 1971 every 3 to 4 years with few exceptions in association with the Congress of the International Union of Physiological Sciences. The first meeting was held in Frankfurt, Germany, at the Max-Planck Institute. Countries hosting the conference have been Germany (twice), Sweden (twice), India, Canada, USA, Great Britain, and now Australia. Over the past 30 years participants at these Proton Transport Conferences have been principal contributors to the major discoveries in the physiology, biochemistry and pharmacology of gastric acid secretion. These include development of the H2-receptor blockers, defining the signaling pathways for the regulation of acid secretion, identifying the gastric proton pump, discovery and development of proton pump inhibitory drugs, and elucidating the physiology and biochemistry of *Helicobacter pylori.*

It is repeatedly mentioned that the year, 2001, is the beginning of 21st century as well as the century of the genome. This does not mean that the role of physiology has arrived at its ultimate end. Rather this is the clarion for the new era of physiology in which the focus is on the functional activity and integration of the gene products. Molecular biology and genetics are both important and convenient tools for any biologist, physiologist or physician, but they do not bring us directly to the answer we really seek. We all know when the mutation in the cystic fibrosis transmembrane regulator gene was discovered. We know when the total genome of *H. pylori* was sequenced. These data only tell us how tough our enemies are. Have we already conquered cystic fibrosis and peptic ulcer? The only way to reach the goal, and the truth, is to clarify the functions of our enemy as well as to refine our weapons.

It now appears that the clinical field of anti-secretory drugs has matured. Proton pump inhibitors promise almost perfect suppression of gastric acid secretion. We have begun to understand the environmental adaptations of that nasty bug that lives on the gastric mucosal surface. Is it useless to continue research on gastric acid secretion? Of course not. All the participants in the Conference have shared a common recognition. This field is still filled with many fascinating riddles as well as a multitude of potential targets for drug development. We now know that the 21st century is the century of physiology.

We thank all the chairpersons, speakers, and delegates. Our special thanks to the sponsors (listed separately) for their generous financial support of the Conference.

Tetsuro Urushidani

John G. Forte

George Sachs

# Contents

## HELICOBACTER PYLORI AND MUCOSAL PROTECTION

## GENE TARGETING MOUSE — MECHANISMS OF ACID SECRETION AND CELL PROLIFERATION



## SIGNAL TRANSDUCTION IN THE PARIETAL CELL — PATHWAY TO ACTIVATION

## CYTOSKELETON AND MEMBRANE TRAFFICKING — THE GASTRIC PARIETAL CELL AS A MODEL

Chapter 1

# The Gastric H,K-ATPase

OLGA VAGIN, KEITH MUNSON, JAI MOO SHIN, NILS LAMBRECHT,

STEVE KARLISH* AND GEORGE SACHS

*UCLA, Los Angeles and *Weizmann Institute, Rehovot, Israel*

This brief review addresses the mechanism of acid secretion and the structure-function relationship of the gastric acid pump, the H,K-ATPase. The growth in knowledge of this enzyme in the last few years, as a result of delineation of the cDNA sequence, determination of the effect of mutations on function, documentation of the site of action of various inhibitors, and recognition of the similarity in 3D structure to the SERCA Ca ATPase has been remarkable.

## Mechanism of Acid Secretion

### (a) The discovery of the acid pump

The discovery of the Na,K-ATPase and definition of its catalytic mechanism in the late 1950's and 1960's (1,2) sparked a conceptual revolution in our understanding of the mechanism of acid secretion. The redox pump theory in which an electrogenic separation of protons and electrons was effected via a redox pump (3), the protons being extruded in parallel with a $Cl^-$ conductance had held sway for several years. This redox theory was undermined by evidence indicating that ATP, similar to the Na,K-ATPase, not redox reactions was the energy source for acid secretion, (4).

This was followed by a preliminary description of a K stimulated ATPase from hog gastric mucosa (5) but the importance of this observation was not recognized until a full length publication describing the presence of a K-stimulated ATPase activity in frog gastric mucosal microsomes appeared (6). A description of the medium alkalinization by this ATPase activity in vesicles (7) then provided the basis for a definition of the gastric acid pump as an electroneutral H for K exchange ATPase that also predicted stimulation of acid secretion would be accompanied by an increased KCl permeability of the ATPase-containing membrane (8). This hypothesis was confirmed by experiments comparing the properties of resting and stimulated rabbit gastric membranes (9).

It is now well accepted that acid secretion by the mammalian stomach is due to the activity of an ATP driven pump that catalyzes electroneutral cytoplasmic H for luminal K exchange, and that stimulation of acid secretion is accompanied by increased access of K to the luminal surface of the pump. Some disagreement still exists as to the precise stoichiometry of the pump. Whereas all agree that, based on thermodynamics considerations, the stoichiometry must be 1H per ATP molecule hydrolyzed at a luminal pH of 1.0, the stoichiometry appears to be 2H per ATP hydrolyzed at a pH of 7.4 (10, 11). The lower value (12) may have been due to unsuspected artifacts such as leakiness or purity of vesicles. As more detailed structure of the enzyme's ion transport sites becomes available, this controversy should become resolved. If the enzyme is shown to have two binding sites which contain carboxylates, protonation of one of the sites at low, but not at neutral pH could shift the stoichiometry from 2H per ATP molecule hydrolyzed to 1H per ATP molecule hydrolyzed.

**(b) Translocation of the H,K-ATPase**

Early electron microscopy suggested stimulation of acid secretion was accompanied by a morphological change in the parietal cell whereby membrane vesicles changed locale from the cytoplasm to microvilli of the secretory canaliculus (13, 14). Disappearance of membrane vesicles from the cytoplasm that is observed by immuno-electron microscopy in association with acid secretion (15) was also consistent with the pump moving into microvilli. Fluorescence microscopy then demonstrated acid secretion takes place into the microvillus-lined secretory canaliculus (16). The association of a cytoskeletal component, ezrin, with stimulated but not with resting membranes (17) and the presence of other SNARE proteins such as Rab 11 (18) in the parietal cell was taken as further support of the vesicle fusion hypothesis.

Freeze fracture images of rapidly frozen fixed tissue indicated the membrane vesicles were in fact tubules (18,19). These data imply the microvilli result from tubule fusion followed by tubule fusion-eversion, rather than vesicle fusion to form the microvillus. The ratio of F to G actin remains constant as the parietal cell changes from the resting morphology to the stimulated state (20). This suggests eversion and extension of the microvillus may occur by removal of monomers from one end of F actin followed by relocation of the G-actin monomers to the other end. Further, maceration of such specimens now reveals the presence of a network of tubules (21). Stimulation of secretion presumably results in eversion of this network to form the microvillar network of the stimulated secretory canaliculus without the need for individual tubule fusion events (22, 23). Much of this is discussed elsewhere in this volume.

## The Gastric H,K-ATPase: Kinetics, Structure/Function

The gastric H,K ATPase is a member of the $P_2$ family of transport ATPases. It undergoes a cycle of phosphorylation and dephosphorylation coupled to the exchange of protons (or more likely hydronium) for potassium ions. It is a heterodimer, consisting of an alpha subunit, 1034 amino acids in length, which contains the catalytic and transport sites, and a beta subunit, 291 amino acids in size which is glycosylated at 6 or 7 sites. The alpha subunit has 75% homology with the alpha subunits of the Na,K-ATPase and the beta subunit has 35% homology with the beta-2 subunit of the Na,K-ATPase at the amino acid level. Additional isoforms of the gastric H,K-ATPase include the colonic H,K-ATPase and ATP1AL1 (24, 25) which share 65% homology with the gastric H,K-ATPase and Na,K-ATPase, respectively. The beta subunits that accompany these particular isoforms remain unknown.

**(a) Kinetics of the H,K-ATPase**

The Post-Albers scheme represents the catalytic cycle of the Na,K-ATPase (26) and applies in general to the reaction cycle of the H,K-ATPase. The H,K-ATPase can transport Na at a pH of 8.5 (27) suggesting similarity to the Na,K-ATPase where ion binding occurs within a pocket lined with ligands from the protein. This would appear to rule out a specialized mechanism for the H,K-ATPase where protons would be shuttled between side chains comprising a "proton wire", and the transported ion is probably hydronium rather than proton. Binding of MgATP and $H_3O^+$ with the ion binding site facing the cytoplasm yields phosphorylation at asp385 to form the Mg.$E_1$-P.$(H_3O^+)_2$ complex.

This form of the enzyme converts spontaneously to the 'ion-binding-site-out' configuration, the Mg.$E_2$-P.$(H_3O^+)_2$ conformation. This releases cation to the outside face and then $2K^+$ are bound from the outside surface forming the Mg.$E_2$-P.$(K^+)_2$ conformer. This is followed by dephosphorylation and transport of $K^+$ inward to form the $E_1$.$(K^+)_2$ form which releases $K^+$ prior to initiating Mg ATP binding. Mg remains bound to the enzyme until the last step of the cycle. These individual steps of the cycle have been established by various rapid kinetic studies (28, 29, 30). The cycle is shown in figure 1.

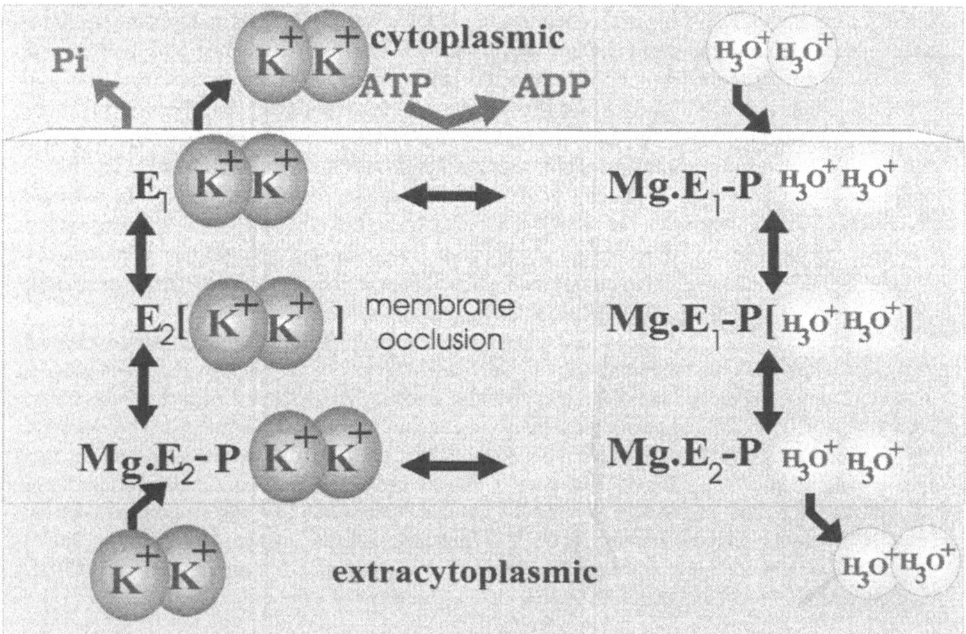


***Figure 1**: The reaction steps of the gastric H,K-ATPase. The three major conformations, $E_1$,$E_0$ (occluded)* and *$E_2$ are shown, and we show that two ions transported per cycle at neutral pH.*

However, there are significant differences in the transport characteristics of the two enzymes. The Na,K-ATPase is electrogenic, since 3Na are extruded in exchange for 2K, but the H,K-ATPase is electroneutral, although each half-reaction may be electrogenic (31, 32).

One explanation for the difference in stoichiometry might be the presence of the lysine at position 791 of the H,K-ATPase instead of a serine in the corresponding position of the Na,K-ATPase. The presence of a lysine in the protonated state adjacent to the ion transport site may substitute for the 3rd Na transported by the Na,K-ATPase (33).

In contrast to the Na,K-ATPase, the H,K-ATPase catalyzes a rapid Rb:Rb exchange (34). Hence, whereas Rb occlusion is easy to demonstrate for the sodium pump, low temperature is required to observe Rb occlusion in the H,K- ATPase (35, 36). This observation implies a more equal total energy for the non- phosphorylated as compared to the Na,K-ATPase-phosphorylated $E_1$ and $E_2$ forms of the gastric enzyme.

**(b)Secondary Structure of the H,K-ATPase**

Various methods have been used and indeed are necessary to demonstrate that the alpha subunit of the H,K-ATPase has ten membrane spanning segments and the beta subunit has a single transmembrane segment. These methods depend on use of the amino acid sequences deduced from the cDNA for each subunit (37, 38). For example, tryptic digestion yielded 4 pairs of membrane segments as defined by N-terminal amino acid sequencing of the fragments separated on tricine-buffered acrylamide gels and visualized either with Coomassie blue staining or fluorescence after labeling with fluorescein maleimide (39, 40). Peculiarly, neither method provided evidence for the C-terminal TM9/TM10 pair of transmembrane segments, but in vitro transcription/translation demonstated their membrane insertion (41). The lack of cysteine labeling in M9 by fluorescein maleimide might indicate some post-translational modification of this region of the catalytic subunit. Identification of inhibitor binding sites on the exoplasmic surface of the pump by peptide isolation and sequencing also identified M1/M2 (SCH28080) M3/M4 (lansoprazole) M5/M6 (omeprazole) and M7/M8 (omeprazole) (42, 43).

The beta subunit was discovered when a peptide generated by tryptic digestion yielded a fragment with high sequence homology to the beta-2 subunit of the Na,K-ATPase (44, 45, 46). Affinity chromatography and yeast-two hybrid analysis have shown that the beta subunit interacts strongly with the amino acid region of the alpha subunit just prior to the entry into TM8 near the exoplasmic surface (47, 48). There are 3 disulfide linkages and 6 or 7 (depending on species) glycosylation sites in this subunit. Disruption of the disulfide linkages or removal of all the glycosylation sites inactivates the enzyme (49, 50) and removal of four sites reduces the enzyme activity (50). The presence of five of the glycosylation sites is necessary for the plasma membrane delivery of the alpha-beta complex in COS cells (50). A two dimensional representation of the H,K-ATPase is shown in figure 2.

## The 3-dimensional structure of the gastric H,K-ATPase.

Insight into the transport mechanism of $P_2$ type ATPases, including the gastric H,K-ATPase, has depended heavily on the high resolution structure of the $E_1 2Ca$ form of the SERCA ATPase (51) derived from X-ray crystallography. This landmark observation was preceded by 2 dimensional crystals in the $E_2$ form whose resolution at 8A confirmed the presence of 10 transmembrane segments in the catalytic subunit. At this resolution, it was difficult to identify the nature of each segment and only the general shape of the cytoplasmic domain could be discerned.

Conservation in the structure of the cytoplasmic domain in $E_1$ was supported by 2 dimensional crystallography of the Neurospora H-ATPase in the $E_1$ configuration where the general shape of the cytoplasmic domain conformed to that of the high resolution SERCA pump. Together the $E_1$ forms showed significant differences when compared to the 8 A SERCA map (52, 53). These results demonstrate large changes in the cytoplasmic domain are associated with the $E_1/E_2$ conformational transition.

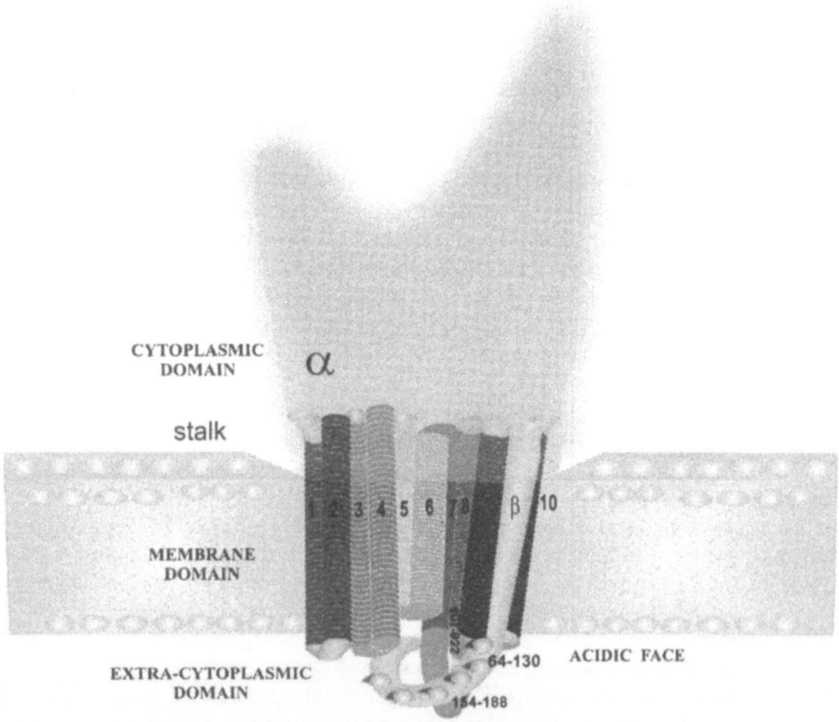


***Figure 2:*** *A general arrangement of the transmembrane segments with the regions of interaction of the alpha and beta subunits numbered. Mutational analysis has shown that carboxylic acids in TM4, 5, 6, and perhaps also TM8, are involved in ion binding in the membrane domain. However, other amino acids are also involved in the ion binding site.*

Although the SERCA and H,K- or Na,K-ATPases are only 20% homologous, there is conservation of sequence in the ATP-binding domains and the phosphorylation domains and general conservation of carboxylic acids in the membrane domain. The amino acid sequences for all or most $P_2$ type ATPases share a common motif with placement of transmembrane segments at opposite ends of functionally homologous cytoplasmic loops containing short stretches of highly conserved sequences. This allows molecular modeling of the H,K-ATPase structure based on the crystals of the SERCA Ca ATPase. Further, the 2 dimensional Ca ATPase crystal has provided a theoretical model for the $E_2$ configuration of the cytoplasmic domain of the SERCA and hence the H,K-ATPase.

**(a) Cytoplasmic Conformational Changes**

The SERCA ATPase has been crystallized in both the $E_1$ and $E_2$ forms but only the structure of the $E_1$ form is available in high resolution (51). However, a model of the cytoplasmic domain in the $E_2$ conformation (PDB ID 1FQU) has been generated from 2D crystals and knowledge of the X-ray structure (PDB ID 1EUL).

Three major domains were defined in the cytoplasmic region of the high resolution structure. The "A" domain contains the amino terminus to residue 95 and the TM2-TM3 loop, "P" contains amino and carboxy terminus regions of the large loop between TM4 and TM5 and the "N" consists of the residues between residues 359 to 605, that contains the nucleotide binding domain. These domains were moved as separate objects to fill the molecular volume and generate a theoretical model for the SERCA pump in the $E_2$ conformation. Alternative experimental methods have sought to define conformational changes induced in the presence of substrate or ligands.

A series of studies on the Na,K-ATPase utilized ascorbate and $H_2O_2$ induced cleavage of the polypeptide backbone catalyzed by Fe bound at the Mg site in the catalytic center (54, 55, 56). The resulting cleavage fragments could be identified using C terminal antibody staining and sequencing.

Cleavages are mediated either by bound Fe ions, or by the bound ATP-Fe complex. Fe binding at single site (defined as site 1) gives cleavage in the A and P domains when the enzyme is in the $E_2$ but not the $E_1$ conformation. With the ATP-Fe complex there is also cleavage in the N domain. However, there is always Fe binding and cleavage at a second site (site 2) near the membrane-water interface, near the entrance to M3 and M1 in the N terminal domain that is therefore independent of conformation (57).

The pattern of fragments generated by similar experiments on the gastric H,K-ATPase was identical (58). Apparently, identical changes occur in the Na,K-ATPase and H,K-ATPases during the transition from $E_1$ to $E_2$. These can be predicted from knowledge of the structure of the SERCA ATPase in the $E_1$ and $E_2$ forms. Thus, a remarkable conservation of the energy transduction mechanism is present in the P-type ATPase pumps.

Figure 3 shows changes in the conformation of the cytoplasmic domain deduced from iron cleavage of the H,K- and Na,K-ATPases (58) and their sequence alignments with the Ca-ATPase. There are large changes in the spatial arrangement of A and P domains with greater separation in $E_1$ as compared to their juxtaposition in $E_2$. After binding of ATP to $E_1$ there must also be movement of the N domain to allow phosphorylation of the P domain from the gamma phosphate of the substrate to give $E_1$-P. This would account for the cleavage in the N domain in the presence of ATP. The currently available crystal structure ($E_1$.2Ca) shows that the N and P domains are too far apart for phosphorylation to occur.

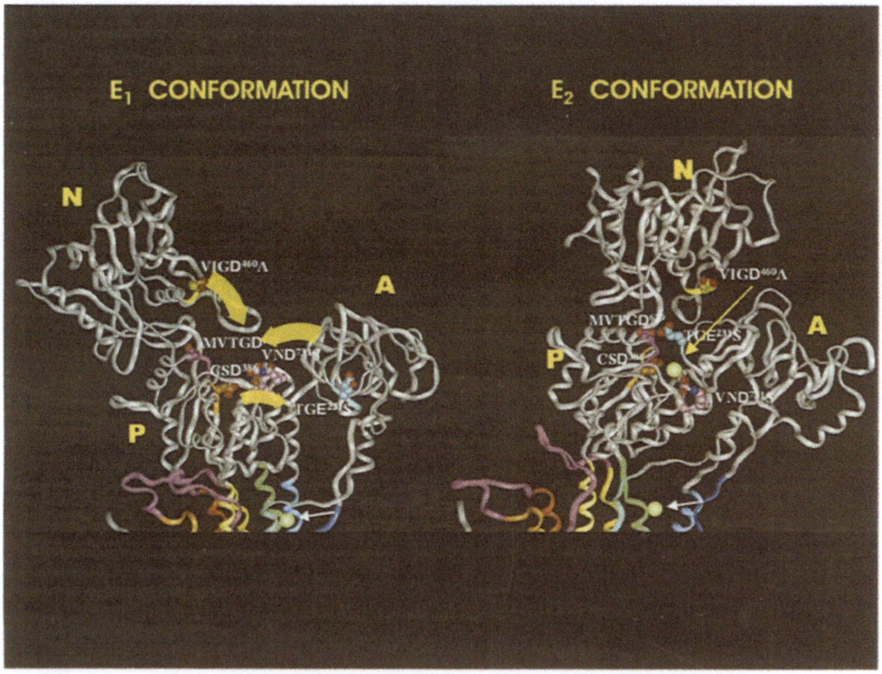


***Figure** 3: A model of the cytoplasmic domain of the gastric H,K-ATPase. The actual structure of the SERCA-ATPase in the $E_1$ and a model of the $E_2$ conformation are used to highlight the amino acids cleaved by exposure to Fe/$H_2O_2$/ascorbate. Illustrated is the separation between the P and A domains in the former and the close association in the latter with the iron-cleavage sites. Highlighted by the white arrow is the amino acid region cleaved in the $E_1$ position (in only the A domain) between M2 and M3, and by yellow arrows the movements that allow cleavage in the N, P and A domains in the $E_2$ position. To allow cleavage at this latter site, site 1, Fe substitutes for Mg at its binding site. A large conformational change occurs in the cytoplasmic domain accompanied by more subtle changes in the membrane domain. The shaded arrows show the direction of movement and the straight yellow arrow highlights the position of Fe in site 1 in the $E_2$ form. In this conformation there is close association between the P and the A domains and a somewhat closer association between the N and the P domains. The cleavage site in the N domain is still quite distant from the P domain but there is likely to be a transient closer association in the $E_1$.ATP form since the N domain shows cleavage in the presence of ATP but not in the presence of K. Site 2 is invariant between the two conformations and is highlighted by the white arrow.*

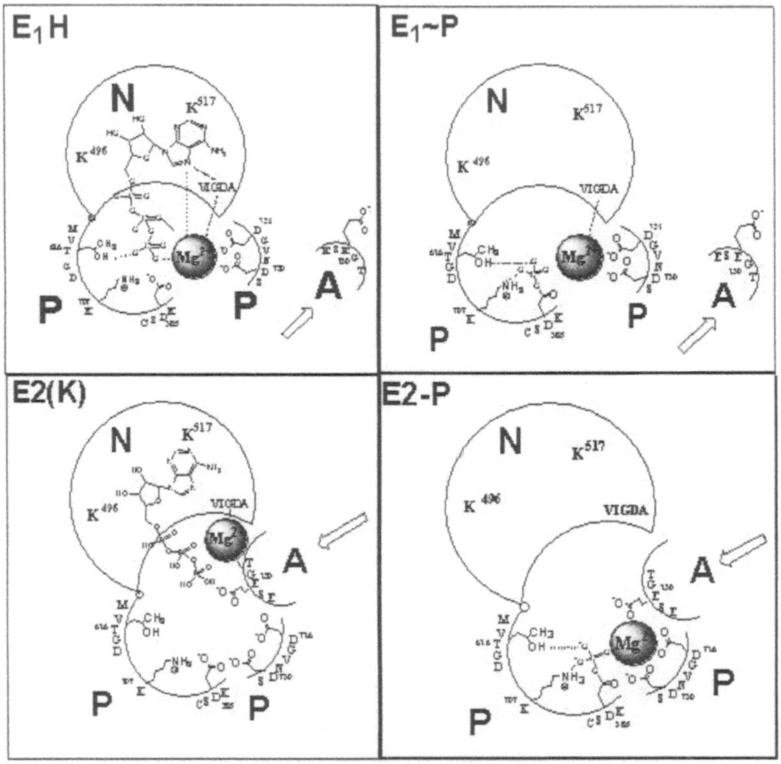


***Figure 4**: Cartoon representation of results obtained with Fe catalyzed cleavage of the H,K-ATPase as visualized in the computer graphic model of figure 3. The amino acids of the H,K-ATPase replace the amino acids in the SERCA-ATPase (ref 59).*

The positions of the Fe catalyzed cleavage can be determined for all four conformational states shown in figure 4, such as $E_1$H (upper left) and $E_2$.K (bottom left) $E_1$-P (top left) and $E_2$P (bottom left) where Fe substitutes for Mg (56, 59). Inspection of figure 3 shows that in the $E_1$ state the N or nucleotide-binding domain is too distant from the P domain to allow phosphorylation. However, when H is bound, the model suggested by the top left diagram of figure 4 indicates an approach of the N and the P domains to allow phosphorylation as diagramed in the top right rendering with ATP bound. In the $E_2$ form (bottom two images) there is close approach of the A and P domains. The binding regions are visualized as including carboxylic acids at the sites of cleavage. These are at ESE in the A domain and at or near MVTGD and VNDS in the P domain. Site 2 near M3 results in cleavage at or near HFVD and is not shown in this representation. In the presence of ATP cleavage occurs to a minor extent at a site near VIGDA in the N domain, allowing the conclusion that there is transient association of the N and P domains between $E_1$.H and $E_1$-P states. There are changes in the relative efficiency of cleavage at site 1 as a function of ligands such as SCH28080. Remarkably in both the $E_1$ and $E_2$ states, the cleavage sites are exactly those predicted from analysis of the SERCA-ATPase structures and are identical to those for the Na,K-ATPase (54, 55).

**(b) Ion binding Domain**

Site-directed mutagenesis studies with both mammalian and baculovirus expression systems (60, 61, 62) have been used to identify the ion binding residues of the H,K-ATPase. The amino acids tested in these studies were suggested by models of both the Ca -ATPase and Na,K-ATPase (51, 63, and PDB ID 1EUL). The orientation of the charged or potentially charged-amino acid side chains in the membrane domain is shown in figure 5.

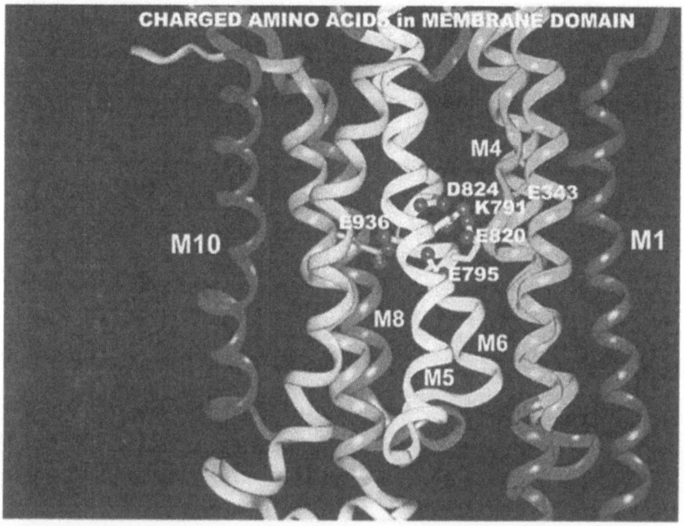


***Figure 5**: The charged amino acids in the E1 configuration of the membrane domain of the gastric H,K-ATPase. Amino acids from the H,K-ATPase sequence have been substituted for the amino acids of the SERCA ATPase. The relevant membrane segments are overlaid. In this view, the loss of helical structure in M4 is evident.*

Figure 6 illustrates the results obtained from our laboratory using HEK 293 cells stably transfected with cDNA of the gastric alpha and beta subunits. Basal wild-type ATPase activity, 10 μmol/mg/hr, is approximately 10% of that noted with purified hog gastric enzyme. Quantitation of the amount of H,K-ATPase alpha subunit in the membrane preparation, however, showed comparable specific activity for the expressed enzyme demonstrating no apparent requirement for proteins exclusive to the parietal cell. Four amino acids appear to be involved in cation binding: glu343, glu795, glu820, and asp824. The mutation of glu936 to aspartate doesn't appear to alter $K_m$app for $NH_4^+$, but changes inhibition by SCH28080 from strictly competitive to non-competitive suggesting the effect of this side chain on the ion access.

The effect of mutation of lys791 on ATPase activity requires special comment. Although mutation of this amino acid results in a large change of $K_m[NH_4^+]$,app/$V_{max}$, this may be due to the removal of a positive charge in the ion-binding domain. This positive charge may compensate for an ion-binding site utilized for Na in the Na,K-ATPase and be critical for the integrity of the ion-binding domain in the H,K-ATPase. The activity of mutants in this region of the H,K-ATPase is low and there is a large rise in the $K_i$ for SCH28080. Hence, mutation of lys791 distorts the structure of the enzyme both at the ion and inhibitor binding sites.

From these mutagenesis studies, it is apparent that M4, 5 and 6 at least, are involved in the movement of ions across the membrane domain of the pump.

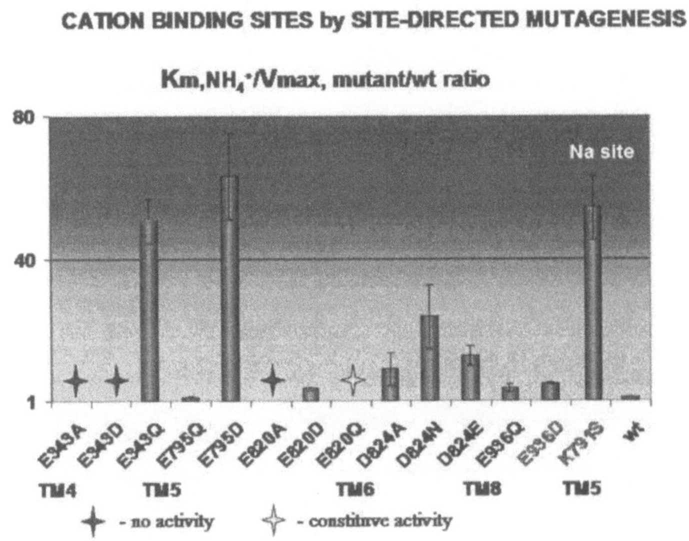


***Figure 6**: The effect of mutation of the carboxylic acids and the lysine predicted to be in the membrane domain. The side chains at positions 343, 795, 820 and 824 play a role in forming the ion-binding sites as do comparable residues in both the Na,K- and Ca-ATPases. E820Q has constitutive activity. The mutation E936D results in non-competitive inhibition. The lysine at position 791 may supply a positive charge in the region of the ion-binding domain of the H,K-ATPase. The dramatic effects of its mutation may therefore relate to distortion of the ion-binding domain rather than implication of the lysine in the ion binding site itself.*

**(c) Inhibitor Binding Sites**

Binding of the covalent class of H,K-ATPase inhibitors (PPI's),such as omeprazole, has been investigated by using labeled inhibitors (39). A common binding site for all of these is cys813, which as can be seen from figure 8, is close to the luminal surface of the putative transport region of the pump at the entry to M6. Binding to this region explains the inability of the pump to generate acid when these inhibitors are covalently bound. The form is likely to be the $E_2$ form of the pump requiring separation of M5/6 loop from M3/4 loop.

A different class of inhibitors is exemplified by SCH28080, a [1,2a] imidazo-pyridine (64). It is strictly K competitive and has stringent structural requirements for activity, in particular, the substituents on the 2 and 3 positions of the imidazole ring are highly restricted. Site directed mutagenesis allows comparison of changes in $K_i$ as compared to cation binding affinity. Some recent results obtained are shown in figure 7.

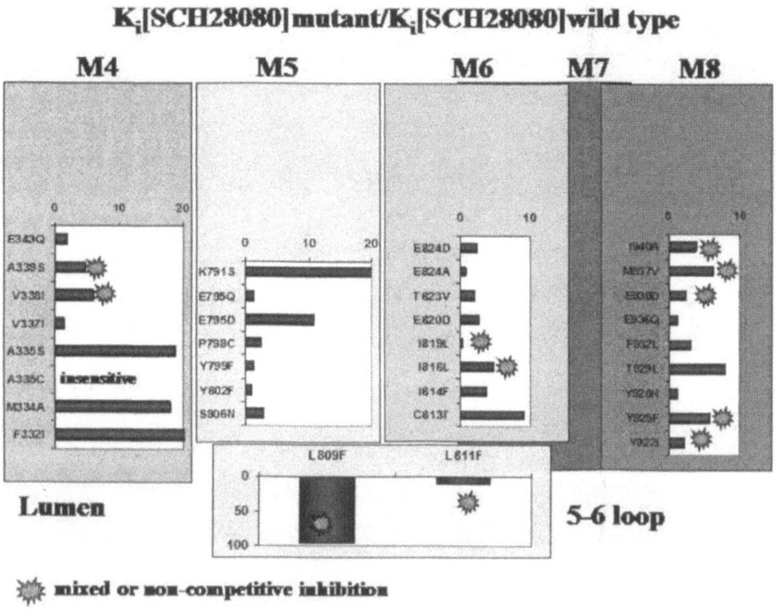


***Figure 7.*** *The effect of mutations of residues in TM4, 5, 6 and 8 on the* $K_i$ *for SCH28080. There are striking effects with mutations towards the outside region of TM4, the loop between TM5 and 6 and amino acids at the entry to TM6. Also illustrated is the finding that some mutations result in mixed or non-competitive inhibition implying an alternative pathway for K access to the cation binding domain.*

The advent of a high-resolution structure for the Ca-ATPase has improved our understanding of how the H,K-ATPase pump works. It is also particularly useful for examining the possible binding site of the imidazo-pyridines to the enzyme. Potential binding sites were first predicted by comparison of the amino acid sequence of the Na,K and H,K-ATPases, since SCH28080 does not inhibit the Na,K-ATPase. With knowledge of the crystal structure of Ca-ATPase and by analyzing the effect of the mutations listed in figure 7 (61, 65,

66) on inhibition of enzyme activity, a working model was developed as depicted in figure 8. This model now allows better predictions of which mutations will alter $K_i$, and the nature of the resulting inhibition which will occur.

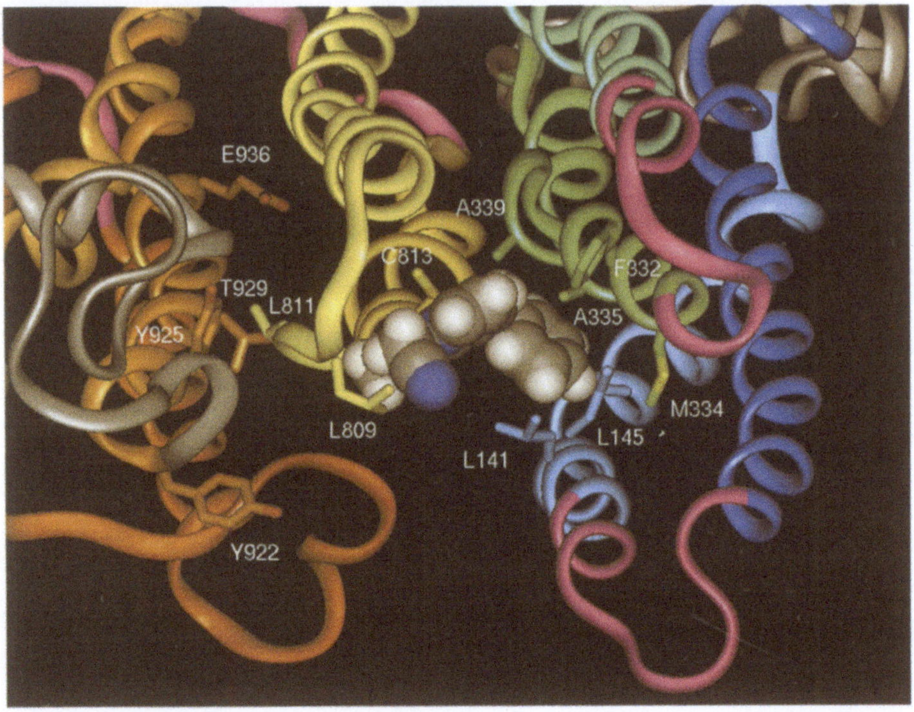


***Figure 8.*** *An illustration of the amino acids (stick form) probably involved in the binding of the K competitive inhibitor SCH28080 (illustrated in CPK) based on the structure of SERCA-ATPase.*

It is really not clear whether such high resolution 3D crystals as obtained from the SERCA-ATPase will be obtained in the near future for the heterodimeric $P_2$ ATPases. The problem is considered to be due to the glycosylation of the beta subunit in both the Na,K- and the H,K-ATPase. The former has 3 N linked sites the latter 6 or 7 depending on species. The exact location of the beta subunit may remain obscure. Nevertheless it is quite remarkable that in spite of sequence heterogeneity, the structural motif of these $P_2$ ATPases is well conserved. The most striking illustration is the consistency of results of Fe cleavage between the Na,K- and H,K-ATPases and their overlap with the model of the SERCA-ATPase. Within the membrane domain, wherein lies the responsibility for ion selectivity and the presence or absence of counter-transport, there may be significant variations that will not be obvious until high resolution images are produced. Yet, with analysis of pump structure-affinity relationships with inhibition by SCH28080, some considerable insight into this limited region of the membrane domain may well be obtained.

Supported in part by USVA and NIHDDK grant #'s DK46917, 53462, 41301 and 17294

Correspondence to gsachs@ucla.edu
fax 310 312 9478

Address
Building 113, Room 324
Wadsworth VA Hospital
Los Angeles Ca 90073

## REFERENCES

1. **Skou, J. C.** (1973) The relationship of the ($Na^+ + K^+$)-activated enzyme system to transport of sodium and potassium across the cell membrane. *J. Bioenerg.*, **4**, 1-30.
2. **Bader, H., Sen, A.K., and Post, R. L.** (1966) Isolation and characterization of a phosphorylated intermediate in the ($Na^+ + K^+$) system-dependent ATPase. *Biochim Biophys Acta*, **118**, 106-115.
3. **Davies R.E.** (1951) The mechanism of hydrochloric acid production by the stomach. *Biol. Rev.*, **26**, 87-120
4. **Sachs, G., Rabon, E., Saccomani, G. and Sarau, H. M.** (1975). Redox and ATP in acid secretion. Ann N Y Adad. Sci 264:456-475.
5. **Sachs, G., Pacifico, A. Collier R.H. and Zweig, R.A** (1968). A possible mechanism for acid secretion by gastric mucosa. *Symposium on Biophysical Aspects of Permeability IUPAB* pp 26-27, 1968
6. **Ganser, A.L. and Forte, J. G.** (1973) $K^+$-stimulated ATPase in purified microsomes of bullfrog oxyntic cells. *Biochim Biophys Acta*, **307**, 169-180.
7. **Lee, J., Simpson, G., and Scholes, P.** (1974) An ATPase from dog gastric mucosa: changes of outer pH in suspensions of membrane vesicles accompanying ATP hydrolysis. *Biochem Biophys Res Commun.*, **60**(2), 825-832.
8. **Sachs, G., H. H. Chang, E. Rabon, R. Schackman, M. Lewin, and G. Saccomani.** (1976) A nonelectrogenic $H^+$ pump in plasma membranes of hog stomach. *J. Biol. Chem.*, **251**, 7690-7698.
9. **Koelz, H. R., G. Sachs, and Berglindh, T.** (1981). Cation effects on acid secretion in rabbit gastric glands. *Am. J. Physiol.*, **241**(5), G431-442.
10. **Reenstra, W. W. and Forte, J. G.** (1981) $H^+$/ATP stoichiometry for the gastric ($K^+ + H^+$)-ATPase, *J. Membr. Biol.*, **61**, 55-60.

11. **Rabon, E. C., McFall, T. L., and Sachs, G.** (1982). The Gastric [H,K]ATPase:$H^+$/ATP Stoichiometry. *J. Biol. Chem.*, **257**, 6296-6299.
12. **Skrabanja, A. T., J. J. De Pont, et al.** (1984). The $H^+$/ATP transport ratio of the ($K^+$ + $H^+$)-ATPase of pig gastric membrane vesicles. *Biochim. Biophys. Acta*, **774**, 91-95.
13. **Forte, T. M. and Forte, J. G.** (1971) A freeze-fracture study of bullfrog gastric oxyntic cells. *J. Ultrastruct. Res.*, **37**, 322-334.
14. **Helander, H. F. and Hirschowitz, B. I.** (1974) Quantitative ultrastructural studies on inhibited and on partly stimulated gastric parietal cells. *Gastroenterology*, **67**, 447-452.
15. **Smolka, A., H. F. Helander, and Sachs, G.** (1983). Monoclonal antibodies against gastric $H^+$ + $K^+$ ATPase. *Am. J. Physiol.*, **245**(4), G589-596.
16. **Mangeat, P., T. Gusdinar, Sahuquet, A., Hanzel, D. K., Forte, J. G., and Magous, R.** (1990). Acid secretion and membrane reorganization in single gastric parietal cell in primary culture. *Biol. Cell*, **69**(3), 223-231.
17, **Hanzel, D., Reggio, H., Bretscher, A., Forte, J. G., and Mangeat, P** (1991). The secretion-stimulated 80K phosphoprotein of parietal cells is ezrin, and has properties of a membrane cytoskeletal linker in the induced apical microvilli. *Embo J.*, **10**(9), 2363-2373.
18. **Jons, T., Lehnardt, S., Bigalke, H., Heim, H. K., and Ahnert-Hilger, G.** (1999). SNARE proteins and rab3A contribute to canalicular formation in parietal cells. *Eur. J. Cell Biol.*, **78**(11), 779-786.
19. **Peranzi, G., D. Bayle, Lewin, M. J., and Soumarmon, A.** (1991). The intramembranous particles of resting and secreting gastric ($H^+$,$K^+$)- ATPase membranes. *Biol. Cell*, **73**(2-3): 163-171.
20. **Wolosin, J. M., Okamoto, C., Forte, T. M., and Forte, J. G.** (1983). Actin and associated proteins in gastric epithelial cells. *Biochim. Biophys. Acta*, **761**(2): 171-182.
21. **Namikawa, T., K. Araki, et al.** (1998). Localization of cytoskeletal filaments during membrane rearrangement in rat parietal cells stimulated with gastrin. *Arch. Histol. Cytol.*, **61**(1): 47-56.
22. **Ogata, T. and Y. Yamasaki** (2000). Morphological studies on the translocation of tubulovesicular system toward the intracellular canaliculus during stimulation of the gastric parietal cell. *Microsc. Res. Tech.*, **48**(5): 282-292.
23. **Ogata, T. and Y. Yamasaki** (2000). Scanning EM of resting gastric parietal cells reveals a network of cytoplasmic tubules and cisternae connected to the intracellular canaliculus. *Anat. Rec.*, **258**(1): 15-24.
24. **Crowson, M. S. and G. E. Shull** (1992). Isolation and characterization of a cDNA encoding the putative distal colon $H^+$,$K^+$-ATPase. Similarity of deduced amino acid sequence to gastric $H^+$,$K^+$-ATPase and $Na^+$,$K^+$-ATPase and mRNA expression in distal colon, kidney, and uterus. *J. Biol. Chem.*, **267**(19): 13740-13748.
25. **Modyanov, N. N., P. M. Mathews, et al.** (1995). Human ATP1AL1 gene encodes a ouabain-sensitive H-K-ATPase. *Am J Physiol* **269**(4 Pt 1): C992-7.
26. **Post, R. L., C. Hegyvary, and Kume, S.** (1972). Activation by adenosine triphosphate in the phosphorylation kinetics of sodium and potassium ion transport adenosine triphosphatase. *J. Biol. Chem.*, **247**(20): 6530-6540.

27. **Polvani, C., Sachs, G., and Blostein, R.** (1989) Sodium ions as substitutes for protons in the gastric H,K-ATPase. J. Biol. Chem., **264**(30), 17854-17859.
28. **Stewart, B., Wallmark, B., and Sachs, G.** (1981). The interaction of $H^+$ and $K^+$ with the partial reactions of gastric H,K-ATPase. *J. Biol. Chem.*, **256,** 2682-2690.
29. **Wallmark, B., Stewart, H. B., Rabon, E., Saccomani, G. and Sachs, G.** (1980). The catalytic cycle of gastric ($H^+$+$K^+$)-ATPase. *J. Biol. Chem.*, **255**, 5313-5319.
30. **Brzezinski, P., Malmstrom, B. G., Lorentzon, P., and Wallmark, B.** (1988). The catalytic mechanism of gastric H/K-ATPase: simulations of pre-steady-state and steady-state kinetic results, *Biochim. Biophys. Acta,* **942**, 215-219.
31. **Thomas, R. C.** (1972) Electrogenic sodium pump in nerve and muscle cells. *Physiol. Rev.*, **52**, 563-594.
32. **Sachs, G., Chang, H. H., Rabon, E., Schackman, R., Lewin, M., and Saccomani, G.** (1976). A nonelectrogenic $H^+$ pump in plasma membranes of hog stomach. *J. Biol. Chem.*, **251**, 7690-7698.
33. **Burnay, M., and Horisberger J.-D.** (2001). A single amino acid mutation in the 5$^{th}$ transmembrane segment of the Bufo Marinus Bladder H,K-ATPase makes its activity electrogenic. *USGEB Young Investigator Meeting*, Lausanne (Dorigny), March 10, 2001
34. **Schackmann, R., Schwartz, A., Saccomani, G., and Sachs, G.** (1977). Cation transport by gastric $H^+$:$K^+$ ATPase. *J. Membr. Biol.,* **32**(3-4): 361-381.
35. **Rabon, E. C., Smillie, K., Seru, V., and Rabon, R.** (1993). Rubidium occlusion within tryptic peptides of the H,K-ATPase. *J. Biol. Chem.*, **268**, 8012-8018.
36. **Soumarmon, A., P. K. Rangachari, and Lewin, M. J.** (1984). Passive transport of $Rb^+$ by hog gastric ($H^+$,$K^+$)-ATPase. *J Biol Chem* **259**(19): 11861-11867.
37. **Bamberg, K., Mercier, F., Reuben, M. A., Kobayashi, Y., Munson, K. B. and Sachs, G.** (1992). cDNA cloning and membrane topology of the rabbit gastric H/K-ATPase α-subunit. *Biochim. Biophys. Acta*, **1131**, 69-77.
38. **Reuben, M. A., Lasater, L. S. and Sachs, G.** (1990). Characterization of a β subunit of the gastric H/K-ATPase. *Proc. Natl. Acad. Sci. U.S.A.*, **87**, 6767-6771.
39. **Besancon, M., Shin, J. M., Mercier, F., Munson, K., Miller, M., Hersey, S., and Sachs, G.** (1993). Membrane topology and omeprazole labeling of the gastric H,K-adenosinetriphosphatase. *Biochemistry*, **32**, 2345-2355.
40. **Shin, J. M., Kajimura, M., Arguello, J. M., Kaplan, J. H., and Sachs, G.** (1994) Biochemical identification of transmembrane segments of the Ca-ATPase of sarcoplasmic reticulum. *J. Biol. Chem.*, **269**, 22533-22537.
41. **Bamberg, K. and Sachs, G.** (1994). Topological analysis of the H,K-ATPase using in vitro translation. *J. Biol. Chem.*, **269**, 16909-16919.
42. **Shin, J.M., Besancon, M., Simon, A., and Sachs, G.** (1993). The site of action of pantoprazole in the gastric H/K-ATPase. *Biochim. Biophys. Acta*, **1148**(2), 223-233.
43. **Besancon, M., Simon, A., Sachs, G., and Shin, J. M.** (1997) Sites of Reaction of the Gastric H,K-ATPase with Extracytoplasmic Thiol Reagents. *J. Biol. Chem.*, **272**, 22438-22446.
44. **Shull, G. E., Lane, L. K., and Lingrel, J. B.** (1986) Amino-acid sequence of the alpha-subunit of the Na K ATPase deduced from a cDNA. *Nature.* **321**(6068):429-431.

45. **Hall, K., Perez, G., Anderson, D., Gutierrez, C., Munson, K., Hersey, S. J., Kaplan, J. H., and Sachs, G.** (1990). Location of the carbohydrates present in the H,K-ATPase vesicles isolated from hog gastric mucosa. *Biochemistry*, **29**, 701-706.
46. **Munson, K. B., Gutierrez, C., Balaji, V. N., Ramnarayan, K., and Sachs, G.** (1991). Identification of an extra-cytoplasmic region of H,K-ATPase labeled by a $K^+$-competitive photoaffinity inhibitor. *J. Biol. Chem.*, **266**, 18976-18988.
47. **Melle-Milovanovic, D., Milovanovic, M., Nagpal, S., Sachs, G., and Shin, J. M.** (1998) Regions of Association Between the Alpha and the Beta Subunit of the Gastric H,K-ATPase, *J. Biol. Chem.*, **273**, 11075-11081.
48. **Shin, J. M. and Sachs, G.** (1994) Identification of a region of the H,K-ATPase α-subunit associated with the β-subunit *J. Biol. Chem.*, **269**, 8642-8646.
49. **Chow, D. C., Browning, C. M., and Forte, J. G.** (1992). Gastric H,K-ATPase activity is inhibited by reduction of disulfide bonds in β-subunit. *Amer. J. Physiol.*, **263**, C39-C46.
50. **Asano S, Kawada K, Kimura T, Grishin AV, Caplan MJ, Takeguchi N.** (2000) The roles of carbohydrate chains of the beta-subunit on the functional expression of gastric H(+),K(+)-ATPase. *J. Biol. Chem.*, **275**(12), 8324-8330.
51. **Toyoshima, C., M. Nakasako, Nomura, H. and Ogawa, H.** (2000). Crystal structure of the calcium pump of sarcoplasmic reticulum at 2.6 A resolution. *Nature* **405**(6787): 647-655.
52. **Zhang, P., Toyoshima, C., Yonekura, K., Green, N. M., and Stokes, D.** (1998) Structure of the calcium pump from sarcoplasmic reticulum at 8-A resolution. *Nature*, **392**, 835-839.
53. **Auer, M., Scarborough, G. A., and Kuhlbrandt, W.** (1998) Three-dimensional map of the plasma membrane $H^+$-ATPase in the open conformation. *Nature*, **392**, 840-843.
54. **Goldshleger, R. and S. J. Karlish** (1999). The energy transduction mechanism of Na,K-ATPase studied with iron- catalyzed oxidative cleavage. *J. Biol. Chem.*, **274**(23): 16213-21.
55. **Goldshleger, R. and S. J. Karlish** (1997). Fe-catalyzed cleavage of the alpha subunit of Na/K-ATPase: evidence for conformation-sensitive interactions between cytoplasmic domains. *Proc. Natl. Acad. Sci., U S A.*, **94**(18): 9596-9601.
56. **Patchornik, G., Goldshleger, R., and Karlish, S. J.** (2000) The complex ATP-$Fe^{2+}$ serves as a specific affinity cleavage reagent in ATP-$Mg^{2+}$ sites of Na,K-ATPase: altered ligation of $Fe^{2+}$ ($Mg^{2+}$) ions accompanies the $E_1P$_$E_2P$ conformational change. *Proc Natl Acad Sci U S A.* **97**, 11954-9.
57. **Tal, D. M., Capasso, J. M., Munson, K., and Karlish, S. J.** (2001) Proximity of Transmembrane Segments M3 and M1 of the alpha Subunit of $Na^+$,$K^+$-ATPase Revealed by Specific Oxidative Cleavage Mediated by a Complex of Cu(2+) Ions and 4,7-Diphenyl-1,10-phenanthroline. *Biochemistry*. **40**, 12505-12514.
58. **Shin, J. M., R. Goldshleger, Munson, K. B., Sachs, G., and Karlish, S. J.** (2001). Selective $Fe^{2+}$-catalyzed oxidative cleavage of gastric $H^+$,$K^+$-ATPase. Implications for the energy transduction mechanism of P-type cation pumps. *J. Biol. Chem.*, **276**(51), 48440-48450.

59. **Patchornik G, Munson,K., Sachs ,G., and Karlish S.J.** (2001) Interactions between N, P and A cytoplasmic domains of Na,K-ATPase studied with ATP-$Fe^{2+}$-catalysed oxidative cleavage and molecular modeling. Evidence for one ATP site. (Manuscript to be submitted to J. Biol. Chem.)
60. **Asano, S., Tega, Y., Konishi, K., Fujioka, M., and Takeguchi, N.** (1996) Functional expression of gastric $H^+$,$K^+$-ATPase and site-directed mutagenesis of the putative cation binding site and catalytic center. *J. Biol. Chem.*, **271**(5), 2740-2745.
61. **Vagin, O., Munson, K., Lambrecht, N., Karlish, S. J., and Sachs, G**. (2001) Mutational analysis of the $K^+$-competitive inhibitor site of gastric H,K-ATPase. *Biochemistry*. 40(25), 7480-7490.
62. **Hermsen, H. P., Swarts, H. G., Wassink, L., Koenderink, J. B., Willems, P. H., and De Pont, J. J.** (2001) Mimicking of $K^+$ activation by double mutation of glutamate 795 and glutamate 820 of gastric $H^+$,$K^+$-ATPase. *Biochemistry*, **40**(21), 6527-6533.
63. **Sweadner, K. J. and C. Donnet** (2001). Structural similarities of Na,K-ATPase and SERCA, the Ca(2+)-ATPase of the sarcoplasmic reticulum. *Biochem. J.*, 356(Pt 3): 685-704.
64. **Wallmark, B., Briving, C., Fryklund, J., Munson, K., Jackson, R., Mendlein, J., Rabon, E., and Sachs, G.** (1987). Inhibition of gastric H,K-ATPase and acid secretion by SCH28080, a substituted pyridyl[1,2α]imidazole. *J. Biol. Chem.*, **262**, 2077-2084.
65. **Lambrecht, N., K. Munson, et al.** (2000). Comparison of covalent with reversible inhibitor binding sites of the gastric H,K-ATPase by site-directed mutagenesis. *J Biol Chem* **275**(6): 4041-8.
66. **Munson, K. B., N. Lambrecht, et al.** (2000). Effects of mutations in M4 of the gastric $H^+$,$K^+$-ATPase on inhibition kinetics of SCH28080. *Biochemistry*, **39**, 2997-3004

Chapter 2

# Human Nongastric H,K-ATPase: Current View On Structure And Functional Properties

GAIL ADAMS, MANORANJANI TILLEKERATNE,
NIKOLAY B. PESTOV, AND NIKOLAI N. MODYANOV
*Department of Pharmacology, Medical College of Ohio, Toledo, OH, 43614 USA*

## INTRODUCTION

The P-type ATPases comprise a large number of highly diverse transport ATPases that are predominantly involved in the active transport of cations across biological membranes. All of these ion pumps share a common feature: formation of a phosphorylated intermediate during the reaction cycle (26,39). Different $K^+$-dependent animal ATPases (X,K-ATPases) are the most closely related among various P-type ATPases. All of the known X,K-ATPases function as cation exchangers that pump $K^+$ into the cell and $Na^+$ or $H^+$ out of the cell. X,K-ATPases exhibit a much higher level of sequence homology between their catalytic α-subunits than with other P-ATPases, and contain a second component, a β-subunit which is absent in other P-ATPases (26,39,53). The catalytic α-subunits are large polytopic proteins (~110 kDa) with 10 transmembrane segments and contain most of the ATPase functional domains such as the ATP-hydrolyzing center and the binding sites for cations and specific inhibitors (8,26,53). The glycosylated β-subunits (core protein ~ 30-35 kDa) have a relatively short cytoplasmic N-terminal domain, a single transmembrane segment, and a large ectodomain containing three conserved disulfide bridges and several carbohydrate chains (11,19). The X,K-ATPase family combines three distinct groups of ion pumps. Two groups, one consisting of the Na,K-ATPase isozymes formed by four α three β isoforms and the second which includes the gastric H,K-ATPase, have long been known and studied extensively (26,53). The recently discovered catalytic α-subunits of nongastric H,K-ATPases encoded by the human ATP1AL1 (alternative name ATP12A) gene and its animal homologues represent the third distinct group (23,27,36). Evidence for the existence of ion-transporting P-ATPases, whose functional properties are closely related, but not identical, to that of the Na,K-ATPase or the gastric H,K-ATPase and which are presumed to be involved in maintenance of electrolyte homeostasis through $K^+$ absorption and proton secretion in kidney and colon being especially essential under pathophysiological conditions, was obtained in many physiological and pharmacological studies (for review, see Refs. 17,46,55). Up to now, many aspects of the structure and function of the nongastric H,K-ATPases remains unknown. The intrinsic subunit composition of the nongastric H,K-ATPases is not yet determined because none of these enzymes has been isolated and analyzed directly by amino acid sequencing, and no β-subunit specific for this ATPase has been described. The basic functional parameters of the human nongastric H,K-ATPase represented by the recombinant complex of the ATP1AL1-encoded protein and gastric H,K-ATPase β-subunit (Atp1al1-βHK) have been analyzed using heterologous expression in *Xenopus laevis* oocytes (20,37), HEK 293 cells (21,22), and Sf-21 insect cells (1). However, some of the results obtained in these experiments are rather controversial, especially in respect to the enzyme cation dependence. In this brief review, we will outline the current status of the studies on the structural and functional characterization of the human nongastric H,K-ATPase and discuss the above mentioned existing discrepancies.

## STRUCTURAL CHARACTERIZATION OF HUMAN NONGASTRIC H,K-ATPase AND ITS RELATIONSHIP WITH OTHER MEMBERS OF X,K-ATPase FAMILY

The study directed towards the molecular characterization of the new human X,K-ATPase that was latter identified as a nongastric H,K-ATPase, was initiated in 1987 when existence of the family of closely related genes in human genome, encoding the catalytic subunits of the X,K-ATPases was demonstrated simultaneously in Moscow and Cincinnati (45,49). Two of the six identified human genes did not correspond to any ion pumps known at that time, and were designated as α-like-genes: ATP1AL1 and ATP1AL2. [The ATP1AL2 has been recently shown to encode a testis-specific isoform of the Na,K-ATPase (56).] In further studies, we demonstrated that the ATP1AL1 is transcriptionally competent, and based on the partial sequence information suggested for the first time that this gene encodes a novel type of human X,K-ATPase rather than an isoform of the known ones (38). [Last year, the Nomenclature Committee of the Human Genome Organization gave this gene a new name "ATP12A", but we are using here the more recognizable old name of "ATP1AL1" that has been accepted and used for many years.]

Expression of the ATP1AL1 was detected at significant levels in human skin and kidney, and at a lower level in colon, brain, placenta, and lung (38,42,50). The cDNA for the ATP1AL1 was cloned from human kidney and skin libraries (24) and closely related cDNAs have been cloned from rat, rabbit and guinea pig distal colons (2,9,15) and mouse and dog kidneys (42). The encoded human Atp1al1 protein and related animal proteins possess all the structural features common to the catalytic subunits of the P-type ATPases. These include several conserved motifs such as a catalytic phosphorylation site (Asp 391), components of the ATP-binding site (Asp 732 and Asp 736), the target of FITC modification (Lys 523) and the disposition of 10 transmembrane domains characteristic of the X,K-ATPase catalytic subunits [see (24) for detailed description]. The human Atp1al1 protein and related animal proteins are structurally equally distant (63-64% amino acid identity) from both the Na,K-ATPase α-subunit isoforms and the gastric H,K-ATPase α -subunit.

It should be noted that sequence diversity between the human Atp1al1 and related animal proteins are significantly higher than divergences between other X,K-ATPase α-subunits from different species (2,9,15,24,26,27,50). For example, the corresponding human and rat Na,K-ATPase isoforms and gastric H,K-ATPases exhibit 96.7-99.1% sequence identity, whereas the human Atp1al1 and rodent sequences share only ~85% identical amino acids. Such a level of similarity is typical for distinct isoforms of the Na,K-ATPase (86-87% identity). Thus, from a structural point of view, the relationship between the nongastric α-subunits resembles the relationship of distinct X,K-ATPase isoforms rather than those of species variants. Therefore, it was suggested that the human Atp1al1 and related animal proteins are products of closely related, but distinct genes and that each of them represents the distinct isoform of the nongastric H,K-ATPase (15,24,27). However, our studies of the ATP1AL1 genomic structure revealed a prominent similarity of the human and rodent DNA sequences that includes not only protein coding regions (~84% identity) but also the 3' untranslated regions (~66-70% identity). A high level of similarity in untranslated regions clearly indicates that the genes for the third group of X,K-ATPases are indeed homologues (50). [Prominent similarity of the expression patterns of the nongastric H,K-ATPase genes in human, mouse, rat, rabbit and dog tissues provide additional support for this conclusion (42).] Therefore, the greater degree of structural diversity may reflect that sequence requirements for the Atp1al1 and its animal homologues were not as strict during evolution as in the case of the extremely well conserved H,K- and Na,K-ATPases (50). Interestingly, most of the dissimilarities between human and rat Atp1al1proteins are concentrated within the N-terminal domain encoded by exons 1 and 2 in the human gene. In this region, the amino acid sequences of human and rat proteins exhibit only 53% amino acid sequence identity. This may indicate that the functional role of the N-terminal domains in nongastric ATPases is not of such significance as in Na,K-ATPases (26,50).

An absolute requirement for a β-subunit to participate in the formation of the active ATPase is a unique feature of the X,K-ATPase family members. The β-subunits play a crucial role in the structural and functional maturation of the active X,K-ATPase molecule. In this process, the β-subunit functions as a specific chaperone whose tight association with the catalytic α-subunit helps the latter in correct membrane insertion and folding, and in translocation to the plasma membrane (8,11,14,19,20). In addition to this important structural role, β-subunits also influence enzymatic and transport functions of the X,K-ATPases through modulation of the enzyme affinities for cations (8,11,19,20). Five closely related genes have been identified as members of the X,K-ATPase β-subunit family in human genome. All of the encoded proteins: three Na,K-ATPase β isoforms (β1NK,β2NK,β3NK), the gastric H,K-ATPase β-subunit (βHK) and the muscle-specific βm protein (41); share a common domain structure, but exhibit a lower degree of sequence similarity (~32 to 48% in the case of human proteins) than the X,K-ATPase α-subunits (11).

Up to now, it remains unclear whether one of the known β-subunits (11) or a hitherto unidentified member of the X,K-ATPase β-subunit family is a real subunit of nongastric H,K-ATPases (22,36,37). The currently available experimental data are rather controversial. Thus, the first expression studies in *Xenopus* oocytes and HEK 293 cells revealed that the βHK is capable of assembling with the Atp1al1 and forming a functionally active ion pump more efficiently than other known X,K-ATPase β-subunits (20,21,37). However, the βHK is probably not the authentic counterpart of the Atp1al1 *in vivo,* because patterns of their tissue-specific expression are different. The presence of βHK mRNA has only been demonstrated in stomach and kidney (for review see 55), whereas RT-PCR revealed significant expression of the ATP1AL1 not only in kidney but also in skin, and at lower levels in colon, brain, placenta, and lung (24,38,42,50). Recent comparative analysis of the effects of different β subunits on the maturation of the ATP1al1expressed in *Xenopus* oocytes revealed that not only gastric βHK, but also β2 -like *Bufo* bladder β and to a lesser extent β2NK are able to protect Atp1al1 from cellular degradation and to produce protease-resistant and correctly folded α-β complexes. For all parameters measured, β1NK is found to be the least efficient partner for Atp1al1 (20). These findings led us to suggest that none of the known β subunits is the real counterpart of Atp1al1 but an as yet unidentified, authentic β should exist that has structural features resembling those of gastric βHK, β2NK or *Bufo* bladder β. One of these essential features might be that these β-subunits, which contain 7 or 8 carbohydrate chains, are more heavily glycosylated than β1NK which has only 3 carbohydrate chains (20).

In contrast to above, it was reported that the β1NK can form a functional ATPase complex with the α subunit of rat nongastric H,K-ATPase expressed in *Xenopus* oocytes (13). Furthermore, recent studies named either β1NK (12,31) or β3NK (44) as a β-subunit of rat nongastric H,K-ATPase because β1NK or β3NK antibodies, respectively, recognized the protein which co-immunoprecipitated with nongastric H,K-ATPase α-subunits from rat colon or kidney microsomes. The immunochemical data have to be confirmed by amino acid sequencing of the α-associated β-subunit. Otherwise, it cannot be excluded that the antibodies just cross-reacted with a hitherto unidentified β-like protein sharing common epitope(s) with β1 or β3 subunits.

It should be noted in conclusion, that the existence of a hitherto unidentified human gene that encodes the authentic β subunit of the nongastric H,K-ATPase cannot be entirely excluded. Also, the possibility of the existence of yet unidentified variants of X,K-ATPase β subunits originated due to alternative splicing, like those found in case of βm (41), or posttranslational modifications can not be ruled out. Thus, considering the collection of the rather controversial data outlined above we have to say that at this point, there is no definite answer to the question: "Which β-subunit is the authentic counterpart of Atp1al1?"

## FUNCTIONAL PROPERTIES OF RECOMBINANT Atp1al1-βHK COMPLEX

For the first time, the protein product of the human ATP1AL1 gene (Atp1al1) was identified as a catalytic α-subunit of the human nongastric H,K-ATPase through its functional expression in *Xenopus* oocytes in combination with the βHK (37). Ion-translocating activity of the Atp1al1-βHK complex in the plasma membrane of cRNAs injected oocytes was evaluated by a Rb uptake assay. It was shown that the Atp1al1-βHK complex is indeed a $K^+$-dependent ATPase (Km ~ 375 μM). The cooperativity coefficient for $K^+$ binding (~1.7) is similar to that for Na,K- and gastric H,K-ATPases (26). This is consistent with the known ion-transport cycle of these enzymes, in which two potassium ions bind at the extracellular face of the enzyme. Most significantly, the Atp1al1-βHK-mediated inward $K^+$ transport was inhibited by the Na,K-ATPase inhibitor, ouabain (Ki ~ 13 μM) and also by high concentrations of the gastric H,K-ATPase inhibitor, SCH28080 (~70% at 500 μM). The Atp1al1-βHK expression produces intracellular alkalinization and ouabain-sensitive proton extrusion as measured with pH-sensitive micro-electrodes. Thus, the recombinant Atp1al1 expressed in *Xenopus* oocytes functions as the catalytic subunit of a novel P-ATPase capable of performing moderately ouabain- and SCH28080-sensitive exchange of extracellular potassium for intracellular protons. (37). Consistent observations have been made when the Atp1al1-βHK complex was expressed in mammalian HEK-293 cells (21).

To gain insight into the enzymatic functions of the human nongastric H,K-ATPase, which until the discovery of the authentic β-subunit, is represented by the recombinant Atp1al1-βHK complex, we have developed a procedure for the preparation of the functionally active Atp1al1-βHK complex in amounts sufficient for extensive biochemical studies. Based on previous experience (58), the baculovirus expression system was selected for these purposes, which we consider a convenient way for large-scale preparation of recombinant proteins.

The three recombinant baculoviruses, containing Atp1al1 cDNA (AcAtp1al1), or ßHK cDNA (AcβHK ), or both these cDNAs together (AcAtp1al1/βHK), were generated and used for infection of Sf-9 and Sf-21 insect cells (1). Substantial synthesis of both recombinant Atp1al1 and βHK proteins was detected upon infection with AcAtp1al1/βHK and co-infection with the AcAtp1al1 and AcβHK viruses of both Sf-9 and Sf-21 cells. However, a significantly higher level of specific ouabain-sensitive ATPase activity indicating formation of the functionally active Atp1al1-βHK complex, was detected only in membranes of Sf21 cells that were co-infected with AcAtp1al11 and AcβHK viruses. It should be noted here that this procedure resulted from an extensive search for the optimal conditions for expression. For example, the particular importance of using a two-stage infection (first with AcbHK ) for efficient formation of the active Atp1al1-βHK complex in Sf-21 membranes was found. Strong differences in the efficiency of formation of the functional Atp1al1-βHK complex in Sf-9 and Sf-21 cells should not be considered as paradoxical, because according to previous observations these cell lines may yield different levels and process recombinant proteins in different ways (30). For example, expression of the SR Ca-ATPase was also much more efficient in Sf-21 cells than in Sf-9 cells (6).

It is important to note that insect cells exhibit very little ouabain-sensitive ATPase activity, which was first revealed upon expression of the Na,K-ATPase in Sf-9 cells (8,57,58). The Sf-9 is a clonal derivative of the Sf-21 cell line (30); therefore it was not surprising that we identified the presence of small but detectable amounts of ouabain-sensitive activity in microsomes of noninfected Sf-21 cells. Although, the endogenous insect ATPase is highly sensitive to ouabain (Ki ~80 nM) and completely resistant to SCH 28080, it is most probably not a Na,K-ATPase because its activity does not change significantly in the presence or absence of $K^+$ and/or $Na^+$. The ratio of the ouabain-sensitive activities of the recombinant Atp1al1-βHK complex to the endogenous activity of the Sf-21 cell membranes varied in the range from 3 to almost 10 in the best preparations.

One of the most interesting observations made in this study was that the Atp1al1-βHK complex exhibits a substantial specific ATPase activity in the absence of added $K^+$, i.e. in nominally $K^+$-free medium. In this case, the $K^+$ concentration in the reaction medium was ~5 μM. The endogenous ATPase activity in the absence of ouabain comprised ~20 % of that of the Atp1al1-βHK complex. The specific ATPase activity of the Sf-21 microsomes used in this study varied in different preparations from 0.8 to 1.65 μmol/mg/h. as measured in the nominally $K^+$-free medium. As shown in Fig.1, under these conditions, the ATPase activity the Atp1al1-βHK complex is moderately sensitive to both ouabain (Ki ~ 55 μM) and SCH28080 (Ki ~95 μM) but much more sensitive to digoxin and digitoxin (Ki ~10 μM) which are more lipophilic than ouabain. The gastric H,K-ATPase also exhibits SCH28080-sensitive ATPase activity in the absence of added K (35,43,54). Although ATP hydrolysis by the Atp1al1-βHK complex in nominally $K^+$-free medium is less sensitive to cardiac glycosides than Na,K-ATPase (26) and much less sensitive to SCH28080 than gastric H,K-ATPase (53), the above observations indicated a certain similarity of the spatial structure of the membrane moieties of the Atp1al1, the Na,K-ATPase and the gastric H,K-ATPase, as was suggested before (24,37). These structural aspects are discussed below.

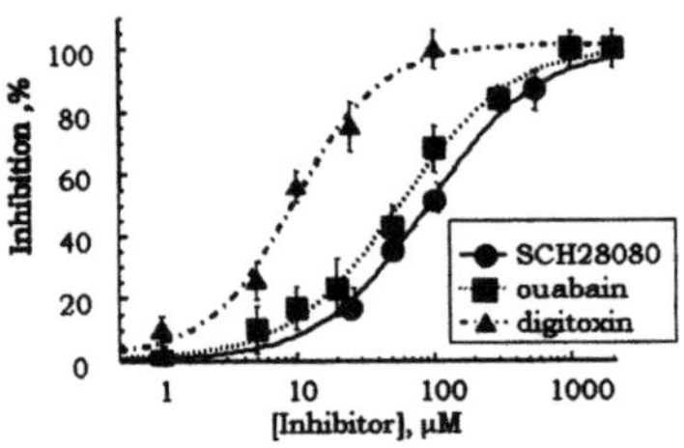


**Fig. 1. Sensitivity of the Atp1al1-βHK complex to ouabain, digitoxin and SCH28080.** ATPase activity was assayed after preincubation in the presence of the indicated concentrations of inhibitors in the nominally $K^+$-free medium containing 1 mM $NaN_3$ according to (1).

Ouabain-sensitive ATP hydrolysis by the Atp1al1-βHK complex was stimulated by increasing the concentrations of added $K^+$ in nominally $Na^+$-free medium in 1.5-3.4 fold (the apparent Km for added $K^+$ activation is ~115 μM). It should be noted that this value is similar to that of the gastric H,K-ATPase that was determined to be within the 0.2-0.4 mM range for the native as well as for the recombinant enzyme (5, 18, 53, 54). In contrast, $K^+$ activates ATP hydrolysis by the Na,K-ATPase with a significantly higher Km that is in the low millimolar range (see, for example Refs. 26, 47, 57, 58). However, this direct comparison is inappropriate, due to the absolute need for $Na^+$ for the Na,K-ATPase activation (47).

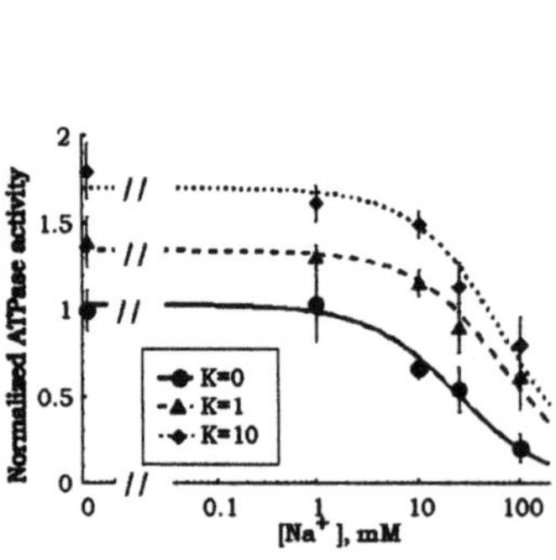


**Fig. 2. Cation dependence of the Atp1al1-ßHK complex: Activation by potassium and inhibitory effect of sodium.** Ouabain-sensitive ATPase activity was measured at the indicated concentrations of added $K^+$ and/or $Na^+$ and normalized to the activity in nominally $K^+$ and $Na^+$ -free medium.

An important goal of this study was to directly measure the effects of different $Na^+$ concentrations on Atp1al1-βHK ATPase activity. Recently, it was reported that expression of the rat nongastric H,K-ATPase in Xenopus oocytes (27) or of the human Atp1al1- βHK in HEK-293 cells (22) lowers intracellular $Na^+$ concentrations. The Atp1al1-βHK complex expressed in Xenopus oocytes in which intracellular $Na^+$ concentrations were increased to more than 80 mM by the co-expression of the amiloride-sensitive epithelial sodium channel, was also found to produce more than a 2-fold decrease in intracellular sodium. The specificity of the effect observed with the Atp1al1-βHK was supported by the fact that oocyte-expressed the Na,K-ATPase but not gastric H,K-ATPase produced a significant decrease in intracellular sodium (36). Therefore, it was suggested that transport modes of the recombinant nongastric ATPases are not limited to the H/K exchange but also include Na/K exchange (22,27,36). That is why it was reasonable to expect that similar to the Na,K-ATPase, $Na^+$ may serve as an activator of ATP hydrolysis by the Atp1al1-βHK complex. However, the opposite effects were revealed in experiments in which ouabain-sensitive ATP hydrolysis was assayed in the presence of $K^+$ and $Na^+$ concentrations varying from 0 to 50 mM and from 0 to 100 mM, respectively. As shown in Fig. 2, the Atp1al1-βHK complex exhibited the highest level of ATPase activity in the absence of added $Na^+$, in nominally $Na^+$-free medium, and an increase in $Na^+$ concentration in the assay medium resulted in a decrease of the ATP hydrolysis rate. The inhibitory effect of $Na^+$ was detected at all $K^+$ concentrations tested, but was the strongest in the absence of added $K^+$ (Ki ~ 24 mM). Potassium partially protected the Atp1al1-βHK complex from this $Na^+$ inhibition and in the presence of 1 or 10 mM $K^+$ the Ki values are ~68 mM and ~70 mM, respectively. The close similarity of these $K_i$ values may be explained by a high affinity of the Atp1al1-βHK complex for $K^+$ due to the fact that concentrations of $K^+$ 1 mM and higher should be considered as close to Vmax conditions. These results clearly indicate that $Na^+$ does not activate ATP hydrolysis by the Atp1al1-βHK complex but has an inhibitory effect on this reaction, and there is an antagonism between $K^+$ stimulation and $Na^+$ inhibition in this reaction (1). It is well established that in the case of Na,K-ATPase for which $Na^+/K^+$ exchange is an authentic function, the presence of $Na^+$ as an ion-activator is an absolute requirement for manifestation of ATPase activity (for reviews, see 8,26,47). Thus, our results do not support the concept that a nongastric H,K-ATPase "is not an H,K-ATPase but rather acts as an Na,K-ATPase" (22). This contradictory issue remains to be explored further. At this point, it cannot be excluded that due to structural and functional similarity between all three groups of X,K-ATPases, the Atp1al1-βHK pump under particular conditions can accept replacement of protons by sodium ions and exhibit mixed specificity with respect to transporting cations.

Previous studies of ion sensitivity of other X,K-ATPases have demonstrated that $Na^+$ also partially inhibits ATP hydrolysis by the gastric H,K-ATPase but at an alkaline pH and has little or no effect on the activity at pH $\leq 7.0$, thus revealing the possibility of competition between protons and $Na^+$ on the cytoplasmic sites (34,53). Sodium ions may also substitute for protons in ion-transporting mechanism of the gastric H,K-ATPase at very low $H^+$ concentration (pH 8.5) but not at a neutral pH. (40,53). In case of the Na,K-ATPase, $K^+$ has been shown to act as a competitive inhibitor of $Na^+$ binding at the cytoplasmic sites, thus suggesting that that $Na^+/K^+$ antagonism could be a plausible mechanism of pump regulation (for review, see 51). Inhibitory effect of sodium ions on the kinetics of the ATP hydrolysis by the Atp1al1-βHK complex reported here, suggest the existence of intracellular $Na^+$-binding site on nongastric H,K-ATPase whose nature is hard to predict at the moment. Upon inhibition, $Na^+$ may directly affect the interaction of $K^+$ with the enzyme and/or may act through blockage of the reaction stages, which require proton participation. The answer cannot be predicted a priori and this issue remains to be examined. It is very unlikely that $Na^+$ may strongly compete with $K^+$ and inhibit ATPase reaction catalyzed by the nongastric H,K-ATPase under normal physiological conditions, since the concentration of $K^+$ in cells is about 10-fold higher than that of $Na^+$ and the Ki value for $Na^+$ inhibition is well above the intracellular $Na^+$ concentration. However, this does not exclude the possibility that this antagonism between intracellular cations might be involved in fine adjustment of the nongastric H,K-ATPase activity *in vivo* in response to changes in ion concentrations.

With respect to sensitivity to the inhibitors, the Atp1al1-βHK complex (and presumably the authentic nongastric H,K-ATPase) occupies an intermediate position between the Na,K-ATPase and the gastric H,K-ATPase. The Na,K-ATPase isoforms (excluding rodent α1NK) have a Ki for cardiac glycosides in the nanomolar range and are resistant to SCH 28080 (8,25,26). In contrast, gastric H,K-ATPase is very sensitive to SCH 28080 with a Ki in the high nanomolar range, but is completely resistant to ouabain (29,35,53).

From the point of view of chemical structure and properties, SCH 28080 and cardiac glycosides, e.g. ouabain, are very different. However all of these compounds exhibit common functional feature: they inhibit the Na,K-ATPase and the gastric H,K-ATPase, respectively, through interaction with the extracytoplasmic surface of the target ion pumps, and both are strong competitors of potassium ions. Strong antagonism between $K^+$ and either ouabain or SCH 28080 was also detected in the reaction of ATP hydrolysis by the Atp1al1-βHK complex. Thus, the Atp1al1-βHK complex exhibits the highest sensitivity to ouabain in the absence of added $K^+$ with an apparent Ki of ~ 55 μM (Fig.1). The presence of 10 mM $K^+$ strongly protects the enzyme from inhibition and the Ki is shifted into the low millimolar range (~1.2 mM). An increase in the $K^+$ concentration in the assay medium also drastically reduces sensitivity to SCH 28080 and the apparent Ki is shifted from ~ 95 μM in the absence of added $K^+$ (Fig.1) to the millimolar range (~1 mM at 5 mM $K^+$; and ~1.6 at 25 mM $K^+$). Studies of ouabain and SCH 28080 effects on other partial reactions of nongastric H,K-ATPase are needed to define which concrete step of the enzyme functioning is affected by each inhibitor.

As it was mentioned above, cardiac glycosides and SCH 28080 are very different in size and shape suggesting different construction of their binding sites formed by extracellular and transmembrane domains of the α-subunits of the Na,K-ATPase and the gastric H,K-ATPase, respectively. However, inhibition of the Atp1al1-βHK complex by both ouabain and SCH28080 indicate that certain essential elements of the binding sites to both inhibitors coexist in the Atp1al1 polypeptide chain. Comparison of the sequences of the Atp1al1 and the Na,K-ATPase α-subunit reveal that the majority of amino acid residues that have been identified by *in vitro* mutagenesis as important for ouabain sensitivity of the Na,K-ATPase (summarized in Ref. 16) are also present at the corresponding positions in the Atp1al1; e.g., Cys-126, Tyr-130, Gln-133, Asn-144, Tyr-330, Leu 815,Thr 819, Tyr 885, Arg-902, Phe 1005 (24). In contrast, the extracellular loop between the first and the second transmembrane segments of the Na,K-ATPase which is thought to be an essential part of the digitalis receptor and the Atp1al1 share only two out of 12 residues. This stretch of high sequence divergence may explain the lower ouabain sensitivity of the nongastric H,K-ATPase.

It was recently shown by mutagenic alterations of the gastric H,K-ATPase that Cys 321, Lys791, Glu795,Cys 813, Ile 816, Glu 820, Thr 823, Pro 827 from the segments TM4 - TM6 and TM8 are involved in determining the affinity for SCH 28080 (3,4,32,52). Glu795, Ile816, Thr823 and Pro827 are conserved in all X,K-ATPases. Cys 813 is substituted in the Atp1al1 and other X,K-ATPases by Thr which is involved in determination of the Na,K-ATPase sensitivity to ouabain (16). Two interesting findings should be noted here: First, the nongastric ATPase also has Lys in the position corresponding to Lys791 of gastric pump and mutation of Lys791Ser resulted in a 20-fold decrease in SCH28080 affinity (52); and second, the position of Glu 820 is occupied by Asp in Atp1al1 and mutation Glu 820Asp in the gastric pump resulted in 8 times lower SCH 28080 sensitivity (3). Both the above mentioned affinities are still much higher than that of the Atp1al1-βHK complex. These observations indicate that determinants of SCH 28080 sensitivity are not limited to the above amino acid residues.

## CONCLUSION

The human nongastric H,K-ATPase is a member of the family of potassium-dependent animal ATPases (X,K-ATPases) which also includes closely related the gastric H,K-ATPase and isoforms of the Na,K-ATPase. The human ATP1AL1 (alternative name ATP12A) gene which encodes the catalytic α-subunit of the nongastric H,K-ATPase (Atp1al1), is expressed at a significant level in human skin and kidney, and at a lower level in colon, brain, placenta, and lung. Up to now, no β-subunit specific for this ATPase has been described. The basic functional parameters of the human nongastric H,K-ATPase represented by recombinant complex of the ATP1AL1-encoded protein and gastric H,K-ATPase β-subunit (Atp1al1-βHK) have been analyzed using different systems of heterologous expression. The recombinant Atp1al1-βHK complex is capable of performing exchange of extracellular potassium for intracellular protons. Its transport modes may also include Na/K exchange. The Atp1al1-βHK complex exhibits enzymatic properties of potassium-dependent ATPase sensitive to cardiac glycosides, SCH 28080 and sodium. Thus, nongastric H,K-ATPase, which differs from Na,K-ATPase in cation-dependence and differs from gastric H,K-ATPase and Na,K-ATPase in sensitivity to inhibitors, represents the third distinct group of X,K-ATPases.

The work in our laboratory is supported by National Institutes of Health Grants HL-36573 and GM-54997 ( to N.N. Modyanov).

## REFERENCES


1. **Adams, G., Tillekeratne, M., Yu, C., Pestov, N. B., and. Modyanov, N. N.** Catalytic functions of nongastric H,K-ATPase expressed in SF-21 insect cells. *Biochemistry* 40: 5765-5776, 2001.
2. **Asano, S., Hoshina, S., Nakaie, Y., Watanabe, T., Sato, M., Suzuki, Y., and Takeguchi, N.** Functional expression of putative H,K-ATPase from guinea pig distal colon. *Am. J. Physiol.* 275 (*Cell Physiol.* 44): C669-C674, 1998.
3. **Asano, S., Matsuda, S., Hoshina, S., Sakamoto, S., and Takeguchi, N.** A chimeric gastric H,K-ATPase inhibitable with both ouabain and SCH 28080 *J. Biol. Chem.* 274: 6848-6854, 1999.
4. **Asano, S., Matsuda, S., Tega, Y., Shimizu, K., Sakamoto, S., and Takeguchi, N.** Mutational analysis of putative SCH 28080 binding sites of the gastric H,K-ATPase. *J. Biol. Chem.* 272: 17668-17674, 1997.
5. **Asano, S., Tega, Y., Konishi, K., Fujioka, M., and Takeguchi, N.** Functional expression of gastric H,K-ATPase and site-directed mutagenesis of the putative cation binding site and catalytic center. *J. Biol. Chem.* 271: 2740-2745, 1996.
6. **Autry, J.M. and L.R. Jones.** Functional co-expression of the canine cardiac Ca2+ pump and phospholamban in Spodoptera frugiperda (Sf21) cells reveals new insights on ATPase regulation. *J. Biol. Chem.* 272: 15872-15880, 1997.
7. **Beil, W., Hackbarth, I., and Sewing, K. F.** Mechanism of gastric antisecretory effect of SCH 28080. *Br. J. Pharmac.* 88:19-23, 1986.
8. **Blanco G. and Mercer, R. W.** Isozymes of the Na-K-ATPase: heterogeneity in structure, diversity in function. *Am. J. Physiol.* 275: F633-F650, 1998.
9. **Campbell, W. G., Weiner, I. D., Wingo, C. S., and Cain, B. D.** H-K- ATPase in the RCCT-28A rabbit cortical collecting duct cell line. *Am. J. Physiol.* 276: F237-F245, 1999.
10. **Chow, D. C. and Forte, J. G.** Characterization of the beta-subunit of the H,K-ATPase using an inhibitory monoclonal antibody. *Am. J. Physiol.* 265: C1562-C1570, 1993.
11. **Chow, D. C. and Forte, J. G.** Functional significance of the beta-subunit for heterodimeric P-type ATPases. *J. Exp. Biol.* 198: 1-17, 1995.
12. **Codina, J., Delmas-Mata, J. T., and DuBose, T. D., Jr.** The alpha-subunit of the colonic H,K-ATPase assembles with beta 1-Na,K-ATPase in kidney and distal colon. *J. Biol. Chem.* 273: 7894-7899, 1998.
13. **Codina, J., Kone, B. C., Delmas-Mata, J. T., and DuBose, T. D., Jr.** Functional expression of the colonic H,K-ATPase alpha-subunit. Pharmacologic properties and assembly with X,K-ATPase beta-subunits. *J. Biol. Chem.* 271: 29759-29763, 1996.
14. **Crambert, G., Hasler, U., Beggah, A.T., Yu, C., Modyanov, N. N., Horisberger, J.-D., Lelievre, L., and Geering, K.** Transport and pharmacological properties of nine different human Na,K-ATPase isozymes. *J. Biol. Chem.* 275: 1976-1986, 2000.
15. **Crowson, M. S. and Shull, G. E.** Isolation and characterization of a cDNA encoding the putative distal colon H,K-ATPase. Similarity of deduced amino acid sequence to gastric H,K-ATPase and Na,K-ATPase and mRNA expression in distal colon, kidney, and uterus. *J. Biol. Chem.* 267: 13740-13748, 1992.

16. **Croyle, M. L., Woo, A. L., and Lingrel, J. B.** Extensive random mutagenesis analysis of the Na,K-ATPase alpha subunit identifies known and previously unidentified amino acid residues that alter ouabain sensitivity--implications for ouabain binding. *Eur. J. Biochem.* 248: 488-495, 1997.
17. **Driel, I. R. and Callaghan, J. M.** Proton and potassium transport by H,K-ATPases.. *Clin. Exp. Pharm. Physiol.* 22: 952-960, 1995.
18. **Forte, J. G., Ganser, A., Beesley, R., and Forte, T. M.** Unique enzymes of purified microsomes from pig fundic mucosa. K-stimulated adenosine triphosphatase and K-stimulated pNPPase. *Gastroenterology* 69: 175-189, 1975.
19. **Geering, K.** The functional role of the beta-subunit in the maturation and intracellular transport of Na,K-ATPase. *FEBS Lett.* 285: 189-193, 1991.
20. **Geering, K., Crambert, G., Yu, C., Korneenko, T. V., Pestov, N. B., and Modyanov, N. N.** Intersubunit interactions in human X,K-ATPases: role of membrane domains M9 and M10 in the assembly process and association efficiency of human, nongastric H,K-ATPase alpha subunits (ATP1al1) with known beta subunits. *Biochemistry* 39: 12688-12698, 2000.
21. **Grishin, A. V., Bevensee, M. O., Modyanov, N. N., Rajendran, V., Boron, W. F., and Caplan, M. J.** Functional expression of the cDNA encoded by the human ATP1AL1 gene. *Am. J. Physiol.* 271: F539-F551, 1996.
22. **Grishin, A. V. and Caplan, M. J.** ATP1AL1, a member of the non-gastric H,K-ATPase family, functions as a sodium pump. *J. Biol. Chem.* 273: 27772-27778, 1998.
23. **Grishin, A. V., Reinhard, J., Dunbar, L. A., Courtois-Coutry, N., Wang, T., Giebisch, G., and Caplan, M. J.** Nongastric H,K-ATPase: cell biologic and functional properties *Semin. Nephrol.* 19: 421-430, 1999.
24. **Grishin, A.V., Sverdlov, V.E., Kostina, M.B., and Modyanov, N. N.** Cloning and characterization of the entire cDNA encoded by ATP1AL1--a member of the human Na,K/H,K-ATPase gene family. *FEBS Lett.* Jul 25: 349: 144-150, 1994.
25. **Hansen, O.** Interaction of cardiac glycosides with Na,K-activated ATPase. A biochemical link to digitalis-induced inotropy. *Pharmacol. Rev.* 36: 143-163, 1984.
26. **Horisberger, J. D.** *The Na,K-ATPase: Structure-Function Relationship*, Austin, TX: R.G. Landes Company, 1994, pp.1-107.
27. **Jaisser, F. and Beggah, A. T.** The nongastric H,K-ATPases: molecular and functional properties. *Am. J. Physiol.* 276: F812-F824, 1999.
28. **Korneenko, T. V., Pestov, N. B., Egorov, M. V., Ivanova, M.V., Kostina, M. B., and Shakhparonov, M. I.** Monoclonal antibodies to the alpha-subunit of the putative human H,K-ATPase encoded by the ATP1AL1 gene. *Russian J. Bioorgan. Chem.* 23: 800-804, 1997.
29. **Keeling, D. J., Taylor, A. G., and Schudt, C.** The binding of a $K^+$ competitive ligand, 2-methyl,8-(phenylmethoxy)imidazo(1,2-a)pyridine 3-acetonitrile, to the gastric H,K-ATPase. *J. Biol. Chem.* 264: 5545-5551, 1989.
30. **King, L. A. and Possee, R. D.** *The Baculovirus Expression System: A Laboratory Guide*, London: Chapman & Hall, 1992, pp. 1-222.

31. **Kraut, J. A., Hiura, J., Shin, J. M., Smolka, A., Sachs, G., and Scott, D.** The Na,K-ATPase beta 1 subunit is associated with the HK alpha 2 protein in the rat kidney. *Kidney Int.* 53: 958-962, 1998.
32. **Lambrecht, N., Munson, K., Vagin, O., and Sachs, G.** Comparison of covalent with reversible inhibitor binding sites of the gastric H,K-ATPase by site-directed mutagenesis. *J. Biol. Chem.* 275: 4041-4048, 2000.
33. **Ljungstrom, M. and Mardh, S.** Kinetics of the acid pump in the stomach. Proton transport and hydrolysis of ATP and p-nitrophenyl phosphate by the gastric H,K-ATPase. *J. Biol. Chem.* 260: 5440-5444, 1985.
34. **Ljungstrom, M., Vega, F.V., and Mardh, S.** Effects of pH on the interaction of ligands with the H,K-ATPase purified from pig gastric mucosa. *Biochim. Biophys. Acta* 769: 220-230, 1984.
35. **Mendlein, J. and Sachs, G.** Interaction of a $K^+$-competitive inhibitor, a substituted imidazo[1,2a] pyridine, with the phospho- and dephosphoenzyme forms of H,K-ATPase *J. Biol. Chem.* 265: 5030-5036, 1990.
36. **Modyanov, N. N., Adams, G., Pestov, N. B., Yu, C., Tillekeratne, M., Korneenko, T. V., Shakhparonov, M.I., Crambert, G., Horisberger, J. D., and Geering, K.** Structural and functional properties of human ouabain-sensitive H,K-ATPase. in *Na/K-ATPase and Related ATPases,* Elsevier, Amsterdam (K. Taniguchi and S. Kanya, eds.) pp.139-146, 2000.
37. **Modyanov, N., Mathews, P., Grishin, A., Beguin, P., Beggah, A., Rossier, B., Horisberger, J.-D., and Geering, K.** Human ATP1AL1 gene encodes a ouabain-sensitive H-K-ATPase. V(1995) *Am. J. Physiol.* 269: C992-C997, 1995.
38. **Modyanov, N. N., Petrukhin, K. E., Sverdlov, V. E., Grishin, A.V., Orlova, M. Y., Kostina, M. B., Makarevich, O. I., Broude, N. E., Monastyrskaya, G. S., and Sverdlov, E. D.** The family of human Na,K-ATPase genes. ATP1AL1 gene is transcriptionally competent and probably encodes the related ion transport ATPase. *FEBS Lett.* 278: 91-94, 1991.
39. **Moller, J.V., Juul, B., and le Maire, M.** Structural organization, ion transport, and energy transduction of P-type ATPases. *Biochim. Biophys. Acta* 1286: 1-51, 1996.
40. **Palvani, C., Sachs, G., and Blostein, R.** Sodium ions as substitutes for protons in the gastric H,K-ATPase. *J. Biol. Chem.* 264: 17854-17859, 1989.
41. **Pestov, N. B., Adams, G., Shakhparonov, M. I., and Modyanov, N. N.** Identification of a novel gene of the X,K-ATPase beta-subunit family that is predominantly expressed in skeletal and heart muscles. *FEBS Lett.* 456: 243-248, 1999.
42. **Pestov, N., Romanova, L., Korneenko, T., Egorov, M., Kostina, M., Sverdlov, V., Askari, A., Shakhparonov, M., and Modyanov, N.** Ouabain-sensitive H,K-ATPase: tissue-specific expression of the mammalian genes encoding the catalytic alpha subunit. *FEBS Lett.* 440: 320-324, 1998.
43. **Reenstra, W. W., Bettencourt, J. D., and Forte, J. G.** Kinetic studies of the gastric H,K-ATPase. Evidence for simultaneous binding of ATP and inorganic phosphate. *J. Biol. Chem.* 263: 19618-19625, 1988.
44. **Sangan, P., Kolla, S. S., Rajendran, V. M., Kashgarian, M., and Binder, H. J.** Colonic H-K-ATPase beta-subunit: identification in apical membranes and regulation by dietary K depletion. *Am. J. Physiol.* 276: C350-360, 1999.

45. **Shull, M. M. and Lingrel, J.B.** Multiple genes encode the human Na,K-ATPase catalytic subunit. *Proc. Natl. Acad. Sci. USA* 84: 4039-4043, 1987.
46. **Silver, R.B. and Soleimani, M.** H,K-ATPases: regulation and role in pathophysiological states *Am. J. Physiol.* 276, F799-F811, 1999
47. **Skou, J.C. and Esmann, M.** The Na,K-ATPase. *J. Bioenerg. Biomembr.* 24: 249-261, 1992.
48. **Stewart, B., Wallmark, B., and Sachs, G.** The interaction of $H^+$ and $K^+$ with the partial reactions of gastric H,K-ATPase. *J. Biol. Chem.* 256: 2628-2690. 1981
49. **Sverdlov, E. D., Monastyrskaya, G. S., Broude, N. E., Ushkaryov, Yu. A., Allikmets, R. L., Melkov, A. M., Smirnov, Yu. V, Malyshev, I.V., Dulobova, I.E., Petrukhin, K.E., Grishin, A.V., Kijatkin, N. I., Kostina, M. B., Sverdlov, V., Modyanov, N., and Ovchinnikov, Yu.A.** The family of human Na,K-ATPase genes. No less than five genes and/or pseudogenes related to the alpha-subunit. *FEBS Lett.* 217: 275-278, 1987.
50. **Sverdlov, V., Kostina, M., and Modyanov, N.** Genomic organization of the human ATP1AL1 gene encoding a ouabain-sensitive H,K-ATPase.. *Genomics* 32: 317-327, 1996.
51. **Therien, A, and Blostein R**. Mechanisms of sodium pump regulation *Am. J. Physiol. Cell. Physiol.* 279: C541-C566, 2000.
52. **Vagin, O., Munson, K., Lambrecht, N., Karlish, S.J., and Sachs G.** Mutational analysis of the K-competitive inhibitor site of gastric H,K-ATPase; *Biochemistry* 40: 7480-7490, 2001
53. **Van Uem, T. J. F., and De Pont, J. J. H. H. M.** Structure and function of gastric H,K-ATPase, in *Molecular Aspects of Transport Proteins* (De Pont, J. J. H. H. M., Ed.) Elsevier: N.Y. 1992, pp 27-55, 1992.
54. **Wallmark, B., Stewart, H. B., Rabon, E., Saccomani, G., and Sachs, G.** The catalytic cycle of gastric H,K-ATPase. *J. Biol. Chem.* 255: 5313-5319, 1980.
55. **Wingo, C. S. and Smolka, A. J.** Function and structure of H-K-ATPase in the kidney. *Am. J. Physiol.* 269: F1-F16, 1995.
56. **Woo, A. L., James, P. F., and Lingrel, J. B.** Characterization of the fourth alpha isoform of the Na, K-ATPase. *J. Membr. Biol.* 169:39-44, 1999.
57. **Xie, Z. J., Wang Y., Liu G., Zolotarjova, N., Periyasamy, S. M., and Askari, A.** Similarities and differences between the properties of native and recombinant Na,K-ATPase. *Arch. Biochem. Biophys.* 330: 153-162, 1996.
58. **Yu, C., Xie, Z., Askari, A., and Modyanov, N.N.** Enzymatic properties of human Na,K-ATPase alpha1beta3 isozyme. *Arch. Biochem. Biophys.* 345: 143-149, 1997.

# Chapter 3

# Functional Consequences of Subunit Interactions in Na,K- and H,K-ATPases

GILLES CRAMBERT, PASCAL BEGUIN, UDO HASLER, AND KATHI GEERING
*Institute of Pharmacology and Toxicology of the University, rue du Bugnon 27, CH-1005 Lausanne, Switzerland*

## INTRODUCTION

In view of the important role of P-type ATPases in many physiological and pathophysiological processes, the elucidation of the mechanisms that regulate their expression and function becomes an important issue. We are interested in the role of protein-protein interactions in the regulation of Na,K-ATPase (NKA) and H,K-ATPases (HKA) and in particular in the elucidation of the basic regulatory effect that results from the interaction of the catalytic α subunit with its partner β subunit and in the identification and characterization of new tissue-specific modulators.

NKA and HKA are the most closely related members of the P-type ATPase family of cation transporters. Among P-type ATPases of higher organisms, only NKA and HKA are oligomeric proteins in which the catalytic α subunit is associated with a second β subunit (Fig.1). Three NKA β isoforms (NKA β1, β2 and β3) and one gastric HKA β subunit (HKA β), but no specific β subunit for non-gastric HKA, have been identified. Increasing experimental evidence suggests that β subunits have two distinct roles (for review see 15). First, β subunits act as specific chaperones which mediate the structural maturation of NKA and HKA α subunits and represent the limiting factor for the cellular expression of NKA and HKA. Individual α subunits expressed in cells without a β subunit are indeed subjected to rapid degradation. Co-expressed in heterologous expression systems, stabilization of the NKA α subunit can be achieved by promiscuous interactions with different NKA β isoforms or HKA β whereas efficient stabilization of the gastric and non-gastric HKA α subunits is only possible with HKA β or NKAβ2. In addition to their chaperone function, β subunits

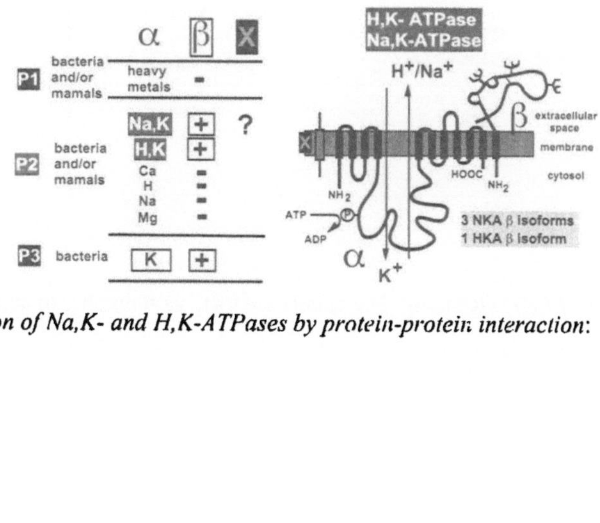


Fig.1: *Regulation of Na,K- and H,K-ATPases by protein-protein interaction*: For details see text

contribute to the intrinsic transport properties of Na,K- and H,K-pumps expressed at the cell surface. This can be revealed by biochemical or mutational perturbations of the β subunit structure or by expression of different β isoforms with NKA or HKA α subunits which result in α-β complexes with different apparent cation affinities (8, 10, 18, 19, 21, 23, 26).

In addition to the very basic and ubiquitous regulation of NKA and HKA expression and function by β subunit association, recent experimental evidence suggests that interaction with other auxiliary subunits may contribute to a more tissue-specific modulation of P-type ATPases adapted to cell-type specific physiological demands. Well defined examples are phospholamban that interacts with and, at low $Ca^{2+}$ concentrations, reversibly inhibits sarco(endo)plasmic Ca-ATPase (SERCA) 2a in cardiac and slow–twitch skeletal muscle (38) and the so-called γ subunit (13, 29) that is expressed in certain renal nephron segments (33, 40), associates specifically with NKA and influences its $Na^+$, $K^+$ and ATP affinities (1, 7, 33, 41) (Fig.2). Significantly, cases of human primary hypomagnesaemia have been linked to a heterozygous mutation in the γ subunit which apparently leads to defective routing of the γ subunit and NKA to the plasma membrane (28). The two identified splice variants of the γ subunit (γa and γb; 24) belong to a recently defined family of small proteins called the FXYD family (39) that includes also phospholemman (30), Mat-8 (mammary tumor marker 8; 31), CHIF (corticosteroid hormone-induced factor; 2), RIC (related to ion channel; 14), and two as yet unknown proteins FXYD 6 and 7 (Fig.2).

| | | aa | tissue distribution | function |
|---|---|---|---|---|
| FXYD1 | phospholemman | 92 | heart | ? |
| FXYD 2a+2b | γa and γb | 66 | kidney | modulator of NKA |
| FXYD3 | Mat-8 (mammary tumor marker) | 87 | | ? |
| FXYD4 | CHIF (corticosteroid hormone-induced factor) | 87 | kidney colon | modulator of NKA ? |
| FXYD5 | RIC (related to ion channel) | 177 | | ? |
| FXYD6 | new | 95 | | ? |
| FXYD7 | new | 77 | | ? |

Fig.2: *The family of FXYD proteins*. For details see text. aa = number of amino acids

The proteins of the FXYD family are characterized by a single transmembrane domain and a signature sequence which includes the FXYD motif, 2 conserved glycines and a serine residue concentrated in and adjacent to the membrane segment. Members of the FXYD family are widely distributed in mammalian tissues, possibly with a prominent expression in epithelial and excitable tissues (39). There is experimental evidence that FXYD proteins may be regulators of ion transporters or may form ion channels by themselves but, with the exception of the γ subunit, their authentic biological function is still unknown.

Based on several indicative observations, we recently became interested in the question whether another member of the FXYD family, namely the corticoidsteroid hormone-induced factor CHIF, may also be a 'γ-like', cell type-specific regulator of NKA. CHIF shows more than 50% sequence similarity to the γ subunit. Like the γ subunit, CHIF is exclusively expressed in basolateral membranes of epithelial cells where it co-localizes with NKA (33, 40). However, whereas the γ subunit is predominantly expressed in the medullary thick ascending limb, and to a lesser extent in other nephron segments of the kidney (33, 40), CHIF is abundant in the collecting duct and in the distal colon (37). Both colon and the collecting duct are target tissues for aldosterone action and it has indeed been shown that aldosterone increases the expression of NKA (12) as well as of CHIF (2).

## EFFECTS OF β SUBUNIT ASSEMBLY ON NKA AND HKA EXPRESSION AND FUNCTION

*The chaperone function of the β subunit*

α subunits of NKA and HKA belong to the subclass of P2-ATPases (27) which are structurally characterized by 10 transmembrane (TM) segments as recently demonstrated by Toyoshima et al. (2000) by high resolution cristallography of SERCA. Compared to α subunits, β subunits are smaller type II proteins with one transmembrane domain, a short cytoplasmic tail and a large ectodomain containing 3 disulfide bridges and several sugar chains (Fig.1).

We have performed a detailed analysis of the membrane topogenesis of NKA and HKA α subunits to get information on the questions why α subunits are degraded when synthesized without a β subunit and how assembly with the β subunit prevents their degradation (3, 5, 6). The results of these studies revealed that during synthesis of NKA and HKA α subunits in the endoplasmic reticulum, 2 N-terminal membrane pairs are produced which form a stable membrane anchor for nascent α subunits which is not recognized by the cellular degradation machinary (Fig.3). Moreover, the large cytoplasmic loop rapidly folds into a correct conformation which prevents exposure of degradation signals during synthesis. On the other hand, membrane insertion of C-terminal membrane segments is inefficient due to the presence of specific sequence information which reduces the interaction with the lipid bilayer. C-terminal membrane pairs are probably in a dynamic equilibrium between the translocon and the cytoplasm and this permits transient exposure of degradation signals to the cytoplasm which mediate degradation of the α subunit by the cytosolic proteasome.

Interaction of the β ectodomain with a short stretch of 10 amino acids in the M7/M8 extracellular loop of the α subunit, which contains an SYGQ motif that was originally shown to be involved in β assembly (9), is the key event for the structural maturation and stabilization of NKA and HKA α subunits (6). β interaction with this domain initially favors the formation of the M7/M8 membrane pair and as a consequence supports intermolecular interactions between other membrane segments. Together, these processes are necessary to complete the proper folding of the α subunit which masks degradation signals and permits the translocation of the C-terminal membrane domain of the α subunit from the translocon into the lipid bilayer (Fig.3).

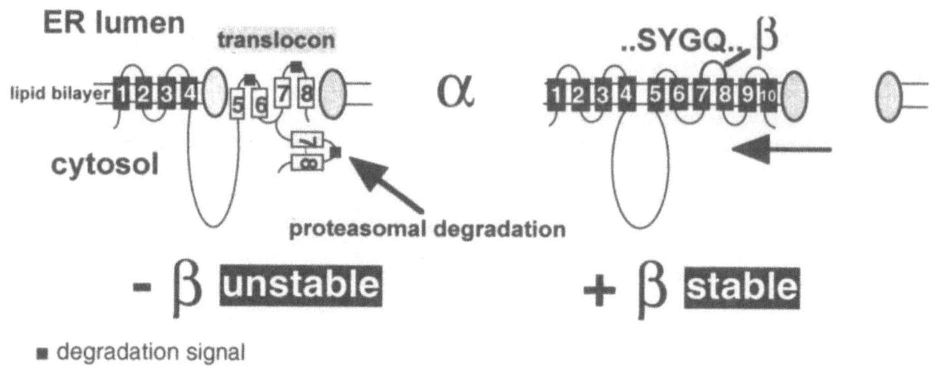


Fig.3: *Topogenesis of α subunits and chaperone function of β subunits of Na,K- and H,K-ATPases.* For details see text.

*β subunit domains and functional effects*

The question arises whether the chaperone and the cation effect of the β subunit are mediated by the same or different interaction sites between α and β subunits. There is evidence that β subunits interact with α subunits in the cytoplasmic, the transmembrane and the extracytoplasmic domain (9, 16, 17, 20, 32, 35). Though the cytoplasmic N-terminal of the β subunit is important for the correct folding of the β subunit itself (18), we could so far not reveal any functional role of cytosolic α-β interactions by mutational analysis (19). This contrasts with a report by Shainskaya et al. (36) suggesting that the NKA β N-terminus interacts with the M6/M7 cytoplasmic loop of the NKA α subunit and contributes to the selectivity for $Na^+$ over $K^+$. As to interactions in the ectodomain, they are important for the stabilization of the α subunit as outlined above. Furthermore, studies with NKA/HKA β chimeras have revealed that ectodomain interactions also contribute to the cation effects of β isoforms (11, 19, 22).

Recently, we have analyzed the possible functional role of α-β interactions in the membrane domain (17). Secondary structure prediction of the putative membrane domain of NKA β1 reveals an α-helical structure which exposes two distinct faces. One face is characterized by the presence of large amino acids e.g. two aligned tyrosine and one phenylalanine residue. The other face of NKA β1 contains 3 aligned glycine residues which form a grove in the α-helix. Interestingly, these glycine residues are part of a G*XXX*G motif which was shown to be involved in dimerization of TM helices (25). To better define the functional relevance of α-β interactions in the TM domain, we replaced each amino acid of the NKA β1 TM domain with tryptophan and then studied the effects of these mutations on the stabilization and the transport properties of the NKA α subunit. Significantly, none of the NKA β1 TM mutants influenced the stabilization of the NKA α subunit and, with one exception, none of them had effects on the pump's transport properties. The only β mutants that produced significant alterations in the transport properties of Na,K-pumps, were those affected in the two aligned tyrosine residues, which are highly conserved in all known β subunits. Single replacements by tryptophan of these tyrosine residues in the NKA β1 TM domain significantly decreased, and double mutations produced additive effects on the apparent $K^+$-affinity of Na,K-pumps compared to that of wild type Na,K-pumps. Altogether, our results indicate that two tyrosine residues in the β's TM domain interact with the α subunit and contribute in concert to the intrinsic transport properties of Na,K-and H,K-pumps.

At present, we have not yet identified the complementary TM segment in the α subunit which interacts with the tyrosine containing face of the β's TM helix but we can put forward an experimentally testable hypothesis. Recent experimental evidence suggests that NKA, and probably HKA α subunits, exhibit a similar spatial organization of their TM domains than SERCA (42), and that the β subunit may be located near TM helices 7 and 10 (34). Provided that the β's TM domain is closely positioned to M8 of the α subunit, as suggested by cross-linking studies (20, 32), then the two tyrosine residues are directed towards M7, making this α TM segment a possible candidate for an interaction site with the tyrosine containing face of the β's TM helix. In such a model, the glycine dimerization motif may be freely exposed away from the α subunit. This would make it accessible for association with another protein, possibly the β subunit itself, and could lead to the formation of hetero-oligomers.

## TISSUE-SPECIFIC MODULATION OF NKA FUNCTION BY CHIF

In addition to the ubiquitous β subunit, which is obligatory for the cellular expression of NKA and HKA and which contributes to their intrinsic transport properties, the γ subunit has been identified as a specific modulator of renal NKA function. Significantly, our recent studies reveal that another member of the FXYD protein family, the corticosteroid hormone-induced factor CHIF, which exhibits a tissue expression distinct from the γ subunit, also fulfills the criteria for an auxiliary, modulatory subunit of NKA (4). When co-expressed with NKA in *Xenopus* oocytes, CHIF can indeed be co-immunoprecipitated with an NKA α antibody over prolonged chase periods, similar to the γ subunit. γ and CHIF interactions with NKA are specific since neither the γ subunit nor CHIF associate with non-gastric H,K-ATPase which, like CHIF, is expressed in kidney and colon. Furthermore, Western blot analysis reveals the presence of γa and γb variants as well as of CHIF in kidney membranes and all 3 peptides can be co-immunoprecipitated with a NKA α antibody indicating the presence of γ-NKA as well as CHIF-NKA complexes in renal tissue.

Comparison of the functional effects of γ splice variants and CHIF on the transport properties of NKA co-expressed in *Xenopus* oocytes revealed that both γ splice variants reduce, whereas CHIF significantly increases, the apparent affinity for intracellular $Na^+$ of Na,K-pumps compared to that of Na,K-pumps expressed alone. The $Na^+$ effect of CHIF is particularly pronounced at physiologically low intracellular $Na^+$ concentrations, e.g. at 5 mM, reaching a 4 times higher Na,K-pump transport activity in the presence than in the absence of CHIF. When measured in the absence of extracellular $Na^+$, a situation in which the $K^+$-activation kinetics most directly reflects the intrinsic binding affinity for external $K^+$, CHIF has no detectable effect on the $K^+$-affinity of Na,K-pumps over a wide range of membrane potentials. On the other hand, γa and γb subunits induce a small but significant decrease of the $K^+$-affinity at very negative membrane potentials. The effects of CHIF and γ subunits on the $K^+$-activation of Na,K-pumps were very different in the presence of extracellular $Na^+$. Under these conditions, the $K^+$-activation kinetics not only depends on the intrinsic binding affinity for $K^+$ but also on the competition between extracellular $Na^+$ and $K^+$. In the presence of external $Na^+$, CHIF induces a large increase in the $K_{1/2}$ value for $K^+$ in the negative potential range which likely reflects an increased affinity for extracellular $Na^+$. γa and γb subunits produce similar changes over the whole potential range but these changes are opposite to those observed with CHIF at low membrane potentials.

Altogether, our results suggest that CHIF and γ subunits influence the transport properties of NKA mainly by affecting steps involved in $Na^+$ transport. This seems to be the only effect for CHIF, while γa and γb have a small additional effect on $K^+$ transport steps.

What is the physiological relevance of these findings? In the kidney, CHIF and γ subunits show a nearly exclusive expression in different segments of the nephron (33, 37, 40). This suggests that the two peptides may have functionally distinct effects which are adapted to nephron segment-specific physiological needs (Fig.4).

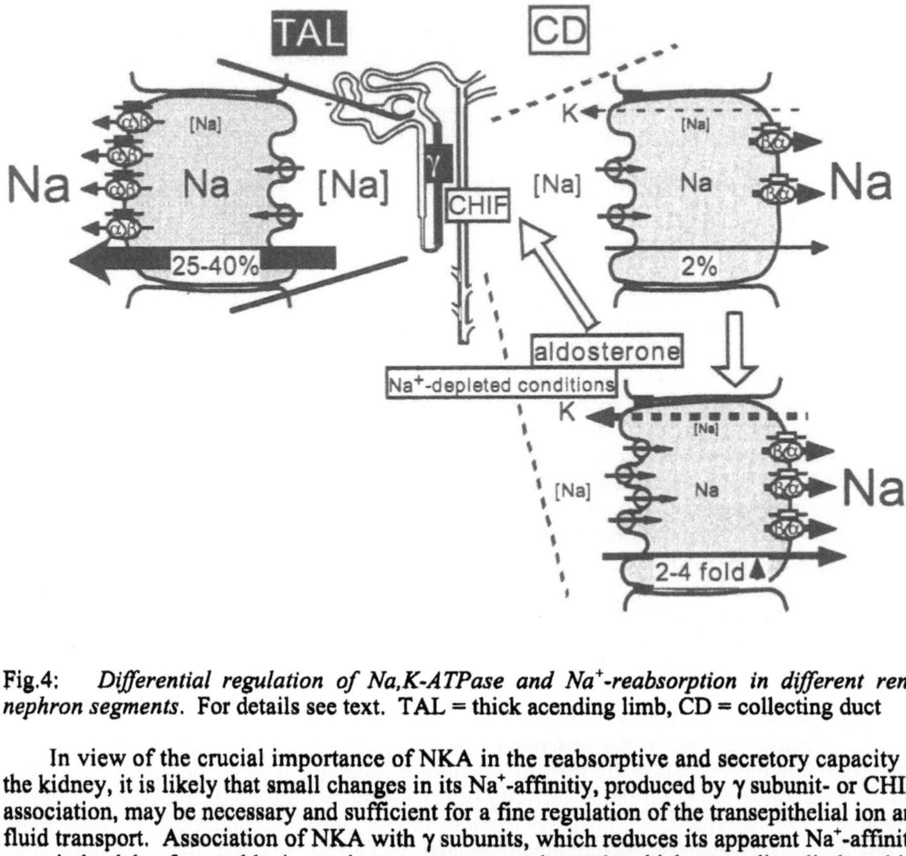


Fig.4: *Differential regulation of Na,K-ATPase and Na⁺-reabsorption in different renal nephron segments*. For details see text. TAL = thick acending limb, CD = collecting duct

In view of the crucial importance of NKA in the reabsorptive and secretory capacity of the kidney, it is likely that small changes in its $Na^+$-affinitiy, produced by γ subunit- or CHIF-association, may be necessary and sufficient for a fine regulation of the transepithelial ion and fluid transport. Association of NKA with γ subunits, which reduces its apparent $Na^+$-affinity, may indeed be favourable in nephron segments such as the thick ascending limb, which reabsorbs as much as 25-40% of the filtered $Na^+$ load. For this process, a high density of Na,K-pumps is needed in the basolateral membrane and the expressed pumps should preferentially have a reduced $Na^+$-affinity in order to be able to cope with and to efficiently extrude the important cellular $Na^+$ load.

In contrast to γ subunits, CHIF increases the $Na^+$-affinity of Na,K-pumps. This may be useful in nephron segments such as the collecting duct or in the distal colon which are ultimate sites for electrolyte conservation. Under conditions of $Na^+$ depletion, luminal $Na^+$ in the collecting duct is decreased to very low values. To favour efficient $Na^+$ reabsorption, the presence of a Na,K-pump with a high $Na^+$-affinity is of great advantage since it permits to maintain low intracellular $Na^+$ concentrations and a high negative, apical membrane potential.

Similar to colon, the collecting duct is a target site for aldosterone which can increase $Na^+$ reabsorption by 2-4 fold when $Na^+$ conservation is required. During the early phase of aldosterone action, apical $Na^+$ entry increases and small changes in the intracellular $Na^+$ concentration are produced, in the range of physiologically low $Na^+$ concentrations. Under these conditions, $Na^+$ reabsorption should considerably improve in the presence of Na,K-pumps associated with CHIF. Finally, it is interesting to note that CHIF and NKA expression is concomittantly increased during the late phase of aldosterone action. This assures that newly synthesized Na,K-pump units can immediately associate with CHIF and that CHIF does not become limiting for the formation of pumps with high $Na^+$-affinity. At last, since $K^+$ secretion in the distal nephron is tightly coupled to $Na^+$ reabsorption and NKA activity, association of CHIF with NKA in the collecting duct may also be favourable to $K^+$ secretion in conditions of a high $K^+$, aldosterone-activated state.

So, in conclusion, our data indicate that CHIF is a new tissue-specific regulator of NKA which modulates its transport properties in an opposite way than γ subunits by increasing its apparent $Na^+$-affinity. It is likely that this regulatory mechanism of NKA plays a crucial role in aldosterone-responsive tissues such as kidney and colon, which are responsible for thc maintenance of body $Na^+$ and $K^+$ homeostasis.

## ACKNOWLEDGMENTS

This work has been supported by grants from the Swiss National Fund for Scientific Research No. 31-53721.98 and from the Roche Research Foundation.

## REFERENCES

1. **Arystarkhova E, Wetzel RK, Asinovski NK, and Sweadner KJ**. The γ subunit modulates $Na^+$ and $K^+$ affinity of the renal Na,K-ATPase. *J Biol Chem* 274: 33183-33185, 1999.
2. **Attali B, Latter H, Rachamim N, and Garty H**. A corticosteroid-induced gene expressing an "IsK-like" $K^+$ channel activity in *Xenopus* oocytes. *Proc Natl Acad Sci USA* 92: 6092-6096, 1995.
3. **Beggah AT, Béguin P, Bamberg K, Sachs G, and Geering K**. β-subunit assembly is essential for the correct packing and the stable membrane insertion of the H,K-ATPase α-subunit. *J Biol Chem.* 274: 8217-8223, 1999.
4. **Béguin P, Crambert G, Guennoun S, Garty H, Horisberger J-D, and Geering K.** CHIF, a member of the FXYD protein family, is a regulator of Na,K-ATPase distinct from the γ subunit. EMBO J: 20: 3993-4002, 2001.
5. **Béguin P, Hasler U, Beggah A., Horisberger J-D, and Geering K**. Membrane integration of Na,K-ATPase α-subunits and β-subunit assembly. *J Biol Chem* 273: 24921-24931, 1998.
6. **Béguin P, Hasler U, Staub O, and Geering K**. Endoplasmic reticulum quality control of oligomeric membrane proteins: Topogenic determinants involved in the degradation of the unassembled Na,K-ATPase α subunit and in its stabilization by β subunit assembly. *Mol Biol Cell* 11: 1657-1672, 2000.
7. **Béguin P, Wang XY, Firsov D, Puoti A, Claeys D, Horisberger J-D, and Geering K.** The γ subunit is a specific component of the Na,K-ATPase and modulates its transport function. *EMBO J* 16: 4250-4260, 1997.

8. **Blanco G and Mercer RW**. Isozymes of the Na-K-ATPase: Heterogeneity in structure, diversity in function. *Am J Physiol* 275: F633-F650, 1998.
9. **Colonna TE, Huynh L, and Fambrough DM.** Subunit interactions in the Na,K-ATPase explored with the yeast two-hybrid system. *J Biol Chem.* 272:12366-12372, 1997.
10. **Crambert G, Hasler U, Beggah AT, Yu C, Modyanov NN, Horisberger J-D, Lelievre L, and Geering K.** Transport and pharmacological properties of nine different human Na, K-ATPase isozymes. *J Biol Chem.* 275: 1976-1986, 2000.
11. **Eakle KA, Kabalin MA, Wang SG, and Farley RA**. The influence of β subunit structure on the stability of $Na^+/K^+$-ATPase complexes and interaction with $K^+$. *J Biol Chem* 269: 6550-6557, 1994.
12. **Feraille E and Doucet A**. Sodium-potassium-adenosinetriphosphatase-dependent sodium transport in the kidney: Hormonal control. *Physiol Rev* 81: 345-418, 2001.
13. **Forbush III B, Kaplan JH, and Hoffman JF.** Characterization of a new photoaffinity derivative of ouabain: Labeling of the large polypeptide and of a proteolipid component of the Na,K-ATPase. *Biochemistry* 17: 3667-3676, 1978.
14. **Fu X and Kamps M.** E2a-Pbx1 induces aberrant expression of tissue-specific and developmentally regulated genes when expressed in NIH 3T3 fibroblasts. *Mol Cell Biol* 17: 1503-1512, 1997.
15. **Geering K.** The functional role of β subunits in oligomeric P-type ATPases. *J Bioenerg Biomembr* 33: 425-438, 2001.
16. **Geering K, Beggah A, Good P, Girardet S, Roy S, Schaer D, and Jaunin P**. Oligomerization and maturation of Na,K-ATPase: Functional interaction of the cytoplasmic $NH_2$ terminus of the β subunit with the α subunit. *J Cell Biol* 133: 1193-1204, 1996.
17. **Hasler U, Crambert G, Horisberger J-D, and Geering K**. Structural and functional features of the transmembrane domain of the Na,K-ATPase β subunit revealed by tryptophan scanning. *J Biol Chem.* 276: 16356-16364, 2001.
18. **Hasler U, Greasley PJ, von Heijne G., and Geering K**. Determinants of topogenesis and glycosylation of type II membrane proteins. Analysis of Na,K-ATPase β1 and β3 subunits by glycosylation mapping. *J Biol Chem.* 275: 29011-29022, 2000.
19. **Hasler U, Wang X, Crambert G, Béguin P, Jaisser F, Horisberger J-D, and Geering K**. Role of β-subunit domains in the assembly, stable expression, intracellular routing, and functional properties of Na,K-ATPase. *J Biol Chem* 273: 30826-30835, 1998.
20. **Ivanov A, Zhao H, and Modyanov NN.** Packing of the transmembrane helices of Na,K-ATPase: Direct contact between β-subunit and H8 segment of α-subunit revealed by oxidative cross-linking. *Biochemistry* 39: 9778-9785, 2000.
21. **Jaisser F, Jaunin P, Geering K, Rossier BC, and Horisberger J-D**. Modulation of the Na,K-pump function by β subunit isoforms. *J Gen Physiol* 103: 605-623, 1994.
22. **Jaunin P, Jaisser F, Beggah AT, Takeyasu K, Mangeat P, Rossier BC, Horisberger J-D, and Geering K**. Role of the transmembrane and extracytoplasmic domain of β-subunits in subunit assembly, intracellular transport, and functional expression of Na,K-pumps. *J Cell Biol* 123: 1751-1759, 1993.
23. **Koenderink JB, Swarts HG, Hermsen HP, and De Pont JJ.** The β-subunits of $Na^+,K^+$-ATPase and gastric $H^+,K^+$-ATPase have a high preference for their own α-subunit and affect the $K^+$ affinity of these enzymes. *J Biol Chem.* 274: 11604-11610, 1999.
24. **Küster B, Shainskaya A, Pu HX, Goldshleger R, Blostein R, Mann M, and Karlish. SJ.** A new variant of the γ subunit of renal Na,K-ATPase. Identification by mass spectrometry, antibody binding, and expression in cultured cells. *J Biol Chem.* 275: 18441-18446, 2000.

25. **Lemmon MA, Treutlein HR, Adams PD, Brünger AT, and Engelman DMA.** Dimerization motif for transmembrane α-helices. *Nat Struct Biol* 1: 157-163, 1994.
26. **Lutsenko S and Kaplan JH.**. An essential role for the extracellular domain of the Na,K-ATPase β subunit in cation occlusion. *Biochemistry* 32: 6737-6743, 1993.
27. **Lutsenko S and Kaplan JH.** Organization of P-type ATPases. Significance of structural diversity. *Biochemistry* 34: 15607-15613, 1995.
28. **Meij IC, Koenderink JB, van Bokhoven H, Assink KF, Groenestege WT, de Pont JJ, Bindels RJ, Monnens LA, van den Heuvel LP, and Knoers NV.** Dominant isolated renal magnesium loss is caused by misrouting of the $Na^+$,$K^+$-ATPase γ-subunit. *Nat Genetics* 26: 265-266, 2000.
29. **Mercer RW, Biemesderfer D, Bliss DP, Collins JH, and Forbush III B.** Molecular cloning and immunological characterization of the γ polypeptide, a small protein associated with the Na, K-ATPase. *J Cell Biol* 121: 579-586, 1993.
30. **Moorman JR, Ackerman SJ, Kowdley GC, Griffin MP, Mounsey JP, Chen ZH, Cala SE, Obrian JJ, Szabo G, and Jones LR.** Unitary anion currents through phospholemman channel molecules. *Nature* 377: 737-740, 1995.
31. **Morrison BW, Moorman JR, Kowdley GC, Kobayashi YM, Jones, LR, and Leder P.** Mat-8, a novel phospholemman-like protein expressed in human breast tumors, induces a chloride conductance in *Xenopus* oocytes. *J Biol Chem* 270: 2176-2182, 1995.
32. **Or E, Goldshleger R, and Karlish SJ.** Characterization of disulfide cross-links between fragments of proteolyzed Na,K-ATPase. Implications for spatial organization of trans-membrane helices. *J Biol Chem* 274: 2802-2809, 1999.
33. **Pu HX, Cluzeaud F, Goldshleger R, Karlish SJ, Farman N, and Blostein R.** Functional role and immunocytochemical localization of the γa and γb forms of the Na,K-ATPase γ subunit. *J Biol Chem* 276: 20370-20378, 2001.
34. **Rice WJ, Young HS, Martin DW, Sachs JR, and Stokes DL.** Structure of $Na^+$,$K^+$-ATPases at 11-A resolution: Comparison with $Ca^{2+}$-ATPase in E1 and E2 states. *Biophys J* 80: 2187-2197, 2001.
35. **Shainskaya A and Karlish SJD.** Chymotryptic digestion of the cytoplasmic domain of the β subunit of Na/K-ATPase alters kinetics of occlusion of $Rb^+$ ions. *J Biol Chem* 271: 10309-10316, 1996.
36. **Shainskaya A. Schneeberger A, Apell HJ, and Karlish SJ.** Entrance port for $Na^+$ and $K^+$ ions on $Na^+$,$K^+$-ATPase in the cytoplasmic loop between trans-membrane segments M6 and M7 of the α subunit. Proximity of the cytoplasmic segment of the β subunit. *J Biol Chem* 275: 2019-2028, 2000.
37. **Shi H, Levy-Holzman R, Cluzeaud F, Farman N, and Garty H.** Membrane topology and immunolocalization of CHIF in kidney and intestine. *Am J Physiol* 280: F505-F512, 2001.
38. **Simmerman HK and Jones LR.** Phospholamban: protein structure, mechanism of action, and role in cardiac function. *Physiol Rev* 78: 921-947, 1998.
39. **Sweadner KJ and Rael E.** The FXYD gene family of small ion transport regulators or channels: cDNA sequence, protein signature sequence, and expression. *Genomics* 68: 41-56, 2000.
40. **Wetzel RK and Sweadner KJ.** Immunocytochemical localization of Na,K-ATPase α- and γ-subunits in rat kidney. *Am J Physiol* 281: F531-F545, 2001.
41. **Thérien AG, Goldshleger R, Karlish SJD, and Blostein R.** Tissue-specific distribution and modulatory role of the γ subunit of the Na,K-ATPase. *J Biol Chem.* 272: 32628-32634, 1997.
42. **Toyoshima C, Nakasako M, Nomura H, and Ogawa H.** Crystal structure of the calcium pump of sarcoplasmic reticulum at 2.6 A resolution. *Nature* 405: 647-655, 2000.

Chapter 4

# HKcα Encodes Ouabain-Insensitive H,K-ATPase in Surface Cells of Rat Distal Colon

HENRY J. BINDER
*Yale University*
*Department of Internal Medicine, 333 Cedar St. / P.O. Box 208019*
*New Haven, CT USA 06520-8019*

The mammalian colon has an important role in total body K homeostasis. The importance of colonic K movement in both subjects and experimental animals with normal renal function and normal dietary intake of sodium and potassium is modest; but in the presence of reduced renal function (e.g., anephric patients on hemodialysis) and during variation of dietary K intake, the colon adapts and become an important site for both K absorption and secretion. Although under normal conditions there is net K secretion in the large intestine that occurs via both potential-dependent and active transport mechanisms, the colon is capable of both active K secretion and active K absorption. Further, these active K transport mechanisms are modified by the dietary K status of the organism.

Several years ago we demonstrated and characterized an active K absorptive process in the rat distal colon and suggested that this active K absorption was mediated by an apical membrane electroneutral, Na-independent H-K exchange, i.e., H,K-ATPase (1, 2). Subsequent physiological studies demonstrated similar active K absorption and proton secretion in guinea pig and rabbit colon (3, 4). Since additional studies also demonstrated that both H,K-ATPase activity (present in apical membranes isolated from the rat distal colon) and active K absorption had both ouabain-sensitive and ouabain-insensitive components (5, 6), we proposed that this ATPase was a member of the family of P-type ATPases with features related to both ouabain-sensitive Na,K-ATPase and ouabain-insensitive gastric parietal cell H,K-ATPase. The cloning of H,K-ATPase α subunit (HKcα)[1] by Crowson and shull in 1992 led to several studies designed to provide increased understanding of the function and regulation of the colonic H,K-ATPase (7). Although the expression and regulation of this ATPase has been extensively studied during the past decade by this (and other) laboratories, understanding of its function remains incomplete.

Active K absorption in the distal colon of the rat is substantially enhanced both by aldosterone (secondary either or dietary sodium restriction or to continuous subcutaneous aldosterone administration ) and by dietary K depletion (1, 14). The mechanism(s) by which active K absorption is increased by aldosterone and by dietary K depletion are dissimilar. Aldosterone is associated with an increase in H,K-ATPase activity as well as increased HKcα message abundance and protein expression. In contrast, dietary K depletion does not appear to alter H,K-ATPase function but does increase: 1) a β subunit

[1]H,K-ATPase α subunit has been designated by some investigators as HKα2.

(HKcβ) that has been cloned from the rat distal colon and that may be the primary β subunit associated with the colonic H,K-ATPase and the 2) basolateral membrane K-Cl cotransport (KCC1). The differential effects of aldosterone and dietary K depletion on apical membrane H,K-ATPase and basolateral membrane KCC1 are summarized in Table 1.

**Table 1**

**Effect of Aldosterone and Dietary K Depletion**

| | Aldosterone | K Depletion |
|---|---|---|
| Active K absorption | ⇑ | ⇑ |
| H,K-ATPase activity | ⇑ | → |
| HKcα message and protein | ⇑ | → |
| HKcβ message and protein | → | ⇑ |
| KCC1 message and protein | → | ⇑ |

The predicted amino acids sequences of HKcα and the α subunits of Na, K-ATPase and gastric parietal cell H,K-ATPase have approximately 50-60% overall homology. Initial studies established that the primary localization of HKcα mRNA was in the distal colon with evidence of its presence in kidney, but not in stomach (8, 9). Further, HKcα mRNA is restricted to surface cells ( and the upper 20% of crypt cells ) with its protein localized to the apical membrane of surface cells (10). Controversy exists as a result of conflicting observations whether HKcα encodes the ouabain-insensitive or the ouabain-sensitive H,K-ATPase isoform. Studies of HKcα in both *Sf*-9 cells and HEK293 cells revealed the expression of ouabain-insensitive H,K-ATPase function and $^{86}$Rb uptake (8, 18). In contrast, evidence of ouabain-sensitive $^{86}$Rb uptake was demonstrated in experiments with HKcα in Xenopus oocytes (12, 13).

This communication summarizes our several observations that support the conclusion that there are **two** H,K-ATPase isoforms in the distal colon ( at least of the rat ) and that HKcα most likely encodes the ouabain-insensitive isoform. If a ouabain-sensitive isoform was encoded by HKcα, it would be necessary to postulate the presence of at least **three** different H,K-ATPase isoforms in the distal colon and would not be compatible with the recent studies that HKcα may also function in the apical membrane as a Na-K exchange. This discussion of multiple H,K-ATPase isoforms must include a detailed analysis of the effect of aldosterone on the ouabain-insensitive and ouabain-sensitive HKcα isoforms.

## Effect of Aldosterone on Active K Absorption and H,K-ATPase

Aldosterone induces an increase in Na,K-ATPase, active $Na^+$ absorption and, as a result of an increase in the electrical potential difference, potential-dependent K secretion in both kidney and colon. When K transport is studied in vitro across mucosa from rat distal colon under voltage clamp conditions, active K absorption is observed in normal rats, but active K secretion is readily demonstrated in rats treated with aldosterone. However, whenever active K secretion is inhibited by one or more experimental maneuvers (e.g., removal of lumen Na, addition of either lumen K channel blockers, serosal ouabain to inhibit

Na,K-ATPase, or serosal bumetanide to inhibit Na,K-2Cl cotransport), a two-three fold increase in active K absorption is observed (i.e., K absorption is unmasked and is substantially enhanced) (15).

Two distinct components of active K absorption have been identified in the rat distal colon: one that is both mucosal Na-sensitive (i.e., the rate of active K absorption is inversely related to the presence of mucosal Na) and ouabain-insensitive and another that is both mucosal Na-insensitive and ouabain-sensitive (6). Aldosterone increases only one of these two components of active K absorption, the ouabain-insensitive component (Figure 1). Similarly, apical membrane H,K-ATPase has two approximately equal components, one that is ouabain-sensitive and the other that is ouabain-insensitive (5). Although aldosterone increases total H,K-ATPase activity in apical membrane, this increase represents primarily a 109% increase in the ouabain-insensitive isoform's activity. In contrast, the activity of the ouabain-sensitive isoform was enhanced by only 42% (16).

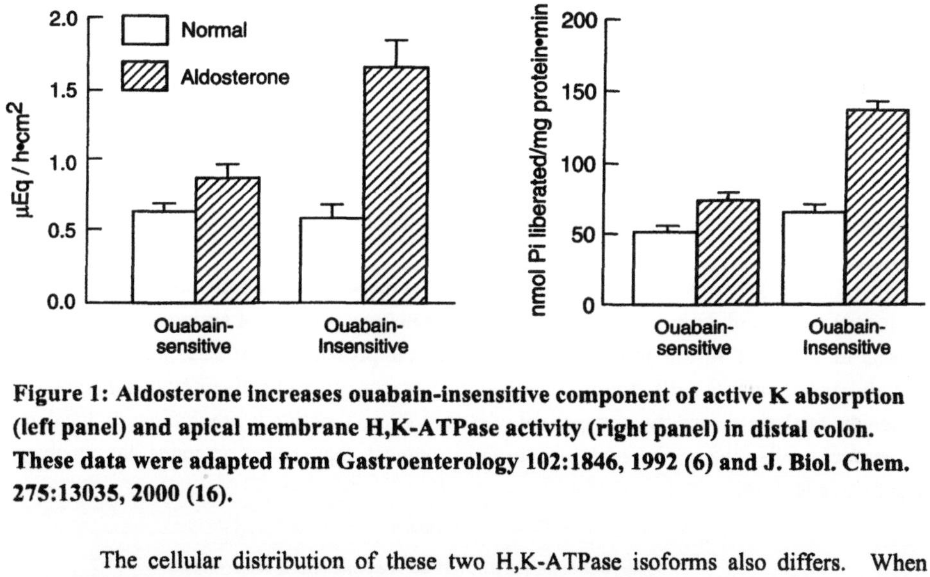


**Figure 1: Aldosterone increases ouabain-insensitive component of active K absorption (left panel) and apical membrane H,K-ATPase activity (right panel) in distal colon. These data were adapted from Gastroenterology 102:1846, 1992 (6) and J. Biol. Chem. 275:13035, 2000 (16).**

The cellular distribution of these two H,K-ATPase isoforms also differs. When H,K-ATPase activity was determined in apical membranes isolated from colonic crypt cells, only ouabain-sensitive H,K-ATPase activity was identified. In contrast, when similar studies were performed with apical membranes isolated from surface cells, both ouabain-sensitive and ouabain-insensitive activities were found (17). Based on studies of intracellular pH (pHi) in which individual cells are analyzed fluorometrically, we believe that the ouabain-sensitive activity represents contamination from crypt apical membranes. These pHi studies provided evidence of a K-dependent recovery to an acid load in both surface and crypt cells and are consistent with an apical membrane H-K exchange (i.e., H,K-ATPase). The K-dependent recovery of pHi to an acid load in crypt cells was completely ouabain-sensitive, while that in surface cells was solely ouabain-insensitive (17) (Table 2).

As a result of these physiological studies of H,K-ATPase isoforms, we believe that the ouabain-insensitive isoform of H,K-ATPase is exclusively localized to apical membranes of surface cells and that the H,K-ATPase isoform present in apical membranes of crypt cells is ouabain-sensitive. Further, these studies indicate that aldosterone increases only the ouabain-insensitive component of active K absorption and that the ouabain-sensitive component of active K absorption is not affected by aldosterone.

## Studies of Cellular and Segmental Localization of HKcα

Analysis of the cellular and segmental distribution of HKcα message and protein and the segmental distribution of active K absorption is consistent with a role of HKcα in active K absorption in the rat colon. First, active K absorption is present solely in the distal colon and has not been identified in the proximal colon even in aldosterone rats (1). Second, although H,K-ATPase activity is readily identified in apical membranes isolated from distal colon, only minimal H,K-ATPase activity can be determined in apical membranes isolated from proximal colon (9). Third, both HKcα message abundance and protein expression are also present in distal colon with but a minimal presence in proximal colon (9). These observations in normal rats are the basis for the thesis that active K absorption in the rat distal colon is most likely closely associated with HKcα.

Evaluation of the spatial distribution of HKcα message and protein as well as that of H,K-ATPase provides additional support for the probability that HKcα encodes the ouabain-insensitive isoform of H,K-ATPase. First, in situ hybridization studies established the localization of HKcα message to surface cells of the distal colon and the upper 20% of crypt cells (10). Second, immunocytochemical studies using a polyclonal antibody that had been raised to a HKcα fusion protein demonstrated that presence of HKcα protein in apical membranes of surface cells, but not in apical membranes of crypt cells in rat distal colon (8). Since the ouabain-insensitive component of H,K-ATPase function is predominantly localized to surface cells, these observations strongly suggest that HKcα encodes the ouabain-insensitive isoform.

**Table 2**

**Spatial distribution of H,K-ATPase in colonic surface and crypt cells**

| | Surface cells | Crypt cells |
|---|---|---|
| HKcα message | 4+ | 0 |
| HKcα protein | 4+ | 0 |
| H,K-ATPase | | |
| Ouabain-insensitive | 3+ | 0 |
| Ouabain-sensitive | 2+ * | 3+ |
| K-dependent pHi recovery | | |
| Ouabain-insensitive | 4+ | 0 |
| Ouabain-sensitive | 0 | 3+ |

* Probably represents contamination from basolateral membranes (5)

### Studies of Aldosterone and HKcα

Recent examination of the effect of aldosterone on HKcα message and protein has also been reported and provide additional support for the role of HKcα in the regulation of the ouabain-insensitive isoform of H,K-ATPase and the ouabain-insensitive component of active K absorption. First, HKcα mRNA abundance was markedly increased in the distal colon of aldosterone treated animals (Figure 2) (9). Second, aldosterone increased HKcα protein in apical membranes isolated from distal colon in western blot studies (9). Third, similar findings were observed in immunocytochemical studies in which there was a substantial increase in the intensity of labeling in apical membranes of the distal colon (Figure 2) (9). Although active K absorption has not been demonstrated in proximal colon of either normal or aldosterone rats, minimal HKcα message abundance and protein expression was observed in the proximal colon of aldosterone (but not normal) rats. In contrast, dietary K depletion which also increases active K absorption did not alter HKcα message or protein in the distal colon (9) (see Table 1). Thus, the results of both the spatial (surface cell vs. crypt cell) and segmental (distal vs. proximal colon) localization and the effect of aldosterone on HKcα message and protein are all consistent with the conclusion that HKcα encodes the ouabain-insensitive H,K-ATPase isoform.

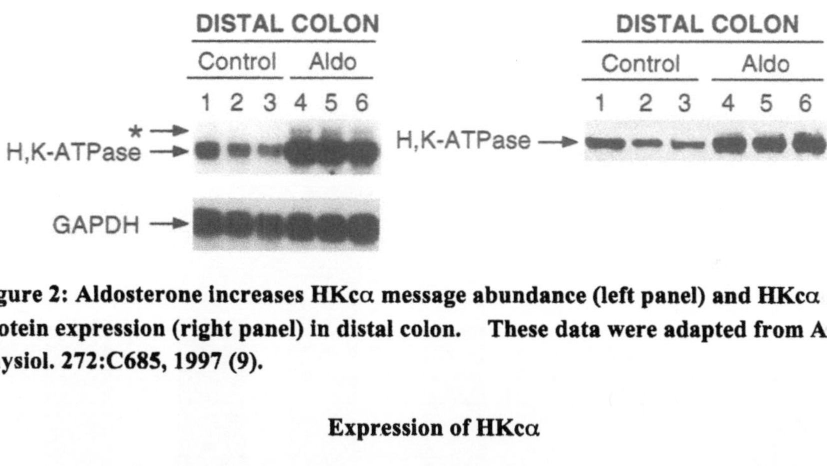


**Figure 2: Aldosterone increases HKcα message abundance (left panel) and HKcα protein expression (right panel) in distal colon. These data were adapted from Am. J. Physiol. 272:C685, 1997 (9).**

### Expression of HKcα

Studies of HKcα expression have been performed with different β subunits in three different cell systems with conflicting observations (11-13). Further, both H,K-ATPase activity and $^{86}$Rb uptake (as a surrogate of K transport) have been determined as parameters of HKcα expression. First, expression of HKcα cDNA in *Sf*-9 cells using baculovirus without the co-infection with any exogenous β subunits resulted in H,K-ATPase activity that was completely ouabain-insensitive (8). It is not known whether *Sf*-9 cells have an endogenous β subunit. Second, HEK293 cells co-transfected with HKcα and with HKcβ also expressed H,K-ATPase activity that was ouabain-insensitive (Figure 3 top) (18). When NaKβ1 was used as the β subunit, ouabain-insensitive H,K-ATPase activity was approximately 80% of that observed in the HKcβ co-transfected cells and when HKcα was singly transfected, H,K-ATPase activity was still present (about 20% of maximal activity). Third, when $^{86}$Rb uptake was determined in parallel studies that were performed in HKcα and HKcβ co-transfected HEK293 cells, there was also expression of stimulated $^{86}$Rb uptake that was ouabain-insensitive (Figure 3 bottom) (18). These several observations provide unequivocal evidence that HKcα expresses an ouabain-insensitive H,K-ATPase.

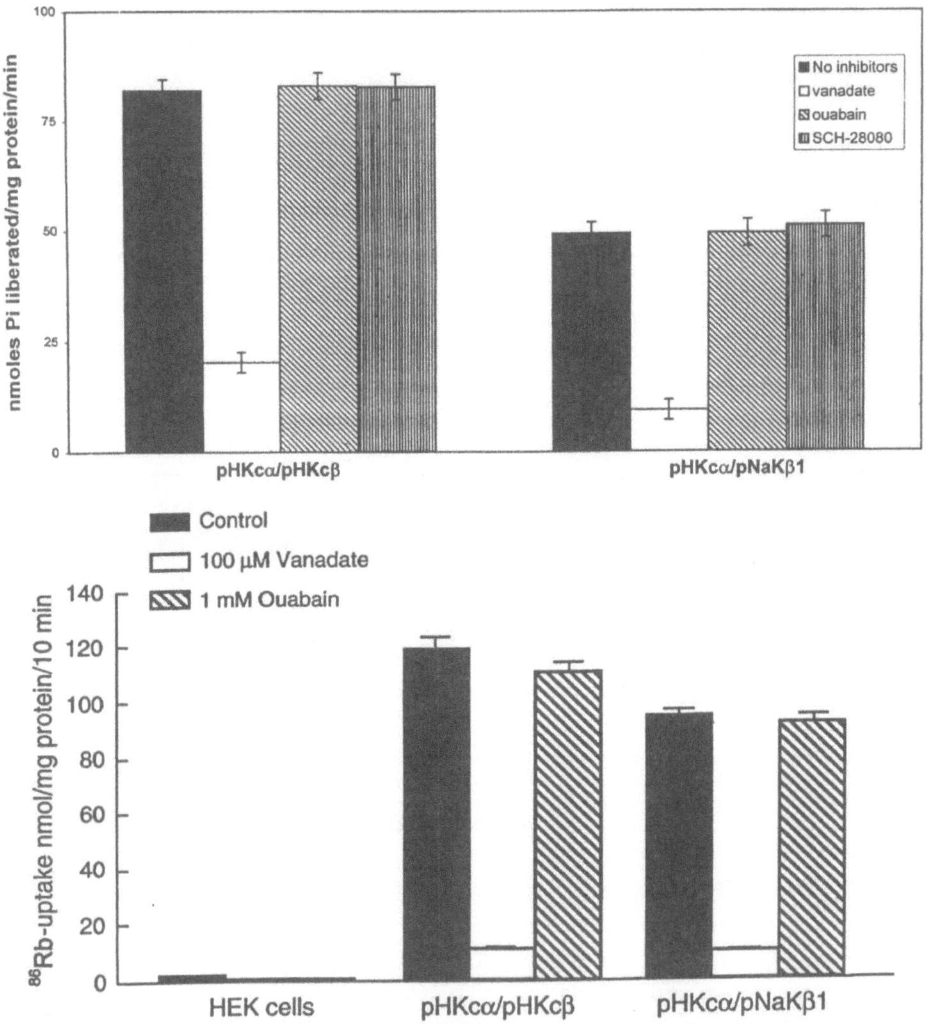


**Figure 3: Expression of HKcα cDNA in HEK293 cells. Expression of HKcα cDNA in PS120 cells as either H,K-ATPase activity (top panel) or $^{86}$Rb uptake is ouabain-insensitive (bottom panel). HKcα was co-transfected with a β subunit, either HKcβ or NaKβ1. (These data were adapted from Am. J. Physiol. 278:C182, 2000)(19).**

In contrast to these several observations, two different groups have performed similar studies in Xenopus oocytes injected with HKcα cRNA and have obtained different results (12, 13). In these studies $^{86}$Rb uptake, but not H,K-ATPase activity was determined. Although $^{86}$Rb uptake was not observed in uninjected oocytes, $^{86}$Rb uptake was present in oocytes co-injected with both HKcα cRNA and that of a β subunit. This $^{86}$Rb uptake was ouabain-sensitive. Since both the physiological studies (Figure 1) and the expression studies in HEK293 and *Sf*-9 cells both suggest that HKcα expresses the ouabain-insensitive component, there is no adequate explanation for these disparate observations. One could speculate that the expression of HKcα in oocytes is associated with the expression of a second protein that modifies the ouabain-sensitivity of H,K-ATPase activity. An alternative suggestion could be that there are three, rather than two H,K-ATPase isoforms in the rat distal colon: both ouabain-sensitive and ouabain-insensitive isoforms in surface cells and an ouabain-sensitive isoform in crypt cells. However, the studies of K-dependent pHi recovery that visualized fluorescence in individual cells and demonstrated only ouabain-insensitive activity in surface cells makes the possibility of three H,K-ATPase isoforms unlikely.

## Expression of H,K-ATPase as a Na-K Exchange

The initial demonstration of H,K-ATPase activity in apical membranes of rat distal colon also reported evidence of Na,K-ATPase activity (5). As neither an independent marker of colonic apical membranes nor an antibody to H,K-ATPase was available at that time, it was assumed that this Na,K-ATPase activity represented basolateral membrane contamination. Three recent studies indicate that this Na,K-ATPase activity likely represents an alternate mode of expression of H,K-ATPase, but do not provide information as to the function of this Na,K-ATPase activity (16, 20, 21).

The studies of HKcα expression in Xenopus oocytes also revealed that $^{86}$Rb uptake that was expressed in oocytes injected with HKcα cRNA also represented Na-K exchange that was ouabain-sensitive (20). These two studies reached the same conclusion that HKcα expresses an ouabain-sensitive ATPase in oocytes that can function either as an H-K or as a Na-K exchange (20, 21). Similar evidence of Na,K-ATPase expression was obtained in studies of native tissue, i.e., apical membranes isolated from rat distal colon, but with a substantially different conclusion (16). First, these studies established that 92% of Na,K-ATPase activity in apical membranes was inhibited by a polyclonal antibody against HKcα protein. Second, this antibody did not alter basolateral Na,K-ATPase activity (Figure 4). Third, HKcα protein expression and ouabain-insensitive H,K-ATPase activity was again shown to be increased by aldosterone by more than 100% (Figures 1, 2). Fourth, the ouabain-sensitive Na,K-ATPase activity and the ouabain-sensitive H,K-ATPase activity were increased by 30% and 42%, respectively. Our interpretation of these results is that Na,K-ATPase activity in apical membranes represents an alternate mode of expression of a ouabain-sensitive H,K-ATPase isoform that has not as yet been identified and cloned (22). The function of this alternate mode of expression is not known.

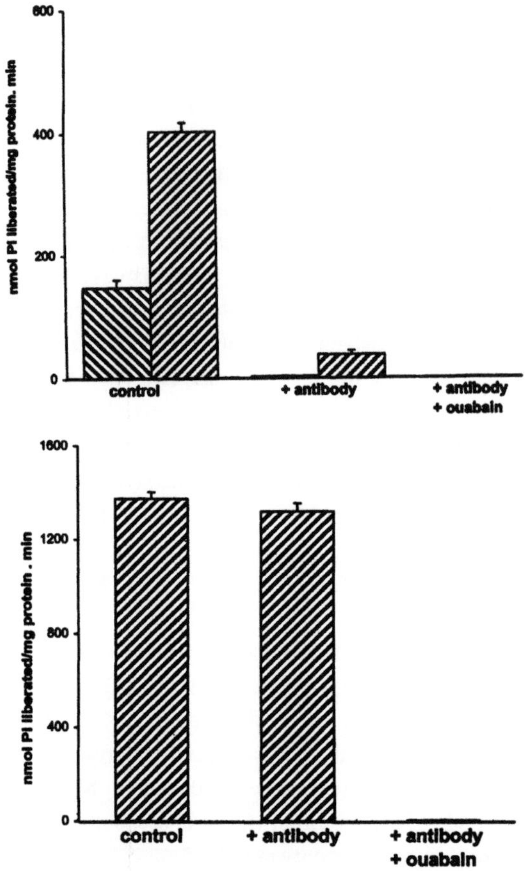


**Figure 4. Effect of rabbit polyclonal antibody (M1) against HKcα on Na,K-ATPase activity in apical and basolateral membranes. Apical membrane activity was inhibited by antibody (top panel) but basolateral activity was unaffected (bottom panel). [Adapted from J. Biol. Chem. 275:113035, 2000 (19)]**

In **conclusion**, we believe that the several observations that are summarized above provide compelling evidence that HKcα expresses the ouabain-insensitive H,K-ATPase isoform that is primarily localized to the apical membrane of surface cells of the rat distal colon and that is up-regulated by aldosterone. Further, the ouabain-sensitive H,K-ATPase is localized to the apical membrane of crypt cells and is not associated with aldosterone enhancement of active K absorption. It is also unlikely that this second H,K-ATPase isoform is linked to the increase in active K absorption induced by dietary K depletion. Finally, although both aldosterone and dietary K depletion increase active K absorption in the distal colon (Table 1), the molecular mechanisms of this enhancement are fundamentally different with aldosterone affecting apical membrane transport processes and dietary K depletion modifying both apical and basolateral membrane transport processes.

**Acknowledgements**: The author is grateful to his colleagues Drs. P. Sangan, V.M. Rajendran and J. Geibel for their active participation in this project. These studies were supported by a USPHS research grant (ROI DK18777) from NIDDK.

## References

1. **Foster, E.S., J. P. Hayslett, and H. J Binder.** Mechanism of active potassium absorption and secretion in the rat colon. *Am. J. Physiol. Gastrointest. Liver Physiol.* 246:G611-G617, 1984.
2. **Sweiry, J., and H. J. Binder.** Active potassium absorption in rat distal colon. *J. Physiol. (Lond.)* 423:155-170, 1990.
3. **Suzuki,Y., and K. Kaneko.** Ouabain-sensitive H-K exchange mechanism in the apical membrane of guinea pig colon. *Am. J. Physiol. Gastrointest. Liver Physiol.* 256:G979-G988, 1989.
4. **Willis N.K, and B. Biagi** Active potassium transport by rabbit descending colon epithelium. *J. Memb. Biol.* 64:195-203, 1982.
5. **Del Castillo, J. R., V. M.Rajendran, and H. J. Binder.** Apical membrane localization of ouabain-sensitive K-activated ATPase activities in rat distal colon. *Am. J. Physiol. Gastrointest. Liver Physiol.* 261:G1005-G1011, 1991.
6. **Pandiyan V., H.J. Binder, and V.M. Rajendran.** Mucosal ouabain and $Na^+$ inhibit active $Rb^+$ ($K^+$) absorption in normal and sodium-depleted rat colon. *Gastroenterology* 102:1846-1853, 1992.
7. **Crowson, M.S. and G.E. Shull.** Isolation and characterization of a cDNA encoding the putative distal colon H-K-ATPase. *J. Biol. Chem.* 267:13740-13748, 1992.
8. **Lee, J., V.M. Rajendran, A.S. Mann, M. Kashgarian, and H.J. Binder.** Functional expression and segmental localization of rat colonic K-ATPase. *J. Clin. Invest.* 96:2002-2008, 1995.
9. **Sangan, P., V.M. Rajendran, A.S. Mann, M. Kashgarian, and H.J. Binder.** Regulation of colonic H-K-ATPase in large intestine and kidney by Na depletion and dietary K depletion. *Am. J. Physiol. Cell Physiol.* 272: C685-C696, 1997.
10. **Jaisser, F., N. Coutry, N. Farman, H.J. Binder, and B.C. Roisser.** Aputative H-K-ATPase is selectively expressed in surface epithelial cells of rat distal colon. *Am. J. Physiol. Cell Physiol.* 265:C1080-C1089, 1993.
11. **Sangan, P., S.S. Kolla, V.M. Rajendran, M. Kashgarian, and H.J. Binder.** Colonic H-K-ATPase β-subunit: identification in apical membranes and regulation by dietary K depletion. *Am. J. Physiol. Cell Physiol.* 276:C350-C360, 1999.
12. **Cougnon, M., G. Planelles, M.S. Crowson, G.E. Shull, B.C. Rossier, and F. Jaisser.** The rat distal colon P-ATPase α-subunit encodes ouabain-sensitive H-K-ATPase. *J. Biol. Chem.* 271: 7277-7280, 1996.
13. **Codina, J., B.C. Kone, J.T. Delmas-Mata, and T.D. DuBose, Jr.** Functional expression of the colonic H,K-ATPase α-subunit. J. Biol. Chem. 271:29759-29763, 1996.
14. **Foster, E.S., H. J Binder , J. P. Hayslett, and J.I. Sandle.** Dietary potassium modulates

active potassium absorption and secretion in rat distal colon. *Am. J. Physiol. Gastrointest. Liver Physiol.* 251: G619-G626, 1986.

15. **Sweiry, J., and H.J. Binder.** Characterization of aldosterone induced potassium secretion in rat distal colon. *J. Clin. Invest.* 83:844-851, 1989.
16. **Rajendran,V.M., H.J. Binder, J. Geibel, and P. Sangan.** Ouabain-sensitive H,K-ATPase functions as Na,K-ATPase in apical membranes of rat distal colon. *J. Biol. Chem.* 275: 13035-13040, 2000.
17. **Rajendran, V.M., S.K. Singh, J. Geibel, and H.J. Binder.** Differential localization of colonic H-K-ATPase isoforms in surface and crypt cells. *Am. J. Physiol. Gastrointest. Liver Physiol.* 274:G424-G429, 1998.
18. **P. Sangan, S. Thevananther, S. Sangan, V.M. Rajendran, and H.J. Binder.** Colonic H-K-ATPase α-and β-subunits express ouabain-insensitive H-K-ATPase. *Am. J. Physiol. Cell Physiol.* 278:C182-C189, 2000.
19. **Rajendran, V.M., H.J. Binder, J. Geibel, and P. Sangan.** Ouabain-sensitive H,K-ATPase functions as Na, K-ATPase in apical membranes of rat distal colon. *J. Biol. Chem.* 275: 13035-13040, 2000.
20. **Cougnon, M., P. Bouyer, G. Planelles, and F. Jaisser.** Does the colonic H-K-ATPase also act as an Na-K-ATPase? *Proc. Natl. Acad. Sci. USA* 95:6516-6520, 1998.
21. **Codina, J., T.D. Dubose, Jr., and T.A. Presslye.** The colonic $H^+K^+$-ATPase functions as a $Na^+$-dependent $K^+$ $(NH_4)$-ATPase in apical membranes from rat distal colon. *J. Biol. Chem.* 274:19693-19698, 1999.
22. **P. Sangan, H.J. Binder, S. Brill, B. Forbush III, and Sasngan S.** Basolateral K-Cl cotransporter regulates colonic K absorption in K depletion. *J. Biol. Chem.* 275:30813-30816, 2000.

Chapter 5

# Molecular Diversity and Regulatory Heterogeneity of the $H^+,K^+$-ATPases in Kidney

THOMAS D. DUBOSE, JR., JUAN CODINA, AND JIAN LI
*Department of Internal Medicine, University of Kansas School of Medicine, 3901 Rainbow Blvd, Kansas City, KS 66160-7350-*

## INTRODUCTION

### Contrasting Features of Gastric and Colonic Isoforms

It is now widely appreciated that two members of the $H^+,K^+$-ATPase family are expressed in kidney: the gastric $H^+,K^+$-ATPase and the colonic $H^+,K^+$-ATPase. The gastric $H^+,K^+$-ATPase α-subunit ($HK\alpha_1$) assembles with a unique β-subunit ($\beta_G$). This heterodimer, when expressed in heterologous systems, or when examined *in vitro,* is insensitive to ouabain, highly sensitive to Sch-28080 and omeprazole, and its abundance in stomach exceeds that in kidney several fold. The gastric $H^+,K^+$-ATPase is localized in kidney predominantly to the collecting tubule apical membrane, where it exchanges $H^+$ for $K^+$ in a stoichiometry of 1:1. Gastric $H^+,K^+$-ATPase does not adapt at the mRNA and protein level to variations in systemic potassium homeostasis (6, 12). In contrast, studies measuring enzyme activity and transport function which was ascribed to $HK\alpha_1$, reported upregulation of $HK\alpha_1$ during metabolic and respiratory acidosis (13, 24). The site of this regulatory response is predominantly in the cortical collecting tubule (1, 2). The α-subunit of the colonic $H^+,K^+$-ATPase ($HK\alpha_2$) assembles with either the $\beta_1$ (5) or $\beta_3$ $Na^+K^+$-ATPase (21). When either β-subunit is expressed with $HK\alpha_2$ heterologously, it is relatively sensitive to ouabain, but resistant to Sch-28080. The colonic $H^+,K^+$-ATPase also exchanges $H^+$ for $K^+$ at the apical membrane, but is capable of transporting $Na^+$ for $H^+$ and $NH_4^+$ for $K^+$ (9-11). This heterodimer adapts robustly to chronic hypokalemia and less dramatically to metabolic acidosis (6, 12, 18). The site of this regulatory response is predominantly in the outer and inner medullary collecting tubules. These functional and pharmacologic differences are contrasted in Table 1.

**TABLE 1**

| | $HK\alpha_1$ | $HK\alpha_2$ |
|---|---|---|
| β-subunit assembly | $\beta_g$ | $\beta_1$, $\beta_3$, $Na^+,K^+$-ATPase |
| Pharmacological sensitivity | Sch-28080 | ouabain |
| Ions transported | $H^+/K^+$ | $H^+,Na^+/K^+,NH_4^+$ |
| Coupling ratio | 1:1 | >1:1 |
| Sch-28080 | 1-10 μM | insensitive |
| Ouabain | insensitive | 1-2 mM |

There is no specific inhibitor for the colonic $H^+,K^+$-ATPase but it, in striking contrast to the gastric isoform, is relatively sensitive to ouabain. Because of the well appreciated specificity of proton pump inhibitors for $HK\alpha_1$, a number of laboratories investigating transport in the kidney and colon, have examined $H^+,K^+$-ATPase function in the presence of high concentrations of Sch-28080. Therefore, it is important to appreciate that Sch-28080, in the rat probe, and that Sp1, NFκB and CREB do not (Figure 9). These deductions are summarized in Figure 10, showing the effects of *H. pylori* on transcription factor binding to the rat homology box I sequence. Thus, we propose that *H. pylori* displaces CREB from CRE and induces AP-2 binding to rat homology box I sequence. Neither of these response elements were predicted by MatInspector V2.2.

**Figure 10.** Effect of *H. pylori* on transcription factor interactions with rat homology box I.

In the case of cold oligo duplex competitions using the $^{32}P$-labeled human probe and uninfected AGS cell extracts, we deduced from the pattern of band attenuations that Sp1, AP-2, AP-1 and NFkB interact with human probe, and that GATA and CRE do not interact (Figure 11). In contrast, with nuclear extract from AGS cells infected with *H. pylori*, the pattern of mobility-shifted bands indicated that Sp1, AP-2 and NFkB continue to interact with human probe, and that AP-1, GATA and CREB do not interact (Figure 12). Thus, AP-1 appears to be the only transcription factor whose interaction with human homology box I is restricted by *H. pylori*. We confirmed this deduction by carrying out a super-shift analysis of infected and uninfected AGS cells nuclear extracts using an AP-1 specific antibody, as shown in Figure 13. In uninfected cells, we detected a prominent super-shifted AP-1 DNA:protein:antibody complex, whose intensity was significantly reduced following infection of the cells with *H. pylori*. These data are summarized in Figure 14, which shows the effects of *H. pylori* on transcription factor binding to the human homology box I sequence. We propose that *H. pylori* infection displaces AP-1 transcription factor from the first homology box, and that all other transcription factor interactions remain essentially unchanged.

**Figure 11.** Un-infected AGS gel shifts with human homology box I DNA probe.

**Figure 12.** *H. pylori*-infected AGS gel shifts with human homology box I DNA probe.

## EVIDENCE FOR SODIUM AND AMMONIUM TRANSPORT BY HK$\alpha_2$

Caplan and associates, demonstrated a coupling ratio of 1 $H^+$:10 $K^+$ when a nongastric $H^+,K^+$-ATPases, ATP1AL1, was expressed in HEK 293 cells (15). Subsequently, studies in our laboratory demonstrated that both $Na^+$-dependent and $Na^+$-independent $K^+$ATPase activities were present in distal colon apical membranes enriched in colonic $H^+,K^+$-ATPase (8). The $Na^+$-dependent $K^+$-ATPase was blocked by a specific anti- HK$\alpha_2$ antibody. In contrast, the anti- HK$\alpha_2$ antibody did not block the $Na^+$-dependent $K^+$-ATPase activity in renal cortex. Furthermore, in this same study, it was demonstrated that the $Na^+$-independent $K^+$ATPase fraction in apical distal colon membranes enriched in HK$\alpha_2$, also had a high affinity for ammonium $NH_4^+$. Therefore, $NH_4^+$ can substitute for $K^+$ efficiently. In summary, the properties of the colonic $H^+,K^+$-ATPase include, the ability to transport $Na^+$ for $H^+$, as well as $NH_4^+$ for $K^+$, resistance to Sch-28080, and sensitive to moderate concentrations of ouabain (Table 1). The physiological significance of $Na^+$ rather than $H^+$ secretion and $NH_4^+$ rather than $K^+$ absorption in either distal colon or medullary collecting duct is not clear. However, a logical model for transport in either alternative mode assumes that the $H^+,K^+$-ATPase would be coupled with other transporters. Such a relationship has not been established at the present time.

## REGULATORY ADAPTATION OF HK$\alpha_2$

Studies from a number of laboratories have shown that bicarbonate absorption and potassium absorption are enhanced by chronic hypokalemia in rabbit (26) and rat collecting tubules perfused *in vitro* (25). Additionally, there is also evidence for regulatory adaptation of the cortical collecting tubule to metabolic acidosis (24) and respiratory acidosis (13). In most transport studies, the Sch 28080-sensitive-$K^+$ dependent component of $HCO_3^-$ absorption in cortical collecting tubule, outer medullary collecting tubule, and terminal inner medullary collecting tubules perfused *in vitro* have been defined as $H^+,K^+$-ATPase transport. Studies in our laboratory (25) investigated the possibility of an ouabain-sensitive component of $HCO_3^-$ absorption in terminal IMCD. In these studies, rats with chronic metabolic acidosis (ammonium chloride added to drinking water for four days) and concomitant chronic dietary potassium depletion, demonstrated a marked up-regulation of the Sch-28080 sensitive component of $HCO_3^-$ absorption ($J_{rCO2}$). While about 70% of $J_{tCO2}$ was inhibited by Sch 28080, these studies demonstrated that ouabain and Sch 28080 have additive inhibitory effects on $J_{tCO2}$ in terminal IMCD during chronic hypokalemia. Similarily, Nakamura and colleagues have demonstrated emergence of a ouabain-sensitive component of $J_{tCO2}$ in OMCD during chronic hypokalemia (20). The demonstration, by these two groups, of a ouabain-sensitive component, suggests the emergence of a unique $H^+,K^+$-ATPase with adaptaton to chronic hypokalemia. Alternatively, post-translational modification or regulation of either the gastric $H^+,K^+$-ATPase or the colonic $H^+,K^+$-ATPase remain possibilities.

### POSSIBLE EXPLANATIONS FOR *EX VIVO/IN VITRO* DIFFERENCES

Two splice variants of HK$\alpha_2$ have been reported recently. The first, (17) described by Kone in rat, is a truncated variant which is missing the first 108 amino acids at the amino terminal end. The second reported splice variant from Wingo and associates (3) in rabbit is an extended protein at the amino terminal. Nevertheless, based on functional data available thus far, neither of these splice variants reveal deviations from the full length HK$\alpha_2$ isoform when expressed heterologously (17). Therefore, possible regulatory expression of splice variants do not offer adequate explanations for functional *ex vivo* and *in vitro* differences noted thus far.

It has also been suggested that HK$\alpha_2/\beta_1$ could assemble with a $\gamma$-subunit or a $\gamma$-like protein (16). Recent studies in our laboratory have demonstrated that $\gamma$-RNA is expressed predominantly in kidney. Therefore, we developed an antibody which recognizes both $\gamma_A$ and $\gamma_B$ subunits. Further studies demonstrated that the anti-$\gamma$ antibody cross-reacted with recombinant and native proteins. In apical-enriched membranes, $\gamma$ protein was shown to be abundant in both renal medulla and renal cortex, but was not detected in distal colon. Nevertheless, these studies demonstrated that $\gamma$-$Na^+K$ associated only with $\alpha_1/\beta_1$ $Na^+$, K-ATPase by coimmunoprecipitation and was not stably associated with HK$\alpha_2$ (unpublished observation). Taken together, $\gamma$-$Na^+K$ ATPase assembly with HK$\alpha_2$ is not demonstrable, so that changes in pharmacologic sensitivity or function *in vitro* as a result of $\gamma$ assembly, is not a viable explanation of these functional differences. In summary, differences in *ex vivo* and *in vitro* function of HK$\alpha_2$ cannot be ascribed to $\gamma$-subunit assembly, expression, or alternative site regulation of truncated isoforms.

The role of $H^+,K^+$-ATPases has been elucidated further from studies by Shull, et al, in a gene-targeted mouse model (19). The HK$\alpha_2$ knockout model does not develop systemic hypokalemia except during dietary $K^+$ restriction. Wasting of potassium from the kidney was not apparent in this knockout model, even during dietary $K^+$ restriction. There was, however, evidence of inappropriate $K^+$ excretion from the colon. Moreover, studies by this group have demonstrated recently that the HK$\alpha_2$-knockout mouse does not display corresponding up-regulation of HK$\alpha_1$. Therefore, upregulation of HK$\alpha_1$ in the HK$\alpha_2$ knockout, is not a possible explanation for the failure of the HK$\alpha_2$ deficient animals to display renal $K^+$-wasting.

Therefore, when taken together, disparities in HK$\alpha_2$ function, when expressed in heterologous systems or cells in culture, compared to native tissues, suggest that a unique and, as yet unidentified $\alpha$-$H^+K^+$ isoform should exist in the collecting duct. This unique transporter functions importantly to increase $K^+$ absorption to protect against further declines in plasma $K^+$ during chronic hypokalemia.

## REFERENCES

1. **Ahn, K. Y., and B. C. Kone.** Expression and cellular localization of mRNA encoding the "gastric" isoform of H,K-ATPase alpha-subunit in rat kidney. *Am J Physiol* 268: F99-109, 1995.
2. **Ahn, K. Y., P. B. Turner, K. M. Madsen, and B. C. Kone**. Effects of chronic hypokalemia on renal expression of the "gastric" H,K-ATPase alpha-subunit gene. *Am J Physiol* 270: F557-66, 1996.
3. **Campbell, W. G., I. D. Weiner, C. S. Wingo, and B. D. Cain.** H-K-ATPase in the RCCT-28A rabbit cortical collecting duct cell line. *Am J Physiol* 276: F237-45, 1999.
4. **Codina, J., J. Cardwell, J. J. Gitomer, Y. Cui, B. C. Kone, and T. D. Dubose, Jr.** Sch-28080 depletes intracellular ATP selectively in mIMCD-3 cells. *Am J Physiol Cell Physiol* 279: C1319-26, 2000.

5. **Codina, J., J. T. Delmas Mata, and T. D. DuBose, Jr.** The alpha-subunit of the colonic H+,K+-ATPase assembles with beta1-Na,K-ATPase in kidney and distal colon. *J Biol Chem* 273: 7894-9, 1998.
6. **Codina, J., J. T. Delmas-Mata, and T. D. DuBose, Jr.** Expression of HKalpha2 protein is increased selectively in renal medulla by chronic hypokalemia. *Am J Physiol* 275: F433-40, 1998.
7. **Codina, J., B. C. Kone, J. T. Delmas Mata, and T. D. DuBose, Jr.** Functional expression of the colonic H,K-ATPase alpha- subunit. Pharmacologic properties and assembly with X, K-ATPase beta-subunits. *J Biol Chem* 271: 29759-63, 1996.
8. **Codina, J., T. A. Pressley, and T. D. DuBose, Jr.** The Colonic H,K-ATPase Functions as a Na-dependent K(NH4)-ATPase in Apical Membranes from Rat Distal Colon. *J Biol Chem* 274: 19693-19698, 1999.
9. **Cougnon, M., P. Bouyer, G. Planelles, and F. Jaisser.** Does the colonic H,K-ATPase also act as an Na,K-ATPase? *Proc Natl Acad Sci U S A* 95: 6516-20, 1998.
10. **Cougnon, M., F. Jaisser, A. Edelman,** T. Anagnostopoulos, and G. Planelles. Ammonium ions are transported by the colonic H,K-ATPase. *J Am Soc Nephrol* 9: 4A, 1998.
11. **Cougnon, M., G. Planelles, M. S. Crowson, G. E. Shull, B. C. Rossier, and F. Jaisser.** The rat distal colon P-ATPase alpha subunit encodes a ouabain-sensitive H,K-ATPase. *J Biol Chem* 271: 7277-80, 1996.
12. **DuBose, T. D., Jr., J. Codina, A. Burges, and T. A. Pressley.** Regulation of H,K-ATPase expression in kidney. *Am J Physiol* 269: F500-7, 1995.
13. **Eiam-ong, S., M. E. Laski, N. A. Kurtzman, and S. Sabatini.** Effect of respiratory acidosis and respiratory alkalosis on renal transport enzymes. *Am J Physiol* 267: F390-9, 1994.
14. **Gallardo, P., L. P. Cid, C. P. Vio, and F. V. Sepulveda.** Aquaporin-2, a regulated water channel, is expressed in apical membranes of rat distal colon epithelium. *Am J Physiol Gastrointest Liver Physiol* 281: G856-63., 2001.
15. **Grishin, A. V., and M. J. Caplan.** ATP1AL1, a member of the non-gastric H,K-ATPase family, functions as a sodium pump [In Process Citation]. *J Biol Chem* 273: 27772-8, 1998.
16. **Jaisser, F., and A. T. Beggah.** The nongastric H,K-ATPases: molecular and functional properties. *Am J Physiol* 276: F812-24, 1999.
17. **Kone, B. C., and S. C. Higham.** A novel N-terminal splice variant of the rat H,K-ATPase alpha2 subunit. Cloning, functional expression, and renal adaptive response to chronic hypokalemia. *J Biol Chem* 273: 2543-52, 1998.
18. **Kraut, J. A., J. Hiura, M. Besancon, A. Smolka, G. Sachs, and D. Scott.** Effect of hypokalemia on the abundance of HK alpha 1 and HK alpha 2 protein in the rat kidney. *Am J Physiol* 272: F744-50, 1997.
19. **Meneton, P., P. J. Schultheis, J. Greeb, M. L. Nieman, L. H. Liu, L. L. Clarke, J. J. Duffy, T. Doetschman, J. N. Lorenz, and G. E. Shull.** Increased sensitivity to K deprivation in colonic H,K-ATPase-deficient mice. *J Clin Invest* 101: 536-42, 1998.
20. **Nakamura, S., H. Amlal, J. H. Galla, and M. Soleimani.** NH4 secretion in inner medullary collecting duct in potassium deprivation: role of colonic H,K-ATPase. *Kidney Int* 56: 2160-7., 1999.
21. **Sangan, P., S. S. Kolla, V. M. Rajendran, M. Kashgarian, and H. J. Binder.** Colonic H,K-ATPase beta-subunit: identification in apical membranes and regulation by dietary K depletion. *Am J Physiol* 276: C350-60, 1999.
22. **Sangan, P., V. M. Rajendran, A. S. Mann, M. Kashgarian, and H. J. Binder.** Regulation of colonic H,K-ATPase in large intestine and kidney by dietary Na depletion and dietary K depletion. *Am J Physiol* 272: C685-96, 1997.

23. **Sangan, P., S. Thevananther, S. Sangan, V. M. Rajendran, and H. J. Binder.** Colonic H,K-ATPase alpha- and beta-subunits express ouabain-insensitive H,K-ATPase. *Am J Physiol Cell Physiol* 278: C182-9, 2000.
24. **Silver, R. B., P. A. Mennitt, and L. M. Satlin**. Stimulation of apical H,K-ATPase in intercalated cells of cortical collecting duct with chronic metabolic acidosis. *Am J Physiol* 270: F539-47, 1996.
25. **Wall, S. M., P. Mehta, and T. D. DuBose, Jr.** Dietary K restriction upregulates total and Sch-28080-sensitive bicarbonate absorption in rat tIMCD. *Am J Physiol* 275: F543-9, 1998.
26. **Wingo, C. S.** Active proton secretion and potassium absorption in the rabbit outer medullary collecting duct. Functional evidence for proton-potassium- activated adenosine triphosphatase. *J Clin Invest* 84: 361-5, 1989.

Chapter 6

# Mutational Analysis of Gastric Proton Pump, α- and β-Subunits

SHINJI ASANO, TOHRU KIMURA & NORIAKI TAKEGUCHI
*Molecular Genetics Research Center, and Faculty of Pharmaceutical Sciences of Toyama Medical and Pharmaceutical University, Toyama, 930-0194, Japan*

## INTRODUCTION

$H^+,K^+$-ATPase is the proton pump responsible for gastric acid secretion (10,16). This pump actively transports proton from inside to outside of gastric parietal cells, and $K^+$ from outside to inside of the cells, coupled with the hydrolysis of intracellular ATP to ADP and inorganic phosphate, resulting in the generation of more than $10^6$ fold proton gradient across the apical membrane. Gastric proton pump consists of two kinds of subunits. One is the catalytic α-subunit, which has 10 transmembrane segments and contains an ATP-binding site, phosphorylation site, binding sites of proton pump inhibitors and ion-binding sites (11). The other is the non-catalytic glycoprotein, β-subunit, which has a single transmembrane segment (12). The β-subunit contains three conserved disulfide bonds and seven N-linked carbohydrate chains in the ectodomain. This subunit is also essential for the functional expression of the proton pump, and is involved in the structural and functional maturation as well as intracellular transport of the functional holoenzyme (15).

$H^+,K^+$-ATPase belongs to the family of P-type ATPases. In the reaction cycle, the P-type ATPases form acid-stable acyl-phosphoenzymes (phosphorylation), and the resulting phosphoenzymes are dephosphorylated (dephosphorylation). These reaction steps are dependent on specific transporting cations. Sarcoplasmic and endoplasmic $Ca^{2+}$-ATPases (SERCA) and $Na^+,K^+$-ATPase also belong to this family.

One of the major research themes of ion-transporting ATPases is to answer the questions where the binding sites of cations are located in the pump molecule, and how the energy of ATP hydrolysis is coupled with active ion transport. To answer these questions, many chemical labeling and mutational studies have been performed especially for SERCA $Ca^{2+}$-ATPases and $Na^+,K^+$-ATPase. As the results, several acidic and hydrophilic amino acids involved in ion binding are identified in their transmembrane segments (8,9). Here, we introduced the mutational study of cation binding site on the $H^+,K^+$-ATPase α-subunit.

Cellular regulation of gastric proton pump is mainly performed by the intracellular localization of the pump molecules. In the resting state, the proton pumps are localized in the intracellular vesicles (tubulovesicles in parietal cells), whereas in the stimulated state they are expressed at cell surface. The β-subunit is very important for the trafficking of gastric proton pump. Here, we also studied the functional role of the β-subunit in the human cell line, and explored the roles of the carbohydrate chains and disulfide bonds of the β-subunit in the trafficking and expression of $H^+,K^+$-ATPase.

## MATERIALS AND METHODS

cDNA Transfection and Preparation of Membrane Fractions — cDNAs of the α- and β-subunits of $H^+,K^+$-ATPase were prepared from rabbit gastric mucosae, and separately cloned in an expression vector, pcDNA3. Introduction of site-directed mutations in the $H^+,K^+$-ATPase α- and β-subunits was carried out by sequential polymerase chain reactions (2).

The α- and β-subunit cDNA transfection in HEK-293 cells was performed by the calcium phosphate precipitation method. Cells were harvested two days after the DNA transfection, and membrane fractions of the cells were prepared as post-nuclear fractions (1).

Assay of $K^+$-ATPase Activity — $K^+$-stimulated ATPase ($K^+$-ATPase) activity was measured in 1 ml of solution comprising 50 μg of membrane protein, 3 mM $MgCl_2$, 1 mM ATP, 5 mM $NaN_3$, 2 mM ouabain and 40 mM Tris-HCl, pH 7.4 in the presence and absence of 15 mM KCl. After incubation at 37 °C for 30 min, the inorganic phosphate released was measured by the method of Yoda and Hokin (17). The $K^+$-ATPase activity was calculated as the difference between activities in the presence and absence of KCl. The expressed $K^+$-ATPase activity in the wild type was inhibited by a proton pump inhibitor, SCH 28080.

Phosphorylation and Dephosphorylation — Fifty micrograms of membrane proteins were phosphorylated in a solution comprising 1 μM ATP ([γ-$^{32}$P]ATP 4 x $10^6$ cpm), 2 mM $MgCl_2$, 1 mM EGTA, 3 mM ouabain and 40 mM Tris-HCl, pH 6.8 at 0 ℃ for 15 s. The reaction was quenched by the addition of ice-cold stop solution containing 10 % trichloroacetic acid and 10 mM inorganic phosphate. The protein was collected at 13,000 x g at 4 ℃ for 3 min, and the pellet was washed with ice-cold stop solution and 30 % sucrose successively, and solubilized in an SDS-PAGE sample buffer comprising 2 % SDS, 2.5 % dithiothreitol, 10 % glycerol and 50 mM Tris-HCl, pH 6.8 and subjected to the SDS-polyacrylamide gel electrophoresis under acidic conditions at pH 6.5. The radioactivity associated with the $H^+,K^+$-ATPase α-subunit separated on the gel was visualized by digital autoradiography (3).

In the dephosphorylation experiments, the membrane protein phosphorylated above was incubated with various concentrations of KCl and 1 mM non-radioactive ATP at 0 ℃ for 15 s. The reaction was stopped as described above, and the samples were run on the SDS-polyacrylamide gel. The radioactivity associated with the $H^+,K^+$-ATPase α-subunit was visualized as described above (3).

Immunofluorescence — COS cells or HEK-293 cells were cultured on glass cover slips and transfected with the α- and β-subunit cDNAs. Within 48 h after transfection, cells were fixed in cold methanol (- 20°C) for 10 min and processed for immunostaining. Intracellular localization of the α- and β-subunits was studied by immunofluorescence with antibodies directed against each subunit. $H^+,K^+$-ATPase α-subunit was detected with rabbit polyclonal antibodies, HK9 or Ab1024, and the β-subunit with monoclonal antibodies, 2G11 or 2B6. Goat anti-mouse IgG conjugated to fluorescein isothiocyanate and goat anti-rabbit IgG conjugated to rhodamine were used as the secondary antibodies. Immunostaining was visualized with a Zeiss LSM 410 laser scanning confocal microscope.

## RESULTS AND DISCUSSION

### 1. The Structure-Function Relationship of Cation Binding Sites on the α-Subunit.

The $H^+,K^+$-ATPase α-subunit contains 10 transmembrane segments (Fig. 1). Several acidic amino acid residues locate in the M4, M5, M6 and M8 segments of the α-subunit. Most of these acidic amino acids are conserved between SERCA $Ca^{2+}$-ATPase and $Na^+,K^+$-ATPase, and involved in binding or determining the affinity for cations. In this study, we especially focused on the results of mutations of Glu-345 in the M4 segment and the mutations of Glu-822 and Asp-826 in the M6 segment.

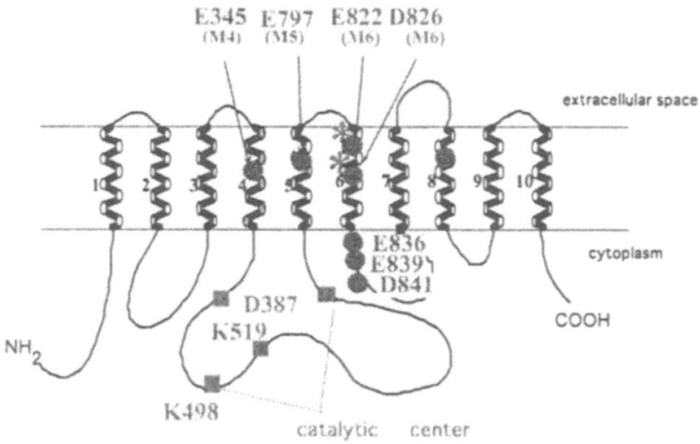


**Fig. 1** A Schematic Model of the Gastric $Na^+$,$K^+$-ATPase α-Subunit. The α-subunit contains 10 transmembrane segments. The catalytic center involved in ATP hydrolysis is primarily localized in the cytoplasmic segment between the M4 and M5 transmembrane segments; D387 is the acyl-phosphorylation site, and K519 is the FITC-binding site. There are several acidic amino acids in the M4, M5, M6 and M8 segments. Asterisks in or close to the M6 segment show the PPI binding sites, C815 and C824. E836, E839 and E841 are the putative sites involved in the interaction with proton.

1-1. Mutational Analysis of Glu-345 in the M4 Segment.

This residue is well-conserved in SERCA $Ca^{2+}$-ATPase and $Na^+$,$K^+$-ATPase. In the case of SERCA $Ca^{2+}$-ATPase, the corresponding amino acid, Glu-309, in the M4 segment forms the $Ca^{2+}$-binding site II, and is not replaceable with other amino acid (14).

1-1-1. Affinity for $K^+$, proton and ATP

In this study, Glu-345 in gastric $H^+$,$K^+$-ATPase was mutated to alanine (Ala), aspartic acid (Asp), asparagine (Asn), glutamine (Gln), isoleucine (Ile), leucine (Leu), lysine (Lys) or valine (Val). Expression levels of the mutant α-subunits were similar with that of the wild-type enzyme. All these mutants except mutant E345Q lost the $K^+$-ATPase activity. The E345Q mutant retained about 50 % of the $K^+$-ATPase activity of the wild-type enzyme. Mutant E345Q showed about 12 times lower affinity for $K^+$ ($K_m$; 2.4 mM) compared with that of the wild-type $H^+$,$K^+$-ATPase ($K_m$; 0.2 mM), indicating that this site is involved in determining the affinity for $K^+$ (1).

Furthermore, pH dependence curve of mutant E345Q shifted alkalineward by pH units of 0.5 compared with that of the wild type, indicating that mutant E345Q shows higher affinity for proton. This mutant also showed higher affinity for ATP compared with the wild type. Therefore, the E→Q mutation of Glu-345 shifted the conformational equilibrium of gastric $H^+$,$K^+$-ATPase toward $E_1$ form (with high-affinity for proton and ATP) (3).

1-1-2. Partial reaction steps, phosphorylation and dephosphorylation

The role of Glu-345 on the partial reaction steps of gastric $H^+$,$K^+$-ATPase was studied. Almost all the mutants except the E345K were phosphorylated from ATP, showing that the ATP-dependent phosphorylation step was not impaired by the mutations. However, the mutations of Glu-345 impaired the $K^+$-dependent dephosphorylation step. The E345Q mutant was dephosphorylated by $K^+$, however, its affinity for $K^+$ was about 10-times lower than that of the wild type. Other Glu-345 mutants were hardly dephosphorylated in the presence of 50-100 mM of $K^+$. From these findings, it is concluded that Glu-345 is involved in the interaction with $K^+$ in the dephosphorylation step of the $H^+,K^+$-ATPase (3).

1-2. Mutational Analysis of the M6 Segment.

The M6 segments of gastric $H^+,K^+$-ATPase, SERCA $Ca^{2+}$-ATPase and $Na^+,K^+$-ATPase are commonly important for interaction with ions (8,9). In gastric $H^+,K^+$-ATPase, there are two acidic amino acid residues, Glu-822 and Asp-826, on this segment (2). In addition, Cys-815 and Cys-824 in or close to the M6 segment are the binding sites of proton pump inhibitors (Fig. 1) (6,7). To study the contribution of each amino acid side chain of the M6 segment to the function of $H^+,K^+$-ATPase, each amino acid residue on this segment (from Cys-815 to Leu-833) was replaced by alanine, and the enzymatic functions of each mutant was observed (5).

**Fig. 2** $K^+$-ATPase activity and ATP-dependent phosphorylation capacity of the M6 mutants.

1-2-1. $H^+,K^+$-ATPase activity.

The expression levels of each mutant α-subunit in the membrane fractions were compared with that of the wild-type α-subunit in the Western blot. The expression levels ranged between 80 % (C824A) and 125 % (I818A) of that of the wild type except for 31 % for mutant L833A. Fig. 2 shows the $K^+$-ATPase activity of the M6 mutants. Mutants C815A, I816A, T817A, F820A, C824A, T825A, F828A and S832A retained 105, 84, 75, 75, 112, 56, 80 and 109 % of the $K^+$-ATPase activity found in the wild type, respectively. Mutants I818A, I821A, E822A, L823A, P829A, S830A and V831A showed 37, 31, 22, 47, 42, 37 and 34 % of the $K^+$-ATPase activity found in the wild type, respectively. Mutants L819A, D826A, I827A and L833A almost lost the $K^+$-ATPase activity.

1-2-2. Phosphorylation capacity.

Mutants C815A, I816A, T817A, I818A, L819A, F820A, E822A, C824A, F828A and S832A retained the ATP-dependent phosphorylation capacity, which was inhibited by SCH 28080 (Fig. 2). Mutants I821A, L823A, T825A, D826A, I827A, P829A, V831A and L833A were hardly phosphorylated from ATP at 0℃. Among them, mutants D826A, I827A and L833A lost the $K^+$-ATPase activity as well as the phosphorylation capacity. The I821A, L823A, T825A and P829A mutants were partially phosphorylated from ATP at 10℃. Their rates of phosphorylation were apparently lower than that of the wild type (data not shown). On the other hand, mutant L819A retained the phosphorylation capacity although this mutant lost the $K^+$-ATPase activity.

1-2-3. Dephosphorylation capacity.

The phosphorylated α-subunit of the wild-type $H^+,K^+$-ATPase was dephosphorylated by the addition of $K^+$ in a concentration-dependent manner. The half-maximally effective concentration of $K^+$ for the dephosphorylation was about 0.1 mM. The phosphorylated α-subunits of mutants C815A and I818A were also dephosphorylated by the addition of $K^+$, although the sensitivity of these phosphorylated mutants for $K^+$ were slightly lower than that of the wild-type enzyme. On the other hand, mutants L819A and E822A were dephosphorylated by only 20 % and 50 % in the presence of 50 mM KCl, respectively (Fig. 3). These results suggest that defect in the $K^+$-dependent dephosphorylation step resulted in the complete or considerable loss of the overall $H^+,K^+$-ATPase activity for mutants L819A and E822A, respectively.

The phosphorylated α-subunit of the wild-type $H^+,K^+$-ATPase was not dephosphorylated by the addition of ADP, indicating that the major phosphorylated intermediate is the $E_2$ form (ADP-insensitive, $K^+$-sensitive form). Mutants L819A and E822A were not dephosphorylated by the addition of ADP, indicating that these mutants have no defect in the process of conversion from $E_1P$ to $E_2P$ (data not shown).

1-2-4. Substitutions of Leu-819, Ile-827 and Leu-833 with other amino acid residues.

Mutant L819A has a defect in the $K^+$-dependent dephosphorylation step, whereas mutants I827A and L833A have defects in the ATP-dependent phosphorylation steps. To study the role of the size of the side chain on the $K^+$-ATPase activity and its partial reactions, Leu-819 was replaced with Gly, Val or Met; Ile-827 was replaced with Val or Met; and Leu-833 was replaced with Gly, Val or Met.

No $K^+$-ATPase activity was observed in mutant L819G, whereas mutants L819V and L819M retained 14 and 26 % of the $K^+$-ATPase activity found in the wild-type enzyme, respectively. These mutants retained the phosphorylation capacity. Mutants L819M and L819V were dephosphorylated by the addition of $K^+$, however, their sensitivity for $K^+$ was 10-times and 100-times lower than that of the wild-type enzyme, respectively, whereas mutant L819G was dephosphorylated by only 30 % in the presence of 50 mM $K^+$. These results suggest that the bulkiness or length of the side chain of the amino acid at the position 819 is critical for determining the affinity for $K^+$ in the dephosphorylation step of the $H^+,K^+$-ATPase.

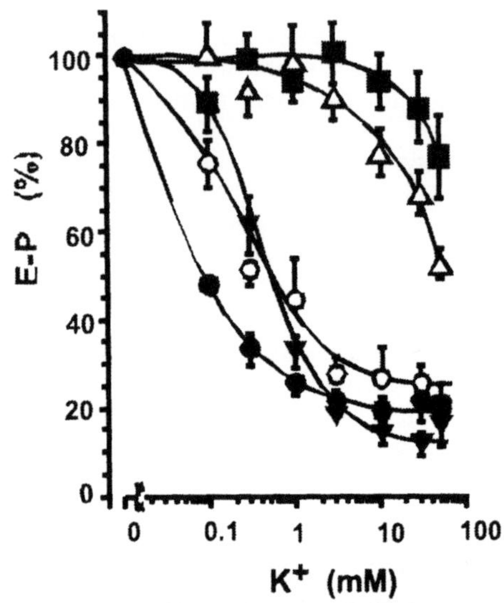


**Fig. 3** Effects of $K^+$ concentrations on the dephosphorylation reaction of the phosphorylated intermediates of the wild-type and the M6 mutants. The radioactivity associated with the phosphorylated $H^+,K^+$-ATPase α-subunit was expressed as the percentage of the control values measured in the absence of KCl: the wild-type $H^+,K^+$-ATPase (●), mutants C815A (▼), I818A (○), L819A (■) and E822A (△).

Mutants I827V and I827M retained 43 % and 49 % of the $K^+$-ATPase activity found in the wild-type enzyme, respectively. These mutants also partially retained the phosphorylation capacity, 20 and 30 % of the phosphorylation levels found in the wild-type enzyme, respectively. These results suggest that the bulkiness or length of the side chain of the amino acid at the position 827 is important in the phosphorylation step of the $H^+,K^+$-ATPase.

The expression level of mutants L833G was 31 % of that of the wild type, which was comparable with that of mutant L833A. However, the expression levels of mutants L833V and L833M were significantly higher than that of mutant L833A. Mutants L833A and L833G lost $K^+$-ATPase activity and phosphorylation capacity whereas mutants L833V and L833M retained them. These results suggest that the bulkiness or length of the side chain of the amino acid at the position 833 is important for the expression of the α-subunit as well as the phosphorylation step of the $H^+,K^+$-ATPase.

1-2-5. Mutations of Glu-822 and Asp-826 in the M6 Segment — There are two acidic amino acid residues, Glu-822 and Asp-826 in the M6 segments. The former residue was replaceable with aspartic acid as well as alanine. Mutant E822D also partially retained the $K^+$-ATPase activity whereas mutants E822Q and E822L lost the activity (2). Constitutive activation of gastric H⁺,K⁺-ATPase was observed by neutralization of this residue (13).

On the other hand, Asp-826 was unreplaceable with other amino acids including Asn and Glu. Double mutant E822D/D826E did not retain the K⁺-ATPase activity and phosphorylation capacity, indicating that the roles of these two negative side chains are not interchangeable.

1-2-6. A helical wheel model of the M6 segment.

If we assume the presence of α-helical structure in the M6 segment, functionally important amino acid residues including Glu-822 and Asp-826 are localized on one side of the α-helix (Fig. 4). These amino acid residues are likely to face ion-binding site or cavity. On the other hand, side chains of amino acids on the opposite side are not directly involved in the function.

Amino acids on the M6 segment were classified in two groups. Amino acid residues on the luminal side (Leu-819 and Glu-822) are involved in K⁺-dependent dephosphorylation or the interaction with K⁺. On the other hand, several amino acids on the cytoplasmic side such as Glu-826 and Ile-827 are involved in proton-dependent phosphorylation.

Recently, precise 3-D structure of SERCA $Ca^{2+}$-ATPase was determined by crystarographic techniques (14). By the alignment of the primary structures, putative 3-D structure of gastric H⁺,K⁺-ATPase can be presumed. Fitting the mutational results shown here to the constructed 3-D structure is to be studied.

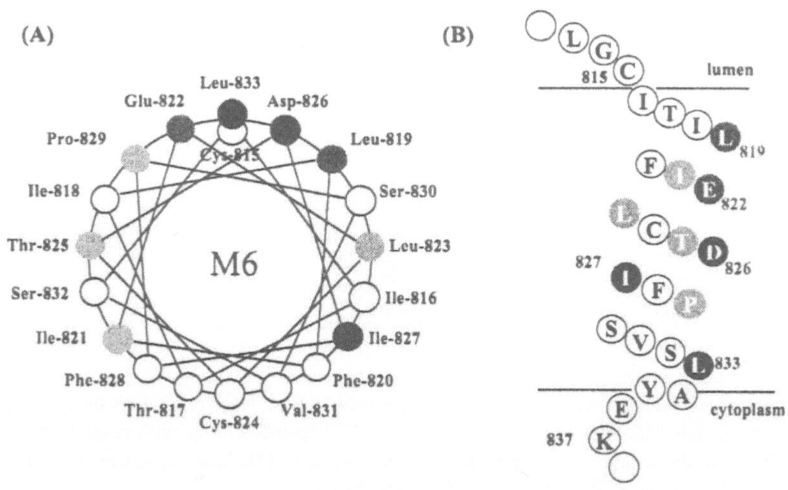


**Fig. 4** A Helical Wheel Model and Putative Membrane Topology of the M6 segment of the H⁺,K⁺-ATPase. (A) Amino acid residues between Cys-815 and Leu-833 are plotted on a helical wheel model. Leu-819 and Glu-822 are involved in K⁺-dependent dephosphorylation. Asp-826, Ile-827 and Leu-833 are involved in phosphorylation. Ile-821, Leu-823, Thr-825 and Pro-829 are involved in the rate of phosphorylation. (B) Putative membrane topology of the M6 segment. Amino acid residues involved in K⁺-dependent dephosphorylation are localized in the luminal half, whereas those involved in phosphorylation are localized on the cytoplasmic half of the M6 segment.

## 2. Site-Directed Mutagenesis of the H⁺,K⁺-ATPase β-Subunit

$H^+,K^+$-ATPase β-subunit is a non-catalytic subunit, however, essential for the functional expression of the ATPase. The β-subunit contains seven carbohydrate chains and three conserved disulfide bonds in the ectodomain (Fig. 5) (12). The molecular mass of the protein core is 32 kDa, whereas the molecular mass of the mature β-subunit is 60-90 kDa because of the attachment of carbohydrate chains. We studied the roles of attachment of carbohydrate chains and formation of disulfide bonds of the β-subunit on the catalytic function and intracellular localization of $H^+,K^+$-ATPase (4).

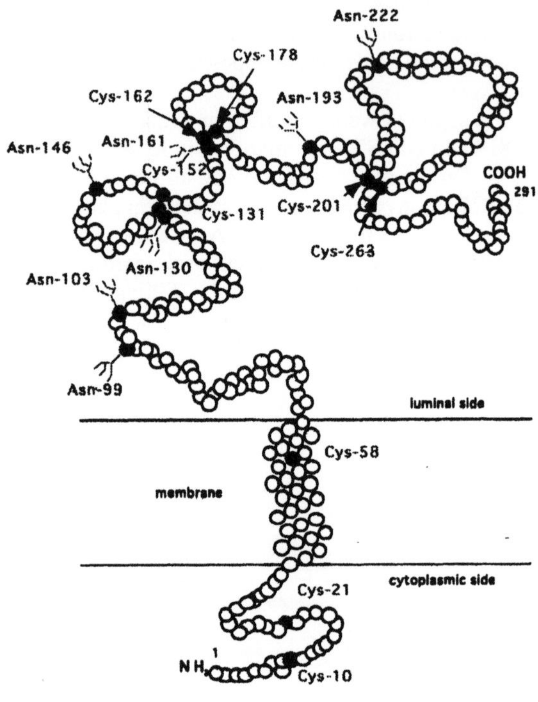


**Fig. 5** Schematic Illustration of the $H^+,K^+$-ATPase β-Subunit. N-glycosylation sites; Asn-99, Asn-103, Asn-130, Asn-146, Asn-161, Asn-193 and Asn-222 (closed circles with sugar chain symbols) and three disulfide bonds; Cys-131–Cys-152, Cys-162–Cys-178, and Cys-201– Cys-263 (closed circles) in rabbit gastric $H^+,K^+$-ATPase β-subunit are shown.

2-1. The Role of Carbohydrate Chains on the Function of Gastric $H^+,K^+$-ATPase.

The β-subunit contains seven carbohydrate chains (Fig. 5). To remove each specific carbohydrate chain, we replaced each asparagine residue in the N-glycosylation site by glutamine ("single-site mutants"). We also replaced several or all seven asparagine residues in the N-glycosylation sites by glutamine ("multi-site mutants").

The expression level of each single-site mutant β-subunit was not significantly different from that of the wild-type β-subunit in the Western blot. When the carbohydrate chains were removed, molecular mass of the mutant β-subunits decreased in a stepwise manner.

2-1-1. Stabilization of the α-subunit

The expression of the α-subunit was increased by the co-expression of the β-subunit. It is due to the stabilization of the α-subunit by the β-subunit in the membrane fraction. Each single-site mutant also stabilized the α-subunit in the membrane. After removal of more than five carbohydrate chains, however, mutant β-subunits did not stabilize the α-subunit.

2-1-2. $H^+,K^+$-ATPase activity

Each single-site mutant retained the $H^+,K^+$-ATPase activity, almost comparable with that of wild-type, indicating that there is no carbohydrate chains specifically important for the $H^+,K^+$-ATPase activity. However, the effect of removing carbohydrate chains was cumulative. When more than three carbohydrate chains were removed (Mutants IV, V, VI and VII), the ATPase activity was decreased in a stepwise manner. The activity was finally lost after removing all the carbohydrate chains (Mutant VII) (Fig. 6).

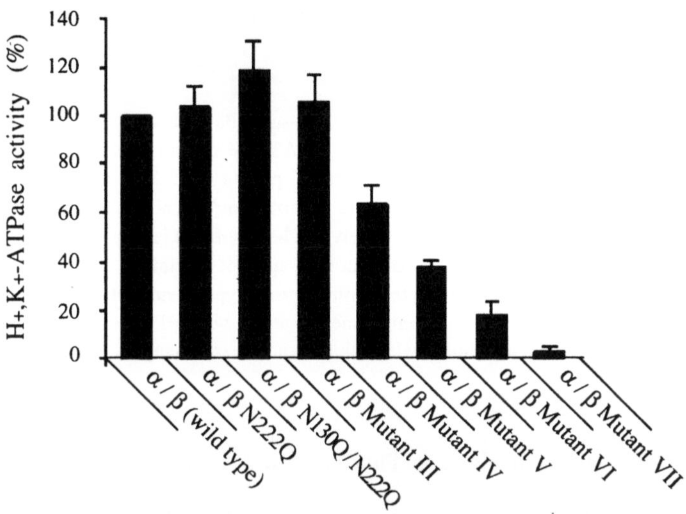


**Fig. 6** $H^+,K^+$-ATPase activities in the membrane fractions of the cells co-transfected with the wild-type α-subunit plus mutant β-subunit cDNAs. The values of the activity are shown as the percentage of the $K^+$-ATPase activity of the wild-type αβ complex.
Mutant III, N99/N130Q/N222Q;
Mutant IV, N99Q/N103Q/N130Q/N222Q;
Mutant V, N99Q/N103Q/N130Q/N146Q/N222Q;
Mutant VI, N99Q/N103Q/N130Q/N146Q/N193Q/N222Q;
Mutant VII, N99Q/N103Q/N130Q/N146Q/N161Q/N193Q/N222Q

2-1-3. Intracellular localization

When the wild-type α-subunit was expressed alone, the α-subunit was expressed only in the intracellular compartment, and no cell surface delivery was observed (B) (Fig. 7). On the contrary, the wild-type β-subunit was observed on the cell surface when the β-subunit expressed alone (A). When the α- and β-subunits were co-expressed, both subunits attained cell surface, indicating that the β-subunit is required for the cell surface delivery of the α-subunit (C, D).

When one or two carbohydrate chains were removed from the β-subunit, the α- and mutant β-subunits attained cell surface (E,F,G,H). However, after removing more than two carbohydrate chains, the α- and mutant β-subunits were retained in the intracellular compartment (I, J).

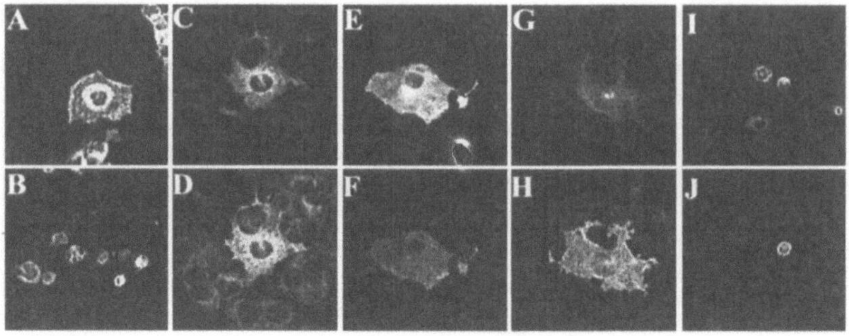


**Fig. 7** Immunolocalization of $H^+,K^+$-ATPase α- and β-subunits expressed in COS cells. COS cells were transfected with the cDNAs encoding the $H^+,K^+$-ATPase α- and β-subunits, either alone or in pairs. The distribution of the β-subunit is depicted in panels A, C, E, G and I, and that of the α-subunit in panels B, D, F, H and J. Panels A and B correspond to cells expressing the wild-type β-subunit alone and the α-subunit alone, respectively. The β-subunit is able to reach the cell surface (panel A), whereas the α-subunit expressed without β-subunit is retained intracellularly (panel B). Panels C and D represent cells co-transfected with the wild-type β-subunit and α-subunit; panels E and F with N99Q β-subunit and the α-subunit; panels G and H with N130Q/N222Q β-subunit and the α-subunit; and panels I and J with Mutant III β-subunit and the α-subunit.

## 2-2. The Role of Disulfide Bonds on the Function of Gastric $H^+,K^+$-ATPase.

The β-subunit contains nine cysteine residues, two in the cytoplasmic domain, one in the transmembrane domain, and six in the ectodomain. The six cysteine residues in the ectodomain form three disulfide bonds. In this study, each cysteine residue of the β-subunit was replaced by serine, and the enzymatic function of the mutants was studied.

### 2-2-1. Stabilization of the α-subunit

When each one of the three disulfide bonds in the ectodomain was cleaved, the mutant β-subunit did not stabilize the α-subunit in the membrane fraction. However, mutant β-subunit stabilized the α-subunit when each free cysteine residues located in the cytoplasmic or transmembrane segments were replaced by serine.

### 2-2-2. $H^+,K^+$-ATPase activity

$H^+,K^+$-ATPase activity was completely lost when one of the three disulfide bonds were removed by mutation (L-1, L-2, L-3), indicating that each disulfide bond is necessary for the expression of catalytic function of $H^+,K^+$-ATPase. The L-1,2,3 mutant, in which all the three disulfide bonds were removed, also showed no $H^+,K^+$-ATPase activity. However, mutants retained $H^+,K^+$-ATPase activity when each free cysteine residues located in the cytoplasmic or transmembrane segments were replaced by serine.

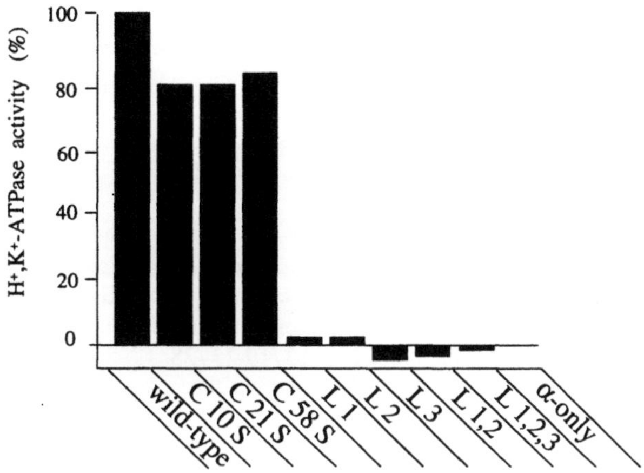


**Fig. 8** $H^+,K^+$-ATPase activities in the membrane fractions of the cells co-transfected with the wild-type α-subunit plus mutant β-subunit cDNAs. The values are shown as the percentage of the $H^+,K^+$-ATPase activity of the wild-type αβ complex. Mutant L1, C131S/C152S; L2, C162S/C178S; L3, C210S/C263S; L1,2, C131S/C152S/C162S/C178S; L1,2,3, C131S/C152S/C162S/C178S/C210S/C263S

2-2-3. Intracellular Localization

When one of the three disulfide bonds was cleaved, the mutant β-subunits were retained in the intracellular compartment. Therefore, each disulfide bond is important for cell surface delivery. However, mutants apparently attained cell surface when each free cysteine residues located in the cytoplasmic or transmembrane segments were replaced by serine.

2-3. Conclusion

Each three disulfide bond is essential for the stabilization of the α-subunit and $H^+,K^+$-ATPase activity. The formation of disulfide bonds is also important for cell surface expression of the $H^+,K^+$-ATPase. On the other hand, each carbohydrate chain is not essential for the catalytic activity nor cell surface delivery. However, the effects of removing carbohydrate chains were cumulative. Cell surface delivery is more dependent on the carbohydrate chains than the catalytic activity.

## REFERENCES

1. **Asano S, Tega Y, Konishi K, Fujioka M, and Takeguchi N**. Functional expression of gastric $H^+,K^+$-ATPase and site-directed mutagenesis of the putative cation binding site and catalytic center. *J Biol Chem* 271: 2740-2745, 1996.

2. **Asano S, Matsuda S, Tega T, Shimizu K, Sakamoto S, and Takeguchi N**. Mutational analysis of putative SCH 28080 binding sites of gastric $H^+,K^+$-ATPase. *J Biol Chem* 272: 17668-17674, 1997.
3. **Asano S, Furumoto R, Tega Y, Matsuda S, and Takeguchi N**. Mutational analysis of theputative cation affinity site on the fourth transmembrane segment of the gastric $H^+,K^+$-ATPase. *J Biochem* 127: 993-1000, 2000.
4. **Asano S, Kawada K, Kimura T, Grishin AV, Caplan MJ, and Takeguchi N**. The role of carbohydrate chains of the β-subunit on the functional expression of gastric $H^+,K^+$-ATPase. *J Biol Chem* 275: 8324-8330, 2000.
5. **Asano S, Io T, Kimura T, Sakamoto S, and Takeguchi N.** Alanine-scanning mutagenesisof the sixth transmembarne segment of gastric $H^+,K^+$-ATPase. *J Biol Chem* 276: 31265-31273, 2001.
6. **Besancon M, Simon A, Sachs G, and Shin JM**. Sites of reaction of the gastric H,K-ATPase with extracytoplasmic thiol reagents. *J Biol Chem* 272: 22438-22446, 1997.
7. **Lambrecht N, Corbett Z, Bayle D, Karlish SJD, and Sachs G.** Identification of the site of inhibition by omeprazole of a α-β fusion protein of the H,K-ATPase using site-directed mutagenesis. *J Biol Chem* 273: 13719-13728, 1998.
8. **Lingrel JB, and Kuntzweiler T.** $Na^+,K^+$-ATPase. *J Biol Chem* 269: 19659-19662, 1994.
9. **MacLennan DH, Rice WJ, and Green NM.** The mechanism of $Ca^{2+}$ transport by sarco(endo)plasmic $Ca^{2+}$-ATPase. *J Biol Chem* 272: 28815-28818, 1997.
10. **Sachs G, Chang HH, Rabon E, Schackman R, Lewin M, and Saccomani G.** A non-electrogenic $H^+$ pump in plasma membranes of hog stomach. *J Biol Chem* 251: 7690-7698, 1976.
11. **Shull GE, and Lingrel JB.** Molecular cloning of the rat stomach $(H^++K^+)$-ATPase. *J Biol Chem* 261: 16788-1679, 1986.
12. **Shull GE.** cDNA cloning of the β-subunit of the rat gastric H,K-ATPase. *J Biol.Chem* 265: 12123-12126, 1990.
13. **Swarts HGP, Hermasen HPH, Koenderink JB, Stekhoven FMAHS, and DePont JJ.** Constituitve activation of gastric $H^+,K^+$-ATPase by single mutation. EMBO J. 17: 3029-3015, 1998.
14. **Toyoshima C, Nakasako M, Nomura H, and Ogawa H.** Crystal structure of the calcium pump of sarcoplasmic reticulum at 2.6 A resolution. *Nature* 405: 647-655, 2000.
15. **Tyagarajan K, Chow DC, Smolka A, and Forte JG.** Structural interactions between α- and β-subunits of gastric H,K-ATPase. *Biochim Biophys Acta* 1236: 105-113, 1995.
16. **Wallmark B, Larsson H, and Humble L.** The relationship between gastric acid secretion and gastric $H^+,K^+$-ATPase activity. *J Biol Chem* 260: 13681-13684, 1985.
17. **Yoda A, and Hokin LE.** On the reversibility of binding of cardiotonic steroids to a partially purified (Na + K)-activated adenosinetriphosphatase from beef brain. *Biochem. Biophys. Res. Commun.* 40: 880-886, 1970.

Chapter 7

# Structure-Activity Relationships of Na,K-ATPase and H,K-ATPase

JAN JOEP H.H.M. DE PONT, JAN B. KOENDERINK, HERMAN G.P. SWARTS AND PETER H.G.M. WILLEMS
*Department of Biochemistry, University of Nijmegen, Nijmegen, The Netherlands*

## INTRODUCTION

The Na,K-ATPase and gastric H,K-ATPase belong to the class of P-type ATPases, which use the energy from ATP for active transport of ions. Although both X,K-ATPases share many similarities, they have different ion specificities ($Na^+$ vs. $H^+$) and different affinities for specific inhibitors (ouabain vs. SCH 28080). How do these enzymes discriminate between the different cations transported? What determines the specific binding of an inhibitor? It is likely that the binding sites for ions and for certain inhibitory drugs are located within or close to transmembrane segments. By making chimeras of the α-subunits of Na,K-ATPase and H,K-ATPase, information could be obtained about transmembrane segments that could be responsible for the inhibitor and cation specificity.

The recombinant baculovirus expression system was used to express Na,K-ATPase, H,K-ATPase, their hybrids and chimeras in insect cells. Basis for this system is the *Autographa californica* multiple nuclear polyhedrosis virus, which infects insect larvae of the fall armyworm *Spodoptera frugiperda*. The genes to be expressed were first cloned into a donor plasmid downstream of a baculovirus promoter. This donor plasmid was then introduced into *E. coli* cells, harbouring the baculovirus genome, as a shuttle vector (bacmid) and a transposition helper vector. Upon site-specific transposition between the donor vector and the bacmid, recombinant bacmids were selected and isolated. Subsequently, insect cells were transfected with these bacmids and the recombinant baculoviruses were harvested. These recombinant viruses could be used directly for production of recombinant proteins. The membrane fractions of the insect cells expressing the recombinant proteins were isolated and Western blot analysis revealed comparable expression patterns. Both Na,K-ATPase and H,K-ATPase α-subunits had an apparent molecular mass of about 100 kDa. The β-subunits possessed a carbohydrate-free and a core-glycosylated form. The carbohydrate-free form of the Na,K-ATPase β-subunit was of a similar molecular mass as the carbohydrate-free H,K-ATPase β-subunit, but the core-glycosylated Na,K-ATPase β-subunit had a lower apparent molecular mass than that of H,K-ATPase, due to the presence of only three glycosylation sites in this subunit in contrast to the seven glycosylation sites present in the H,K-ATPase β-subunit. The molecular mass of all α-subunits expressed was similar to that of the isolated α-subunits. However, the recombinant β-subunits were less glycosylated than the isolated β-subunits, as has been reported before (7,5,16). Expression levels of the subunits of each hybrid or chimerical ATPase were in general comparable to those of Na,K-ATPase and H,K-ATPase.

## HYBRIDS

In order to investigate α- and β-subunit interactions, cation-binding sites and inhibitor-binding sites more than 40 different recombinant baculoviruses were generated and the expressed ATPases were analysed. Assembly of α- and β-subunits is a crucial step in the formation of active X,K-ATPases. Although the similarity of the α-subunits is almost 70%, the β-subunits are only 30% identical. The β-subunits of Na,K-ATPase and H,K-ATPase are involved in correct folding (11) of the α-subunits. In addition the β-subunit of H,K-ATPase plays a role in endocytosis of the enzyme (4), that overcomes continuous gastric acid secretion. In order to investigate the interaction between the α- and the β-subunits of these transport systems more closely we expressed two hybrid enzymes (NaKαHKβ, HKαNaKβ in insect cells (19). In the hybrid ATPases the subunits cross-assembled as shown in co-immunoprecipitation experiments. However, for both hybrid ATPases there seemed to be less co-precipitated β-subunits compared to the wild type enzymes, indicating a less efficient assembly between the α- and β-subunits of the hybrid ATPases. This could be due to differences in the amino acid composition or to differences in glycosylation levels between the two β-subunits. Furthermore, we determined that both hybrid ATPases possessed a low

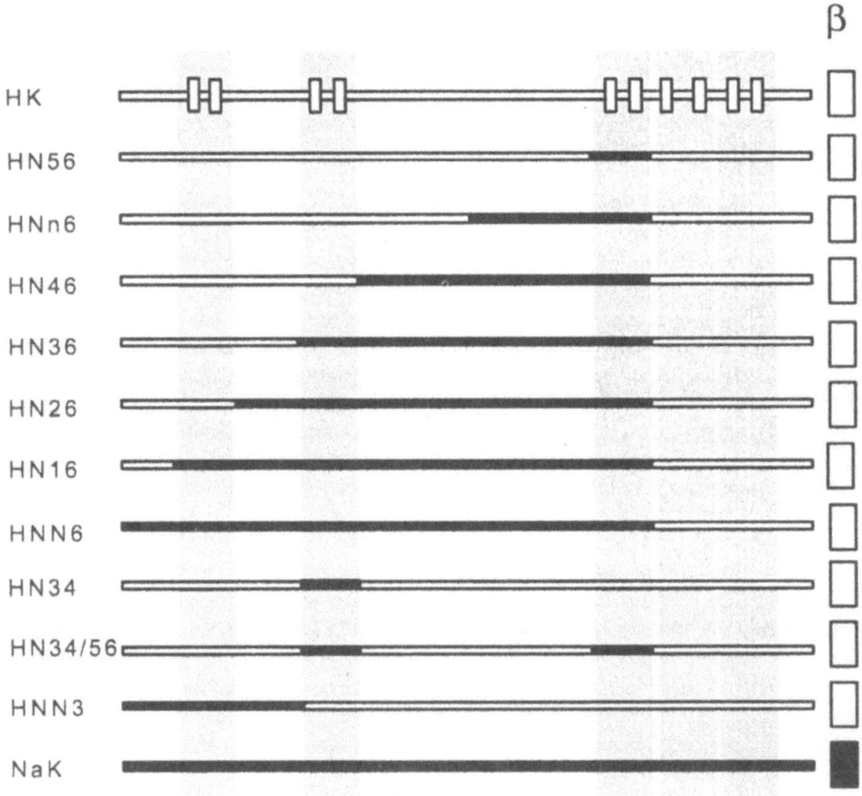


*Figure 1* *Schematic presentation of the constructed H,K-ATPase (white)/ Na,K-ATPase (black) chimeras.*

ATPase activity which was about 10% of the activity of recombinant Na,K-ATPase or H,K-ATPase. A less efficient subunit assembly is probably the reason for this lower ATPase activity. When the Na,K-ATPase ß-subunit was replaced by the H,K-ATPase ß-subunit the apparent $K^+$-affinity in the ATPase reaction decreased similarly as has been reported in ouabain binding experiments by Eakle et al. (8). In contrast, the apparent $K^+$-affinity of the complementary hybrid ATPase was increased as compared to that of H,K-ATPase. The hybrid ATPase NaKαHKβ was phosphorylated to 21% of the Na,K-ATPase phosphorylation level. Surprisingly, we were not able to measure any specific phosphorylation with the hybrid HKαNaKβ. Indeed, HKαNaKβ has an increased apparent $K^+$ affinity, which could reduce the amount of phosphoenzyme. Thus both Na,K-ATPase and H,K-ATPase require their own β-subunits for optimal activity. When the β-subunits were exchanged, the enzyme activity decreased and the apparent $K^+$-affinity of both hybrid ATPases was modified. It has been demonstrated that the binding region for the β-subunit is located in the C-terminal region of both the Na,K-ATPase and H,K-ATPase α-subunits (5,21,23,35). Hence, an optimal assembly between chimerical α-subunits and the H,K-ATPase β-subunit could be ensured through the presence of the C-terminal 187 amino acids of the H,K-ATPase α-subunit. This finding has been applied in all chimeras presented in this chapter.

## SODIUM-BINDING SITES

Ten chimerical constructs were produced after introduction of six unique restriction sites in both the cDNAs of the rat gastric H,K-ATPase α-subunit and the rat Na,K-ATPase $\alpha_1$-subunit (17,18) (Figure 1). First, the transmembrane hairpin M5-M6 of H,K-ATPase was exchanged by the similar region of Na,K-ATPase, generating the chimerical ATPase HN56. Since most of the polar amino acids in this hairpin are similar in the two enzymes, we expected to keep the $K^+$-sensitivity and we hoped to gain the $Na^+$-sensitivity. However, chimera HN56 could be phosphorylated in the absence of $Na^+$ and the dephosphorylation reaction was not stimulated by $K^+$. Next, the part originating from Na,K-ATPase was progressively increased in the N-terminal direction, generating the chimerical ATPases HNn6, HN46, HN36, HN26, HN16,

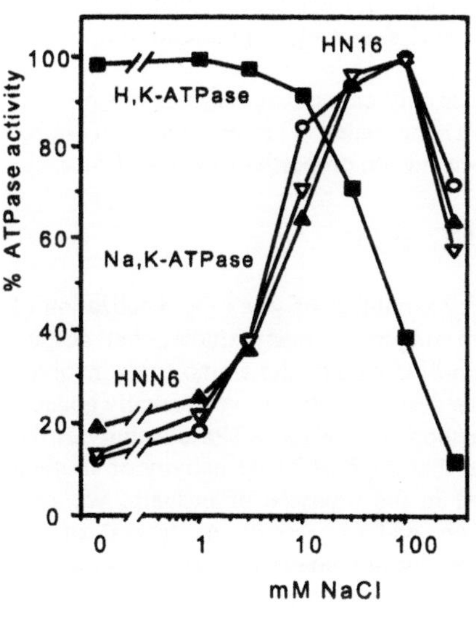


*Figure 2*
*$Na^+$-dependence of $K^+$-ATPase activity of the H,K-ATPase/Na,K-ATPase chimeras HNN6 and HN16, Na,K-ATPase and H,K-ATPase.*

and HNN6. In addition, we prepared the chimeras HN34, HN34/56 and HNN3 (Figure 1). The chimeras HN26, HN36 and HN46 could not be phosphorylated. All other chimeras could be phosphorylated, but only in case of chimeras HN16 and HNN6 $Na^+$ enhanced the phosphorylation level. The latter two chimeras also showed $Na^+$-dependent ATPase activity in the presence of $K^+$ (Figure 2). The $Na^+$-activation curves of these chimeras superimposed that of Na,K-ATPase. At very high $Na^+$ concentrations the $K^+$-stimulated ATPase activities of Na,K-ATPase and the chimeras HN16 and HNN6 were inhibited. $Na^+$ did not stimulate the H,K-ATPase activity. In contrast, increasing concentrations of $Na^+$ gradually inhibited the activity of this enzyme. Thus the $Na^+$-activation curves of HN16 and HNN6 are rather similar to that of Na,K-ATPase. Of the other chimeras tested only the chimeras HN34 and HN34/56 showed $K^+$-stimulated ATPase activity, but $Na^+$ did not stimulate this activity.

These studies indicate that the presence of the M1-M2 hairpin of Na,K-ATPase (in combination with M3-M4 and M5-M6) is essential for obtaining $Na^+$-activated reactions. The origin of the N-terminal tail was not important in this respect. The M1-M2 hairpin alone is not sufficient for $Na^+$-activated reactions, since $Na^+$ did not increase the phosphorylation level of chimera HNN3. It is tempting to speculate how this fits with the Toyoshima model for $Ca^{2+}$-ATPase (33), that shows that only amino acids from M4, M5, M6 and M8 are directly involved in $Ca^{2+}$ binding. It is known that, whereas $Ca^{2+}$-ATPase binds and transports two $Ca^{2+}$ ions, Na,K-ATPase transports three $Na^+$ ions and binding of all three ions is needed for ATPase activity. It could theoretically be that one of the $Na^+$ ions binds to a region of the protein that includes the M1-M2 hairpin. Alternatively, the interaction of the M1-M2 hairpin with the M3-M4 and M5-M6 hairpins through the hydrogen bond network might be necessary to obtain a functional $Na^+$-binding pocket.

Studies with vanadate indicated that the chimeras HN16 and HNN6 have in contrast to Na,K-ATPase a very low affinity for this compound (18), suggesting that these chimeras have a high preference for the $E_1$ conformation, whereas Na,K-ATPase has a preference for the $E_2$ conformation. The $E_2K$ conformer of the Na,K-ATPase is 3 orders of magnitude more stable than $E_1K$, while the $E_1K$ and $E_2K$ conformations of the H,K-ATPase are energetically nearly equivalent (9,28). Chimeras HNN6 and HN16 have also a high apparent affinity for ATP compared to Na,K-ATPase. This is attributable to the relative rates of interconversion of the $E_1/E_2$ enzyme conformations which yield an equilibrium that is probably more close to that of H,K-ATPase than to that of Na,K-ATPase. Consequently, the chimeras HNN6 and HN16 most likely possess the $K^+$ de-occlusion properties of the H,K-enzyme, and not those of Na,K-ATPase.

In summary, chimeras HNN6 and HN16 are not only chimerical enzymes in structure, but also in function. On one hand they possess the $Na^+$-stimulated ATPase reaction of Na,K-ATPase, while on the other hand they have the $K^+$ de-occlusion properties of H,K-ATPase.

## OUABAIN-BINDING SITE

Na,K-ATPase is specifically inhibited by ouabain. For a number of years the localization of the ouabain-binding site on Na,K-ATPase has been subject of many studies, but despite attempts to unravel the precise location of the ouabain-binding site, the amino acids involved in direct binding are still unknown. In the experiments described above, we generally added a low concentration of ouabain in order to inhibit the endogenous Na,K-ATPase activity of the Sf9 cells. When we once omitted ouabain we noticed that the $K^+$-ATPase activity of chimera HN34/56 was considerably higher than as measured in the presence of ouabain. We next systematically investigated the effect of ouabain on ATPase activity of several of the chimeras mentioned above (Figure 3). The ATPase activity of chimera HN34/56 was the only

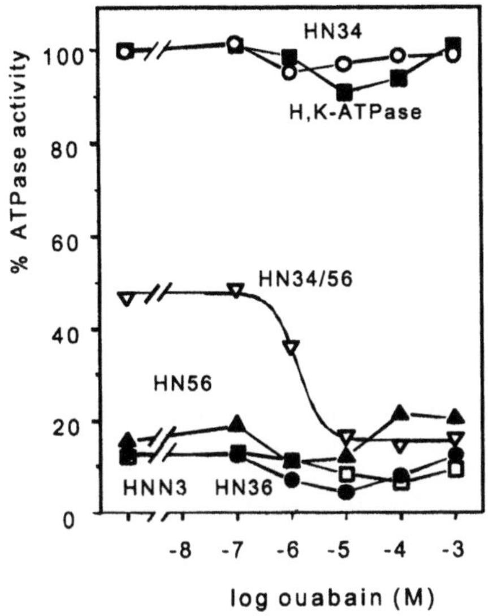


*Figure 3*
*The effect of ouabain on the $K^+$-ATPase activity of several H,K-ATPase/Na,K-ATPase chimeras and H,K-ATPase.*

one that was inhibited by ouabain (17). The inhibition was found at very low ouabain concentrations (IC50 = 2 μM).

Some of the chimeras tested in this study (HN56, HNN3) had no ATPase activity, but could be phosphorylated. Ouabain had no effect on the phosphorylation level of these mutants. It increased, however, the phosphorylation level of mutant HN34/56. This increase could be explained by a stabilization of the $E_2P$ intermediate by ouabain (17). The high-affinity binding of ouabain by chimera HN34/56, was also indicated by the low ouabain concentration required to replace 50% of bound [$^3$H]ouabain. This concentration is similar to that for replacement of [$^3$H]ouabain in purified rabbit kidney Na,K-ATPase.

The finding that this chimera was sensitive to ouabain was surprising for two reasons. All chimeras used by us originated from the rat enzyme, which is very insensitive toward ouabain. In addition, Lingrel and co-workers had presented strong evidence that the extracellular domain between M1 and M2 is involved in the binding of ouabain (26,27). When they replaced Arg and Asp, present at the borders of M1 and M2 in the rat Na,K-ATPase, by Gln and Asn (the amino acids that are present in the highly sensitive sheep enzyme) the rat Na,K-ATPase became sensitive to ouabain. The extracellular loop between M1 and M2 in chimera HN34/56 originated from rat H,K-ATPase, in which enzyme on these crucial positions Gln and Asn are present. We therefore replaced these amino acids by Arg and Asp, respectively. This indeed led to a marked reduction in the affinity for ouabain on the ATPase activity, indicating that Gln and Asn are responsible for the high-affinity inhibitory effect of ouabain.

It has been suggested that the overall conformation of the first extracellular loop may influence ouabain sensitivity indirectly by altering the stability or structure of the intermediate of the Na,K-ATPase catalytic cycle which is competent to bind ouabain. Transmembrane segments M4, M5, and M6 are thought to establish the cation-binding pocket, in which movement of hairpin M5-M6 may play a role in active transport (10,22). It has been proposed that binding of ouabain to this hairpin inhibits cation transport by immobilizing these transmembrane domains (25). In the last decade many amino acid substitutions in Na,K-

ATPase have been shown to decrease the ouabain sensitivity (6). The interpretation of this loss of function is fraught with difficulties. The major reason for this problem is the question whether the substitution is really directly affecting the ouabain binding. We observed a gain of function, which is likely to be due to a direct effect. We can conclude that introduction of transmembrane hairpins M3-M4 and M5-M6 of Na,K-ATPase into H,K-ATPase results in the formation of a high affinity ouabain-binding site. Future studies in which amino acids in these regions will be substituted can more clearly localize the ouabain-binding site.

## CONSTITUTIVE ATPase ACTIVITY

Negatively charged residues within transmembrane segments of P-type ATPases are assumed to be involved in cation binding and transport (1,24). This is also the case for the gastric H,K-ATPase. Site-directed mutagenesis of residue $Glu^{343}$ in transmembrane segment four (2,3), $Glu^{795}$ in transmembrane segment five (12,32), and $Glu^{820}$ in transmembrane segment six (13,32) provided strong evidence that these three residues are involved in $K^+$-stimulation of the ATPase activity. The specificity of these residues is different. $Glu^{820}$ can only be replaced by Asp, indicating that the negative charge of this residue is essential for $K^+$-activation (13). On position 795, Gln, but not Asp or Asn, can replace Glu resulting in a small increase in $K^+$-affinity, suggesting that a side chain with a carbonyl residue and with the right length is necessary and sufficient (12). On position 343 only Gln can replace Glu, but this results in a reduction in $K^+$-affinity, suggesting that the Glu on this position is the preferable amino acid residue (2,3). Interestingly, several other mutants of $Glu^{343}$, like E343D, could still be phosphorylated, but no $K^+$-dependent stimulation of the dephosphorylation of the phosphorylated intermediate was found (2).

In previous studies we observed that some mutants of $Glu^{820}$ (E820Q, E820A, E820N) (30) and $Glu^{795}$ (E795L, E795A) (12) possessed a relatively high ATPase activity in the absence of $K^+$. This $K^+$-independent ATPase activity (also named constitutive ATPase activity) was caused by an enhanced spontaneous dephosphorylation rate of the phosphorylated intermediate. We postulated that the altered cation-binding site mimics the stimulatory effect of the $K^+$-filled binding pocket on the hydrolysis of the aspartyl phosphate in the phosphorylated enzyme.

The ATPase activity of all mutants was, like that of the wild type enzyme, very sensitive to inhibition by SCH 28080. In addition, the steady-state ATP-dependent phosphorylation level of the wild-type enzyme as well as of $K^+$-sensitive mutants was decreased by SCH 28080 (31). In contrast, the steady-state phosphorylation level of the constitutively active mutants was only partly decreased by SCH 28080 and that of the mutant E820Q was even enhanced (31). The latter increase was also found in five double mutants of $Glu^{795}$ and $Glu^{820}$ that displayed $K^+$-independent ATPase activity (14).

This anomalous behavior of constitutively active mutants was explained by differences in the $E_2 \leftrightarrow E_1$ equilibrium between the enzymes with $K^+$-dependent and $K^+$-independent ATPase activity (31). $K^+$-dependent enzymes are assumed to be present in the $E_2$-form. Upon pre-incubation of these enzymes with SCH 28080 a complex is formed and ATP-dependent phosphorylation can no longer occur. Mutants with $K^+$-independent ATPase activity are assumed to be mainly present in the $E_1$-form. Pre-incubation with SCH 28080 does not result in an $E_2$.SCH complex and upon addition of ATP phosphorylation occurs, finally resulting in formation of the phosphorylated intermediate $E_2$-P. Next, $E_2$-P forms a stable complex with SCH 28080 and is accumulated since its dephosphorylation rate is slower than that of the phosphorylated intermediate formed in the absence of SCH 28080 (31).

The preference of mutants with constitutive ATPase activity for the $E_1$-form was confirmed by recent studies with vanadate (14). We found that the ATPase activity of mutants with $K^+$-independent ATPase activity was much less sensitive toward vanadate than that of enzymes with $K^+$-dependent ATPase activity. The $K^+$-independent mutants have thus a relatively high preference for the $E_1$-form.

We next investigated whether the preference of $K^+$-independent mutants for the $E_1$-conformation and the high spontaneous dephosphorylation rate are two coupled processes (29). We therefore prepared a series of single, double and triple mutants of glutamates 343, 795 and 820. At position 343 either an Asp (no $K^+$-stimulated activity) or a Gln (reduced $K^+$-affinity) was used, whereas on positions 795 and 820 Glu was replaced by a Gln. For clarity these mutants are written with their single letter code. The first letter indicates the amino acid at position 343, the second at position 795 and the third at position 820 (e.g. DQQ is mutant E343D/E795Q/E820Q).

All mutants were functionally active as indicated by ATP-dependent phosphorylation. After determining (i) the steady-state ATP-dependent phosphorylation level, (ii) the $K^+$-dependency of the ATPase reaction, (iii) the effect of $K^+$ on the dephosphorylation process and (iv) the effect of vanadate on the phosphorylation level, we could divide the mutants and the wild type enzyme in three groups. The $K^+$-sensitive mutants (QEE, EQE, QQE) including the wild type enzyme (EEE) are shown in Figure 4, the constitutively active mutants (EEQ, QEQ, EQQ, QQQ) are presented in Figure 5 and the non-dephosphorylating mutants (DEE, DQE, DEQ, DQQ) are given in Figure 6.

Figure 4A shows that the wild type enzyme (EEE) and the three mutants QEE, EQE, and QQE were phosphorylated by ATP and that SCH 28080 (0.1 mM) completely prevented ATP-dependent phosphorylation of these preparations. The EQE mutant had a slightly higher $K^+$-sensitivity in the ATPase reaction than the wild type enzyme (Figure 4B). The apparent $K^+$-affinities of the QQE and QEE mutants were reduced by approximately one log unit. The data from the dephosphorylation experiments were in line with those of the ATPase assays, showing a slightly higher apparent $K^+$-affinity for the EQE mutant and a reduced apparent affinity for the QEE and QQE mutants, compared to the wild type enzyme (Figure 4C). Figure 4D shows that the phosphorylation levels of all these mutants were very sensitive toward vanadate suggesting that they have a preference for the $E_2$ conformation.

The phosphorylation levels of the four mutants with a Gln on position 820 and either a Glu or a Gln on positions 343 and 795 were lower than those of the $K^+$-sensitive preparations (cf. Figure 5A with Figure 4A). The EQQ and the QQQ mutants had especially low EP levels. Pre-incubation of the membranes with SCH 28080 did not prevent ATP-dependent phosphorylation of these mutants. Conversely, it enhanced the steady-state phosphorylation levels. Of note, the phosphorylation level measured in the presence of SCH 28080 was of the same order of magnitude as that of $K^+$-sensitive mutants measured in the absence of SCH 28080, suggesting that they have similar intrinsic capacities. The constitutively active mutants showed an SCH 28080-sensitive ATPase activity in the absence of $K^+$ (Figure 5B). This activity was not affected by $K^+$ concentrations up to 10 mM and decreased at higher concentrations. Figure 5C demonstrates that the phosphorylated intermediates of the EEQ, EQQ, and QQQ mutants were hydrolysed in 3s at 21°C to 10% or less of the original level. The phosphorylated intermediate of the QEQ mutant was reduced to 30% of the original level in this time period. The residual phosphorylation levels were not affected by $K^+$. Figure 5D shows that these mutants were very insensitive toward vanadate suggesting that they have a preference for the $E_1$ conformation.

Figure 6A shows that the steady-state phosphorylation level of the four mutants with an Asp at position 343 and either a Glu or a Gln at positions 795 and 820 was rather similar to

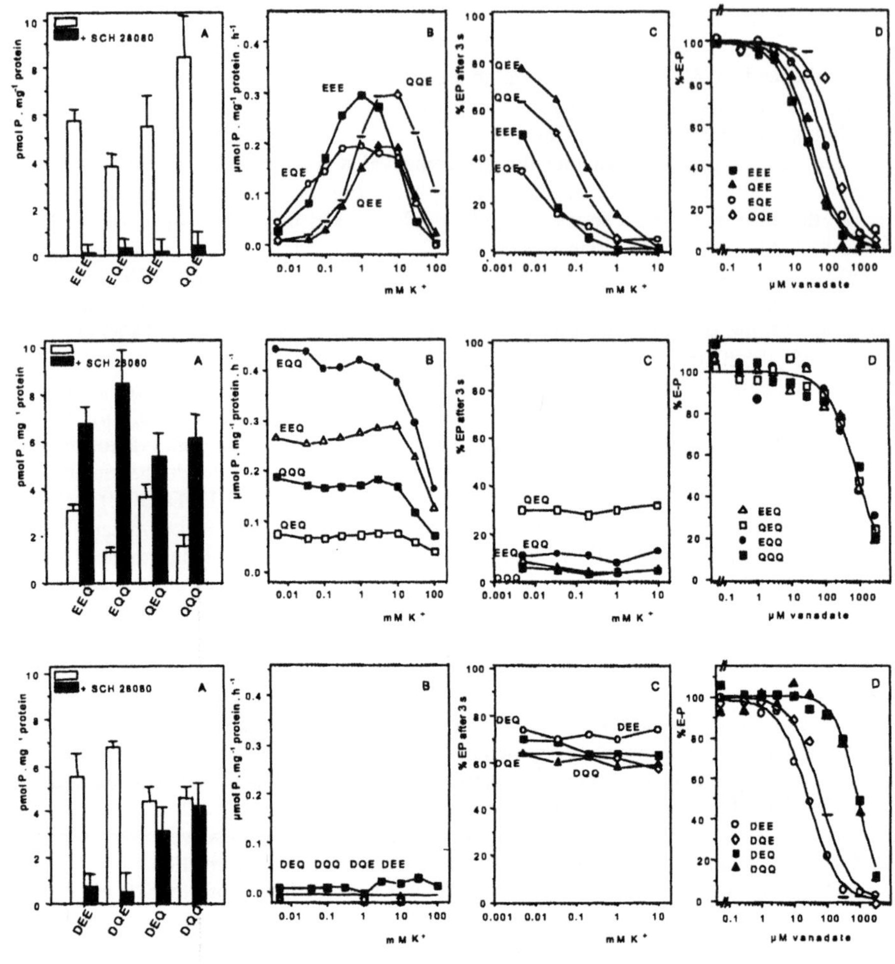


*Figures 4, 5 and 6.*
*The effect of SCH 28080 on the ATP-phosphorylation capacity (A), the K⁺-sensitivity of the ATPase activity (B), the K⁺-sensitivity of the dephosphorylation reaction (C) and the vanadate sensitivity of the level of the phosphorylated intermediate (D) of the K⁺-sensitive HK-ATPase mutants (Figure 4), the K⁺-independent H,K-ATPase mutants (Figure 5), and the K⁺-insensitive H,K-ATPase mutants (Figure 6).*

that of the wild type enzyme (cf. Figure 6A with Figure 4A). SCH 28080 prevented ATP-dependent phosphorylation of the DEE and DQE mutants. This inhibition, however, did not occur in the DEQ and DQQ mutants. No K⁺-activated or constitutive ATPase activity could be demonstrated in any of these mutants (Figure 6B). The dephosphorylation measurements (Figure 6C) showed a rather low spontaneous dephosphorylation and no K⁺-activation of the dephosphorylation process. Figure 6D shows the vanadate sensitivity of the phosphorylation level of these mutants. The two mutants with a Glu on position 820 had similar $IC_{50}$ values as the $K^+$-sensitive mutants. The two mutants with a Gln on residue 820 had similar $IC_{50}$ values as the mutants with constitutive ATPase activity. This confirms the experiments with SCH 28080 and indicates that the presence of a Gln on position 820 resulted in a preference for the $E_1$-conformation, whereas a Glu on this position favored the $E_2$-conformation.

Thus the two properties of constitutively active mutants of gastric H,K-ATPase: a high $K^+$-independent dephosphorylation rate and preference for the $E_1$-conformation can be uncoupled. The E343D mutant alone had no ATPase activity, but could be phosphorylated by ATP and had an $E_2$-preference. On the other hand, the E820Q mutant possessed $K^+$-independent ATPase activity, a high spontaneous dephosphorylation rate and an $E_1$-preference. Combination of these two mutations did no longer result in an enhanced $K^+$-independent dephosphorylation rate. However, the $E_1$-preference, the other property of constitutively active mutants, was kept.

In the Toyoshima model for $Ca^{2+}$-ATPase (33) two $Ca^{2+}$ binding sites were found. The residue homologous to $Glu^{795}$ participates in $Ca^{2+}$ binding site-I, whereas the residues homologous to $Glu^{343}$ and $Glu^{820}$ are involved in $Ca^{2+}$ binding site-II. Using this model it is not yet possible to understand why constitutive ATPase activity in H,K-ATPase is found with some mutants of $Glu^{820}$ and $Glu^{795}$, but not with mutants of $Glu^{343}$. Furthermore, $Ca^{2+}$ ions stimulate phosphorylation and the constitutive ATPase activity primarily reflects an increase in the spontaneous dephosphorylation rate. Even if the structure of the catalytic subunit of H,K-ATPase would be rather similar to that of $Ca^{2+}$-ATPase, one has to realize that the published structure is of the enzyme in the $E_1(Ca^{2+})$-form, whereas $K^+$-stimulated or spontaneous dephosphorylation takes place with the enzyme present in the $E_2$-P-form. There is quite some evidence that there are large structural changes in P-type ATPases during the catalytic cycle (24). As long as we do not know how the amino acid residues in the membrane are positioned in the $E_2$-conformation, the mechanism of spontaneous activation of the dephosphorylation process is unclear.

Although a direct mimicking of the $K^+$-filled binding pocket by mutagenesis is possible, it might also be that some of these mutations simulate the effect of $K^+$-binding without a direct effect on the $K^+$-binding site. It could for instance be that they change the mobility of the transmembrane domains five and six (10,22), and so mimic the $K^+$-activation. Mutation of $Glu^{820}$ may also have unexpected effects elsewhere in the protein. In T4 lysozyme it has been reported that insertion of three alanine residues at position 73 results in the conversion of two elsewhere located α-helices, that form a bend of 105°, into a single straight α-helix (34). Another possibility is that the binding pocket is modified in such a way that $H^+$ can replace $K^+$ in the dephosphorylation reaction. We recently observed a similar change in cation selectivity ($Na^+$ vs. $K^+$) in the D804A mutant of Na,K-ATPase (20).

ATP could phosphorylate all single, double and triple mutants used in the present study. It is generally assumed that binding of $H^+$-ions is necessary for phosphorylation and that most of the amino acid residues involved in $H^+$-binding have a similar function in $K^+$-binding. Nevertheless, the specific orientation of these residues must be different in the $E_1$-form that binds protons compared to the $E_2$-form that binds $K^+$-ions.

Why is the dephosphorylation in the E343D mutant blocked? One explanation is that $Glu^{343}$ is indeed directly involved in $K^+$ binding, in which interaction both oxygens of the carboxyl group are involved. Asp cannot replace Glu in $K^+$-binding, apparently because of its shorter length. Gln has the right length, but can only partially replace Glu, probably because it has only one of the two binding oxygens left, which explains the decrease in affinity by one order of magnitude. The alternative explanation for the lack of dephosphorylation of the E343D mutant is that all mutations other than Gln block the conformational change needed for communication with the phosphorylation domain. In that sense it is interesting that in the Toyoshima model of $Ca^{2+}$-ATPase, transmembrane segment 4 is unwound around $Glu^{309}$. Mutation of the latter residue might affect the unwinding of the α-helix, so disrupting the coupling to the phosphorylation site.

The starting point for the present investigation was our finding that mutant E820Q showed constitutive ATPase activity associated with $K^+$-independent dephosphorylation and a preference for the $E_1$-conformation. These properties remained when in addition to a Gln on position 820 Gln residues were present on positions 343 and/or 795. We assume that constitutive ATPase activity requires a permanent signaling between the cation-binding site and the phosphorylation domain, so that spontaneous dephosphorylation occurs. This signaling is not negatively influenced by the presence of Gln instead of Glu residues in transmembrane segments four and five. We could, however, block the constitutive ATPase activity of mutant E820Q when, in addition, an Asp residue replaced Glu343. This occurred with either a Glu or a Gln on position 795. Strikingly, the tendency of the E820Q mutation toward the $E_1$-form was kept in these mutants.

This leads to the conclusion that the high spontaneous dephosphorylation rate and the preference for the $E_1$-conformation are two separate effects of the E820Q mutation and that the preference for the $E_1$-conformation is not the result of the increased dephosphorylation rate. The $K^+$-independent dephosphorylation in the E820Q mutant could be blocked when in this mutant $Glu^{343}$ was additionally replaced by an Asp. The latter mutation only reduces the length of this side chain by a single $CH_2$-group. This additional mutation left the second effect (preference for the $E_1$-conformation) intact. It is likely that both effects were due to conformational changes caused by the E820Q mutation. The enhanced spontaneous dephosphorylation rate must be due to a better accessibility to water of the aspartyl phosphate bond located in the P-domain. The E343D mutation would then physically block the conformational change between the cation-binding pocket and the phosphorylation domain. According to the Toyoshima model (33) this could be possible since $Glu^{343}$ is located between $Glu^{820}$ and the P-domain. The change in the $E_2 \leftrightarrow E_1$ equilibrium in the rightward direction by the E820Q mutation might be caused by a change in the cation-binding domain itself, resulting in a change in the cation selectivity of this pocket in favor for $H^+$. The latter change is probably not affected by the additional E343D mutation. This speculative explanation for the findings can be tested when detailed atomic structures of gastric H,K-ATPase in several conformational states are available.

## REFERENCES

1. **Arguello JM, and Kaplan JH.** Glutamate 779, an intramembrane carboxyl, is essential for monovalent cation binding by the Na,K-ATPase. *J Biol Chem* 269: 6892-9, 1994.
2. **Asano S, Furumoto R, Tega Y, Matsuda S, and Takeguchi N.** Mutational analysis of the putative $K^+$-binding site on the fourth transmembrane segment of the gastric $H^+,K^+$-ATPase. *J Biochem (Tokyo)* 127: 993-1000, 2000.
3. **Asano S, Tega Y, Konishi K, Fujioka M, and Takeguchi N.** Functional expression of gastric $H^+,K^+$-ATPase and site-directed mutagenesis of the putative cation binding site and the catalytic center. *J Biol Chem* 271: 2740-5, 1996.
4. **Caplan MJ.** Ion pumps in epithelial cells: sorting, stabilization, and polarity. *Amer J Physiol* 272: G1304-G1313, 1997.
5. **Colonna TE, Huynh L, and Fambrough DM.** Subunit interactions in the Na,K-ATPase explored with the yeast two-hybrid system. *J Biol Chem* 272: 12366-72, 1997.

6. **Croyle ML, Woo AL, and Lingrel JB.** Extensive random mutagenesis analysis of the $Na^+/K^+$-ATPase α subunit identifies known and previously unidentified amino acid residues that alter ouabain sensitivity - implications for ouabain binding. *Eur J Biochem* 248: 488-95, 1997.
7. **DeTomaso AW, Xie ZJ, Liu G, and Mercer RW.** Expression, targeting, and assembly of functional Na,K-ATPase polypeptides in baculovirus-infected insect cells. *J Biol Chem* 268: 1470-8, 1993.
8. **Eakle KA, Kabalin MA, Wang SG, and Farley RA.** The influence of β subunit structure on the stability of $Na^+/K^+$-ATPase complexes and interaction with $K^+$. *J Biol Chem* 269: 6550-7, 1994.
9. **Faller LD, Diaz RA, Scheiner-Bobis G, and Farley RA.** Temperature dependence of the rates of conformational changes reported by fluorescein 5'-isothiocyanate modification of $H^+,K^+$- ATPases and $Na^+,K^+$-ATPases. *Biochemistry* 30: 3503-10, 1991.
10. **Gatto C, Lutsenko S, Shin JM, Sachs G, and Kaplan JH.** Stabilization of the H,K-ATPase M5M6 membrane hairpin by $K^+$ ions. Mechanistic significance for $P_2$-type ATPases. *J Biol Chem* 274: 13737-40, 1999.
11. **Geering K.** The functional role of the β-subunit in the maturation and intracellular transport of Na,K-ATPase. *FEBS Lett* 285: 189-93, 1991.
12. **Hermsen HPH, Koenderink JB, Swarts HGP, and De Pont JJHHM.** The carbonyl group of glutamic acid-795 is essential for gastric $H^+,K^+$-ATPase activity. *Biochemistry* 39: 1330-7, 2000.
13. **Hermsen HPH, Swarts HGP, Koenderink JB, and De Pont JJHHM.** The negative charge of glutamic acid 820 in the gastric $H^+,K^+$-ATPase α-subunit is essential for $K^+$ activation of the enzyme activity. *Biochem J* 331: 465-72, 1998.
14. **Hermsen HPH, Swarts HGP, Wassink L, Koenderink JB, and Willems PHGM, De Pont JJHHM.** Mimicking of $K^+$-activation by double mutation of glutamate-795 and glutamate-820 of gastric $H^+,K^+$-ATPase. *Biochemistry* 40, 6527-33, 2001.
15. **Klaassen CHW, Fransen JAM, Swarts HGP, and De Pont JJHHM.** Glycosylation is essential for biosynthesis of functional gastric $H^+,K^+$-ATPase in insect cells. *Biochem J* ; 321: 419-24, 1997.
16. **Klaassen CHW, Van Uem TJF, De Moel MP, De Caluwé GLJ, Swarts HGP, and De Pont JJHHM.** Functional expression of gastric H,K-ATPase using the baculovirus expression system. *FEBS Lett* 329: 277-82, 1993.
17. **Koenderink JB, Hermsen HPH, Swarts HGP, Willems PHGM, and De Pont JJHHM.** High-affinity ouabain binding by a chimeric gastric $H^+,K^+$-ATPase, containing transmembrane hairpins M3-M4 and M5-M6 at the α1-subunit of rat $Na^+,K^+$-ATPase. *Proc Nat Acad Sci USA* 97: 11209-14, 2000.
18. **Koenderink JB, Swarts HG, Stronks HC, Hermsen HP, Willems PH, and De Pont JJ.** Chimeras of $X^+$, $K^+$-ATPases. The M1-M6 region of $Na^+$, $K^+$-ATPase is required for $Na^+$-activated ATPase activity, whereas the M7-M10 region of $H^+,K^+$-ATPase is involved in $K^+$ de-occlusion. *J Biol Chem* 276: 11705-11, 2001.
19. **Koenderink JB, Swarts HGP, Hermsen HPH, and De Pont JJHHM.** The beta-subunits of $Na^+,K^+$-ATPase and gastric $H^+,K^+$-ATPase have a high preference for their own alpha-subunit and affect the $K^+$ affinity of these enzymes. *J Biol Chem* 274: 11604-10, 1999.
20. **Koenderink JB, Swarts HGP, Hermsen HPH, Willems PHGM, and De Pont JJHHM.** Mutation of aspartate 804 of $Na^+,K^+$-ATPase modifies the cation binding pocket and thereby generates a high $Na^+$-ATPase activity. *Biochemistry* 39: 9959-66, 2000.

21. **Lemas MV, Takeyasu K, and Fambrough DM.** The carboxyl-terminal-161 amino acids of the Na,K-ATPase α-subunit are sufficient for assembly with the β-subunit. *J Biol Chem* 267: 20987-91, 1992.
22. **Lutsenko S, Anderko R, and Kaplan JH.** Membrane disposition of the M5-M6 hairpin of $Na^+$,$K^+$-ATPase α–subunit is ligand dependent. *Proc Nat Acad Sci USA* 92: 7936-40, 1995.
23. **Melle-Milovanovic D, Milovanovic M, Nagpal S, Sachs G, and Shin JM.** Regions of association between the α and the β subunit of the gastric H,K-ATPase. *J Biol Chem* 273: 11075-81, 1998.
24. **Møller JV, Juul B, and Le Maire M.** Structural organization, ion transport, and energy transduction of P-type ATPases. *Biochim Biophys Acta* 1286: 1-51, 1996.
25. **Palasis M, Kuntzweiler TA, Arguello JM, and Lingrel JB.** Ouabain interactions with the H5-H6 hairpin of the Na,K- ATPase reveal a possible inhibition mechanism via the cation binding domain. *J Biol Chem* 271: 14176-82, 1996.
26. **Price EM, Rice DA, and Lingrel JB.** Site-directed mutagenesis of a conserved, extracellular aspartic acid residue affects the ouabain sensitivity of sheep Na,K-ATPase. *J Biol Chem* 264: 21902-6, 1989.
27. **Price EM, Rice DA, and Lingrel JB.** Structure-function studies of Na,K-ATPase. Site-directed mutagenesis of the border residues from the H1-H2 extracellular domain of the alpha subunit. *J Biol Chem* 265: 6638-41, 1990.
28. **Rabon EC, Bassilian S, Sachs G, and Karlish SJD.** Conformational transitions of the H,K-ATPase studied with sodium ions as surrogates for protons. *J Biol Chem* 265: 19594-9, 1990.
29. **Swarts HGP, Koenderink JB, Hermsen HPH, Willems PHGM, and De Pont JJHHM.** $K^+$-independent gastric $H^+$,$K^+$-ATPase: Dissociation of $K^+$-independent dephosphorylation and preference for the $E_1$-conformation by combined mutagenesis of transmembrane glutamate residues. *J Biol Chem* 2001 published on the Web July 24th.
30. **Swarts HGP, Hermsen HPH, Koenderink JB, Schuurmans Stekhoven FMAH, De Pont JJHHM.** Constitutive activation of gastric $H^+$,$K^+$-ATPase by a single mutation. *EMBO J* 17: 3029-35, 1998.
31. **Swarts HGP, Hermsen HPH, Koenderink JB, Willems PHGM, De Pont JJHHM.** Conformation-dependent inhibition of gastric $H^+$,$K^+$-ATPase by SCH 28080 demonstrated by mutagenesis of glutamic acid 820. *Mol Pharmacol* 55: 541-7, 1999.
32. **Swarts HGP, Klaassen CHW, De Boer M, Fransen JAM, De Pont JJHHM.** Role of negatively charged residues in the fifth and sixth transmembrane domains of the catalytic subunit of gastric $H^+$,$K^+$- ATPase. *J Biol Chem* 271: 29764-72, 1996.
33. **Toyoshima C, Nakasako M, Nomura H, Ogawa H.** Crystal structure of the calcium pump of sarcoplasmic reticulum at 2.6 A resolution. *Nature* 405: 647-51, 2000.
34. **Vetter I, Baase W, Heinz D, Xiong J, Snow S, Matthews B.** Protein structural plasticity exemplified by insertion and deletion mutants in T4 lysozyme. *Protein Sci* 5: 2399-415, 1996.
35. **Wang SG, Eakle KA, Levenson R, Farley RA.** $Na^+$-$K^+$-ATPase α-subunit containing Q905-V930 of gastric $H^+$- $K^+$-ATPase α preferentially assembles with $H^+$-$K^+$-ATPase β. *Amer J Physiol* 272: C923-C930, 1997.

Chapter 8

# The Effect of Mutations Within a Cluster of Homologous Amino Acids Present in the M5, M6 and M8 Transmembrane Regions of the H,K-ATPase

E.C.Rabon, S.J. Rulli and E. V. Skripnikov
*Tulane University and the Veterans Administration, New Orleans, LA 70112*

## ABSTRACT

Five conserved amino acids, asn-794, glu-797, thr-825, asp-826 and glu-938, in transmembrane regions M5, M6 and M8 of the P2 family of ATPases have been identified as a putative cation binding site. To test if these conserved amino acids influence the $K^+$ -stimulated ATPase activity of the H, K -ATPase, site mutations were introduced into the alpha-subunit of the rabbit gastric H, K -ATPase and co-expressed with the rabbit gastric beta-subunit in HEK 293T cells. In transmembrane domain M5, mutations of asn-794 (leu, ala, asp), but not glu-797 (gln) significantly attenuated or inactivated the H,K-ATPase. Several mutations of lys-793 (ser, glu, arg, ala, leu) also partially inactivated the H,K-ATPase and is consistent with a role for this residue in H,K-ATPase function. Substitution of aliphatic (ala, leu) but not charged residues (ser, asp, lys) was tolerated in M6 at thr-825 but all substitutions of asp-826 (leu, ala, asn, thr) inactivated the H,K-ATPase. Aliphatic substitutions of glu-938 (gln, ala) in transmembrane region M8 were tolerated. The data utilizing single aliphatic substitutions present clear evidence of the importance of asn-794 and asp-826 in cation dependent function. H,K-ATPase inactivation dependent upon charge substitutions at thr-825 implicates thr-825, but evidence for the involvement of the carboxyl side chain at glu-797 and glu-938 in cation dependent function is lacking. To test whether the co-ordination of shared sidechain oxygens from multiple residues might mask the potential contribution of a single residue, double mutations were introduced into residues of M5/ M6, M5/ M8 and M6/ M8. The H,K-ATPase containing double substitutions in M5/ M6 at glu-797 (gln) / thr-825 (ala, leu) and glu-797 (gln) / glu-938 (gln, ala) was fully functional while the H,K-ATPase containing double substitutions in M5/ M8 at glu-797 (gln) / glu-938 (gln, ala) and in M6/ M8 at thr-825 (ala, leu) / glu-939 (gln, ala) was partially inhibited. These data are consistent with a role for glu-938, but not glu-797 in cation dependent ATPase activity. Finally, the double mutation thr-825 (asp)/ asp-826-(thr) failed to restore function lost with single substitutions (thr-825 (asp) and asp-826 (thr)) at this position. This is consistent with a strict requirement for the charged side chain geometry at this position rather than a requirement simply for the conservation of charge between the charged pair. Overall, these data are consistent with a model utilizing 4 residues, asn-794, thr-825, asp-826 and glu-938 in the cation binding site located in M5/ M6 and M8 and suggests that this cation binding site is mostly conserved within the P2 family.

## INTRODUCTION

The gastric H,K-ATPase is a P2 ion translocation pump responsible for $H^+/K^+$ exchange across the gastric parietal cell secretory membrane. A variety of biochemical and site mutational data within the P2 ATPase family is consistent with the idea that cation binding sites of the transported ligands are composed of amino acids found within the M4/M5/M6 and M8 transmembrane helices (3)(9)(6)(2)(12)(5). The structural solution of the SR Ca-ATPase identifies 11 amino acid residues organized in 2 clusters that define the cation binding sites for $Ca^{2+}$ binding in this ATPase (13). The conservation of these residues within the H,K- and Na,K-ATPases suggests that these residues may also be involved in cation binding. In a previous study, we introduced mutations into several of the amino acids homologous to the putative $Ca^{2+}$ ATPase "site 1" to investigate if these conserved amino acids strongly influenced cation dependent functions of the H,K ATPase (11). This study placing isosteric or aliphatic substitutions into glu-797 (gln), thr-825 (leu, ala) and glu-938 (gln) showed that sidechain removal of a single residue did not disrupt the $K^+$-dependent activation or SCH28080-dependent inhibition of the H,K-ATPase. This observation suggests either that the cation binding domain in the H,K-ATPase is divergent from that of the SR Ca-ATPase or that multiple site coordination masks the functional role of a single residue. Figure 1 presents a homologous model of the putative cation binding site 1 of the H,K-ATPase. This site is composed of 5 residues present in transmembrane helices M5, M6 and M8. The present investigation introduces single mutations into asn-794 (M5) and asp-826 (M6), to investigate the influence of single mutations of these residues. In addition, double mutations were introduced in residues in transmembrane pairs M5/ M6, M5/ M 8 and M6/ M8 to evaluate whether a 2nd mutation in the putative cation binding site of the H,K-ATPase might unmask the contribution of a residue not apparent from a single mutation.

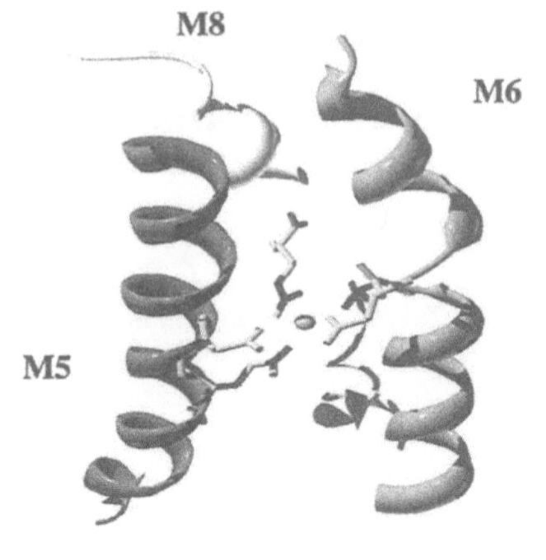


Figure 1 *Model of putative H,K-ATPase cation binding site 1.* Homology model of the putative cation binding site 1 utilizing contributions of amino acid residues in transmembrane domains M5 (asn-794, glu-797) M6 (thr-825, Asp-826) and M8 (glu-938).

## METHODS

*Plasmid constructs and site-directed mutagenesis:* Plasmids containing the α- (accession # X64694) and β-subunits (accession # M35544) of the rabbit (Oryctolagus cuniculus) gastric H,K-ATPase were constructed in pcDNA3 (Invitrogen) as described previously (10). Site directed mutations were introduced into pcDNA3-α using the Quick Change Site Directed Mutagenesis Kit (Stratagene) according to the manufacture's instructions with minor modifications. Following the selection and expansion of single, satellite-free colonies, plasmids were isolated with the Wizard Midi Prep Kit (Promega), re-purified on Quiagen Spin Mini Prep columns and submitted for sequence analysis.

*Cell culture and transient transfection:* HEK 293T cells cultured in Dulbecco's modified Eagle Medium (DMEM), 10 % fetal bovine serum, 2.0 mM glutamine and 0.1 mg/mL penicillin / streptomycin at 37°C in 95 / 5% $CO_2$ atmosphere were co-transfected with 9 µg pcDNA3-

α,18 μg pcDNA3-β and 1.3 μL LipofectAMINE™ per μg DNA. The transfected cells were cultured overnight, then supplemented with 20 mL of the standard culture medium containing 10% fetal calf serum and grown for an additional 48 hr.

*Membrane Preparation:* Membranes were prepared from the transiently co-transfected cells 72 hr. post transfection by Dounce homogenization utilizing a 7 mL vessel and tight fitting pestle. The membrane homogenate was resolved by centrifugation over a 44 % sucrose column, resuspended in 10 mM Pipes/ Tris, pH 7.4 and frozen at -80°C until assay.

*Western Analysis*: The total membrane-bound protein was determined by the Pierce BCA assay using BSA as a standard. Aliquots containing 12-25 μg of total membrane protein were resolved by SDS PAGE run under reducing conditions according to the method of Laemmli and transferred to nitrocellulose for Western analysis (7). The nitrocellulose membranes were blocked in 1% Carnation instant nonfat dry milk and probed with anti-α MAb, 12.18. The probed membranes were incubated with secondary anti-mouse horseradish peroxidase (Sigma) and visualized using ECL horseradish peroxidase substrate. Chemilumenesence was captured on ECL Hyperfilm. The α-subunits were estimated by quantitative densitometry of 2 or 3 film exposures for each n = 3 aliquots of 12-20 μg of total membrane-bound protein. A maximal estimate of the alpha subunit in each mutant preparation was obtained by comparison of the transfected membrane preparation to the "G1" fraction purified from hog gastric mucosa. Chemilumenisence was captured as an 8 bit tiff file using Biosoft's ScanAnalysis (ver 2.56). A linear standard curve was obtained from the integrated density of background corrected aliquots containing from 20 to 60 ng of the H,K ATPase.

*ATPase Analysis:* ATPase activity was measured in membrane preparations of non transfected (control) and transiently co-transfected HEK 293T cell using the malachite green assay of Lancetta (8). Typically, 4 μg of membrane-bound protein was assayed using 0.1 mL of ATPase reaction buffer containing; 1 mM ATP, 3 mM $MgCl_2$, 0.1 mM ouabain, 1.5 mM EGTA, 60 mM Tris-HCl, 1 μg/ mL oligomycin, 1 μg/ mL nigericin ± 10 mM KCl with a final assay pH of 7.4. The reaction, started by the addition of ATP, was incubated at 37°C for 30 min. and stopped with the addition of 45 μL of malachite green reagent and 45 μL of citrate buffer. Inorganic P was measured by its absorption at 650 nm using a Molecular Devices VERSAmax multiwell plate reader. The expressed H,K-ATPase was differentiated from an endogenous, nonspecific ATPase by its characteristic SCH28080-inhibited, KCl-stimulated activity. Each average measurement reports the analysis of 2 or more membrane preparations of n = 4 samples in each assay. The KCl-stimulated activity was derived by subtracting the ATPase activity measured in $Mg^{2+}$ medium alone ($Mg^{2+}$ ATPase activity) from the ATPase activity at 10 mM KCl. The SCH28080-inhibitible activity was derived by subtracting the residual activity measured in the presence of 55 μM SCH 28080 from the total activity measured in the presence of 10 mM KCl.

*Modeling the H,K-ATPase:* The rabbit gastric H,K-ATPase alpha subunit (swissprot. accession # P27112) was modeled to the template structure of SERCA 1a (pdb entry 1EUL.pdb) using the Swiss-PdbViewer (4). The initial Blast sequence alignment utilized an open gap penalty of 3 and an extended gap penalty of 2 (1).

*Cell line / Antibodies / Materials:* HEK293T cells were a kind gift of Dr. Jamboor K. Vishwanatha. Anti-α, MAb 12.18, was a kind gift from Dr. A. Smolka. The rabbit gastric alpha- and beta-subunits cDNA were a kind gift of George Sachs.

## RESULTS

*The influence of lys-793 in the cytoplasmic proximity of putative site 1.*

The model of Panel A, figure 2 positions lys-793 adjacent to the cytoplasmic side of putative cation site 1. Due to its proximity and potential for interaction with the cation site residues, several mutations were introduced to alter the sidechain structure and charge at this position. The Western analysis of typical membrane preparations shown in Panel B indicates that the alpha-subunit is expressed in preparations where charged residues such as arg, ser and glu are substituted at this position. Previous Western analysis has established that aliphatic substitutions such as ala and leu are well expressed. Panel C provides a summary of H,K-ATPase activity averaged from 2 or more membrane preparations in each of the lys-793 mutations. All mutations, independent of the charge polarity or the side chain size reduced both the $K^+$-stimulated and SCH28080-inhibited components of the H,K-ATPase activity. The SCH28080-sensitive component in each mutant was approximately equal to the $K^+$-stimulated activity as expected from $K^+$ competitive inhibition kinetics.

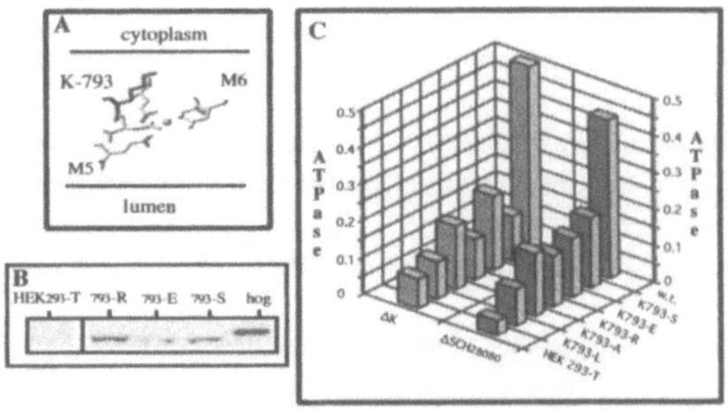


*Figure 2. H,K-ATPase activity profile of lys-793 mutations.*
*Panel A.* Carbon bond structure of site 1. The dark carbon bonds indicate the position of lys-793 mutations relative to the cytoplasmic and lumenal sides of the membrane.
*Panel B.* Western analysis of alpha-subunit expressing specific mutations of lys-793. HEK 293-T ; untransfected membrane preparation. hog ; 50 ng of porcine gastric H,K-ATPase.
*Panel C.* H,K-ATPase activity profile of mutations of lys-793. The measured activity is reported from n=4 replicates in µmol/ mg (membrane protein) / hr.

*Mutation of asn-794 in M5 or thr-825 and asp-826 in M6.*

Asn-794 is an unexpected cation site candidate identified in the crystalline structure of SERCA. The model in panel A of figure 3 shows the position of this conserved residue in the H,K-ATPase. Several mutations were

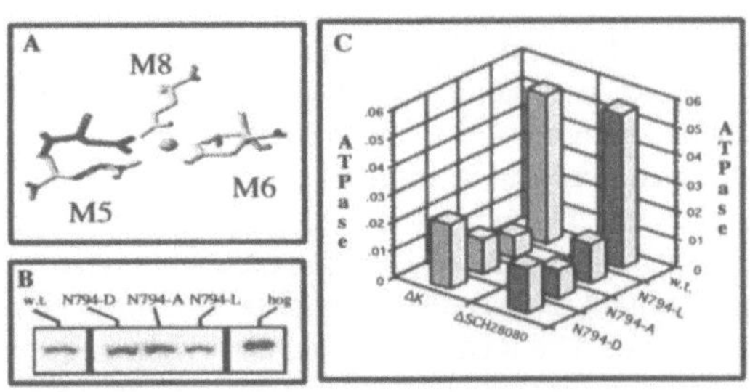


Figure 3 *H,K-ATPase activity profile of asn-794 mutations.*
*Panel A.* Carbon bond structure of site 1 indicating the position of asn-794 mutations. The darkened bond identifies asn-794 in the M5 transmembrane domain.
*Panel B.* Western analysis of alpha-subunit expression of asn-794. w.t. (wild type) ; untransfected. hog ; porcine gastric H,K-ATPase (50 ng).
*Panel C.* H,K-ATPase activity profile of mutations of asn-794. The $K^+$-stimulated and SCH28080-inhibitable activity is reported from n=4 replicates in µmol/ mg (membrane protein) / hr.

introduced at this position to either modify the side chain bulk (ala, leu) or provide an isosteric charged replacement (asp). The Western analysis of a typical preparation shown in Panel B indicates that the alpha subunit carrying each of the mutations is expressed. The averaged ATPase activity shown in panel C indicates that all mutations introduced at this position reduce both the $K^+$-stimulated and SCH28080-inhibited components of the ATPase activity. Interestingly, H,K-ATPase activity lost due to aliphatic substitution (ala, leu) at this position is partially restored by the isosteric substitution of asp for asn.

**Table I**
**Measured H,K-ATPase in single mutations**

| mutant | $\Delta K^+$ | ΔSCH |
|---|---|---|
| thr 825-leu | - | *0.89 ± .15 |
| thr 825-ala | - | *0.91 ± .36 |
| thr 825-ser | 0.24 ± .06 | 0.17 ± .05 |
| thr 825-asp | 0.09 ± .02 | 0.09 ± .03 |
| thr 825-lys | 0.07 ± .05 | 0.07 ± .04 |
| | | |
| asp 826-leu | 0.06 ± .03 | 0.09 ± .02 |
| asp 826-ala | 0.08 ± .01 | 0.11 ± .00 |
| asp 826-asn | 0.17 ± .01 | 0.15 ± .01 |
| asp 826-thr | 0.06 ± .00 | 0.07 ± .03 |

Values represent average activity from 2 or more preparations ± std. dev. * data taken from Rulli et al (11). Activity reported in µmol/ mg (membrane protein) / hr.

The data in Table I summarize the H,K-ATPase present in single mutations of contiguous residues in M6, thr-825 and asp-826. Mutations introduced in thr-825 yielded divergent results. Aliphatic substitutions (leu and ala) spared the H,K-ATPase function while charged substitutions(ser, asp and lys) inactivated it. Within the charged substitutions, mutations at thr-825 involving either the substitution of a larger charged group (lys for thr) or a carboxy group (asp for thr) reduced the activity of the expressed ATPase more than that of the conservative substitution of ser for thr. All mutations introduced in asp-826 inactivated the H,K-ATPase. Little difference was noted between the loss of activity with aliphatic substitutions (ala and leu) or a charge substitution of thr for asp. Preparations of the ATPase containing the isosteric substitution of asn for asp demonstrated the highest partial activity of this mutation series.

*Introduction of double mutations into the putative cation binding domain.*

*Double mutations of M5/ M6 and M6/ M8.*
The ATPase activity of preparations containing single mutations of glu-797 (gln), thr-825 (ala, leu) and glu-938 (gln, ala) contrasts markedly to that observed with aliphatic mutations in asn-794 (ala, leu) and asp-826 (ala, leu). To test the possibility that ATPase function is maintained in the former mutants because of ligand co-ordination between the remaining residues of the site, double mutations were introduced to eliminate the side chain contribution of each residue of a candidate site pair. Accordingly, mutations incorporating ala or leu at thr-825 and glu-938 were introduced into the H,K-ATPase alpha-subunit containing the isosteric substitution, gln for glu, at residue-797. Panel B, of figure 4 reports the average measured

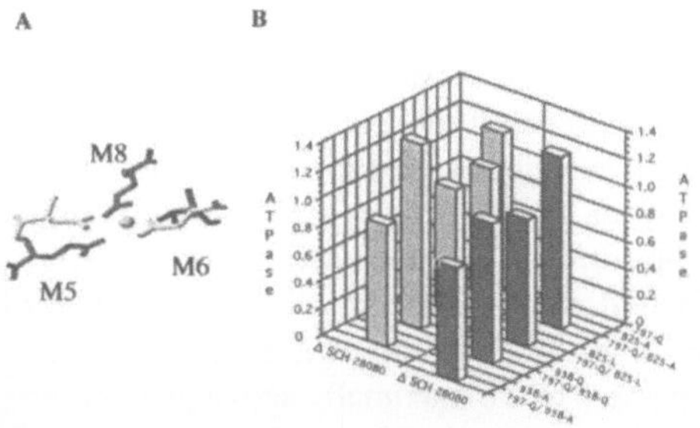


Figure 4 *H,K-ATPase activity profile of double mutations in M5/ M6 and M5/ M8.*
*Panel A.* Carbon bond structure of site 1 indicating the positions of paired, modified residues in M5/ M6 and M5/ M8. The darkened carbon bonds indicate the positions of residues in M5 (glu-797), M6 (thr-825) and M8 (glu-938).
*Panel B.* H,K-ATPase activity profile of single and paired mutations. The SCH28080 inhibitable activity is reported in µmol/ mg (membrane protein) / hr. for the identified substitution(s) using n=4 replicates.

activity of 2 or more preparations of the double mutant constructs and the H,K-ATPase activity of each single mutant present in the double mutant constructs. The data show that the average activity of H,K-ATPase preparations containing double mutations in M5/ M6 and M5/ M8 are similar to those of the H,K-ATPase preparations containing only single mutations in either transmembrane domain.

In contrast to the H,K-ATPase activity exhibited by the double mutants in M5/ M6 and M5/ M8, double mutations in M6/ M8 at thr-825 (ala) / glu-938 (ala) and thr-825 (leu) / glu-938 (ala) reduced the measured activity below that measured in each single mutant. As indicated in panel B of figure 5, the double mutants exhibit approximately half the measured ATPase of preparations containing single mutations at each position.

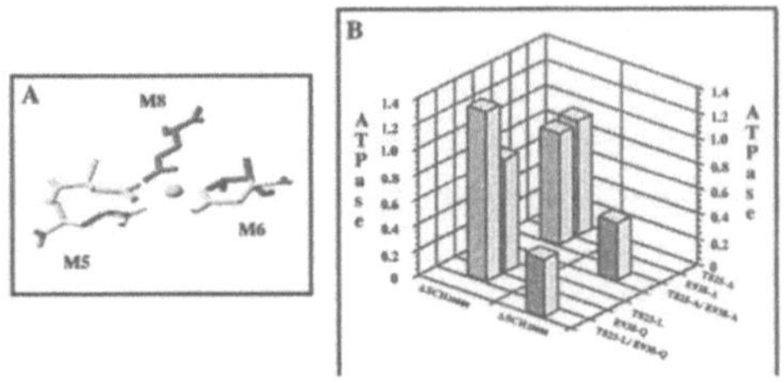


Figure 5 *H,K-ATPase activity profile of double mutations in M6/ M8.*
*Panel A.* Carbon bond structure of site 1 indicating the positions of paired modified residues in M6/ M8. The darkened carbon bonds identify residues in M6 (thr-825) and M8 (glu-938).
*Panel B.* H,K-ATPase activity profile of preparations containing single and double mutations. The SCH28080-inhibitable activity of single and double mutants of the H,K-ATPase is reported from n=4 replicates.

The level of alpha subunit expression was measured by Western analysis of representative preparations in order to determine the specific activity of the double mutant preparations independent of the alpha subunit expression level. Quantitative densitometry of Western assays shown in Table II indicates that the expression level of the double mutations varied about 4-fold from 0.86 ± .37 to 3.51 ± 1.3 µg alpha subunit/ mg preparation. The relative specific activity, reported as the measured activity corrected for the level of alpha subunit expression was compared to the maximally active preparation of glu-797 (gln) / thr-825 (ala) in Table II. Both M5/ M6 double mutants were fully active while the H,K-ATPase of both the M5/ M8 and the M6/ M8 double mutants was inhibited 50 - 80%.

**Table II**
**Specific Activity of Double Mutants**

| Double mutant | Measured ATPase (µmol/mg/hr) | Expression (µg/ mg) | Relative specific act. |
|---|---|---|---|
| glu-797 (gln) / thr-825 (ala) | 1.25 ± .04 | 1.66 ± .45 | 1 |
| glu-797 (gln) / thr-825 (leu) | 0.59 ± .05 | 0.86 ± .37 | 0.91 |
| glu-797 (gln) / glu-938 (gln) | 1.02 ± .02 | 2.75 ± .65 | 0.49 |
| glu-797 (gln) / glu-938 (ala) | 0.88 ± .04 | 3.51 ± 1.3 | 0.33 |
| thr-825 (ala) / glu-938 (ala) | 0.38 ± .13 | 2.52 ± .73 | 0.20 |
| thr-825 (leu) / glu-938 (gln) | 0.39 ± .04 | 1.13 ± .92 | 0.46 |
| thr-825 (asp) / asp-826 (thr) | 0.04 ± .01 | 2.60 ± .93 | 0.02 |

Measured ATPase reports SCH28080-sensitive activity of n = 4 replicates of total membrane protein. Specific activity, normalized to the level of alpha-subunit expression in aliquots of the preparations used for ATPase analysis was compared to that obtained in the glu-797 (gln) / thr-825 (ala) preparation.

*Double mutation of M6/ M6.*
M6 contains the charged doublet thr-825/ asp-826 indicated in Panel A of figure 6. In contrast to single mutations in other residues of the putative cation binding site and the tolerance to aliphatic substitutions at thr-825, charge substitutions at thr-825 and asp-826 are lethal to the H,K-ATPase. To investigate if the ATPase function was related to the conservation of the side chain charge in this doublet or to the geometry

of the charged residues, the side chains were switched to obtain the double mutant, thr-825-asp/ asp-826-thr. As indicated in figure 6, the charge switch in the amino acid pair failed to restore activity lost in each of the single mutations. Table II indicates that the loss of function was not due to the loss of ATPase expression.

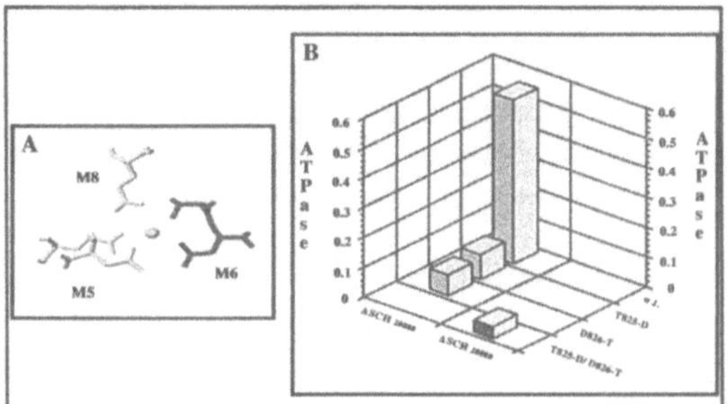


Figure 6 *H,K-ATPase activity profile of a double mutation in M8.*
*Panel A.* Carbon bond structure of site 1 indicating the position of the modified pair of residues in M8. The darkened carbon bonds indicate the positions of residues in M8 (thr-825) and (asp-826).
*Panel B.* H,K-ATPase activity profile of single and paired mutations. The SCH28080-inhibitable activity is reported from n=4 replicates.

## DISCUSSION

Biochemical and structural investigations have identified 11 residues within 4 transmembrane domains of the P-2 ATPase family that may function as cation binding site(s). The crystalline structure of the SR Ca-ATPase proposes that these residues define 2 sites, site1 composed of 5 residues in M5, M6 and M8 and site 2 composed of 6 residues in M4 and M6. Site 1 is comprised predominently of charged residues while site 2 is composed predominently of uncharged residues from an unwound portion of the M5 helix and two acidic residues in M6. This investigation placed single or double mutations into each of the residues of putative cation binding site 1 to determine if substitutions of side chain charge or geometry at these positions would influence cation-dependent function of the H,K-ATPase.

The results presented herein indicate that side chain substitutions in 4 of 5 potential residues in putative cation site 1 influence cation-dependent function. 2 residues, asn-794 and asp-826, are essential for H,K-ATPase function as none of the amino acids substitutions at these positions support a fully functional ATPase. 2 other residues, thr-825 and glu-938 are also likely to be involved in H,K-ATPase function as selective, single substitutions in thr-825 and paired substitutions in glu-797/ thr-825 and thr-825/ glu-938 significantly inhibit function. In contrast, the single substitution of glu-797 (gln) and the double mutations involving glu-797/ thr-825 paired substitutions provided little evidence of carboxyl sidechain involvement of glu-794 in H,K-ATPase function.

The data here show that both M5 and M6 contain important adjacent residues. Lys-793, on the cytoplasmic side of asn-794 in M5 is clearly important for function as all substitutions, (ser, glu, arg, leu and ala ) significantly reduce the H,K-ATPase activity. This residue is positioned to hydrogen bond with any of the putative site 1 residues and could potentially serve an important function by bridge formation between these residues. Single substitutions in both M5 (thr-825) and M6 (asp-826) inhibit H,K-ATPase function. Double mutations were introduced to restore the charge, but reverse the side chain position. The restoration of the conserved but reversed charged pair failed to salvage the H,K-ATPase function indicating the strict geometrical requirement of the sidechain position.

The basic premise of this investigation was that substitutions in cation site residues would be poorly tolerated because they would disrupt the co-ordination needed for cation binding. This premise was found to be true for asn-794 and asp-826, but the residues glu-797, thr-825 and glu-938 were surprisingly tolerant to substitutions. This tolerance suggests that either the cation binding site geometry is different from the Ca-ATPase or that the loss of a single co-ordinating residue is masked by the contribution of the remaining residues. The present interpretation is that the sidechains of thr-825 and glu-938 are important for cation binding , but the contribution is masked in single mutations while the side chain of glu-797 does not contribute to the cation binding site.

In summary, the present data provide evidence of the functional involvement of 4 residues in putative cation binding site1, asn-794, thr-825, asp-826 and glu-938.

## REFERENCES


1. **Altschul, S. F., W. Gish, W. Miller, E. W. Myers, and D. J. Lipman.** Basic local alignment search tool. *J.Mol.Biol.* 215: 403-410, 1990.
2. **Asano, S., Y. Tega, K. Konishi, M. Fujioka, and N. Takeguchi.** Functional expression of gastric $H^+,K^+$-ATPase and site-directed mutagenesis of the putative cation binding site and catalytic center. *J Biol Chem* 271: 2740-2745, 1996.
3. **Clarke, D. M., T. W. Loo, G. Inesi, and D. H. MacLennan.** Location of high affinity $Ca^{2+}$-binding sites within the predicted transmembrane domain of the sarcoplasmic reticulum $Ca^{2+}$-ATPase. *Nature* 339: 476-478, 1989.
4. **Guex, N., and M. C. Peitsch.** SWISS-MODEL and the Swiss-PdbViewer: An environment for comparative protein modeling. *Electrophoresis* 18: 2714-2723, 1997.
5. **Hermsen, H. P., J. B. Koenderink, H. G. Swarts, and J. J. de Pont.** The carbonyl group of glutamic acid-795 is essential for gastric $H^+,K^+$-ATPase activity. *Biochemistry 2000 Feb 15;39(6):1330-7* 39: 1330-1337, 2000.
6. **Jewell-Motz, E. A., and J. B. Lingrel.** Site-directed mutagenesis of the Na,K-ATPase: Consequences of substitutions of negatively-charged amino acids localized in the transmembrane domains. *Biochemistry* 32: 13523-13530, 1993.
7. **Laemmli, U. K.** *Nature* 227: 680-685, 1970.
8. **Lancetta P., A. L., Reinach P. and Candia O.** An Improved Assay for Nanomole Amounts of inorganic Phosphate. *Anal Biochem.* 100: 95-97, 1979.
9. **le Maire, M., S. Lund, A. Viel, P. Champeil, and J. V. Mueller.** Ca $^{2+}$ -induced conformational changes and location of Ca $^{2+}$ transport sites in sarcoplasmic reticulum $Ca^{2+}$-ATPase as detected by the use of proteolytic enzyme (V8). *JBC* 265: 1111-1123, 1990.
10. **Rulli, S. J., M. N. Horiba, E. Skripnikova, and E. C. Rabon.** Glu-857 Moderates $K^+$-stimulation and SCH 28080 Inhibition of the Gastric H,K-ATPase. *JBC* 274: 15245-15250, 1999.
11. **Rulli, S. J., N. M. Louneva, E. Skripnokova, and E. C. Rabon.** Site-directed Mutagenesis of a Cation Binding Sites of the Gastric H,K-ATPase. *JBC*:, 2000.
12. **Swarts, H. G., C. H. Klaassen, M. de Boer, J. A. Fransen, and J. J. de Pont.** Role of negatively charged residues in the 5th and 6th transmembrane domains of the catalytic subunit of gastric $H^+,K^+$-ATPase. *JBC* 271(47): 29764-29772, 1996.
13. **Toyoshima, C., M. Nakasako, and H. Ogawa.** Crystal structure of the calcium pump of sarcopasmic reticulum at 2.6Å resolution. *Nature* 405: 647-655, 2000.

Chapter 9

# Host-Specific *H. pylori* Inhibition of H,K-ATPase α Subunit Gene Expression

ADAM SMOLKA and MONIKA Gö õ z
*Department of Medicine, Gastroenterology and Hepatology Division*
*Medical University of South Carolina, Charleston, SC*

## INTRODUCTION

*Life is a struggle, not against sin, not against the Money Power, not against animal magnetism, but against hydrogen ions.*

H.L. Mencken, 1919

With this sentence, in a book review on the mechanisms of death, Mencken wrote a concise biography of the bacterium we now know as *Helicobacter pylori*. Mencken was a widely-read literary critic, but even he was unlikely to have read the Italian description 27 years earlier of gastric mucosal spiral bacteria (1). For *H. pylori*, life is indeed a struggle against hydrogen ions. The strategies undertaken by *H. pylori* against the lethal effects of low gastric pH are well known. The spiral highly motile morphology of the organism allows penetration into the gastric mucin layer where a pH gradient up to neutral pH in the immediate vicinity of the surface epithelial cells affords some measure of protection. Highly specific interactions between bacterial and epithelial apical membrane glycoproteins and glycolipids ensure bacterial adhesion in this relatively high pH zone. Most prominently, *H. pylori* takes up urea through proton-regulated mechanisms elegantly described by George Sachs (14); urease activity produces $CO_2$ and ammonia, neutralizing hydrogen ions in the immediate vicinity of the organism.

In this study, we present evidence for yet another strategy by *H. pylori* to counter low gastric pH, namely that *H. pylori* infection down-regulates H,K-ATPase gene expression. In keeping with human gastric mucosa being the major reservoir of *H. pylori* colonization, we reasoned that transcriptional down-regulation would be less efficient with non-human H,K-ATPase promoters, and that species differences in H,K-ATPase transcriptional down-regulation may pin-point *H. pylori*-sensitive response elements. Down-regulation of H,K-ATPase gene expression would lead to fewer proton pumps being expressed in parietal cells, and thus result in lower capacity to secrete protons into the gastric lumen.

Clinical evidence for our hypothesis comes from retrospective analyis of iatrogenic mini-epidemics of hypochlorhydria during acid secretory studies (7) and volunteer experiments which showed that acute infection leads to neutrophilic gastritis and decreased gastric acid secretion (9, 10). Also, a recent clinical study showed that normal acid secretion is restored in patients with hypochlorhydria when their *H. pylori* infection is eradicated (4). The pathogenesis of this hypochlorhydria remains unclear. Bacterial products have been implicated (2, 8), and *H. pylori*-induced inflammation increases epithelial IL-1ß secretion, which may in turn inhibit acid secretion (13). *H. pylori* urease activity cannot account for hypo- or achlorhydria since that mechanism is finely regulated to titrate only the bacterial periplasm to a pH of 6 or higher (14).

In previous studies, we showed that gastric adenocarcinoma (AGS) cells possessed functional histamine $H_2$ receptors, and that human HKα 5'-flanking sequence transfected into AGS cells was activated by histamine and inhibited by *H. pylori* (6). Briefly, treatment of AGS cells with histamine induced transient dose-dependent elevations of intracellular calcium AGS cells, as measured by fluorescence changes of the non-ratiometric calcium probe Fluo-3. The $Ca^{2+}$ transients were suppressed by addition of EDTA to the cell culture medium, or by pre-incubation of the cells with the $H_2$ receptor antagonist cimetidine (6). Also, histamine treatment of AGS cells induced a dose-responsive increase in intracellular cyclic AMP, counteracted by pre-incubation of cells with cimetidine (6). In the same study, AGS cells were transiently transfected with reporter gene constructs containing 2.2 kb of human HKα 5'-flanking sequence (6). Physiological concentrations of histamine caused dose-responsive activation of the HKα promoter, which was suppressed by pre-incubation of the cells with cimetidine. Additionally, human HKα promoter activity stimulated by histamine and phorbol ester (PMA) was also dose-dependently inhibited by *H. pylori* infection of the cells for 24 hours (6). Phorbol ester and oleoylacetylglycerol activation of HKα promoter was inhibited by staurosporine and calphostin C respectively. These data implicate protein kinase C as a potential downstream target of *H. pylori* down-regulation of H,K-ATPase gene expression.

## MATERIALS and METHODS

Expression of proton pumps in normal or *H. pylori*-infected gastric mucosa was assessed by immunohistochemistry. Sections of human gastric mucosal biospies obtained endoscopically in the MUSC Gastroenterology Clinic were immunolabeled with a monoclonal antibody (HK 12.18) directed against a cytoplasmic epitope of the H,K-ATPase α subunit (12). Proton pump α subunit mRNA levels were measured in gastric biopsies from normal and *H. pylori*-infected rhesus monkeys by ribonuclease protection assay. Biopsy RNA preparations were incubated with $^{32}P$-labeled antisense riboprobes for HKα mRNA and 18S ribosomal RNA and then treated with ribonuclease to remove single-stranded RNA. Radiolabelled RNA duplexes were detected by phosphor imager gel analysis. For the study of *H. pylori* effects on H,K-ATPase α subunit gene expression, our experimental model consisted of human gastric adenocarcinoma (AGS) cells growing in culture, transient lipid-mediated transfection of these cells with pGL2 luciferase reporter constructs containing human or rat HKα 5'-flanking sequence, and lastly *H. pylori* inocula (*vac*A$^+$, *cag*A$^+$, urease$^+$) with which we infected the transfected cells over a wide range of multiplicities of infection. The activity of HKα promoter constructs was measured by luminometric assay of luciferase-driven light emission from luciferin (6).

## RESULTS and DISCUSSION

If *H. pylori* infection leads to down-regulation of H,K-ATPase gene expression, then one would expect that gastric parietal cells of *H. pylori*-infected patients would express fewer copies of H,K-ATPase than normal controls. We tested this hypothesis by direct visualization of proton pumps in gastric biopsy sections from individuals undergoing endoscopy at the MUSC Gastroenterology Clinic. Apparently healthy parietal cells were visible in hematoxylin/eosin-stained sections from normal and *H. pylori*-infected patients in whom the presence of acute neutrophil infiltration was scored as grade 2, although it is impossible to assess in such cases the duration of *H. pylori* infection (data not shown). Immunolabeling of these sections with a monoclonal antibody against the H,K-ATPase α subunit (Figure 1) revealed normal numbers and intensity of H,K-ATPase labeling in three uninfected patients (upper panels), and somewhat reduced numbers of parietal cells and a significantly reduced intensity of proton pump fluorescence in three *H. pylori*-infected patients (lower panels). We repeated this experiment in the rhesus monkey model of *H. pylori* infection (3), one advantage of this approach being that the duration of *H. pylori* infection is known exactly. The results were found to be qualitatively similar to the human gastric biopsy experiment (data not shown).

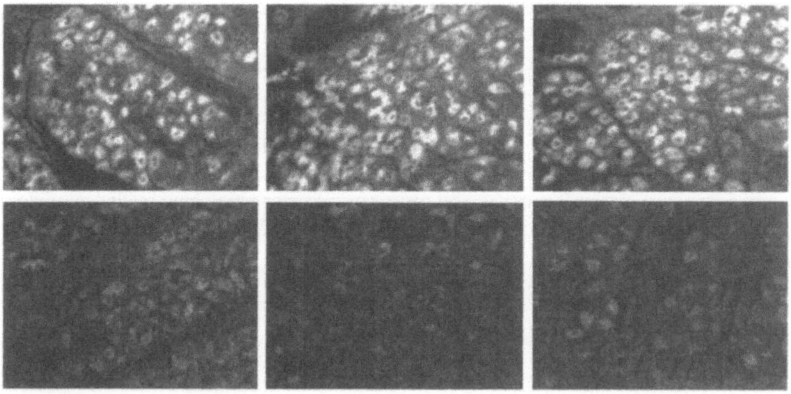

**Figure 1.** *H. pylori* infection reduces parietal cell H,K-ATPase expression in human gastric biosies.

Clearly, given lower levels of proton pump expression, we would also expect to find lower levels of mRNA coding for H,K-ATPase in gastric mucosa. We measured HKα mRNA levels in gastric biopsies from normal and *H. pylori*-infected rhesus monkeys by ribonuclease protection assay, as shown in Figure 2. Comparison of the signal intensity ratios of HKα mRNA to constitutively-expressed 18S RNA (as a control for sample loading) showed that *H. pylori*-infected monkeys expressed lower levels of HKα mRNA than uninfected monkeys. These data are consistent with those of Furuta *et al*, who showed that H,K-ATPase mRNA levels in *H. pylori*-infected patients as measured by RT-PCR were lower than mRNA levels in patients where *H. pylori* had been eradicated (5).

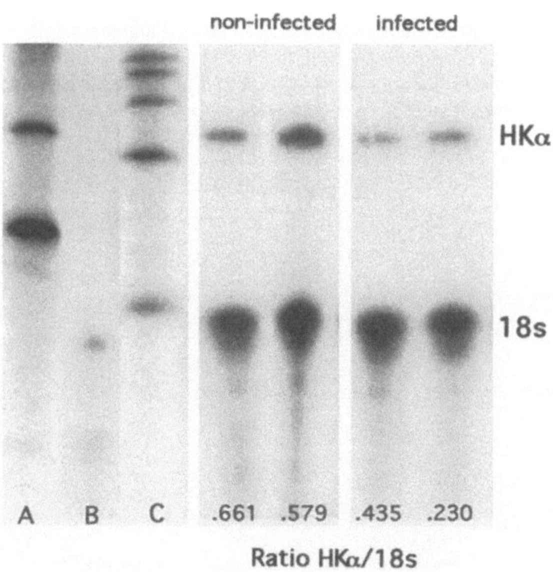


**Figure 2.** A, HKα and 18S RNA riboprobes alone; B, riboprobes + ribonuclease; C, molecular markers.

In the course of earlier studies of human HKα promoter inhibition by *H. pylori* (6), we had noted that rat HKα 5'-flanking sequence was significantly less sensitive to inactivation by *H. pylori* than the corresponding human HKα sequence. This observation was consistent with human gastric mucosa being the only major reservoir of *H. pylori* colonization, so we examined more closely the differential sensitivity of rat and human promoters to *H. pylori* infection. Human and rat HKα 5'-flanking sequences were originally cloned and sequenced by Gary Shull, who noted three regions of close sequence homology, shown in Figure 3 as homology boxes I, II, and III (11). In this diagram, we show all putative response elements predicted by consensus sequence matching by the computer algorithm MatInspector V2.2. We cloned these 2.2 kb human and rat HKα 5'-flanking sequences into pGL2

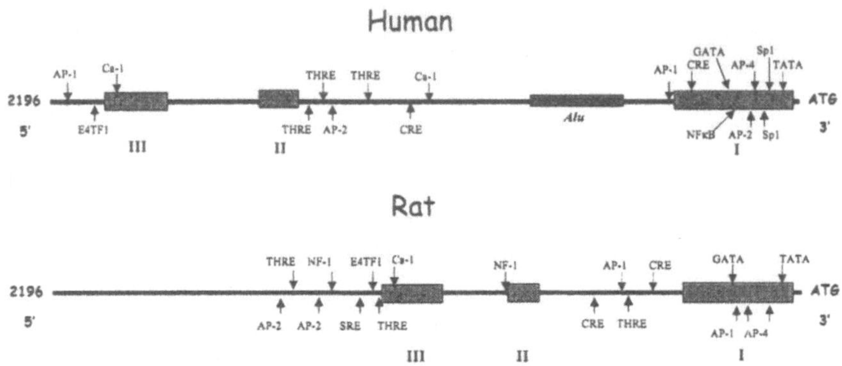


**Figure 3.** 5'-flanking sequences of human and rat H,K-ATPase α subunit genes.

luciferase reporter vectors, transiently transfected AGS cells and measured HKα promoter activity in response to graded doses of *H. pylori*. The data showed that the human HKα promoter was inhibited half-maximally by *H. pylori* infection at an MOI of 150, whereas the half-maximal inhibition for the rat HKα promoter occurred at an MOI of 250 (Figure 4).

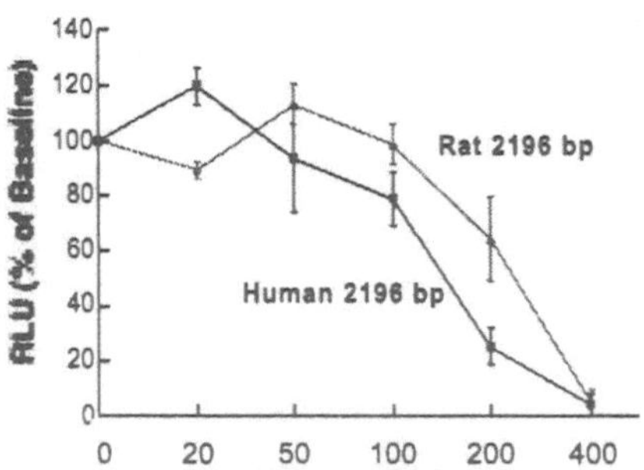


**Figure 4.** Differential inhibition of human and rat HKα 5'-flanking sequence activity by *H. pylori*.

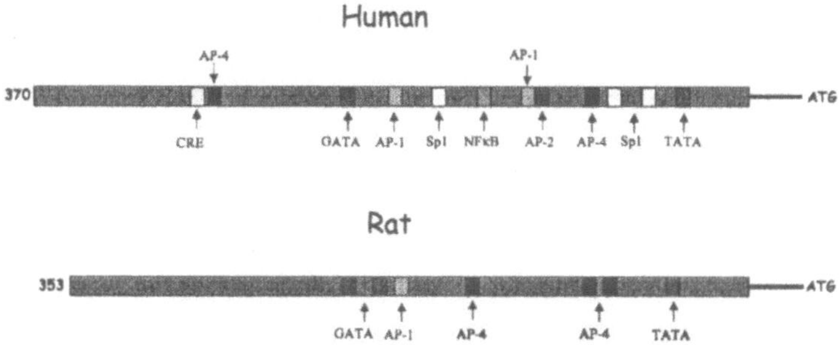


**Figure 5.** Human and rat H,K-ATPase α subunit promoter; DNA probes of homology box I.

This differential *H. pylori* sensitivity of human and rat HKα promoter sequences was preserved when AGS cells were transfected with successively truncated portions of the respective rat and human 5'-flanking sequences. This study focuses on data acquired using homology box I, the first homologous region of both promoters. As shown in Figure 5, in which the relative locations of putative response elements are indicated as predicted by the MatInspector V2.2 program, the human homology box I has three Sp1 sites, an NFκB site, and a CRE (cyclic AMP response element), none of which appears to be in the homologous rat sequence. The human sequence also has two AP-1 sites compared to one in the rat. Both sequences have homologously located GATA and AP-4 binding sites. When we transfected AGS cells with these homology box I constructs, and measured responses to *H. pylori*, we again saw differential sensitivity to *H. pylori*, with half-maximal inhibitory MOIs of 100 for the human and 250 for the rat promoters (Figure 6).

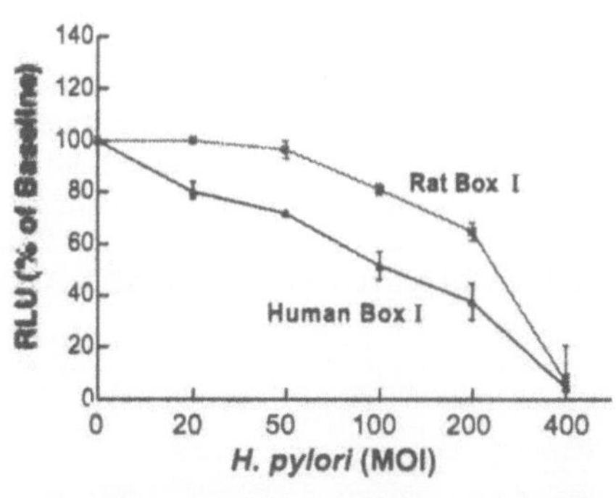


**Figure 6.** Differential inhibition of human and rat HKα homology box I promoter activity by *H. pylori*.

In order to correlate the species-specific promoter down-regulation by *H. pylori* with interactions of specific transcription factors with putative response elements in homology box I, we turned to electrophoretic mobility shift analysis. Firstly, we studied the full spectrum of retarded complexes formed between $^{32}$P-labeled rat and human homology box I DNA probes and AGS cell nuclear protein extracts. With nuclear extract from uninfected AGS cells, at least five discrete DNA:protein

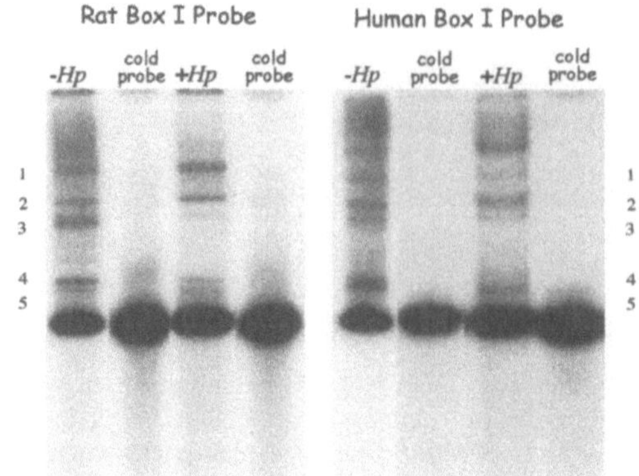


**Figure 7.** AGS cell nuclear extract gel shifts with rat and human HKα homology box I DNA probes.

complexes or bands were formed by the rat probe; with extracts from AGS cells infected with *H. pylori* at MOIs of 50 for 24 hours, band 3 was significantly attenuated, indicating *H. pylori*-mediated interference in formation of this particular DNA:protein complex (Figure 7). Probe specificity was confirmed by addition of a 200x molar excess of unlabeled probe, which eliminated all signal. With the human probe, interacting with uninfected AGS cell nuclear extracts, again at least five discrete DNA:protein complexes were formed; with *H. pylori*-infected AGS cell nuclear extracts, changes in DNA:protein complexes were more subtle, with some degree of attenuation of complexes at bands 2, 4 and 5 (Figure 7).

In order to identify the transcription factors responsible for retardation of rat and human homology box I probes, we used a panel of unlabelled oligonucleotide duplexes reproducing the consensus sequences for six response elements predicted to be present on the human and rat homology box I sequences. These were Sp1, AP-2, AP-1, NFκB, GATA, and CRE. Two hundred times molar excess of cold AP-1, GATA, and CRE oligo duplexes eliminated band 1 from uninfected AGS extracts probed with $^{32}P$-labeled rat DNA, and the GATA oligo also eliminated the band 3 complex. Cold Sp1, AP-2 and NFκB oligos had no effect on the banding of DNA:protein complexes (Figure 8). We conclude that AP-1, GATA and CREB interact with rat probe and that Sp1, AP-2 and NFκB do not interact. Similar analysis of *H. pylori*-infected AGS extracts revealed that AP-2, AP-1 and GATA interact with

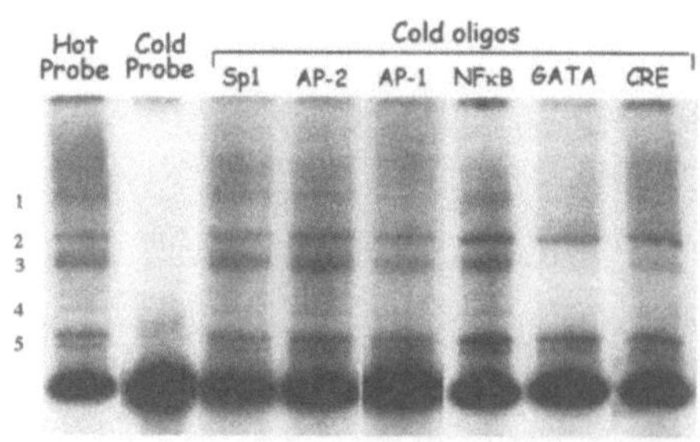


**Figure 8.** Un-infected AGS nuclear extract gel shifts with rat HKα homology box I DNA probe.

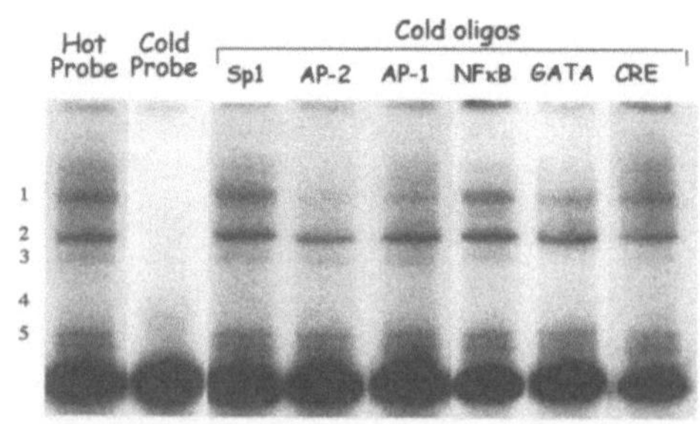


**Figure 9.** *H. pylori*-infected AGS nuclear extract gel shifts with rat HKα homology box I DNA probe.

the rat probe, and that Sp1, NFκB and CREB do not (Figure 9). These deductions are summarized in Figure 10, showing the effects of *H. pylori* on transcription factor binding to the rat homology box I sequence. Thus, we propose that *H. pylori* displaces CREB from CRE and induces AP-2 binding to rat homology box I sequence. Neither of these response elements were predicted by MatInspector V2.2.

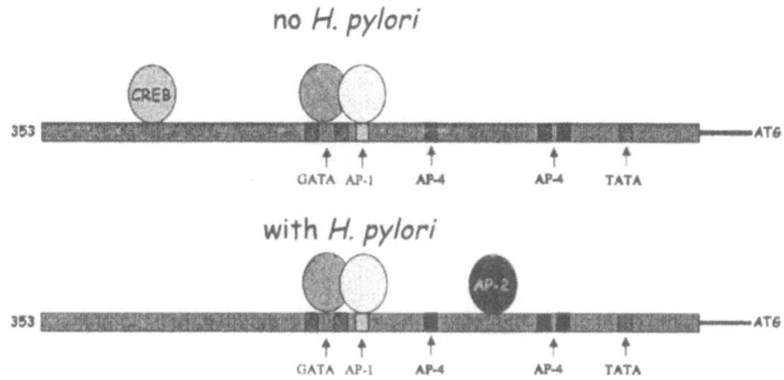


**Figure 10.** Effect of *H. pylori* on transcription factor interactions with rat homology box I.

In the case of cold oligo duplex competitions using the $^{32}P$-labeled human probe and uninfected AGS cell extracts, we deduced from the pattern of band attenuations that Sp1, AP-2, AP-1 and NFkB interact with human probe, and that GATA and CRE do not interact (Figure 11). In contrast, with nuclear extract from AGS cells infected with *H. pylori*, the pattern of mobility-shifted bands indicated that Sp1, AP-2 and NFkB continue to interact with human probe, and that AP-1, GATA and CREB do not interact (Figure 12). Thus, AP-1 appears to be the only transcription factor whose interaction with human homology box I is restricted by *H. pylori*. We confirmed this deduction by carrying out a super-shift analysis of infected and uninfected AGS cells nuclear extracts using an AP-1 specific antibody, as shown in Figure 13. In uninfected cells, we detected a prominent super-shifted AP-1 DNA:protein:antibody complex, whose intensity was significantly reduced following infection of the cells with *H. pylori*. These data are summarized in Figure 14, which shows the effects of *H. pylori* on transcription factor binding to the human homology box I sequence. We propose that *H. pylori* infection displaces AP-1 transcription factor from the first homology box, and that all other transcription factor interactions remain essentially unchanged.

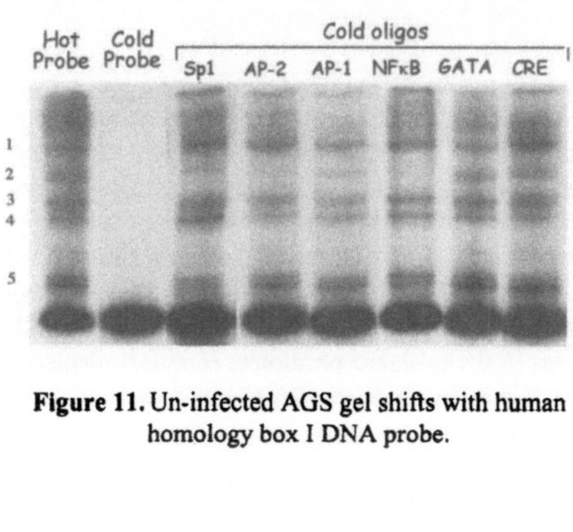


**Figure 11.** Un-infected AGS gel shifts with human homology box I DNA probe.

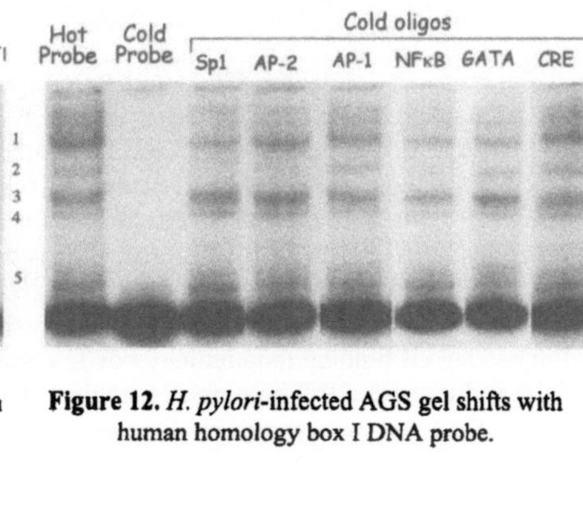


**Figure 12.** *H. pylori*-infected AGS gel shifts with human homology box I DNA probe.

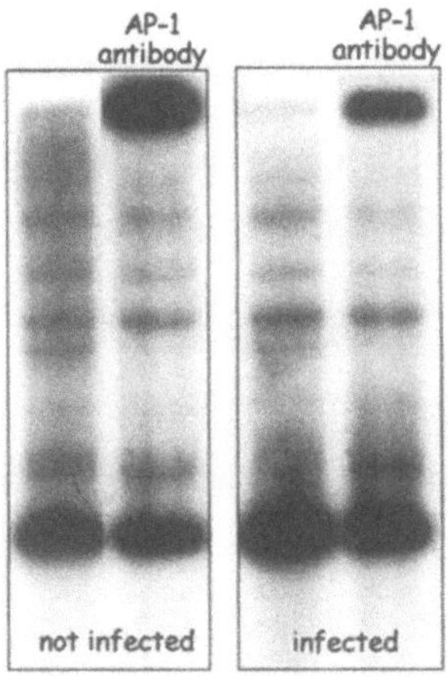


**Figure 13.** Super-shift analysis of *H. pylori*-infected AGS cells with AP-1 antibody.

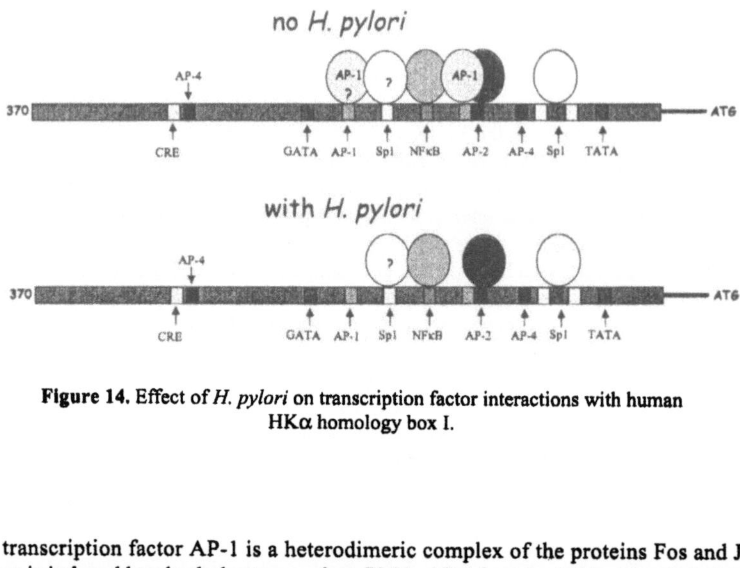


**Figure 14.** Effect of *H. pylori* on transcription factor interactions with human HKα homology box I.

The transcription factor AP-1 is a heterodimeric complex of the proteins Fos and Jun, whose expression is induced by phorbol esters such as PMA. AP-1 is activated by phosphorylation of Jun, which is promoted by activation of PKC by phorbol esters. Our earlier demonstration that *H. pylori* inhibits PMA-induced activation of human HKα subunit promoter is therefore consistent with our finding that AP-1 transcription factor activation is down-regulated in gastric epithelial cells infected with *H. pylori*. The data in this study are consistent with *H. pylori* inhibiting basal and stimulated transcription of HKα gene in gastric epithelial cells. Furthermore, the HKα promoter is more sensitive to *H. pylori* than rat promoter; human HKα promoter basal transcription is associated with Sp1, AP-1, AP-2 and NF κB binding to homology box I, whereas in the rat HKα promoter, *H. pylori* displaces CREB and induces AP-2 binding to homology box I. We conclude that *H. pylori* down-regulates human HKα basal transcription by displacement of the AP-1 transcription factor from the homology box I region of HKα 5'-flanking sequence.

## ACKNOWLEDGEMENTS

We thank Dr. David Lewin for providing human gastric biopsy sections, Drs. Andre Dubois and Marjorie Walker for providing rhesus monkey gastric biopsies, and Charles Hammond, Kellie Larsen, and April Angel for expert technical assistance. This study was supported by NIH DK 43138 and NASA NAG8-1385 awards to A.S.

## REFERENCES

1. **Bizzozero G**. Sulle ghiandole tubulari del tubo gastro-enterico e sui rapporti del loro epitelio coll'epitelio di rivestimento della mucosa. *Atti della Reale Accademia delle Scienze di Torino* 28: 233-251, 1892.
2. **Cave DR, King WW, and Hoffman JS.** Production of two chemically distinct acid inhibitory factors by *Helicobacter pylori. Eur J Gastroenterol Hepatol* 5: S23-S27, 1995
3. **Dubois A, Berg DE, Incecik ET, Fiala N, Heman-Ackah LM, Perez-Perez GI, and Blaser MJ.** Transient and persistent experimental infection of nonhuman primates with *Helicobacter pylori*: Implications for human disease. *Infect Immun* 64: 2885-2891, 1996.
4. **El-Omar EM, Oien K, El-Nuiumi A, Gillen D, Wirz A, Dahill S, Williams C, Ardill JE, and McColl KE.** *Helicobacter pylori* infection and chronic gastric acid hyposecretion. *Gastroentrology* 113: 15-24, 1997.
5. **Furuta T, Baba S, Takashima M, Shirai N, Xiao F, Futami H, Arai H, Hanai H, and Kaneko E.** H,K-adenosine triphosphatase mRNA in gastric fundic gland mucosa in patients infected with *Helicobacter pylori. Scand J Gastroenterol*, 34: 384-390, 1999.
6. **Gooz M, Hammond CE, Larsen K, Mukhin YV, and Smolka A.** Inhibition of human gastric H,K-ATPase α-subunit gene expression by *Helicobacter* pylori. *Am J Physiol Gastrointest Liver Physiol* 278: G981-G991, 2000
7. **Graham DY, Alpert LC, Lacey-Smith J, and Yoshimura HH.** *Campylobacter pylori* infection is a cause of epidemic achlorhydria. *Am J Gastroenterol* 83: 974-980, 1995.
8. **Kobayashi H, Kamiya S, Suzuki T, Kohda A, Muramatsu S, Kuramada T Ohta U, Miyazawa N, Kimura N, Mutoh T, Shirai A, Takagi S, Harasawa S, Tani N, and Miwa T.** The effect of *Helicobacter pylori* on gastric acid secretion by isolated parietal cells from a guinea pig. *Scand J Gastroenterol* 31: 428-433, 1996
9. **Marshall BJ, Armstrong JA, McGechie DB, and RJ Glancy.** Attemp to fulfill Koch's postulates for pyloric *Campylobacter. Med J Australia* 142: 436-439, 1985.
10. **Morris A and Nicholson G.** Experimental and accidental *C. pylori* infection in humans. In: Blaser MJ (ed) *Campylobacter pylori in Gastritis and Peptic Ulcer Disease*. New York: Igaku-Shoin, 1989, pp 61-72.
11. **Newman PR, Greeb J, Keeton TP, Reyes AA, and Shull GE.** Structure of the human gastric H,K-ATPase gene and comparison of the 5'-flanking sequences of the human and rat genes. *DNA and Cell Biology*, 9: 749-762, 1990
12. **Smolka AJ, Larsen KA, and Hammond CE.** Location of a cytoplasmic epitope for monoclonal antibody HK 12.18 on H,K-ATPase α subunit. *Biochem Biophys Res Commun* 273: 942-947, 2000
13. **Wallace JL, Cucala M, Mugridge K,** Secretagogue-specific effects of interleukin-1 on gastric acid secretion. *Am J Physiol Gastrointest Liver Physiol* 261: G559-G564, 1991.
14. **Weeks DL, Eskandari S, Scott DR, and Sachs G.** A $H^+$-gated urea channel: the link between *Helicobacter pylori* urease and gastric colonization. *Science* 287:482-485, 2000.

Chapter 10

# $H^+K^+$-ATPase, the Dominant Gastric Autoantigen in *Helicobacter pylori* Infection

B. J. APPELMELK[1], M. P. BERGMAN[1], A. V. D. AAR[1], W. LAKHAI[1], M. M. D'ELIOS[2], A. AZZURRI[2], A. AMEDEI[2], M. BENAGIANO[2], F. CIANCHI[2], S. ROMAGNANI[2], J. G. FORTE[3], G. F. DEL PRETE[2], J. J. DE PONT[4], C. M. J. E. VANDENBROUCKE-GRAULS[1]

[1]*Dept. Med. Microbiology, Vrije Universiteit Med. Center, Amsterdam, The Netherlands.*
[2]*Depts. Int. Med. and Surgery, Univ. Florence, Italy.*
[3]*Dept. Mol. and Cell. Biol, Univ. California, Berkeley, USA.*
[4]*Dept. Biochem., Univ. Nijmegen, The Netherlands.*

## INTRODUCTION

The pathogen *Helicobacter pylori* colonizes the gastric mucosa of half of mankind and causes a variety of diseases. While in the early days of *H . pylori* research most attention was given to its ability to cause peptic ulcers, much of the current focus is on its causative role in gastric cancer. Gastric cancer ranks second in malignancy-associated mortality, and in certain regions in The People's Republic of China this mortality amounts to 70/100.000/year.

The research on *H. pylori*-induced gastric carcinogenesis encompasses several aspects:

1. How does *H. pylori* induce gastric cancer? It is becoming clear that infection-associated inflammation is the key. Patients with certain gene polymorphisms in IL-1 and IL-1 receptor antagonist (Il-1ra) are at increased risk of developing gastric cancer (6). The most likely mechanism is the relatively high ratio of IL-1/IL-1ra in those patients, which leads to a pro-inflammatory environment in the gastric mucosa. Inflammation is associated with the formation of oxygen and nitrogen radicals that cause cancer through mutation of key host genes.
2. How does *H. pylori* cause atrophy? It has been known for a long time that atrophic gastritis precedes gastric cancer (10). A recent study shows that *H. pylori*-infected patients with *corpus predominant* gastritis have a 34.5 fold increased risk of developing gastric cancer (9).
3. Does eradication of *H. pylori* decrease the risk of subsequent development of gastric cancer? Several randomized, placebo controlled longitudinal studies are being performed to answer this question. As antibiotic treatment of all *H. pylori*-infected patients is not a realistic option, an important question is whether it is possible to identify patients that should be treated, i.e. those at risk of developing gastric cancer. Of potential interest in this respect are recently commercialized Elisa assays for Pepsinogen I and II and Gastrin 17; based on the outcome these serological determinations patients with moderate to severe corpus atrophy can be identified (12).

We will focus on topic 2 and demonstrate that gastric $H^+K^+$-ATPase is the dominant gastric autoantigen in *H. pylori* infection (1), and that $H^+K^+$-ATPase-specific T-cells can be isolated from patients with autoimmune gastritis (4). Likely, these autoreactive T-cells play a crucial role in *H. pylori*-induced corpus inflammation and atrophy.

## MATERIALS AND METHODS

**Immunoblotting.** Purified pig gastric $H^+K^+$-ATPase (10 μg per lane) was dissolved (15 min at 37 degrees Celsius) in SDS-PAGE sample buffer without SH-reagent. After electrophoresis and electro-blotting with conventional procedures, blot strips were incubated with sera (diluted 1:100-1:500) from *H. pylori*-infected patients or healthy individuals (as negative controls). As positive controls served sera from patients with autoimmune gastritis as well as the $H^+K^+$-ATPase β-chain specific monoclonal antibody 2G11(3). Mab 2G11 was used at a dilution of $1:10^6$. Binding of antibodies was detected by chemoluminescence.

**Cloning of T-cells.** Gastric biopsies from corpus and antrum were kept in tissue culture medium for 10-14 days with IL-2 (30 U/ml). Lymphocytes were harvested and redistributed by limiting dilution on heterologous antigen presenting cells (APC), i.e. PBMC from buffycoats after irradiation, together with + IL-2/PHA. Clones were scored after 10 days. Antigen-specific proliferation was tested in the presence of autologous APC and $^3$H-labelled thymidine incorporation (cpm) was measured. Antigens used were purified pig $H^+K^+$-ATPase and a *H. pylori*-cell lysate. The results are expressed as mitogenic index, i.e., cpm with antigen/cpm without antigen.

**Determination of perforin-mediated cytotoxic activity.** Chromium-labeled autologous Epstein-Barr virus transformed B-cells were laden with purified pig gastric $H^+K^+$-ATPase. Then they were brought into close contact through centrifugation with $H^+K^+$-ATPase-specific T-cells and co-incubated for 8 hours after which the percentage specific lysis was calculated.

## RESULTS AND DISCUSSION

In 1991, Riccardo Negrini (Brescia, Italy) was the first to report that a considerable percentage of *H. pylori*-infected patients develop serum autoantibodies to gastric parietal cells, in particular the canaliculi (8) (Fig. 1).

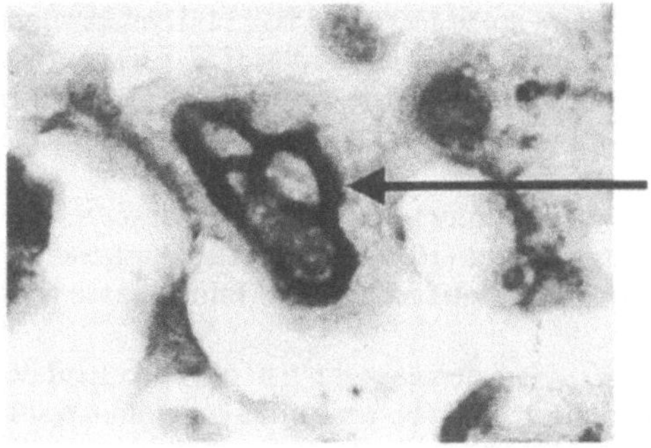

**Fig. 1.** *Detection of anti-gastric autoantibodies by immunohistochemistry (original magnification x100). The outline of two parietal cells (grey) is visible; the arrow indicates the intense staining of the canaliculi.*

Two other research groups (Gerhard Faller and Thomas Kirchner in Erlangen Germany, and Vorobjova and colleagues from Tartu, Estonia) have reported similar data (13). It is estimated that 30-50 % of *H. pylori*-infected patients develop gastric autoantibodies. Predominantly through the work of Faller it has become clear that infected patients developing gastric anti-canalicular autoantibodies have very clear characteristics (Table 1), i.e., they have an increased corpus atrophy and also in other respects are similar to patients with autoimmune gastritis. Among patients with longstanding infection the number of patients with gastric autoantibodies increases with the duration of the infection (13). Eradication of the bacterium through antimicrobial therapy leads to loss of autoantibodies (7).

**Table 1. Characteristics of *H. pylori*-infected patients developing anti-canalicular autoantibodies**

| | |
|---|---|
| Increased | |
| | -Corpus atrophy |
| | -Influx of periglandular and intraepithelial lymphocytes in corpus |
| | -Serum gastrin levels |
| | -Apoptosis in corpus epithelium |
| Decreased | |
| | -Duodenal ulcer |
| | -Acid output |
| | -Pepsinogen I/II ratio |

Both by immunoprecipitation and by SDS-PAGE it has been shown that the epitopes that are recognized are located on the α and β chain of gastric $H^+K^+$-ATPase. An example is given in Fig. 2.

Gastric $H^+K^+$-ATPase is also the autoantigen recognized by sera from patients with AIG and pernicious anemia (AIG/PA)(11). This and other findings (2) have led to the two

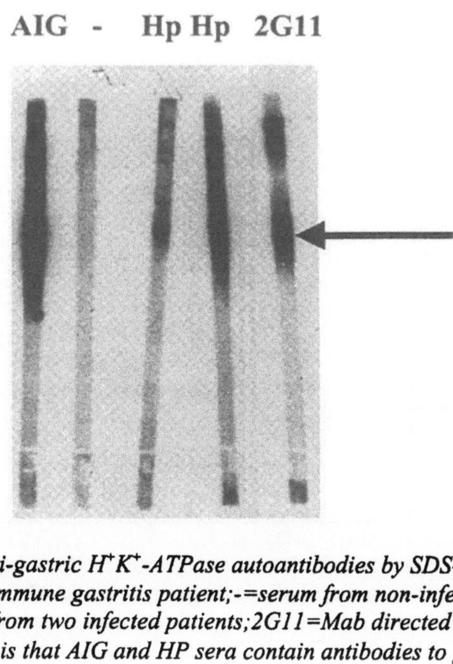


**Fig. 2.** *Detection of anti-gastric $H^+K^+$-ATPase autoantibodies by SDS-PAGE immunoblot. AIG=serum from autoimmune gastritis patient;-=serum from non-infected healthy individual;Hp=serum from two infected patients;2G11=Mab directed to β-chain of gastric $H^+K^+$-ATPase. Evident is that AIG and HP sera contain antibodies to gastric $H^+K^+$-ATPase β-chain (arrow)*

following questions: Is *H. pylori*-associated corpus atrophy an autoimmune process? Does *H. pylori*-associated corpus atrophy precede *H. pylori*-negative AIG/PA?

Murine experimental AIG is mediated by T-cells, not by antibodies specific for gastric $H^+K^+$-ATPase. Hence, we hypothesized that chronic infection with *H. pylori* induces the expansion of autoreactive gastric $H^+K^+$-ATPase specific T-cells. We also hypothesized that this type of T-cells are present in AIG, and started T-cell cloning from gastric biopsies of AIG patients as well as from "classical" *H. pylori*-infected patients showing no signs of corpus involvement. From 5 AIG patients 46 of the 280 isolated T-cell clones proved specific for gastric $H^+K^+$-ATPase, the remaining ones of an as yet unidentified specificity. None of the 259 clones isolated from "classical" *H. pylori* patients was specific for gastric $H^+K^+$-ATPase, while 50 of them proved specific for *H. pylori* antigens. Fig 3 shows that gastric $H^+K^+$-ATPase-specific T-cells recognize their autoantigen antigen in a HLA-restricted manner.

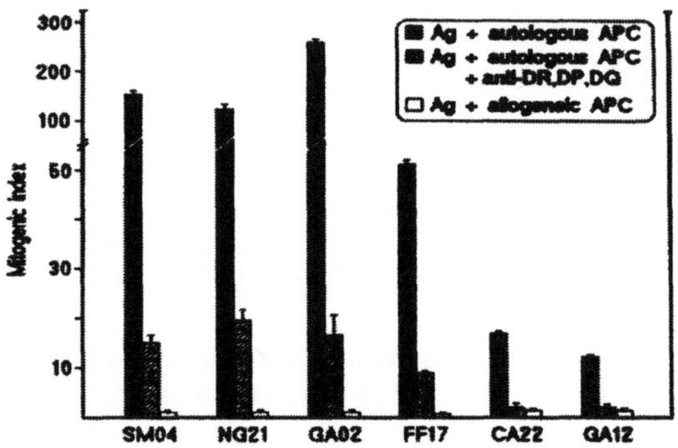


**Fig 3.** *HLA-restricted proliferation of gastric $H^+K^+$-ATPase-specific T-cell clones. T-cells proliferate well to gastric $H^+K^+$-ATPase presented by autologous APC but not by allogeneic APC. Mab specific for HLA-DR,DP,DQ strongly reduces mitogenic index (From ref. 4).*

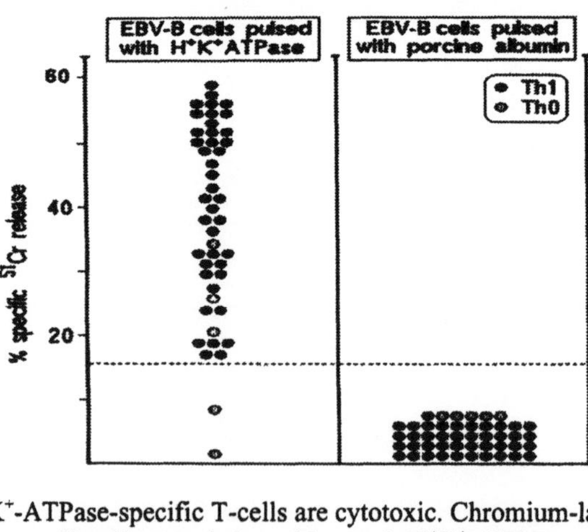


**Fig 4.** Gastric $H^+K^+$-ATPase-specific T-cells are cytotoxic. Chromium-labeled EBV-transformed autologous B-cells were laden with gastric $H^+K^+$-ATPase and co-incubated with gastric $H^+K^+$-ATPase-specific T-cells. The high chromium release demonstrates a T-cell and autoantigen dependent lysis of the B-cells; lysis in this assay system is perforin dependent (From ref. 4).

Fig 4 shows that autoreactive T-cells are capable of target cell lysis. From fig 4 we conclude that these autoreactive T-cells have the potential to damage gastric mucosa through elimination of parietal cells, i.e. the hallmark of corpus atrophy. It was also shown that the gastric $H^+K^+$-ATPase-specific T-cells express CD4 and display the Th1 phenotype (Interferon γ high, IL-4 low); furthermore they were found to kill Jurkat cells through a Fas-FasL dependent mechanism (not shown)

The proposed pathogenesis of *H. pylori*-induced gastric autoimmunity is as follows (2): *H. pylori* infection induces HK-ATPase-specific T cells. T-cells kill HK-ATPase-presenting parietal cells, which leads to corpus atrophy. Severe atrophy leads to loss of *H. pylori*, and finally, prolonged corpus inflammation leads to continuous exposure of gastric $H^+K^+$-ATPase, and independence of the autoimmune process from *H. pylori* as a stimulus.

## SUMMARY AND OUTLOOK

We have demonstrated that *H. pylori* induces gastric autoimmunity directed at gastric $H^+K^+$-ATPase. T-cells specific for this antigen were isolated from AIG patients, and were able to kill target cells by apoptosis and perforin-mediated lysis. A role for *H. pylori* infection in initiating AIG is likely.

Future studies have to address at least the following questions:

1. Which are the epitope(-s) recognized by the gastric $H^+K^+$-ATPase-specific T-cells. To answer this question the synthesis of overlapping peptides of both α and β- chain is needed. In the past, epitope-scanning has been used with success for the identification of antigenic peptides recognized by murine gastric $H^+K^+$-ATPase-specific T-cells (5).
2. Are gastric $H^+K^+$-ATPase-specific T-cells present in *H. pylori*-infected patients before overt AIG is diagnosed? Answering this question requires T-cell cloning in the early stages of *H. pylori*-associated corpus atrophy.
3. To which degree does the atrophic corpus mucosa return to normal after *H. pylori* eradication? This is a hotly debated topic; it is likely that atrophy may be partially reversible. A recent paper showed that even some premalignant gastric mucosal lesions disappeared after antibiotic treatment
4. Does measurement of anti-parietal autoantibodies contribute to defining a sub-group of patients that are most at risk for developing gastric cancer? Possibly, but not proven yet, measurement of pepsinogen I/II and gastrin levels may define such a treatment subgroup, and the question is whether inclusion of data on presence of gastric autoantibodies improves identifation. Elisa screening (PG I/II, gastrin, anti- $H^+K^+$-ATPase) in high incidence area's, followed by antibiotic treatment of selected patients might be a valid approach for decreasing the very high mortality due to gastric cancer.

## REFERENCES

1. **Appelmelk, B. J., G. Faller, D. Claeys, T. Kirchner, and C. M.** Vandenbroucke-Grauls. Bugs on trial: the case of Helicobacter pylori and autoimmunity. *Immunol Today* 19: 296-299, 1998.
2. **Bergman, M. P., Faller, G., D'Elios, M. M., Del Prete, G., Vandenbroucke-Grauls, C. M. J. E., and Appelmelk, B. J.** The role of *Helicobacter pylori* in gastric autoimmunity. In: *Helicobacter pylori*:physiology and genetics (ed. H.L.T. Mobley, G.L. Mendz, and S. L. Hazell), ASM press, 2001.
3. **Chow, D. C. and J. G. Forte.** Characterization of the beta-subunit of the H(+)-K(+)-ATPase using an inhibitory monoclonal antibody. *Am.J Physiol* 265: C1562-C1570, 1993.
4. **D'Elios, M. M., M. P. Bergman, A. Azzurri, A. Amedei, M. Benagiano, J. J. De Pont, F. Cianchi, C. M. Vandenbroucke-Grauls, S. Romagnani, B. J. Appelmelk, and G. Del Prete.** H(+),K(+)-atpase (proton pump) is the target autoantigen of Th1-type cytotoxic T cells in autoimmune gastritis. *Gastroenterology* 120: 377-386, 2001.
5. **De Silva, H. D., P. A. Gleeson, B. H. Toh, I. R. Van Driel, and F. R. Carbone.** Identification of a gastritogenic epitope of the H/K ATPase beta-subunit. *Immunology* 96: 145-151, 1999.
6. **El Omar, E. M., M. Carrington, W. H. Chow, K. E. McColl, J. H. Bream, H. A. Young, J. Herrera, J. Lissowska, C. C. Yuan, N. Rothman, G. Lanyon, M. Martin, J. F. Fraumeni, Jr., and C. S. Rabkin.** Interleukin-1 polymorphisms associated with increased risk of gastric cancer. *Nature* 404: 398-402, 2000.
7. **Faller, G., M. Winter, H. Steininger, N. Lehn, A. Meining, E. Bayerdorffer, and T. Kirchner.** Decrease of antigastric autoantibodies in Helicobacter pylori gastritis after cure of infection. *Pathol.Res.Pract.* 195: 243-246, 1999.
8. **Negrini, R., L. Lisato, I. Zanella, L. Cavazzini, S. Gullini, V. Villanacci, G. Poiesi, A. Albertini, and S. Ghielmi.** Helicobacter pylori infection induces antibodies cross-reacting with human gastric mucosa. *Gastroenterology* 101: 437-445, 1991.
9. **Ohkusa, T., K. Fujiki, I. Takashimizu, J. Kumagai, T. Tanizawa, Y. Eishi, T. Yokoyama, and M. Watanabe.** Improvement in atrophic gastritis and intestinal metaplasia in patients in whom Helicobacter pylori was eradicated. *Ann.Intern.Med.* 134: 380-386, 2001.
10. **Sipponen, P. and B. J. Marshall.** Gastritis and gastric cancer. Western countries. *Gastroenterol.Clin.North Am.* 29: 579-5vi, 2000.
11. **Toh, B. H., J. W. Sentry, and F. Alderuccio.** The causative H+/K+ ATPase antigen in the pathogenesis of autoimmune gastritis. *Immunol Today* 21: 348-354, 2000.
12. **Varis, K., P. Sipponen, F. Laxen, I. M. Samloff, J. K. Huttunen, P. R. Taylor, O. P. Heinonen, D. Albanes, N. Sande, J. Virtamo, and M. Harkonen.** Implications of serum pepsinogen I in early endoscopic diagnosis of gastric cancer and dysplasia. Helsinki Gastritis Study Group. *Scand.J Gastroenterol.* 35: 950-956, 2000.
13. **Vorobjova, T., G. Faller, H. I. Maaroos, P. Sipponen, K. Villako, R. Uibo, and T. Kirchner.** Significant increase in antigastric autoantibodies in a long-term follow-up study of H. pylori gastritis. *Virchows Arch.* 437: 37-45, 2000.

Chapter 11

# The Gastric H/K ATPase in the Pathogenesis of Autoimmune Gastritis

BAN-HOCK TOH, MARK BIONDO, AIDEN MARSHALL, DEANNE GREENWOOD, JOHN SENTRY, AND FRANK ALDERUCCIO

*Department of Pathology and Immunology, Monash University Medical School, Prahran, Victoria 3181, Australia*

## ABSTRACT

Pernicious anaemia is the commonest cause of vitamin B12 deficiency in man affecting ~2% of the population. Autoimmune gastritis is the underlying cause of this disease. The gastritis results in loss of parietal cells that secrete intrinsic factor required for vitamin B12 absorption in the small intestine. The target antigen of these parietal cell antibodies is the gastric H/K ATPase responsible for secretion of HCl. To examine the role of the gastric H/K ATPase in the pathogenesis of autoimmune gastritis, we adopted and established mouse models of autoimmune gastritis. We showed that the gastritis is initiated by the β subunit of the ATPase because transgenic expression of this subunit in the thymus of gastritis-susceptible BALB/c or BALB/cCrSlc mice renders the mice resistant to development of autoimmune gastritis. We showed that immunological tolerance to the β subunit was established by expression of this subunit in the thymus because thymocytes from transgenic mice failed to transfer gastritis to naïve recipients. The mouse gastritis is mediated entirely by TH1-type CD4 T cells. We showed that early lesions of gastritis are characterised by influx of CD4 T cells that secrete a mix of TH1 and TH2 cytokines with the exception of IL-4. A single injection of antibodies to interferon γ prevented the development of gastritis. An unresolved question is the nature of the trigger that initiates gastritis in genetically-susceptible mice. Microbial organisms may initiate gastritis by acting as molecular mimics or through by-stander activation following the generation of pro-inflammatory cytokines. We investigated the role of the pro-inflammatory cytokine GM-CSF in the initiation of gastritis by the transgenic production of this cytokine in the stomach. We selected this cytokine because we have previously shown that it is generated in gastritic lesions and because GM-CSF is a potent activator of antigen-presenting cells that links innate to adaptive immunity. We showed that transgenic expression of GM-CSF in the stomach was alone sufficient to initiate gastritis in genetically-susceptible mice. Our studies support the role of pro-inflammatory by-stander activation in the initiation of gastritis.

Autoimmune gastritis belongs to a group of organ-specific autoimmune diseases characterised by organ-restricted pathology and circulating organ-specific autoantibodies (1, 4-6). In autoimmune gastritis, the gastric pathology is associated with autoantibodies to gastric parietal cells and to intrinsic factor, itself a secretory product of human parietal cells. The pathologic lesion of autoimmune gastritis results in loss of gastric parietal cells and zymogenic cells. In turn, the loss of gastric parietal cells together with autoantibodies to intrinsic factor in gastric juice results in intrinsic factor deficiency leading to failure of intrinsic factor-dependent absorption of vitamin B12 (cobalamin) from the terminal ileum and subsequent pernicious anemia.

Pernicious anemia is associated with significant morbidity in the general population. It is the commonest cause of cobalamin deficiency in humans and remain undiagnosed in almost 2% of persons over the age of 60yr. It comprises about a third of patients with neurologic disorders due to cobalamin deficiency without overt signs of anemia.

## GASTRIC PATHOLOGY

Autoimmune gastritis or Chronic Atrophic Gastritis Type A is restricted to the fundus and body of the stomach with sparing of the antrum. The gastric lesion is distinguished from Chronic Atrophic Gastritis Type B that involves the antrum and is associated with Helicobacter pylori infection. The gastritis is characterised by a chronic inflammatory infiltrate in the submucosa extending into the lamina propria with loss of gastric parietal cells and zymogenic cells and their replacement by mucous-containing cells, a change known as "intestinal metaplasia".

## GENESIS OF AUTOIMMUNE GASTRITIS

A genetic predisposition to the development of autoimmune gastritis is supported by an increased incidence of pernicious anemia and of gastric autoantibodies in first degree relatives of patients with pernicious anaemia, clustering of related organ-specific autoimmune diseases and organ-specific autoantibodies in first degree relatives and in twin studies. The demonstration that the gastric H/K ATPase is the major autoantigen reactive with parietal cell autoantibodies in chronic Helicobacter pylori-associated gastritis with corpus atrophy has led to the suggestion that the atrophy is a result of an autoimmune reaction to the gastric H/K ATPase. In turn, this has led to speculation that Helicobacter pylori is the environmental trigger for the development of autoimmune gastritis (2).

## IDENTIFICATION OF GASTRIC H/K ATPASE AS THE MOLECULAR TARGET IN AUTOIMMUNE GASTRITIS

Identification of the autoantigen(s) targeted by the immune system constitutes the first step in any concerted study directed towards a better understanding of the pathogenesis of autoimmunity. In organ-specific autoimmune diseases, organ-specific autoantibodies provide a convenient starting point for identification of molecular targets. In the case of autoimmune gastritis and pernicious anemia, circulating parietal cell autoantibodies have been shown to react with two polypeptides- the catalytic α and the glycoprotein β subunits of the gastric H/K ATPase, the enzyme located on gastric secretory membranes and responsible for acid secretion in the stomach.

## STUDIES OF IMMUNOPATHOGENESIS USING MOUSE MODELS OF HUMAN AUTOIMMUNE GASTRITIS

Identification of the gastric H/K ATPase as the molecular target of parietal cell autoantibodies raises the question of whether this gastric enzyme initiates the lesion of autoimmune gastritis. To address this question, we established a number of mouse models of human autoimmune gastritis. The most commonly used mouse model of gastritis is that induced by day 3 neonatal thymectomy (Figure 1).

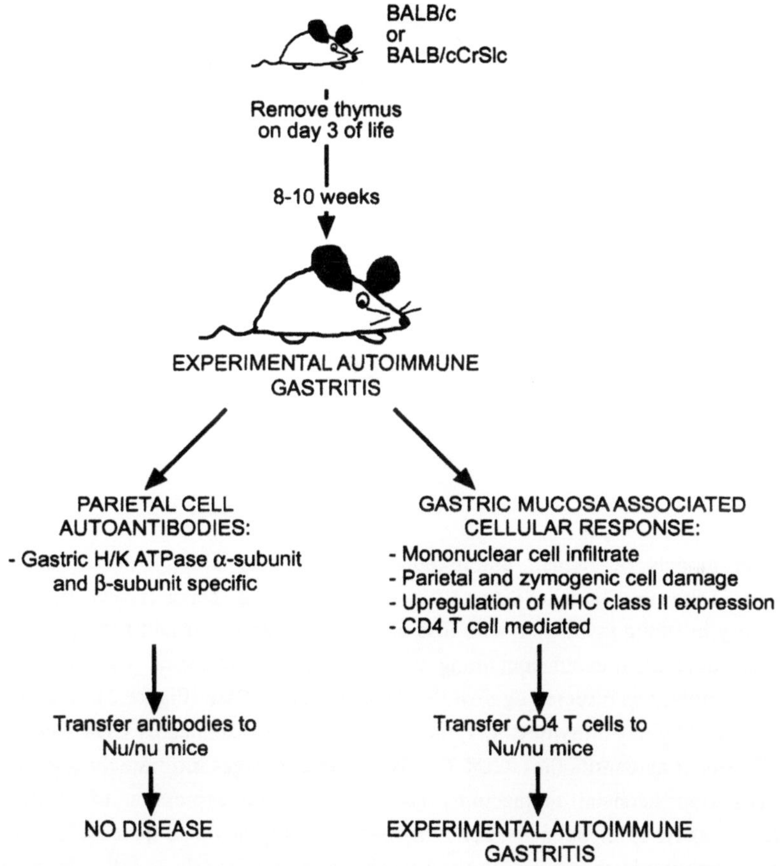


**Figure 1** Mouse autoimmune gastritis initiated by neonatal thymectomy. Gastritis is induced by surgical removal of the thymus from three day old BALB/c or BALB/cCrSlc mice. Eight to ten weeks after thymectomy, gastritic mice are identified by circulating parietal cell autoantibodies reactive with gastric H/K ATPase and histological evidence of a destructive gastritis. Experimental data indicate that gastritis is mediated by TH1 CD4+ T cells and not by CD8 T cells or autoantibodies.

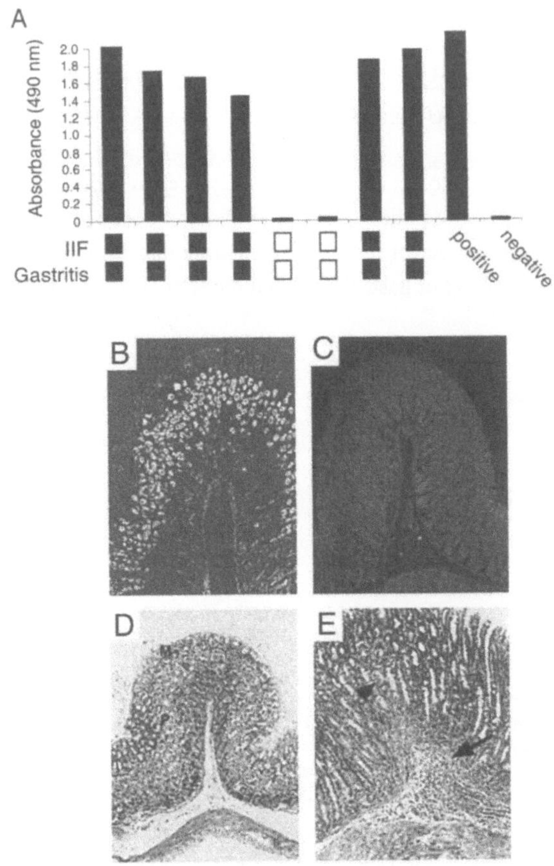


**Figure 2**. Development of parietal and H/K ATPase reactive autoantibodies correlates with the presence of gastritis. Autoantibodies reactive with gastric H/K ATPase produced in mice following neonatal thymectomy are identified by (A) ELISA and (B) indirect immunofluorescence (IIF). Sera from normal mice are non-reactive (C). D. Normal stomach illustrating typical gastric glands and the three major cell types, parietal cells (P), zymogenic cells (Z) and surface mucous cells (M). E. Gastritic stomach showing mononuclear cells infiltrating gastric mucosa (arrow) and between gastric glands. Parietal and zymogenic cells are lost and replaced by mucus secreting cells (arrowhead).

The mouse models of human autoimmune gastritis very closely mirror their human counterpart. Like the human disease, mouse gastritis is characterised by a chronic inflammatory infiltrate associated with loss of gastric parietal cells and zymogenic cells and their replacement by mucous-containing cells. The gastric lesion is also associated with parietal cell antibodies directed against the gastric H/K ATPase (Figure 2). In autoimmune gastritis induced by day 3 thymectomy, the gastric lesion is mediated by CD4 T cells and not by CD8 T cells or autoantibodies. CD4 T cells and macrophages infiltrate the gastric mucosa at 4 weeks after neonatal thymectomy accompanied with expression of MHC class II molecules in gastric epithelial cells. In contrast, there is only marginal increase in CD8 T cells with B cells arriving later and peaking at 8 weeks appearing largely as follicular aggregates. The CD4 T cells produce a mix of Th1 and Th2 type cytokines including IFN$\gamma$, IL-10, TNF$\alpha$ and GM-CSF but not IL-4. Gastritis can be prevented by treatment with neutralising antibodies to IFN$\gamma$ indicating that the pathogenic CD4 T cells belong to the TH1 subset.

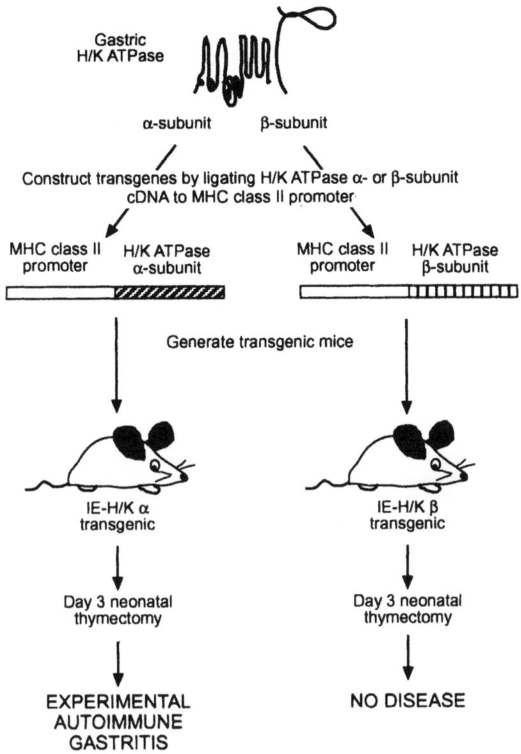


**Figure 3.** Transgenic expression of the gastric H/K ATPase β-subunit in the thymus abrogated autoimmune gastritis induction. Transgenic mice were generated expressing the α- or β-subunit of gastric H/K ATPase under control of a MHC class II promoter. Transgenic and non-transgenic littermates were neonatally thymectomised at day 3 and assessed for experimental autoimmune gastritis. Mice expressing transgenic α-subunit developed gastritis similar to non-transgenic mice. In contrast, transgenic mice expressing the β-subunit were completely protected from disease induction.

The demonstration that TH1 CD4 T cells mediate the lesion of autoimmune gastritis raises the question whether these T cells are also directed against the gastric H/K ATPase. To address this question, we used a transgenic approach to express either one of the 2 subunits of the gastric H/K ATPase ectopically in the thymus in an attempt to render developing CD4 T cells tolerant to these proteins (Figure 3). Ectopic expression in mouse thymus of the H/K ATPase α subunit but not the β subunit prevented the development of mouse autoimmune gastritis induced by a variety of approaches. The observations provide compelling evidence that autoimmune gastritis is initiated by pathogenic CD4 T cells directed towards the β subunit, thus identifying the β subunit as the causative or initiating autoantigen.

## MODEL FOR THE INITIATION AND PROPAGATION OF AUTOIMMUNE GASTRITIS

Based on the foregoing experimental data, we propose that autoimmune gastritis is the consequence of activation of antigen presenting cells in the stomach following uptake of gastric antigen triggered by a local pro-inflammatory milieu (Figure 4). Activated antigen presenting cells then migrate to the draining gastric lymph node where they present a gastritogenic β subunit peptide by MHC Class II molecules to naïve precursors of pathogenic CD4 T cells resulting in their activation through their T cell receptor (TCR). In turn, the activated CD4 T cells migrate to the gastric mucosa resulting in the lesion of autoimmune gastritis.

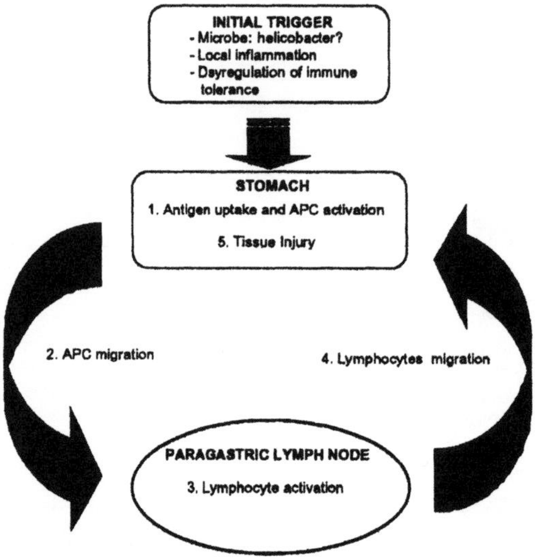


**Figure 4.** Model for initiation and propagation of gastric autoimmunity. The initial trigger may be provided by tissue specific pathogens, local inflammation and/or dysregulation of immune tolerance mechanisms. This is followed sequentially by (1) uptake of autoantigen and activation of antigen presenting cells (APC); (2) migration of APC to the local draining lymph node where antigen is presented, leading to (3) activation of antigen specific autoreactive T cells. (4) Following activation, lymphocytes migrate to the site of tissue inflammation through homing and chemokine signaling where they participate in destruction of epithelial cells in gastric mucosa.

## INITIATION OF AUTOIMMUNE GASTRITIS

Microbial organisms may trigger autoimmunity either by acting as molecular mimics and/or through the generation of a local pro-inflammatory milieu. To test the hypothesis that autoimmunity can be initiated by the presence of a pro-inflammatory milieu, we generated a transgenic mouse that expressed GM-CSF in the stomach (3) (Figure 5). We selected GM-CSF as the pro-inflammatory cytokine for local gastric expression because we have previously shown that it is expressed in the stomachs of gastritic mice and because of its potent capacity to stimulate antigen-presenting cells. Expression of this cytokine in the stomach did indeed initiate gastric autoimmunity with features indistinguisable from that induced by other methods. These studics support the contention that autoimmunity can be initiated by by-stander activation of antigen-presenting cells. In the context of the stomach it lends credence to the proposal that it can be initiated by a gastric pathogen such as Helicobacter pylori (2).

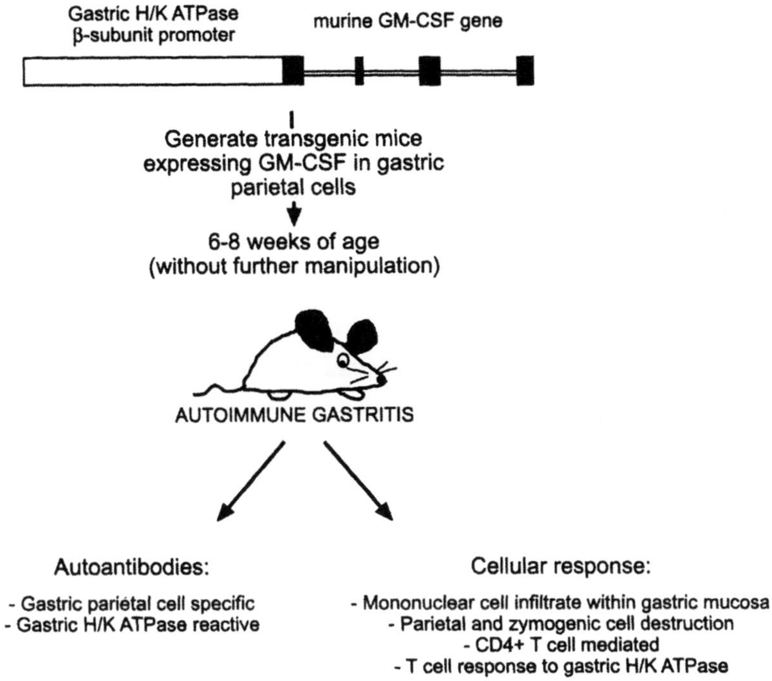


**Figure 5.** Initiation of autoimmune gastritis by ectopic expression of GM-CSF in the stomach. Transgenic mice were generated expressing the pro-inflammatory cytokine GM-CSF in parietal cells under control of the gastric H/K ATPase β-subunit promoter. Without any further manipulation, GM-CSF transgenic mice spontaneously develop autoimmune gastritis characterised by circulating autoantibodies specific for parietal cells and the gastric H/K ATPase and a destructive gastritis.

## ACKNOWLEDGEMENTS

We acknowledge support from the National Health and Medical Research Council of Australia and the Alfred Foundation.

Address for correspondence: Ban-Hock Toh

Department of Pathology and Immunology, Monash University Medical School, Prahran, Victoria, 3181, Australia

Telephone: (613) 99030713

Fax: (613) 99030731

Email: Ban-Hock.Toh@med.monash.edu.au

## REFERENCES

1. **Alderuccio F and Toh BH.** Immunopathology of autoimmune gastritis. *Histo Histopath* 15, 869-879, 2000.
2. **Appelmelk BJ, Faller G, Claeys D, Kirchner T, Vandenbroucke-Grauls CM.** Bugs on trial: the case of Helicobacter pylori and autoimmunity. *Immunol Today* 19: 296-299, 1998.
3. **Biondo M, Nasa Z, Marshall A, Toh BH, and Alderuccio F.** Local transgenic expression of granulocyte-monocyte colony stimulating factor (GM-CSF) initiates autoimmunity. *J. Immunol* 166, 2090-2099, 2001.
4. **Toh, BH, Van Driel IR, Gleeson PA.** Pernicious anemia. *N Engl J Med* 337: 1441-1448, 1997.
5. **Toh, BH, Gleeson PA, Whittingham S, Van Driel IR.** Autoimmune gastritis and pernicious anemia. In The Autoimmune Diseases, NR Rose and IR Mackay, editors. Academic Press, 459-476, 1998.
6. **Toh BH, Sentry J and Alderuccio F.** The causative H/K ATPase antigen in the pathogenesis of autoimmune gastritis. *Immunol today* 21, 352-358, 2000.

Chapter 12

# Duodenal Protection: Influence of Melatonin and other Intestinal Transmitters on Duodenal Bicarbonate Secretion and Enterocyte Signaling

GUNNAR FLEMSTRÖM and MARKUS SJÖBLOM
*Department of Physiology, Uppsala University*
*PO Box 572, SE-751 23 Uppsala, Sweden*

## INTRODUCTION

The duodenal epithelium secretes $HCO_3^-$ at higher rates than does the stomach (or more distal small intestine) and the duodenal secretion is currently accepted as the most important defence mechanism against acid discharged from the stomach (4,6). $HCO_3^-$ entering the continuous layer of visco-elastic mucus gel on top of the epithelial surface maintains pH in its cell-facing portion at neutrality at acidities encountered in the healthy duodenum.

The mucosal $HCO_3^-$ secretion is under central as well as local intestinal neurohumoral influence. This made it of interest to study effects of melatonin on mucosal $HCO_3^-$ secretion and duodenal enterocyte intracellular calcium ($[Ca^{2+}]_i$) (18,19). The neurohormone melatonin is synthesized and released from the pineal gland in a circadian fashion and exerts its action within the central nervous system as well as in peripheral tissues. Circadian rhythms in pain and discomfort are frequently observed in duodenal ulcer and other gastrointestinal diseases. Knowledge about the neurohumoral mediation of these variations may provide interesting information about intestinal pathophysiology and provide clues to future therapy. Importantly, melatonin is also produced by the enterochromaffin cells (EC-cells) of the intestinal mucosa (16) and some studies suggest that nutrients in the lumen influences the intestinal release of melatonin and other EC-cell products.

## METHODS

### Rat Duodenum in Vivo

Male outbred Sprague-Dawley rats weighing 190-260 g or F1-hybrids of Lewis x Dark Agouti rats, weighing 200-260 g were placed in a conditioning unit under standardized temperature and light conditions (21-22C, 12:12 h light-dark cycle) for at least four days after purchase. Animals were deprived of food for 20-24 h before the experiments, but had free access to drinking water. Experiments were started by anaesthetizing the animal close to 9 am with 5-ethyl-5-(1'-methyl-propyl)-2-thiobarbiturate (Inactin®, RBI), 120 mg/kg body weight intraperitoneally. Anesthesia was performed within the Animal Department by the person who had previously handled the animals. Subsequently, the rats were tracheotomized with a tracheal tube to facilitate respiration and body temperature was maintained at 37-38C throughout experiments by a heating pad controlled by a rectal thermistor probe. The surgical and experimental procedures have been described previously (5,12,18). Ringer solution was continuously infused into a femoral vein (1.0 ml/h) to compensate for fluid loss and to avoid acid-base changes during the experiments. Blood acid/base balance was controlled (AVL Compact 3, Graz, Austria) taken at the start and end of experiments. Values of blood pH (all experiments included) were between 7.28 and 7.45, and changes in base excess within ± 3 mmol/l. The abdomen was opened by a midline incision and the gastric pylorus was ligated with a suture. For measurement of duodenal mucosal $HCO_3^-$ secretion, a 12 mm segment of duodenum with its blood supply intact, starting 10-12 mm distal to the pylorus and thus devoid of Brunner's glands, was cannulated in situ between two glass tubes connected to a reservoir. Fluid (10 ml of 154 mM NaCl), maintained at 37°C by a water jacket, was rapidly circulated bya gas lift of 100% oxygen. $HCO_3^-$ secretion into the luminal perfusate was continuously titrated with 50 mM HCl at pH 7.4 under automatic control of a pH-stat system (Radiometer, Copenhagen).

In view of the central nervous as well as peripheral synthesis and actions of melatonin, melatonin and receptor ligands were administered close intra-arterially (i.a.) to the duodenum, thus minimizing any central nervous action of the compounds in these experiments. The effects of central nervous melatonin were studied by intracerebroventricular (i.c.v.) administration of melatonin. For duodenal i.a. infusions, the hepatic artery was cannulated, tied 3-4 mm proximal to its entrance into the liver, and perfused in the retrograde direction at 17 µl/min. The distribution was checked visually at the start of experiments by i.a. injection of a small amount (~0.1 ml) of a marker dye (Evans Blue). For i.c.v. infusions, a metal cannula was inserted into the right lateral cerebral ventricle using a stereotaxic instrument. Artificial cerebrospinal fluid alone or this fluid containing melatonin was infused. The location of the end of the cannula within the i.c.v. space was tested at the end of most experiments by adding Evans blue solution to the infusate, followed by dissection of the brain.

**Duodenal enterocytes**

A 3-cm segment of duodenum, starting 2-3 mm distal to the pylorus was promtly excised via an abdominal midline incision and freed from mesentary. The segment was opened along the antemesenterial axis and the luminal surface rinsed in a (NRM) solution containing (mmoles/l): $Na^+$ 114.4; $K^+$ 5.4; $Ca^{2+}$ 1.0; $Mg^{2+}$ 1.2; $Cl^-$ 121.8; $SO_4^{2-}$ 1.2; phosphate 6.0; Hepes 15.0; pyruvate 1.0; and glucose 10 *plus* 10 mg/l phenol red, 0.1 mg/ml gentamicin and 2.0 % fetal calf serum (1,2). The pH was adjusted to 7.40 immediately before use and temperature was maintained at 37°C. The sheet of duodenal wall was then put on a precleaned glass slide (lumen side up) and the mucosa was gently scraped off and cut into into 0.3-0.8 mm in diameter pieces which were dispersed and briefly shaken in NRM solution also containing 0.5 mM dithiothreitol (DTT). After sedimentation for 2-3 min, the supernatant was removed and the tissue fragments (in the sediment) washed three times in NRM solution (not containing DDT). Following brief gassing with 100% $O_2$, the tissue fragments (15-20 µl) were then exposed to mild digestion for 3 min by inoculation in 10 ml NRM solution containing 0.1 mg/ml collagenase type H (Sigma Co.) and 0.1 mg/ml dispase II (Mannheim). Digestion was stopped by adding DTT (to a concentration of 0.3 mM) and the solution centrifuged (3 min at 1,000 g). The pellet was washed three times by suspension in 10 ml DME/F12 (with 15 mM Hepes and 2.5 mM glutamine) followed by centrifugation (3 min at 1,000 g). $HCO_3^-$ (1 mM), gentamicin (0.01 mg/ml), fetal calf serum (2.0 %) and were allwas added to the DME/F12 and pH adjusted to 7.40. The preparation procedure yielded aggregates (10-200 cells) of interconnected duodenal enterocytes as well as (smaller amounts of) single cells. The aggregates were composed predominately of cells with rounded, vacuolated morphologies characteristic of crypt cells (2,21).

For measurement of $[Ca^{2+}]_i$, 70 µl of the cell aggregate suspension was loaded at 37°C with Fura-2 acetoxymethyl ester for 45 - 60 min in an electrolyte solution (mmoles/l: $Na^+$ 141.2; $K^+$ 5.4; $Ca^{2+}$ 1.0; $Mg^{2+}$ 1.2; $Cl^-$ 146.4; phosphate 0.4; TES 20.0; and glucose 10; pH 7.40) found appropriate for studies of cell aggregates from other tissues (17) The fura-2 loaded cell aggregates were spun down and placed on an uncoated and precleaned circular glass cover slip at the bottom of a temperature-controlled (37°C) perfusion chamber and fixed on top of the slip by cover of a uniformly sized pore polycarbonate membrane filter (17). The covering filter and the cell preparation were perfused (1 ml/min) with the electrolyte solution and receptor ligands to be tested added by inclusion in the perfusate. Changes in $[Ca^{2+}]_i$ in the fura-2-loaded cells were measured by the dual-wavelength excitation ratio technique by exposure to alternating 340 and 380 nm light with the use of a filter changer under the control of an InCytIM-2 system (Intracellular Imaging).

# RESULTS

Close intra-arterial (i.a.) infusion of melatonin (20, 200 and 2000 nmol/kg, h) to the duodenum significantly increased mucosal $HCO_3^-$ secretion (Fig. 1) but did not affect the transmucosal electrical potential difference. The secretion started to increase 20 min after the start of infusion of the lowest dose tested (20 nmol/kg, h) and peaked after 60 min. The rise was from 10.6 ± 1.1 to 20.0 ± 2.3 μEq/cm, h; secretion did not increase further with the higher doses. In contrast, $HCO_3^-$ secretion tended to decline with 200 and 2,000 nmol/kg, h, but was always significantly greater than the basal secretion. Prostaglandin $E_2$ (20 μM luminally) added at the end of experiments as a test of the viability of the preparation increased the $HCO_3^-$ secretion and also caused a rise in transmucosal PD. As found with melatonin, close i.a. infusion of the full melatonin receptor agonist 2-iodo-N-butanoyl-5-methoxytryptamine (2-IbMT) significantly stimulated the secretion.

The melatonin $MT_1/MT_2$ ($MT_2 > MT_1$) receptor antagonists luzindole (600 nmol/kg) did not affect on spontaneous (basal) $HCO_3^-$ secretion but abolished the response to melatonin. In contrast, Prazosin, which is an antagonist at melatonin $MT_3$ receptors, did not affect the rise in secretion in response to melatonin and neither luzindole nor prazosin inhibit the response to prostaglandin $E_2$.

The effects of central nervous administration of melatonin were tested by i.c.v. infusion of the compound. Infusion of 20 nmol/kg, h of melatonin for 110 min into a lateral brain ventricle affected neither the duodenal $HCO_3^-$ secretion nor the mean arterial blood pressure.

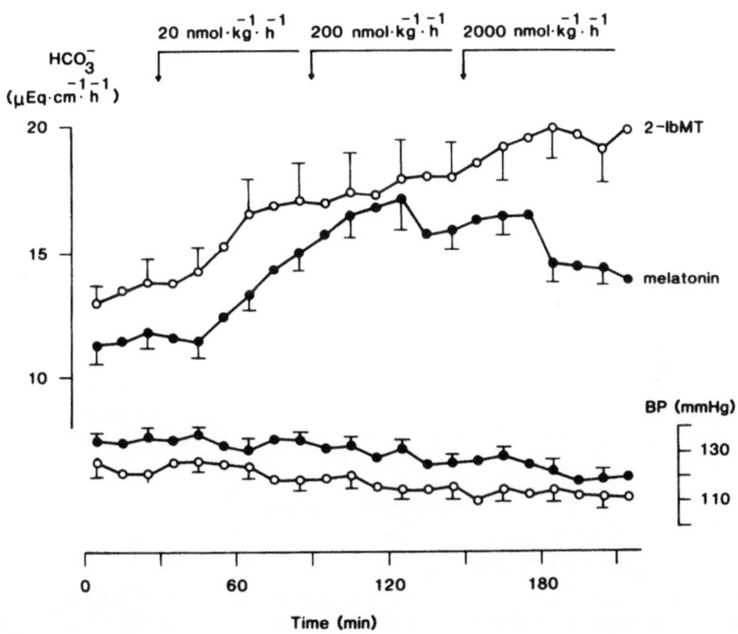


**Figure 1.** Melatonin and the full melatonin receptor agonist 2-IbMT administered to the duodenum by close intra-arterial infusion significantly ($P<0.05$) increased mucosal $HCO_3^-$ secretion. Means ± SE of $HCO_3^-$ secretion and mean arterial blood pressure are shown (n=6 in both groups).

Aggregates of duodenal enterocytes provided more stable preparations than single enterocytes. Superfusion with cholecystokinin octapeptide (CCK-8; 1-50 nM), caerulin (100 nM), ADP (100 μM) or carbachol (1-100 μM) induced transient rises in $[Ca^{2+}]_i$, similar to those observed with CCK-8 and carbachol in duodenal enterocytes in primary culture (2). Galanin (10-100 nM), in contrast, caused a slowly developing and more long-lasting rise in $[Ca^{2+}]_i$ and also potentiated the responses to CCK-8 (Fig. 2) and carbachol (not shown). Responses to melatonin (5-100 nM) were similar to those of galanin.

One or two cells in an aggregate were often the first to respond to CCK-8 (or carbachol) with a rise in $[Ca^{2+}]_i$. This was followed by spread of the $[Ca^{2+}]_i$ signal within the aggregate, suggesting that interconnecting enterocytes respond to stimuli increasing $[Ca^{2+}]_i$ as a syncytium.

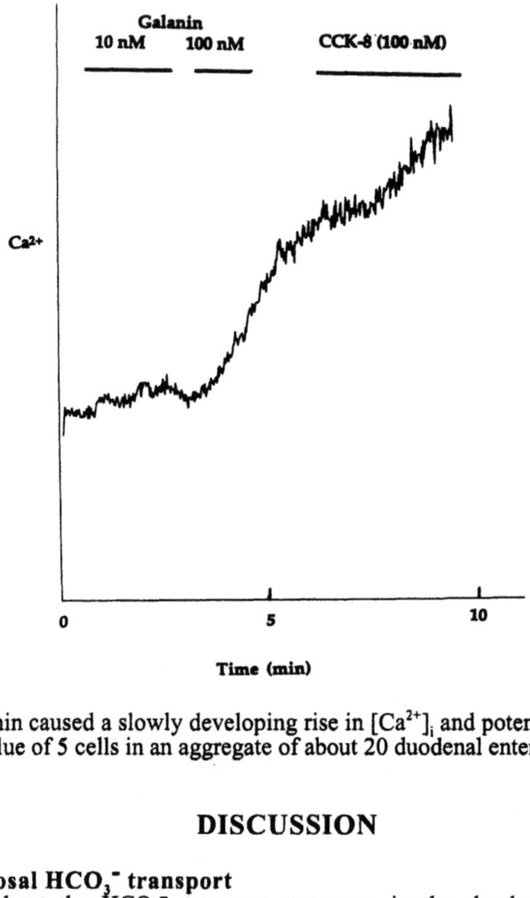


**Figure 2.** Galanin caused a slowly developing rise in $[Ca^{2+}]_i$ and potentiated the response to CCK-8. Mean value of 5 cells in an aggregate of about 20 duodenal enterocytes with crypt cell morphology

## DISCUSSION

### Duodenal mucosal $HCO_3^-$ transport

Knowledge about the $HCO_3^-$ transport processes in the duodenal epithelium *per se* originates from numerous studies involving amphibian mucosa *in vitro*, isolated mammalian duodenal enterocytes and cell membranes; and, more recently, from studies of genetically modified mice. Microfluorospectrophotometry of mixed villous and crypt enterocytes isolated from rat and human proximal duodenum and loaded with the pH-sensitive fluoroprobe 2',7'-*bis*(2-carboxyethyl)-5(6)-carboxyfluorescein (BCECF), confirmed previous hypotheses that duodenal enterocytes possess at least three mechanisms for acid/base transport (9) An amiloride-sensitive $Na^+/H^+$ exchanger that extrudes acid. Duodenal enterocytes import $HCO_3^-$ at the basolateral membrane by $Na^+(n)$-$HCO_3^-$ cotransport and export $HCO_3^-$ by $Cl^-/HCO_3^-$ exchange as well as via an apical anion conductive pathway. Recent evidence indicates that the cystic fibrosis transmembrane conductance regulator (CFTR) is the ubiquitous membrane spanning conductance that transports $HCO_3^-$ as well as $Cl^-$ (3,7,20). CFTR-/- mice, as well as patients with cystic fibrosis, have decreased resting and stimulated duodenal $HCO_3^-$ secretion (15).

Cyclic AMP stimulated $HCO_3^-$ transport across murine duodenum *in vitro* involves electrogenic transport via CFTR-channel as well as electroneutral transport via a CFTR-dependent $Cl^-/HCO_3^-$ exchange process that is associated with epithelial carbonic anhydrase activity (3). The latter finding may be of particular interest with respect to the mucosal location (crypt and/or villus) of the transporters. Suppression of carbonic anhydrase activity decreases duodenal mucosal $HCO_3^-$ secretion. Furthermore, the carbonic anhydrase isoenzyme (CA II) associated with $HCO_3^-$ transport, is located mainly in the villi and not in the duodenal crypts (14). This suggests that part of duodenal alkaline secretion originates from the villi; in contrast to the earlier hypothesis that intestinal secretions were of crypt origin while absorptive function resided in the villi. Fig. 3 presents a model of the $HCO_3^-$ transport by duodenal epithelium.

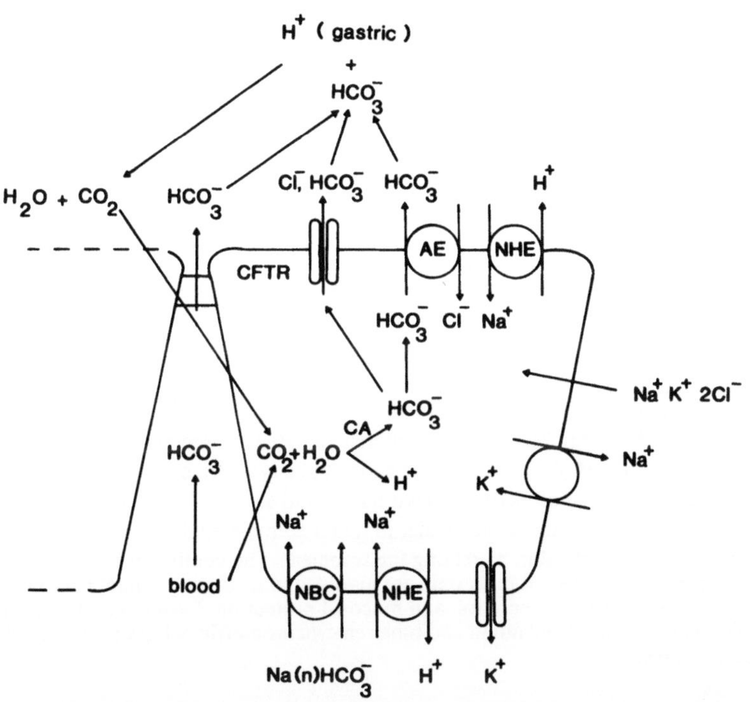


**Figure 3.** Mechanisms of $HCO_3^-$ transport by duodenal enterocytes. The model is based on studies of isolated cells, membrane vesicles and duodenal mucosa invitro and in vivo in several species. Species differences, if any, seem small but the mucosal location of the duodenal ion transport processes (villus and/or crypt cell) has not yet been determined fully. $HCO_3^-$ reaches the epithelium via blood as well as seconday to the intracellular carbonic anhydrase conversion of $CO_2$ + $H_2O$ to $HCO_3^-$. In addition, $HCO_3^-$ is imported via $NaHCO_3$ transport and extruded via $Cl^-/HCO_3^-$ exchange and ionic conductance.

**Melatonin and other intestinal transmitters**

The mucosal $HCO_3^-$ secretion is under central as well as local intestinal neurohumoral influence. A number of common diseases in humans, including gastroduodenal ulcer and irritable bowel syndrome, are well known for showing circadian rhythms in pain and discomfort. Further, studies in fasting animals have demonstrated that the gastric secretions of $HCO_3^-$ and mucus, both important in mucosal protection, shows day and night rhythms with peak times different from that of the mucosa-aggressive H+ secretion (11). This phase shift in secretory rhythms may, in theory, result in circadian variations in mucosal vulnerability to acid injury. Fuller information about circadian rhythms and, in particular, about the neurohumoral pathways from internal clocks to tissues and ion transporting cells, should provide relevant information about intestinal pathophysiology and may provide clues to future therapy.It is thus of interest to study the effects the pineal gland neurohormone melatonin on the mucosa-protective $HCO_3^-$ secretion in the duodenum. In humans and other mammals, including rodents, melatonin secretion from the pineal gland peaks at darkness (night), independently of species differences in day or night activity (22). Melatonin is synthesized from tryptophan, with serotonin as an intermediate, and is released from the pineal gland in mammals into the circulation. Being a non-polar and lipid-soluble indoleamine, melatonin crosses the blood-brain barrier and acts at melatonin-specific receptors in the central nervous system as well as such receptors in peripheral tissues.Three G-protein coupled melatonin receptor subtypes are reported and the mammalian $MT_1$ and $MT_2$ subtypes, but not yet the MT3 subtype, have been cloned (22).

Importantly, melatonin is also one of the transmitters produced by the enterochromaffin cells of the intestinal mucosa (16) and the total amount of melatonin in the alimentary tract is considerably higher than that in the central nervous system (8). A physiological role of this source of melatonin has not been clarified before, nor has any interaction between pineal gland and intestinal mucosal melatonin production. Like the enterochromaffin cell products uroguanylin and guanylin (10) and serotonin, intestinal melatonin may have a role in the reaction of the mucosa to luminal contents.

The results of the present work demonstrates that melatonin stimulates duodenal mucosal secretion of $HCO_3^-$, very likely via action at enterocyte $MT_2$-receptors, and that stimulation occurs without significant changes in the transepithelial electrical potential difference. The highest concentration of 2-iodomelatonin binding sites are, as proposed for the electroneutral $HCO_3^-$ exporter, located in the villi of the rat small intestine (13). Melatonin may thus stimulate the mucosa-protective $HCO_3^-$ secretion in the most acid-exposed (villus) portion of the duodenal epithelium.

Luzindole similarly inhibits the marked rise in $HCO_3^-$ secretion induced by central nervous (i.c.v.) infusion of the adrenoceptor agonist phenylephrine, a response also abolished by sub-laryngeal ligation of all nerves around the carotid arteries (18). The response in pineal gland ectomized animals is the same as that in intact animals, making it unlikely that central nervous melatonin is involved in mediating the response. The combined results thus strongly suggest that intestinal sources of melatonin are involved in mediating central nervous influence on duodenal epithelial $HCO_3^-$ secretion and mucosal protection. Further knowledge about the release and local actions of melatonin and other enterochromaffin cell products would seem of considerable interest.

## REFERENCES

1. **Chew CS, Ljungström M, Smolka A, and Brown MR.** Primary culture of secretagogue-responsive parietal cells. *Am J Physiol Gastrointest Liver Physiol* 256: G254–G263, 1989.
2. **Chew CS, Säfsten B, and Flemström G.** Calcium signaling in cultured human and rat duodenal enterocytes. *Am J Physiol Gastrointest Liver Physiol* 275: G296-G304, 1998.

3. **Clarke LL, and Harline MC.** Dual role of CFTR in cAMP-stimulated $HCO_3^-$ secretion across murine duodenum. *Am J Physiol Gastrointest Liver Physiol* 274: G718-G726, 1998.
4. **Flemström G.** *Gastric and duodenal mucosal bicarbonate secretion.* In: *Physiology of the Gastrointestinal Tract* (3rd ed.), edited by Johnson LR, Jacobson ED, Christensen J, Alpers D, and Walsh JH, New York: Raven, 1994, p. 1285-1309.
5. **Flemström G, and Jedstedt G.** Stimulation of duodenal mucosal bicarbonate secretion in the rat by brain peptides. *Gastroenterology* 97: 412-420, 1989.
6. **Flemström G, and Isenberg, JI.** Gastroduodenal mucosal alkaline secretion and mucosal protection. *News Physiol Sci* 16: 23-28, 2001.
7. **Hogan DL, Crombie DL, Isenberg JI, Svendsen P, Schaffalitsky de Muchadell OB, and Ainsworth MA.** CFTR regulates cAMP- and $Ca^{2+}$-mediated duodenal epithelial secretion. *Am J Physiol Gastrointest Liver Physiol* 272: G872-G876, 1997.
8. **Huether G, Poeggeler, B, Reimer A., and George A.** Effect of tryptophan administration on circulating melatonin levels in chicks and rats: evidence for stimulation of melatonin synthesis and release in the gastrointestinal tract. *Life Sci* 51: 945-953, 1992.
9. **Isenberg JI, Ljungström M, Säfsten B, and Flemström G.** Proximal duodenal enterocyte transport: evidence for $Na^+$-$H^+$ and $Cl^-$-$HCO_3^-$ exchange and $NaHCO_3$ cotransport. *Am J Physiol Gastrointest Liver Physiol* 265: G677-G685, 1993.
10. **Joo NS, London RM, Kim HD, Forte LR, and Clarke LL.** Regulation of intestinal $Cl^-$ and $HCO_3^-$ secretion by uroguanylin.*Am J Physiol Gastrointest Liver Physiol* 274: G633-G644, 1998.
11. **Larsen KR, Moore JG, and Dayton MT.** Circadian rhythms of acid and bicarbonate efflux in fasting rat stomach. *Am J Physiol Gastrointest Liver Physiol* 260: G610-G614, 1991.
12. **Larson GM, Jedstedt G, Nylander O, and Flemström G.** Intracerebral adrenoceptor agonists influence rat duodenal mucosal bicarbonate secretion. *Am J Physiol Gastrointest Liver Physiol* 271: G831-G840, 1996.
13. **Lee PP, and Pang SF.** Melatonin and its receptors in the gastrointestinal tract. *Biol Signals* 2:181-193, 1993.
14. **Lönnerholm GL, Knutson L, Wistrand PJ, and Flemström G.** Carbonic anhydrase in the normal rat stomach and duodenum and after treatment with omeprazole and ranitidine. *Acta Physiol Scand* 136: 253-262, 1989.
15. **Pratha VS, Hogan DL, Mårtensson BA, Bernard J, Zhou R, and Isenberg JI.** Identification of transport abnormalities in duodenal mucosa and duodenal enterocytes from patients with cystic fibrosis. *Gastroenterology* 118: 1051-1060, 2000.
16. **Raikhlin NT, and Kvetnoy IM.** Melatonin and enterochromaffine cells. *Acta Histochem* 55:19-24, 1976.
17. **Shariatmadari R, Lund PE, Krijukova E, Sperber GO, Kukkonen JP, and Åkerman KEO.** Reconstitution of neurotransmission by determining communication between differentiated PC12 pheochromocytoma and HEL 92.1.7 eythroleukemia cells. *Pflügers Arch-Eur J Physiol* 442: 312-320, 2001
18. **Sjöblom M, Jedstedt G, and Flemström G.** Peripheral melatonin mediates neural stimulation of duodenal mucosal bicarbonate secretion. *J Clin Invest* 108: 625-633, 2001
19. **Sjöblom M, Jedstedt G, and Flemström G.** Effects of melatonin on duodenal enterocyte calcium signaling and bicarbonate secretion (abstr). *Gastroenterology* 120: A-312, 2001.
20. **Seidler U, Blumenstein I, Kretz A, Viellard-Baron D, Rossmann H, Colledge WH, Evans M, Ratcliff R, and Gregor M.** A functional CFTR protein is required for mouse intestinal cAMP-, cGMP- and $Ca^{(2+)}$-dependent $HCO_3^-$ secretion. *J Physiol (Lond)* 505: 411-423, 1997.

21. **Trier JS.** Studies of the small intestinal crypt epithelium. II. Evidence for and mechanisms of secretory activity by undifferentiated crypt cells of the human small intestine. *Gastroenterology* 47: 480-495, 1964.
22. **Vanecek J.** Cellular mechanisms of melatonin action. *Physiol Rev* 78: 687-721, 1998.

# Chapter 13

# Role of Cytokines in *Helicobacter pylori*-Induced Gastric Epithelial Cell Matrix Metalloproteinase Secretion and Activation

MONIKA GÖÖZ, MARIA SHAKER, PAL GÖÖZ, ADAM J. SMOLKA
*Medical University of South Carolina, Charleston, SC*

## INTRODUCTION

Initial responses to gastric colonization by *Helicobacter pylori* include a transient period of achlorhydria and activation of neutrophils (1). In our previous work (2) we showed that acute bacterial infection of human gastric adenocarcinoma cells (AGS) inhibits the activity of human H,K-ATPase α subunit promoter, which may contribute to lowered acid output found in acute *H. pylori* infection in vivo. The inflammatory response is hypothesized to be induced both directly by bacterial factors like LPS, urease, porins, etc and indirectly by induction of the chemokine IL-8 and cytokines (IL-1, TNFα), leading to mucosal damage and ulceration (1). Matrix metalloproteinases play major roles in tissue degradation, cell migration and also inflammatory processes. We have previously shown that *H. pylori* infection of AGS cells increased secretion of epithelial MMPs-1 and –3 and bacterial MMP-3-like activity (3). Our aim in the present study was to characterize the relationship between activation of matrix metalloproteinases and IL-1ß and TNFα, potent stimulators of IL-8 secretion in gastric epithelia during *H. pylori* infection.

## EXPERIMENTAL

Human gastric adenocarcinoma cells (AGS) in serum-free medium were treated for 24 and 48 hr with IL-1ß (10 ng/ml) or TNFα (10 ng/ml) and MMP activities in concentrated conditioned media were assessed by zymography using non-reducing 10% or 4-16% acrylamide gels containing 0.1% gelatin, or 0.1% casein (NuPAGE, Novex). IL-1ß and TNFα both stimulated gelatinolytic activity (Fig.1), but IL-1ß also stimulated caseinolytic activity (data not shown), prompting further study of the IL-1ß effect. (Untreated cells supernatant is shown in lane C.)

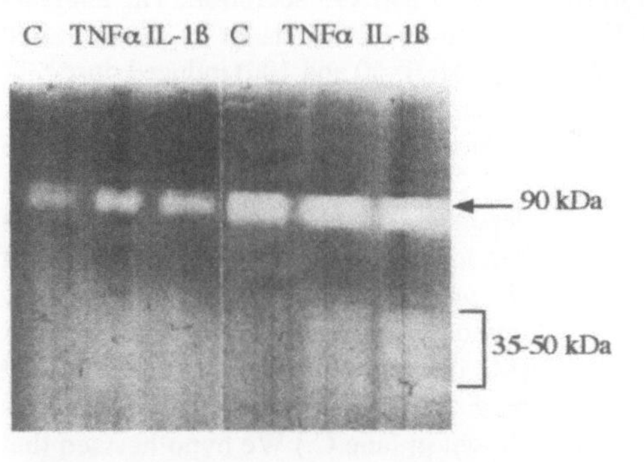


**Fig. 1** Gelatin zymography of TNFα and IL-1ß treated AGS cells supernatant.

As both *H. pylori* and IL-1ß were previously shown to increase secretion of IL-8 from gastric epithelial cells, we determined the IL-1ß concentrations that would induce comparable IL-8 secretion to *H. pylori* concentrations used in our earlier metalloproteinase study (3). AGS cells were treated either with IL-1ß (0.2-

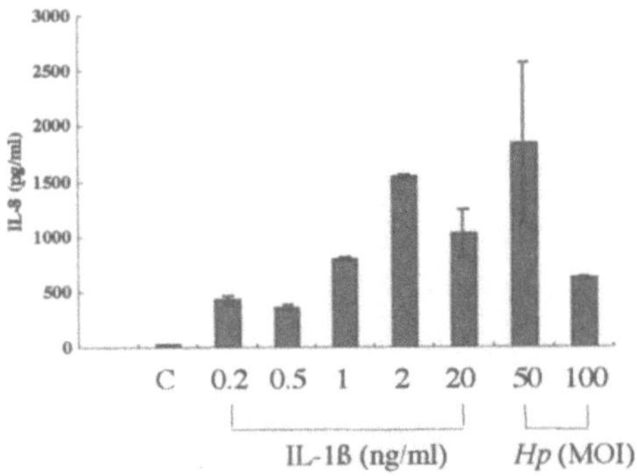


**Fig. 2** IL-8 secretory response induced by IL-1ß and *H. pylori* (ELISA).

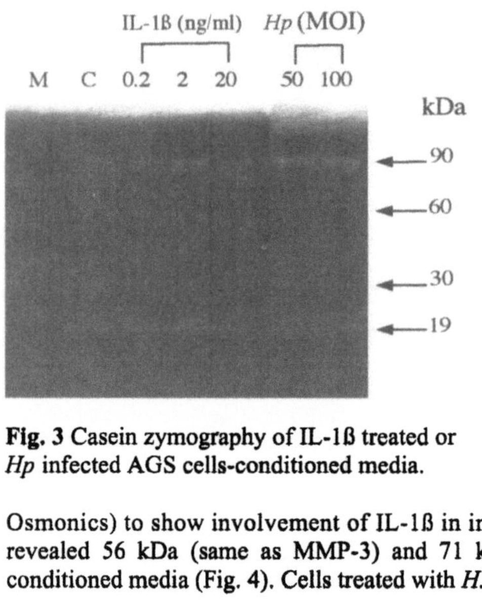


**Fig. 3** Casein zymography of IL-1ß treated or *Hp* infected AGS cells-conditioned media.

20 ng/ml) or infected with *H. pylori* (multiplicities of infection of 50 and 100) in serum-free Ham's F-12 medium. IL-8 content of unconcentrated supernatants was measured by ELISA (R&D System). As shown in Figure 2, comparable IL-8 secretory responses (~1000ng /ml) were induced by 2.0 ng/ml IL-1ß and by *H. pylori* at MOI=50. (Medium from untreated cells is shown as C.)

MMP activity secreted by AGS cells during IL-1ß treatment or *H. pylori* infection was assessed by caseinolytic zymography. Aliquots of concentrated conditioned serum-free media were electrophoresed on casein-containing zymogram gels (for details see Fig 1). Densitometric analysis of lanes in Fig 3 using NIH Image showed that IL-1ß and *H. pylori* concentrations (2ng/ml and MOI=50, respectively) inducing similar IL-8 secretory response induced comparable increase in AGS cell MMP activity at 90 kDa (1302 vs 1351) and 60 kDa (330 vs 244). (Medium alone is shown in lane M and medium from untreated AGS cells is shown in lane C.) Because we had previously shown (3) that MMP-3 production of AGS cells is increased during *H. pylori* infection, we used MMP-3 monoclonal antibody (Chemicon) in immunoblot analysis (Bis-tris NuPAGE gels, Novex, transferred to nitrocellulose membrane, Osmonics) to show involvement of IL-1ß in induction of MMP-3 secretion. The analysis revealed 56 kDa (same as MMP-3) and 71 kDa immunoreactive bands in concentrated conditioned media (Fig. 4). Cells treated with *H. pylori* (MOI=50 and 100) induced dose-dependent increases in both immunoreactive bands, whereas IL-1ß (0.5-2ng/ml) treatment induced dose-dependent increase only in the 71 kDa band. Concentrated serum-free media of *H. pylori* alone did not show any immunoreactivity (data not shown). (Medium alone is shown in lane M and medium from untreated AGS cells is shown in lane C.) We hypothesized that the 71 kDa band represented a MMP-3/TIMP-3 heterodimer and that, like *H. pylori* infection, IL-1ß treatment

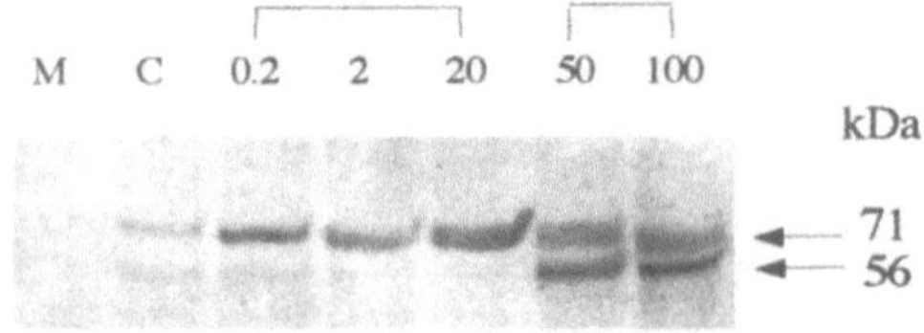


**Fig 4.** Immunoblot analysis of IL-1ß treated or *Hp* infected AGS cells-conditioned media using MMP-3 monoclonal antibody.

increased both MMP-3 and TIMP-3 secretion by AGS cells. To test this hypothesis we used polyclonal TIMP-3 antibody (Chemicon) in immunoblot analysis of IL-1ß (0.2-20 ng/ml) or *H. pylori* (MOI=50 and 100) treated AGS cells-conditioned media (for details see Fig 4). The 71 kDa band shown in Figure 4 was identified as a MMP-3/ TIMP-3 complex in Figure 5. Using NIH Image analysis we found dose-dependent increase in immunoreactivity of the complexes formed by AGS TIMP-3/MMP-3 during IL-1ß treatment. We have previously shown that the antibody also recognizes a bacterial TIMP-3-like factor. Thus the 71 kDa band shown in *H. pylori* infected cells' supernatant reflects both human and bacterial TIMP-3 complexes. The 24 kDa and 17 kDa bands represent bacterial TIMP-3-like immunoreactivity.

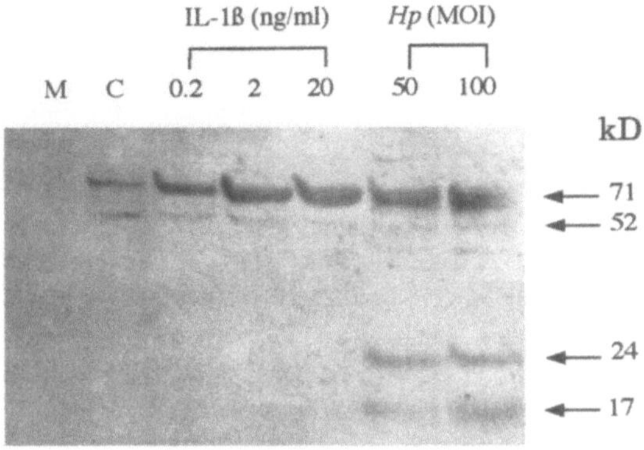


**Fig. 5** Immunoblot analysis of IL-1ß treated or *Hp* infected AGS cells-conditioned media using TIMP-3 polyclonal Ab.

To further characterize the functional property of secreted MMP-3, we measured enzymatic activity in AGS/IL-1ß and AGS/ *H. pylori* conditioned media using a synthetic, fluorogenic specific MMP-3 substrate (M2110, Chemicon). Although IL-1ß and *H. pylori* both elicited dose-dependent increases in evoked fluorescence, *H. pylori*-treated media expressed 30 times more MMP-3 activity than IL-1ß- treated media (note the difference in scale between the two bar graphs; M refers to medium alone and C refers to untreated AGS cells- conditioned medium (**p<0.01, ***p<0.001 vs untreated/ uninfected.)

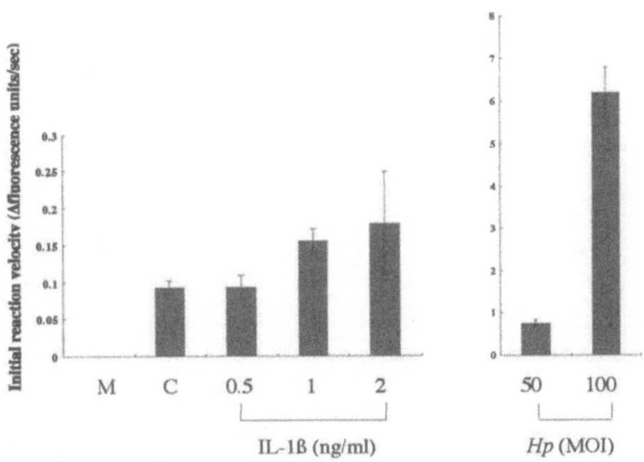


**Fig. 6** MMP-3 activity of (A) IL-1ß treated or (B) *H. pylori* infected AGS cells-conditioned media M2110 synthetic substrate.

## CONCLUSIONS

We conclude that both TNFα and IL-1ß increase active MMP secretion of gastric epithelial cells, thus contributing to the epithelial tissue destruction during *H. pylori* infection. However, other bacterial factors, including bacterial MMP-3 are needed to mediate the full gastric epithelial cell MMP activation response induced by *H. pylori* infection.

## REFERENCES

1. **Crabtree JE**. Immune and inflammatory responses to *Helicobacter pylori* infection. *Scand J Gastroenterol* 31 Suppl 215:3-10, 1996.

2. **Gööz M, Hammond C, Larsen K, Mukhin Y, and Smolka AJ.** Inhibition of human gastric H,K-ATPase α subunit gene expression by *H. pylori*. Am J Physiol 278:G981-991,2000.

3. **Gööz M, Gööz P, and Smolka AJ.** Epithelial and bacterial metalloproteinases and their inhibitors in *H. pylori* infection of gastric epithelial cells. Am J Physiol (Accepted for publication).

Chapter 14

# HCl Causes Less Intracellular Acidification In Surface Epithelial Cells Of Isolated Necturus Gastric Mucosa Than Other Inorganic Acids

KIVILUOTO T, NYLANDER-KOSKI O, MUSTONEN H, KIVILAAKSO E
*Department of Surgery, Helsinki University Central Hospital, Helsinki, Finland,*

## ABSTRACT

Luminal acid causes intracellular acidification in the gastric epithelium, but the mechanism how $H^+$ enters surface cells remains obscure. This study addresses the problem by assessing how different acids affect intracellular pH in gastric surface cells. Isolated Necturus antral mucosa was exposed to HCl, $HNO_3$, $H_2SO_4$ and $H_3PO_4$ at pH 2.30. Intracellular pH was measured with microelectrodes. Exposure to luminal $HNO_3$, $H_2SO_4$ or $H_3PO_4$ caused significantly larger intracellular acidification than exposure to HCl. The degree of acidification was not dependent on the valence or nature of the anionic counterion of the acid, but significantly correlated with the amount of undissociated molecular form of the acid in the solution.

## INTRODUCTION

The gastric mucosa is continuously exposed to relatively strong luminal acid, which implies effective mechanisms to protect the mucosa against acid induced damage and to maintain intracellular pH ($pH_i$) within physiologic range in the surface epithelial cells. Direct measurements with microelectrodes indicate that physiologic intragastric concentrations of HCl acidify slightly but significantly $pH_i$ in surface epithelial cells (3,4), thus implicating that some luminal $H^+$ does penetrate inside the cells. However, the mechanism how luminal acid enters the surface cell is not known, since the phospholipid bilayer cell membrane is virtually impermeable to $H^+$. The apical membrane of surface cells is known to contain amiloride sensitive cation ($Na^+$) channels, which may be permeant also to $H^+$, but in the presence of luminal acid these channels are closed thus precluding major transport of $H^+$ inside the cell (5).

Gutknecht and Walter (2) have proposed that the entry of luminal $H^+$ inside the gastric mucosa might occur in the form of molecular HCl, since phospholipid bilayers are, in general, highly permeable to small uncharged molecules. The same authors also proposed that monovalent acids are more permeant in phospholipid bilayers than multivalent acids (2). On the other hand, Barreto and Lichtenberger (1) have presented that it is the anionic counterion which dictates the permeation of $H^+$ in phospholipid bilayers, $Cl^-$ permitting much higher

$H^+$ permeation than $SO_4^{2-}$ or $PO_4^{3-}$. In this study we have investigated acid movement across the apical cell membrane of surface epithelial cells in isolated Necturus antral mucosa by exposing the mucosa to various inorganic acids with different valence and anionic counterion and by comparing their effects on $pH_i$ to that of HCl.

## MATERIALS AND METHODS

Isolated Necturus antral mucosa was mounted in a perfusion chamber and perfused with Ringer's solution, gassed with 95%$O_2$/5%$CO_2$ on both sides. Serosal side was buffered to pH 7.25 with $HCO_3$. The mucosal (luminal) side was perfused with HCl and one of the following acids: $HNO_3$, $H_2SO_4$, $H_3PO_4$ at equal acidity (pH 2.30). In the mucosal HCl, $HNO_3$, $H_2SO_4$, $H_3PO_4$ solutions the respective anion was substituted for all other anions in the Ringer's solution. The mucosas were exposed to the two acids in a random order, the HCl exposure step serving as the control. Between the exposures to the different acids the mucosa was allowed to recover for 10-15 minutes in Ringer's solution buffered to pH 6.0 with MES. Intracellular pH ($pH_i$) was measured with double-barrelled proton-selective microelectrodes positioned under microscopic control in a surface epithelial cell. The signals of the pH-selective barrel ($H^+$ activity plus apical transmembrane potential) and the voltage barrel (apical transmembrane potential) were subtracted to give the net $H^+$ activity signal. The results are given as mean±SE. Statistical analysis was performed with Student's t-test for paired variates. In correlation analysis Pearson's correlation coefficient R was used.

## RESULTS AND DISCUSSION

Exposure of the gastric mucosa to the "natural" gastric acid, HCl, caused significantly ($p<0.05$) less intracellular acidification in the surface epithelial cells than the "unnatural" inorganic acids $HNO_3$, $H_2SO_4$ or $H_3PO_4$ (n=11 for each group) (Fig. 1). No correlation was observed between the valence of an acid and the degree of intracellular acidification provoked by the same acid (Pearson's R=0.32, p=0.1)(Fig. 2). The monovalent $HNO_3$ provoked significantly larger acidification than the likewise monovalent HCl, but also somewhat larger acidification than the divalent $H_2SO_4$, indicating that the valence of an acid is not the critical determinant of acid permeation inside the cell. Nor is acid permeation dependent on the nature of anionic counterion in the way suggested by Barreto and Lichtenberger (1), since $Cl^-$ did not permit higher $H^+$ permeation than $SO_4^{2-}$ or $PO_4^{3-}$. In contrast, our findings are explicable in terms of the relative amount of molecular (unionised) acid in the perfusate. Unlike monovalent acids, multivalent acids dissociate in several steps. Knowing the dissociation constants it is possible to calculate the amount of the acid, which is in the molecular form in an aqueous solution. At pH 2.3 in aqueous solution the molecular concentrations for the acids used in the present experiments are as follows: HCl 49 pM, $HNO_3$ 25 μM, $H_2SO_4$ 161 nM and $H_3PO_4$ 39 mM. Thus, the molecular concentration of an acid (in logarithmic scale) in aqueous solution seems to correlate with the degree of intracellular acidification provoked by it (Pearson's R=0.48, p=0.003, log(molecular concentration) vs $\Delta pH_i$)(Fig. 3). This suggests that gastric acid (and other acids) enters the surface epithelial cell, at least in part, in an unionised molecular form.

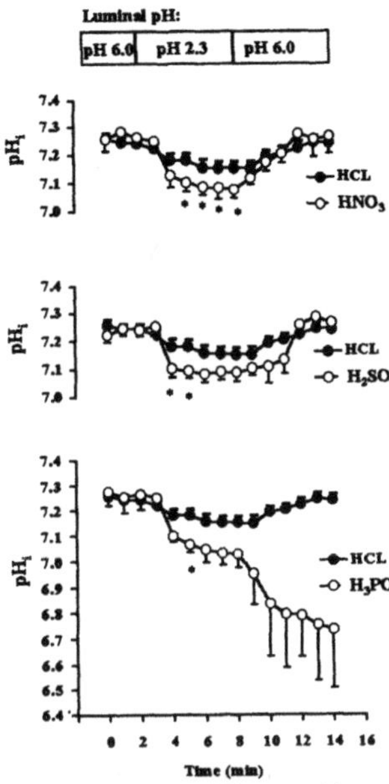


Fig. 1. Effect of different acids on intracellular pH.

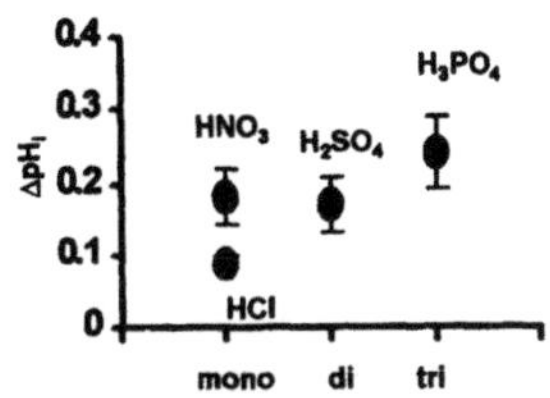


Fig. 2. Relationship between the valence of an acid and degree of intracellular acidification

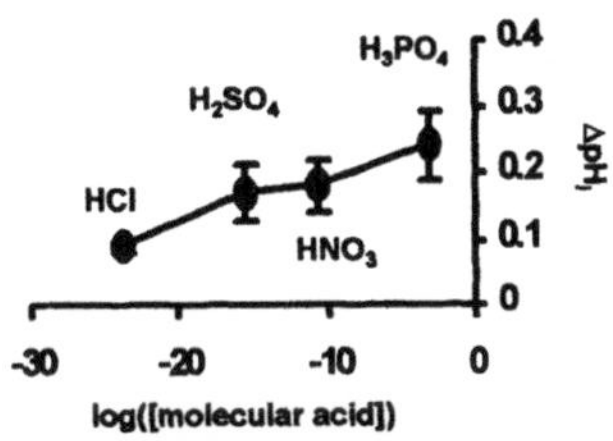


Fig. 3.Relationship between the concentration of molecular acid and the degree of intracellular acidification

## REFERENCES

1. **Barreto J, Lichtenberger L.** A phospholipid bilayer model for gastric membranes: vesicle acidification driven by a million fold proton gradient. *Gastroenterology* 98(5): A19,1990

2. **Gutknecht J, Walter A.** Transport of protons and hydrochloric acid through lipid bilayer membranes. *Biochimica et biophysica acta* 641(1): 183–188, 1981

3. **Kivilaakso E, Kiviluoto T.** Intracellular pH in isolated Necturus antral mucosa in simulated ulcerogenic conditions. *Gastroenterology* 95: 1198–1205, 1988

4. **Kiviluoto T, Mustonen H, Kivilaakso E.** Effect of barrier–breaking agents on intracellular pH and epithelial membrane resistances: studies in isolated necturus antral mucosa exposed to luminal acid. *Gastroenterology* 96: 1410–1418, 1989

5. **Mustonen H, Kivilaakso E.** Luminal acid increases apical cell membrane resistance in isolated necturus antral mucosa. *Gastroenterology* 113: 875-883, 1997

**Supported** by grants from the Research Foundation of Helsinki University Central Hospital, Sigrid Juselius Foudation and Academy of Finland

Chapter 15

# Ethanol Increases Intracellular Free Calcium, Closes Gap Junctions And Decreases Cell Volume In Primary Cultured Rabbit Gastric Epithelial Cell Monolayers

MUSTONEN H, KIVILUOTO T, KIVILAAKSO E.
*Department of Surgery, Helsinki University Central Hospital, Helsinki, Finland*

## ABSTRACT

The roles of cell calcium, gap junctions and cell volume in ethanol induced gastric damage were investigated by assessing intracellular free calcium levels, the patency of gap junctions and cell volume changes in ethanol exposed gastric epithelial cell monolayers. Intracellular free calcium concentration was increased, gap junctions were closed and cell volume was decreased after exposure to 5% (vol/vol) ethanol. Since gap junctions are known to be calcium gated, it is likely that their closure is secondary to the elevated cytosolic calcium in ethanol injured cells and may have a protective function by limiting intercellular spread of impending cell injury. The mechanism of cell shrinkage and its potential contribution to ethanol injury remains to be worked out.

## INTRODUCTION

Luminal ethanol is a well established ulcerogenic "barrier breaker" in gastric mucosa. Many efforts have been made to clarify the mechanisms of ethanol injury in gastric mucosa (1;3-5;7;8), but its effects at cellular or epithelial level remain to be detailed. In Necturus antrum artificially increased intracellular free calcium concentration induces similar changes in epithelial cell membrane resistances and potentials as do low concentrations (0.5-4%) of luminal ethanol (5), suggesting a role for cytosolic calcium in ethanol induced injury. Although ethanol is often used in hyperosmolar concentrations which should allegedly promote cell shrinkage, its influence on cell volume seems to be relatively weak due to its high permeability in cell membranes. In canine gastric mucosa, exposure to 12.5% ethanol produced widened and irregular intracellular spaces (1) suggesting cell volume decrease, but the role of cell shrinkage in ethanol injury remains obscure. Cell shrinkage might also perturb intercellular gap junctions, but the behaviour of these junctions in ethanol injury is not known. In order to further elucidate the mechanism of ethanol induced epithelial injury, we have measured intracellular free calcium concentrations, assessed the patency of gap junctions and estimated cell volume changes in primary cultured gastric epithelial cell monolayers exposed to ethanol.

## METHODS

Gastric epithelial cells were isolated from male white rabbits. The oxyntic mucosa was minced into small pieces and incubated in a medium containing 0.07% collagenase. Isolated rabbit gastric epithelial cells were cultured on collagen type I coated glass cover slips in Ham's F12 medium with 10% FBS. The cells were incubated in 5% $CO_2$ at 37° C, forming a complete polarized monolayer in 48h. For the measurement of intracellular calcium the cells were loaded with the acetoxymethyl ester of fura-2. Emission intensity (510 nm) was measured with cooled digital camera at 340 nm and 380 nm excitation. The 340/380 ratio image was calculated. Intracellular free calcium concentration ($[Ca^{2+}]_i$) was calculated according to the Grynkiewicz equation (2). The patency of gap junctions was assessed by measuring fluorescence recovery after photobleaching (FRAP). The cells were loaded with acetoxymethyl ester of 5-carboxyfluorescein diacetate and in a small area of the monolayer the fluorescence signal was photobleached with a laser. The recovery of fluorescence intensity was measured with a confocal microscope with or without 5% ethanol in Ham's F12. For cell volume measurements the monolayer was loaded with Calcein-AM and imaged along Z-axis with confocal microscope. The change in depth of a cell was intercepted as a measure of change in cell volume.

## RESULTS AND DISCUSSION

The basal intracellular free calcium concentration was 65±9 nM and after 10 minutes exposure to 5% ethanol it increased to 140±17 nM ($p=0.03$, $N=5$) (Fig. 1). Even though ethanol significantly increased cytosolic free calcium level, it still remained much lower compared to those seen generally in dying cells ($>10^{-5}$M). In assessing the patency of gap junctions with FRAP experiments, the recovery at 10 minutes after the bleaching was without ethanol 52.8±11.1% and with 5% ethanol 9.4±3.2% ($p=0.01$, $N=6$), indicating partial closure of gap junctions (Fig. 2). Since high calcium concentration is known to close gap junctions, it is likely that their closure is secondary to the elevated cytosolic calcium in ethanol injured cells. It is possible that this mechanism may contribute to epithelial protection by impeding the spread of dissipated ionic gradients and other cytosolic derangements from injured cells to neighbouring vital cells. After 5 minutes exposure to 5% ethanol the cell volume decreased by −9.4±2.7% ($p=0.003$, $N=18$) (Fig. 3). The baseline permeability (permeability without transporters or channels) of ethanol through cell membranes is large (6), even larger than the

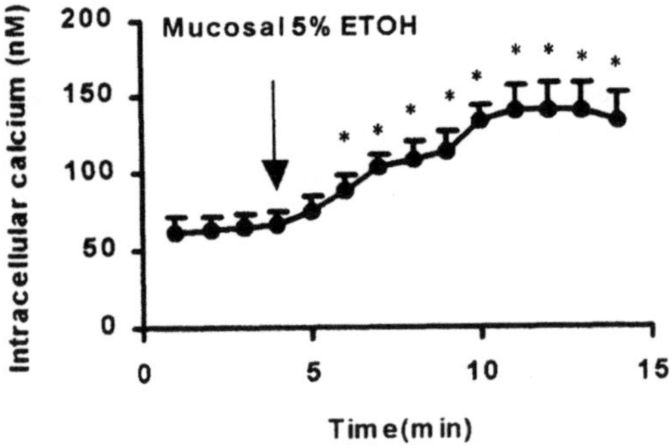


**Figure 1.** Intracellular free calcium concentration before and after 5% ethanol exposure. Intracellular calcium is increased by 115%. * ($p<0.05$) as compared to 0% ethanol exposure.

basal permeability of water. This is due to the high lipid partition ratio of ethanol (0.48 for octanol/water (6)). This indicates that even though the osmolarities of ethanol solutions are high they should not exert strong osmotic pressure to the cells and therefore they should not cause substantial volume changes. It also suggests that the volume decrease observed in the present study is not due to osmotic pressure but is caused by some more specific ethanol-linked mechanism. Whether it has a role in ethanol induced gastric damage remains to be worked out.

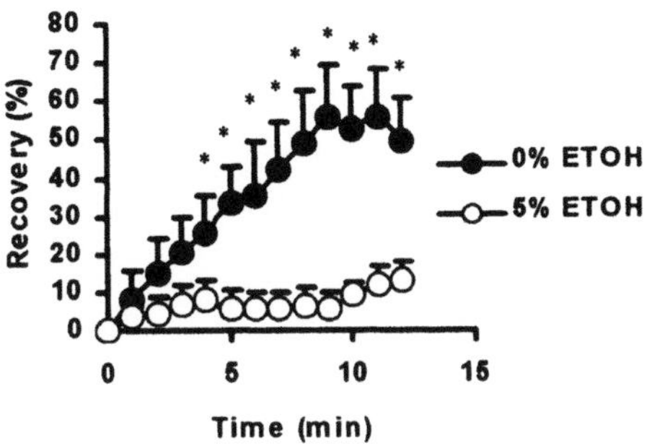


**Figure 2.** Fluorescence recovery after photobleaching with or without 5% ethanol. Gap junctions are partially closed by ethanol. * (p<0.05) as compared to 0% ethanol exposure.

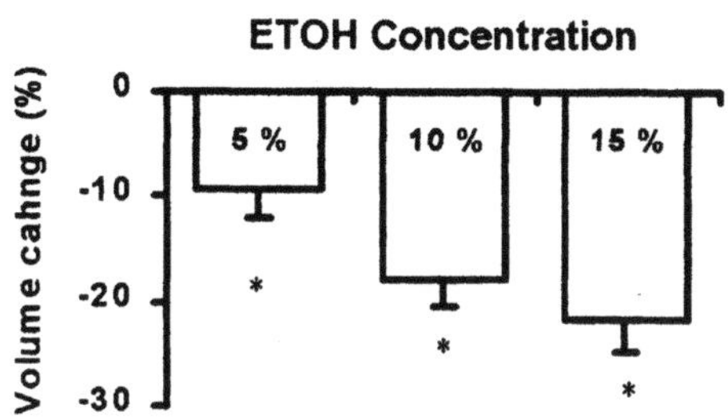


**Figure 3.** Cell volume changes following exposure to different ethanol concentrations as compared to control (* p<0.05)

## REFERENCES


1. **Dinoso, V. P. J., S. Ming, and J. McNiff.** Ultrastructural changes of the canine gastric mucosa after topical application of graded concentrations of ethanol. *Am J Dig Dis* 21: 626-632, 1976.
2. **Grynkiewicz, G., M. Poenie, and R. Y. Tsien.** A new generation of $Ca^{2+}$ indicators with greatly improved fluorescence properties. *Journal of Biological Chemistry* 260: 3440-3450, 1985.
3. **Kiviluoto, T., H. Mustonen, and E. Kivilaakso.** Effect of barrier-breaking agents on intracellular pH and epithelial membrane resistances: studies in isolated Necturus antral mucosa exposed to luminal acid. *Gastroenterology* 96: 1410-1418, 1989.
4. **Ranta-Knuuttila, T., H. Mustonen, and E. Kivilaakso.** Topical prostaglandin E2 protects isolated gastric mucosa against acidified taurocholate-, but not ethanol- or aspirin-induced injury. *Dig Dis Sci* 45: 99-104, 2000.
5. **Rutten, M. J. and C. D. Moore**. Low doses of ethanol have $Ca^{2+}$ ionophore-like effects on apical membrane potential of in vitro Necturus antrum. *Am J Physiol* 261: G92-103, 1991.
6. **Stein, W.** Transport and Diffusion Across Cell Membranes. San Diego, Academic Press. 1987.

7. **Sue, M. W. and P. H. Guth.** A fluorescent in vivo microscopic method to assess surface mucosal integrity in the rat stomach. Effects of ethanol and prostaglandin. *Gastroenterology* 89: 415-420, 1985.
8. **Wilkes, J. M., H. J. Ballard, D. T. Dryden, and B. H. Hirst**. Proton permeability and lipid dynamics of gastric and duodenal apical membrane vesicles. *Am J Physiol* 256: G553-G562, 1989.

**Supported** by grants from the Research Foundation of Helsinki University Central Hospital, Sigrid Juselius Foudation and Academy of Finland

Chapter 16

# Induction of Small Intestinal Damage by Inhibition of Both NO Synthase and COX-2

AKIKO TANAKA, SHOKO HASE, TOHRU MIYAZAWA
KENJI KANATSU AND KOJI TAKEUCHI
*Department of Pharmacology and Experimental Therapeutics*
*Kyoto Pharmaceutical University*
*Misasagi, Yamashina, Kyoto 607-8414, Japan*

## INTRODUCTION

Non-steroidal antiinflammatory drugs (NSAIDs) such as indomethacin cause intestinal ulceration as an adverse effect (2,3). Prostaglandin (PG) deficiency caused by cyclooxygenase-1 (COX-1) is considered to play a critical role in the pathogenesis of these lesions. However, recent studies showed that inhibition of both COX-1 and COX-2 is required for induction of damage in the gastrointestinal mucosa following administration of NSAIDs (4-6). Since inhibition of COX-1 was accompanied by the mucosal expression of COX-2 (4), it is assumed that PGs produced by COX-2 contribute to the mucosal integrity by antagonizing deleterious events caused by inhibition of COX-1.

On the other hand, nitric oxide (NO) plays a dual role in the pathogenesis of NSAID-induced intestinal damage; a protective role of constitutive NO synthase (cNOS)/NO and a proulcerogenic role of inducible NO synthase/NO (2). Indeed, indomethacin-induced intestinal lesions were prevented or worsened by prior administration of the NO donor NOR-3 or the NO synthase inhibitor $N^G$-nitro-L-arginine methyl ester (L-NAME), respectively (5). Thus, cNOS/NO is important for maintaining mucosal integrity of the small intest ine, just mimicking the role for COX-1/PGs. If so, intestinal damage might also be induced by concurrent inhibition of NO synthase and COX-2.

In the present study, we therefore examined whether intestinal damage was provoked in rats following the combined administration of L-NAME and the COX-2 selective inhibitor rofecoxib. We also examined the mucosal expression of COX-2 mRNA after administration of L-NAME.

## MATERIALS AND METHODS

**Induction of Intestinal Lesions:** Male Sprague Dawley rats (200-220 g) were used. L-NAME (50 mg/kg; Sigma Chemicals) the NO synthase inhibitor or SC-560 (10 mg/kg; Cayman Chemicals) the selective COX-1 inhibitor was given p.o., either alone or in combination with rofecoxib (10 mg/kg; Synthesized in our laboratory) the selective COX-2 inhibitor, and the animals were killed 24 hr after the administration. The small intestines were excised and treated with 2% formalin for fixation of the tissue walls, then they were opened along the mesenteric attachment and examined for damage under a dissecting microscope with square grids (x10). The area ($mm^2$) of macroscopically visible lesions was measured, summed per small intestine, and used as a lesion score. The person measuring the lesions did not know the treatments given to the animals.

**Analysis of COX-2 mRNA:** The animals were killed under deep ether anesthesia 6 hr after p.o. administration of L-NAME (50 mg/kg), SC-560 (10 mg/kg) or rofecoxib (10 mg/kg),

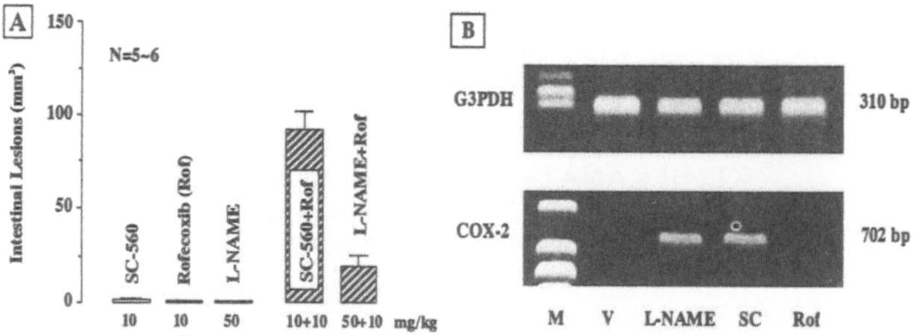


Figure 1. A. Intestinal ulcerogenic property of L-NAME, SC-560 and rofecoxib, either alone or in combination, in rats. L-NAME (50 mg/kg), SC-560 (10 mg/kg) or refecoxib (10 mg/kg) was administered p.o., and the animals were killed 24 hr after the administration. Data are presented as the mean±SE from 5~6 rats. B. Gene expression of COX-2 in the rat intestinal mucosa following administration of L-NAME, SC-560 or rofecoxib. The animals were administered L-NAME (50 mg/kg), SC-560 (SC: 10 mg/kg) or rofecoxib (Rof: 10 mg/kg) p.o., and killed 6 hr later. M: marker, V: vehicle.

and the small intestines were removed, frozen in liquid nitrogen, and stored at -80℃ until use. Analysis of COX-2 mRNA was performed by RT-PCR, according to a previous paper (4).

## RESULTS

### Induction of Intestinal Damage

Oral administration of neither SC-560 (10 mg/kg) nor rofecoxib (10 mg/kg) alone provoked damage in the small intestinal mucosa within 24 hr (Figure 1A). However, SC-560 produced hemorrhagic lesions in the small intestine at an incidence of 100% when administered together with rofecoxib, the damage score being 95.6±9.7 mm². Likewise, L-NAME (50 mg/kg) did not by itself cause any gross damage in the small intestinal mucosa. When L-NAME was administered together with the selective COX-2 inhibitor rofecoxib (10 mg/kg), this treatment, however, provoked damage in the intestine at an incidence of 100%, the damage score being 21.5±5.7 mm².

### Expression of COX-2 mRNA in the Intestinal Mucosa

The gene expression of COX-2 was negligible in the small intestinal mucosa of normal rats. However, it was clearly up-regulated in the rat 6 hr after administration of L-NAME (50 mg/kg) or SC-560 (10 mg/kg) but not rofecoxib (10 mg/kg)(Figure 1B). In contrast, G3PDH mRNA was observed in the intestinal mucosa of rats, irrespective of whether they were treated with vehicle, L-NAME, SC-560 or rofecoxib.

## DISCUSSION

The present study confirmed our previous findings that neither the selective COX-1 inhibitor (SC-560) nor COX-2 inhibitor (rofecoxib) alone caused intestinal damage, yet their combination provoke apparent damage in the small intestine (4). In addition, we also found that inhibition of NO production by L-NAME did not induce gross damage in the intestine whereas the agent when administered with the selective COX-2 inhibitor did produce damage, similar to that induced by inhibition of both COX-1 and COX-2. Furthermore, we showed that the COX-2 expression was up-regulated in the intestinal mucosa after administration of L-NAME as well as SC-560.

Wallace et al. (6) recently reported that neither SC-560 nor celecoxib induced gastric lesions but their combination caused damage in the stomach, suggesting that inhibition of both COX-1 and COX-2 is required for the occurrence of NSAID-induced gastric injury. We also showed that thse two agents when administered together produced damage in the intestine, although neither of these agents alone induced any damage (4). Thus, the present findings together with those data indicate that COX-2 as well as COX-1 play a "housekeeping" role in the gastrointestinal mucosa and that the adverse reaction of NSAIDs is not accounted for solely by inhibition of COX-1.

Of interest, we found in this study that intestinal damage was provoked after the combined administration of L-NAME and rofecoxib at the incidence of 100%, although the severity of damage was less than that observed after the combination of the selective COX-1 and COX-2 inhibitors. In addition, we observed the up-regulation of COX-2 expression in the intestinal mucosa after administration of L-NAME as well as the selective COX-1 inhibitor SC-560. We previously reported that the mucosal $PGE_2$ content of the small intestine was markedly decreased by SC-560, yet gradually recovered from 6 hr after the administration, in a rofecoxib-sensitive manner (4). A rapid up-regulation of COX-2 expression might represent a compensatory response to inhibition of PG or NO biosynthesis and contributes to maintenance of the mucosal integrity under such adverse conditions. Indeed, we have previously reported that the blockade of NO production by L-NAME decreased intestinal mucus secretion and resulted in enhancement of bacterial translocation and aggravation of NSAID-induced intestinal lesions, the events similar to those observed after administration of SC-560 (3,5).

In conclusion, the present study suggests that intestinal damage occurs by the inhibition of both COX-1 and COX-2 as well as both NO synthase and COX-2. Inhibition of COX-1 or NO synthase up-regulates the COX-2 expression, and this may counteract the deleterious events caused by endogenous NO or PG deficiency. Finally, the present findings strongly suggest a "housekeeping" role for COX-2 in the small intestine, especially under the adverse conditions.

## REFERENCES


1. **O'Neill GP and Ford-Hutchinson AW.** Expression of mRNA for cyclooxygenase-1 and cyclooxygenase-2 in human tissues. FEBS 1993; 330: 156-160.
2. **Robert A and Asano T.** Resistance of germ-free rats to indomethacin-induced intestinal inflammation. Prostaglandins 1977; 14: 333-341.
3. **Tanaka A, Kunikata T, Konaka A, Kato S and Takeuchi K.** Dual action of nitric oxide in pathogenesis of indomethacin-induced small intestinal ulceration in rats. J Physiol Pharmacol 1999; 50: 405-417.
4. **Tanaka A, Mizoguchi H, Araki H, Komoike Y and Takeuchi K**: Is intestinal toxicity of nonsteroidal anti-inflammatory drugs accounted for by cyclooxygenase-1 inhibition? Gastroenterology 118, A826 (2000)
5. **Tanaka A, Mizoguchi H, Kunikata T, Miyazawa T, Takeuchi K.** Protection by constitutively formed nitric oxide of intestinal damage induced by indomethacin in rats. J Physiol Paris, 2001; 95: 35-41,
6. **Wallace JL, McKnight W, Reuter BK and Vergnolle N.** NSAID-induced gastric damage in rats: Requirement for inhibition of both cyclooxygenase 1 and 2. Gastroeneterology 2000; 119: 706-714.

Chapter 17

# Role of Protein Kinases on Acid-Induced Duodenal Bicarbonate Secretion in Rats

OSAMU FURUKAWA, MASAHIKO HIROKAWA, PAUL H. GUTH, ELI ENGEL, JONATHAN D. KAUNITZ

*GLAVAHS & CURE / UCLA School of Medicine, Los Angeles, CA,*

## INTRODUCTION

Luminal acid is the most potent physiologic stimulus for duodenal bicarbonate secretion (DBS) (1). Indomethacin inhibits acid-induced DBS, consistent with the involvement of cyclooxygenase (COX) (2). Also, protein kinase A (PKA) is thought to be involved, since cAMP stimulates the cystic fibrosis transmembrane conductance regulator (CFTR), a candidate apical $HCO_3^-$ channel and the basolateral sodium bicarbonate cotransporter (NBC) (3, 4). DBS also cAMP dependent. Nevertheless, little is known about the involvement of other kinases, such as protein kinase C (PKC) and tyrosine kinase (TK), in acid-induced DBS. We thus tested the hypothesis that acid-induced DBS requires intracellular signaling pathways by testing the effects of protein kinase inhibitors, staurosporine for PKC, H-89 for PKA and genistein for TK on acid-induced DBS in rats.

## MATERIALS AND METHODS

Male Sprague-Dawley rats fasted for 18 hours were used in all experiments. Experiments were carried out under urethane-anesthetized condition (1.25 g/kg, i.p.). A 2 cm duodenal loop was prepared and perfused in order to measure duodenal $HCO_3^-$ secretion (2). The duodenal loop was perfused with $O_2$ gas bubbled and prewarmed saline using a peristaltic pump at 1 ml/min. For pH-stat measurement, 10 mM HCl was added to the reservoir to maintain pH at 7.0. Mucosal acidification was performed by perfusing the duodenal loop for 10 min with pH 2.2 HCl made isotonic with NaCl (2). Indomethacin (0.1 μM), H-89 (PKA inhibitor; 30 μM), staurosporine (PKC inhibitor; 0.1 μM) and genistein (TK inhibitor and CFTR activator; 1 μM) were added to acidified saline and then perfused, respectively. In some cases, genistein (1, 10 μM) and phorbol 12-myristate 13-acetate (PMA, 1 μM), were added to the luminal solution to examine their effects on basal DBS.

## RESULTS

After a 10 min luminal acidification, DBS gradually increased to a peak of 1.5~2 times basal values, occurring 30 min after acidification. Indomethacin almost completely inhibited acid-induced DBS. Intravenous injection of $PGE_2$ (0.1 mg/kg) reversed the inhibitory effects of indomethacin. Delta increases in DBS after $PGE_2$ injection for 30 min in control (non-treated group) and indomethacin treated group being $0.292 \pm 0.088$ and $0.293 \pm 0.050$, respectively. H-89 inhibited acid-induced DBS, and also inhibited $PGE_2$ induced DBS. Delta increases in DBS after 0.3 mg/kg $PGE_2$ injection for 30 min in control and H-89 group being 0.585 ±

0.057 and 0.203 ± 0.128, respectively. Between both groups, differences in delta increases were statistically significant. Staurosporine also almost completely inhibited acid-induced DBS, delta increase in DBS after 0.1 mg/kg $PGE_2$ administration for 30 min being 0.224 ± 0.065. Genistein, but not genstin, completely inhibited acid-induced DBS, which was reversed with $PGE_2$. Delta increase in DBS in genistein treated group being 0.333 ± 0.029. Addition of 0.1 μM of PMA in the luminal solution (reservoir solution) gradually increased basal DBS, 25 and 30 min after addition, the increases were statistically significant. However, in contrast with PMA, genistein did not stimulate basal DBS at 1 and 10 μM.

## DISCUSSION

It is well known that basal DBS is ruled by both humoral factor, such as $PGE_2$, VIP, glucagon, others and nerve system (1). Moreover, it is reported that acid-induced DBS is controlled by nerve system and indomethacin is also known to inhibit this response (2), suggesting the involvement of COX, cAMP and PKA pathway, little is known about the involvement of other kinases, such as PKC and tyrosine TK in acid-induced DBS. Intracellular calcium plays an important role in the intestinal epithelial transport (5,6). Hogan et. al. (5). reported that A23187, a calcium ionophore, increased rabbit DBS. They also reported this was not affected pretreatment with tetrodotoxin, indicating A23187 acted to the epithelial cells directly, not mediated by nerve system. They also mentioned verapamil, a Ca channel antagonist, inhibited VIP stimulated-DBS, but not cAMP- or $PGE_2$-stimulated DBS. Our results regarding PMA stimulates basal DBS and staurosporine inhibits acid-induced DBS suggest that not only on basal DBS but also on acid-induced response, $Ca^{2+}$ and PKC pathway are involved in this response. Since H-89, a specific PKA inhibitor, inhibited both acid-induced and $PGE_2$-stimulated DBS, it is obvious that COX and PKA pathway are involved in acid-induced response. However, staurosporine also showed the inhibitory effect on acid-induced response, it is very clear that PKC also plays an important role on acid-induced DBS. It is likely that other mediators except $PGE_2$, which induce $Ca^{2+}$ entry into the epithelial cells might be released in response to acid challenge.

Genistein is known to have several different properties. In general, this compound is a classical TK inhibitor, however, this shows stimulatory effect on CFTR, candidate apical $HCO_3^-$ channel (3). Thus, the action of genistein on DBS is greatly complicated. According our result as for the effect of genistein on basal DBS, it seems that its action on CFTR can be excluded in our experimental condition. The mechanism of inhibitory effect of genistein on acid-induced DBS is still unclear at present. However, Ruiz et. al. (7) demonstrated $CO_2$ gas can stimulate SRK, a non receptor type TK, leading to stimulate NBC activity which is thought the main molecule for uptake bicarbonate ion into the cell from extra cellular fluid in the kidney cells. In our experiment, it is very easy to conjecture the production of $CO_2$ into the duodenal epithelial cells, the same pathway is likely to come out into duodenal cells.

## CONCLUSION

As is the case with basal secretion, the $Ca^{2+}$-PKC pathway and cAMP-PKA pathway are involved in acid-induced DBS. Under basal conditions, PKA and PKC activate DBS. Following luminal acidification, however, activation of both PKA, PKC and TK are required for acid-induced DBS response.

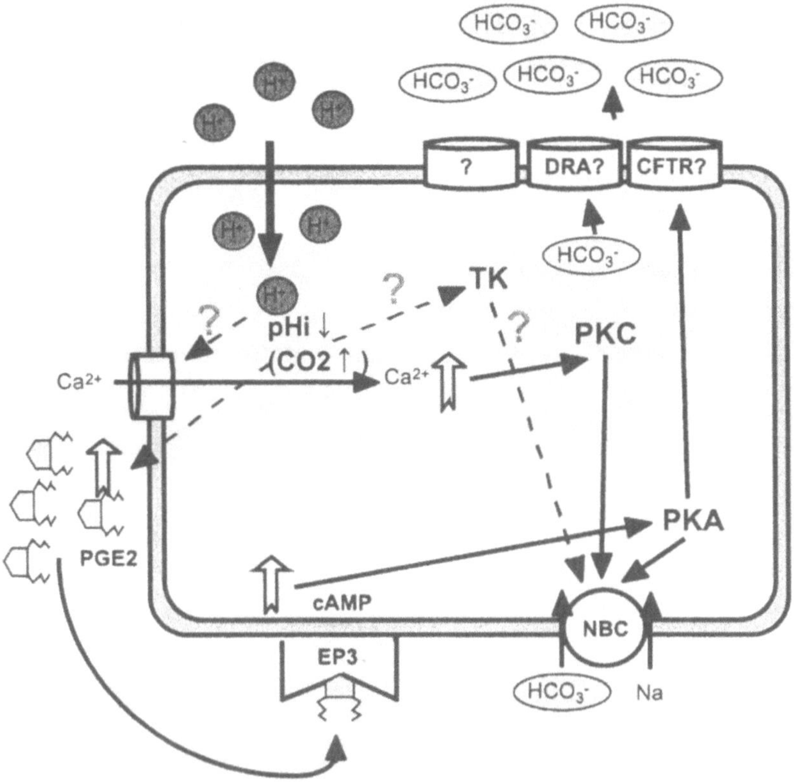


## REFERENCES


1 **Allen, A., G. Flemstruöm, A. Garner and E. Kivilaakso.** Gastroduodenal mucosal protection. *Physiol. Rev.* **73**: 823-857, 1993.

2 **Takeuchi,K., H. Tanaka, O. Furukawa and S. Okabe.** Gastroduodenal $HCO_3^-$ secretion in anesthetized rats: effects of 16,16-dimethyl $PGE_2$, topical acid, and acetazolamide. *Jpn. J .Pharmacol.* **41**: 87-99, 1986.

3 **Illek, B., J. R. Yankaskas and T. E Machen.** cAMP and genistein stimulate $HCO_3^-$ conductance through CFTR in human airway epithelia. *Am. J. Physiol.* **272**: L752-L761, 1997.

4 **Jacob P, S. Christiani, H. Rossmann, G. Lamprecht, D. Vieillard-Baron, R. Muller, M. Gregor and U. Seidler.** Role of Na(+)HCO(3)(-) cotransporter NBC1, Na(+)/H(+) exchanger NHE1, and carbonic anhydrase in rabbit duodenal bicarbonate secretion. *Gastroenterology* **119**: 406-419, 2000.

5 **Hogan, D. L., Yao, B. and Isenberg, J. I.** Modulation of bicarbonate secretion in rabbit duodenum. *Dig. Dis. Sci.* **43**: 120-125, 1998.

6 **Odes, H. S., R. Reimer, R. Muallem, M. Schwenk, W. Beil,. and K. -F Sewing.** Role of protein kinases C in duodenal mucosal bicarbonate secretion in the guinea pig. *Pharmacology* **53**: 60-65, 1996.
7 **Ruiz, O. S., R. B. Robey, Y. Y Qiu., L. J. Wang, C. J. Li, J. Ma and J. A. Arruda.** Regulation of the renal Na-HCO3 cotransporter. XI. Signal transduction underlying $CO_2$ stimulation. *Am. J. Physiol* **277**: F580-F586, 1999.

Chapter 18

# Acid Induces Duodenal Epithelial Acidification and $HCO_3^-$ Secretion via Activation of the $Na^+$:$HCO_3^-$ Cotransporter and CFTR in Mice

MASAHIKO HIROKAWA, OSAMU FURUKAWA, PAUL H. GUTH, ELI ENGEL, SHAOYOU CHU, MARSHALL H. MONTROSE, AND JONATHAN D. KAUNITZ

*VAGLAHS & CURE/UCLA School of Medicine, Los Angeles, CA, 90073, USA*

## INTRODUCTION

Rapid variation of duodenal luminal pH between pH 2 and 7 in humans due to the opposing effects of pancreatic alkaline and gastric acid secretions exposes the duodenal mucosa to profound pH stress [1]. With the measurement of robust epithelial $HCO_3^-$ secretion and neutral pH in the juxtamucosal mucus gel, despite the presence of luminal acid, the 'bicarbonate hypothesis' was developed, wherein $HCO_3^-$, secreted by the epithelial cells, completely neutralizes luminal acid in-diffusing through the mucus gel [2]. Duodenal $HCO_3^-$ secretion (DBS) is thought to be the single most important duodenal defense mechanism. We hypothesized that cellular base uptake precedes DBS and that DBS requires CFTR activation. The primary aim of this study is to characterize a mouse model of intracellular pH ($pH_i$) measurement and DBS that could then be used in mice in which key proteins related to $HCO_3^-$ transport are deleted. To achieve this goal, we used techniques that had been used successfully top measure $pH_i$ and DBS in the rat and adapted them for use in a mouse model. We have devised an *in vivo* microscopy system in mice for measurement of the fluorescence ratio (495nm/450nm), and we tested the role of the $Na^+$:$HCO_3^-$ cotransporter (NBC) and CFTR in DBS using acid perfusion.

## MATERIALS AND METHODS

All studies were performed on male C57Bl/6 mice fasted for 18 hours before the experiments. After urethane (1.25g/kg) anesthesia, the abdomen was opened. **$pH_i$ studies**: An intestinal loop between pylorus and duodenum was made after the pancreaticobiliary duct was ligated. The anterior wall of the duodenum was incised to expose the mucosa. A perfusion chamber, fashioned out of a bacteriological loop, was placed over the exposed mucosa, which was then incubated with 2',7'-bis(2-carboxyethyl)-5(6)-carboxyfluorescein (BCECF-AM) for 15min to load the duodenal epithelial cells with a pH sensitive fluorescent dye. Duodenal mucosa was superfused with pH 7.0 or 2.2 buffers via PE50 tubing and a Harvard infusion pump. The fluorescence ratio (495nm/450nm) was measured at baseline with pH 7.0 perfusion, and during and after a 10 min pH 2.2 acid exposure. For the confocal images, performed by S.

Chu and M. Montrose at the University of Indiana, mice were placed prone on the microscope stage with a portion of the duodenal mucosa protruding down to a chamber kept at 37°C (Figure 1). Images were obtained as a Z-stack with illumination by the 488 nm Ar laser. **DBS studies**: The duodenal loop was perfused with 100% $O_2$ gassed saline heated to 37°C through two polyethylene catheters. DBS was measured by pH-stat and back-titration with duodenal perfusion with or without the $Na^+$:$HCO_3^-$ inhibitor DIDS or the CFTR inhibitor NPPB. DBS was measured at baseline with pH 7.0 perfusion, and after 10min pH 2.2 acid or $PGE_2$ (0.1mg/ml) challenges. In some experiments, indomethacin (5mg/kg) was administered subcutaneously 1 hour before the experiment.

## RESULTS

We established an *in vivo* microscopy system for measurement of the fluorescence ratio in mice. Examination of mucosa loaded with BCECF *in vivo* with confocal microscopy revealed that the fluorescence was limited to the epithelial cells. pH 2.2 perfusion for 10 min rapidly reduced the fluorescence ratio (495nm/450nm). After acid removal, the fluorescence ratio increased to the baseline 10 min after perfusate pH was returned to 7.0. After 10min luminal acid challenge (pH2.2), DBS gradually increased to the peak of 1.5-2 times basal value 40min after acidification (pH 2.2) of the duodenal loop. The $Na^+$:$HCO_3^-$ inhibitor DIDS (0.5mM) and the CFTR inhibitor NPPB (0.1mM) inhibited acid-induced DBS increase. Pretreatment with indomethacin also inhibited acid-induced DBS increase. $PGE_2$ (0.1mg/ml) luminally applied to the duodenal loop rapidly enhanced DBS in perfused mouse duodenum.

## DISCUSSION

Using a technique modeled on our previous studies of duodenal surface cell $pH_i$ in rat, we loaded the dye BCECF into duodenal epithelium *in vivo* in mice. The presence of intercellular dye enabled the fluorescence ratio (495nm/450nm) to be measured *in vivo* with a fluorometric technique. The fluorescence ratio is related to $pH_i$, therefore luminal pH change leads to a change of $pH_i$ and corresponding fluorescence ratio. The confocal images confirmed intracellular dye localization with BCECF in epithelial cells, we established a novel *in vivo* microscopy system for measurement of the fluorescence ratio in mice.

Bicarbonate is secreted in response to a number of agonists that increase intracellular adenosine 3', 5'-cyclic monophosphate (cAMP) and calcium concentration including vasoactive intestinal polypeptide (VIP), carbachol and enteric nerve activation [3)4)]. Our results are consistent with prior studies demonstrating that $PGE_2$ stimulates DBS in mice [5)6)]. CFTR is an apical membrane channel for $Cl^-$, regulated by cAMP-dependent protein kinase A. We demonstrated that the $Na^+$:$HCO_3^-$ inhibitor DIDS and the CFTR inhibitor NPPB inhibited acid-induced DBS increase. Basal DBS and acid- and $PGE_2$-stimulated DBS increase were diminished in CFTR knockout mice [6)], consistent with a major functional role of CFTR in mediating DBS. We previously showed that the $Na^+$:$HCO_3^-$ inhibitor DIDS abolished overshoot over baseline after acid exposure in rats [7)], and that DIDS inhibited acid-induced

DBS increase in mice. We propose that intercellular $HCO_3^-$ uptake via a DIDS-sensitive basolateral $Na^+$:$HCO_3^-$ cotransporter, which temporally precedes $HCO_3^-$ secretion in response to rapid shifts of luminal pH, is an early acute duodenal response to luminal acid. We intend to repeat these studies in mice in which key $HCO_3^-$ transport genes have been deleted.

## CONCLUSION

We have established a novel *in vivo* microscopy system for measurement of $pH_i$ and DBS in mice. DBS is in large part a result of acid/base transport mediated by the $Na^+$:$HCO3^-$ cotransporter and the CFTR.

## REFERENCES


1. **Ovesen,L., Bendtsen F., Tage-Jensen U., Pedersen NT., Gram BR., and Rune SJ.** Intraluminal pH in the stomach, duodenum, and proximal jejunum in normal subjects and patients with exocrine pancreatic insufficiency. Gastroenterology: 90958-962, 1986
2. **Garner A., Flemström G., Allen A., Heylings JR., and McQueen S.** Gastric mucosal protective mechanism: roles of epithelial bicarbonate and mucus secretion. Scand J Gastroenterol Supple 101:79-86, 1984
3. **Hogan DL., Yao BG., Steinbach JH., Isenberg JI.** The enteric nervous system modulates mammalian duodenal mucosal bicarbonate secretion. Gastroenterology 105: 410-417, 1993
4. **Yao BG., Hogan DL., Bukhave K., Koss MA., Isenberg JI.** Bicarbonate transport by rabbit duodenum in vitro: effect of vasoactive intestinal peptide, prostaglandin $E_2$ ,and cyclic adenosine monophosphate. Gastroenterology 104: 732-740, 1993
5. **Takeuchi K., Ukawa H., Kato S., Furukawa O., Araki H., Sugimoto Y., Ichikawa A., Ushikubi F., and Narumiya S.** Impaired duodenal bicarbonate secretion and mucosal integrity in mice lacking prostaglandin E-receptor subtype $EP_3$ .Gastroenterology 117: 1128-1135, 1999
6. **Hogan DL., Crombie DL., Isenberg JL., Svendsen P., Schaffalitzky de Muckadell OB., and Ainsworth MA.** Acid-stimulated duodenal bicarbonate secretion involves a CFTR-mediated transport pathway in mice. Gastroenterology 113: 533-541, 1997
7. **Akiba Y., Furukawa,O., Guth PH., Engel E., Nastaskin I., and Kaunitz JD.** Acute adaptive cellular base uptake in rat duodenal epithelium. Am J Physiol. 280: G1083-1092, 2001

Chapter 19

# Insights Into the Cell Biology, Development and Pathology of the Gastric Mucosa Revealed in Gastric H/K ATPase and Gastrin-Deficient Mice

IAN R. VAN DRIEL[1], TEO V. FRANIC[1], LOUISE M. JUDD[2],
SIMON P. BARRETT, KATRINA L. SCARFF[3],
LINDA C. SAMUELSON[4] AND PAUL A. GLEESON[1]

*Department of Pathology and Immunology, Monash University Medical School,*
*Alfred Hospital, Melbourne, Victoria, AUSTRALIA 3181*
*1 Current address: The Russell Grimwade School of Biochemistry and Molecular Biology,*
*The University of Melbourne, Melbourne, Vic 3010, AUSTRALIA*
*2 Current address: Department of Molecular Genetics, Biochemistry and Microbiology,*
*University of Cincinnati, 231 Bethesda Ave, Cincinnati, OHIO 45267, USA*
*3 Current address: Department of Biochemistry and Molecular Biology,*
*Monash University, Clayton, Melbourne, Victoria, AUSTRALIA 3800*
*4 Department of Physiology, The University of Michigan,*
*Ann Arbor, MI 48109-0622, USA*

## INTRODUCTION

The gastric $H^+/K^+$-ATPase, present in gastric parietal (oxyntic) cells, exchanges luminal $K^+$ for cytoplasmic $H^+$ and is the enzyme principally responsible for gastric luminal acidification (6, 25, 32). The gastric $H^+/K^+$-ATPase is composed of two non-covalently linked subunits, an α (H/Kα) and a β (H/Kβ), with apparent molecular weights of 95 kDa and 60-90 kDa, respectively (1, 25). H/Kα (encoded by the mouse gene *Atp4a*) contains the catalytic sites of the enzyme and has 10 transmembrane domains. The highly glycosylated H/Kβ_(2, 31)(encoded by the mouse gene *Atp4b*) has one transmembrane domain (30), is required for the transport of an active enzyme complex from the ER to apical membranes (9, 11), and for gastric acid secretion (26). More detail on the structure and function of the gastric H/K ATPase subunits is supplied elsewhere in this volume.

Gastric parietal cells contain an extensive network of secretory membranes. These membranes contain the gastric $H^+/K^+$-ATPase protein and are the site of gastric acid production (6, 24, 28). In the resting state the membranes appear as a network of cytoplasmic helically coiled tubules often termed tubulovesicles (6, 22). Upon stimulation with secretagogues the membranes are transformed into a secretory canaliculus, which is an expansion of the apical membrane and composed of long microvilli. These actively secreting membranes are associated with the actin cortical cytoskeleton probably via associations mediated by a member of the ERM family of proteins, ezrin (5, 10).

Recently, mice in which the genes encoding the gastric $H^+/K^+$ ATPase subunits were mutated have been produced (26, 29). Analysis of these mice has allowed delineation of the role that these molecules play in acid secretion, parietal cell membrane biogenesis, and development of the gastric mucosa. Here we review work on the H/Kβ-deficient mice and the role that the constitutively elevated levels of gastrin found in the H/Kβ-deficient mice play in determining phenotype. We also discuss the H/Kα-deficient mice highlighting differences between the mutants.

## PRODUCTION OF H/Kβ-DEFICIENT MICE AND GROSS PHENOTYPE

We produced mice in which exon 1 of the Atp4b gene was interrupted after the third codon so a functional protein could not be produced (26). The mutation was designated the *Atp4b*$^{kls}$ allele. PCR analysis of cDNA, immunoblotting experiments and immunohistology clearly demonstrated that no H/Kβ mRNA or protein was produced in mice homozygous for the *Atp4b*$^{kls}$ allele.

The $H^+/K^+$-ATPase-deficient mice were born at the expected Mendelian ratio, thrived normally and were fertile. As noted elsewhere in this volume the gastric H/K ATPase is expressed in a number of organs particularly the kidney. We have not observed any symptoms that would suggest a deficit in kidney function although more detailed analysis of ion transport in this organ is currently underway and may reveal more subtle deficits in H and K homeostasis.

## H/Kβ-DEFICIENT MICE ARE ACHLORHYDRIC AND HYPERGASTRINEMIC

The stomachs of all H/Kβ-deficient mice examined had a neutral pH compared to the acidic pH observed in littermates expressing normal levels of H/Kβ. This data confirmed the functional requirement of the β subunit in parietal cells *in vivo*. A number of studies involving expression of the subunits of the $H^+/K^+$-ATPase in heterologous cell lines had demonstrated that while the catalytic α subunit required association with a β subunit to attain enzymatic activity and to be transported to the cell surface (9, 11) the β subunit of the $Na^+/K^+$-ATPase (Na/Kβ) could substitute for H/Kβ (11, 13). Gastric parietal cells express large amounts of $Na^+/K^+$-ATPase, hence, it was possible that in the absence of the H/Kβ an Na/Kβ subunit would be able to substitute and maintain acid secretory activity. However, this does not appear to be the case. The reason why the H/Kα did not pair with the Na/Kβ (or why Na/Kα does not pair with H/Kβ) in parietal cells is not clear. In normal parietal cells the huge abundance of the H/Kβ may conceivably out compete the Na/Kβ, but obviously is not a possibility in H/Kβ-deficient mice.

Gastrin is the major hormone controlling gastric acid secretory levels and is released from gastric G cells in response to elevated gastric luminal pH. Therefore, it was not surprising that serum gastrin levels in the H/Kβ-deficient mice were elevated relative to normal mice. We did not observe a correlation between gastrin levels and the age of the mice indicating that prolonged achlorhydria did not exacerbate the hypergastrinemia.

## PARIETAL CELL ONTOGENY IN H/Kβ-DEFICIENT MICE

Correct expression of the H/Kβ is required for normal development of parietal cells and normal biogenesis of their secretory membranes (26). There was a delay in the normal morphological development of the mucosa and a delay in the appearance of parietal cells in H/Kβ-deficient mice (Fig. 1). There was a clear deficit in the number of parietal cells in the mucosa of 15 and 17 days day old mice. EM observation of mucosa from 17 day-old mice revealed mitochondria-rich cells with an abnormal canaliculus which were the immature parietal cells. At older ages the number of $H^+/K^+$-ATPase α subunit expressing cells in H/Kβ-deficient mice was similar to littermates (7) (see below).

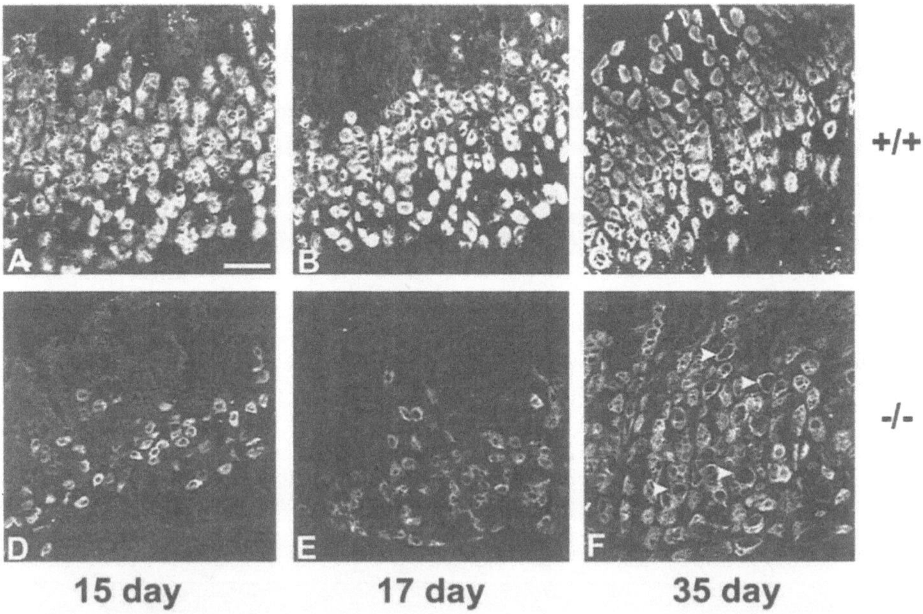


**Figure 1. Reactivity of H/Kα–specific antiserum to frozen stomach sections from H/Kβ–deficient mice.** Sections of frozen stomach from (A and D) 15-day-old, (B and E) 17-day-old, and (C and F) 35-day-old wild-type (A–C) and H/Kβ–deficient (D–F) mice were incubated with a polyclonal rabbit serum specific for H/Kα subunit followed by anti–rabbit immunoglobulin–FITC and examined by using confocal microscopy. Arrowheads indicate some cells that contain dilated canaliculi. At 15 and 17 days the number of parietal cells is significantly reduced in H/Kβ-deficient mice. At 35 days the number of parietal cells in mice of both genotypes is similar (Bar in A=50 μm and applies to all panels.)

## THE GASTRIC H/K ATPase IS REQUIRED FOR THE FORMATION OF TUBULAR SECRETORY MEMBRANES OF PARIETAL CELLS

Parietal cells in H/Kβ-deficient mice have an abnormal membrane structure (Fig 2) (7, 26). The tubular membranes corresponding to the resting state of the membranes were absent. Instead, an abnormal canaliculus was present. These compartments were however, significantly different in morphology to the secretory canaliculus of normal parietal cells. The structures were far more open and the microvilli shorter. The size of these membranous compartments in the parietal cells appears to increase with age. In the parietal cells of older animals these structures were often over 25 μm in diameter. These dilated canaliculi appeared to be totally intracellular.

As we did not observe the typical tubulovesicles in any cell found in the H/Kβ-deficient mice at any age it is thus possible that the $H^+/K^+$-ATPase is required for the formation of the tubulovesicular membranes. This possibility is also raised by the work of Courtois-Coutry *et al.* (4) that suggested H/Kβ contains a signal required for endocytosis and retrieval of the $H^+/K^+$-ATPase from the active canaliculi membranes to the tubulovesicle compartment. If the $H^+/K^+$-ATPase contains the predominant internalisation signal for the parietal cell secretory membrane then, in the absence of the β subunit, the secretory membrane may be

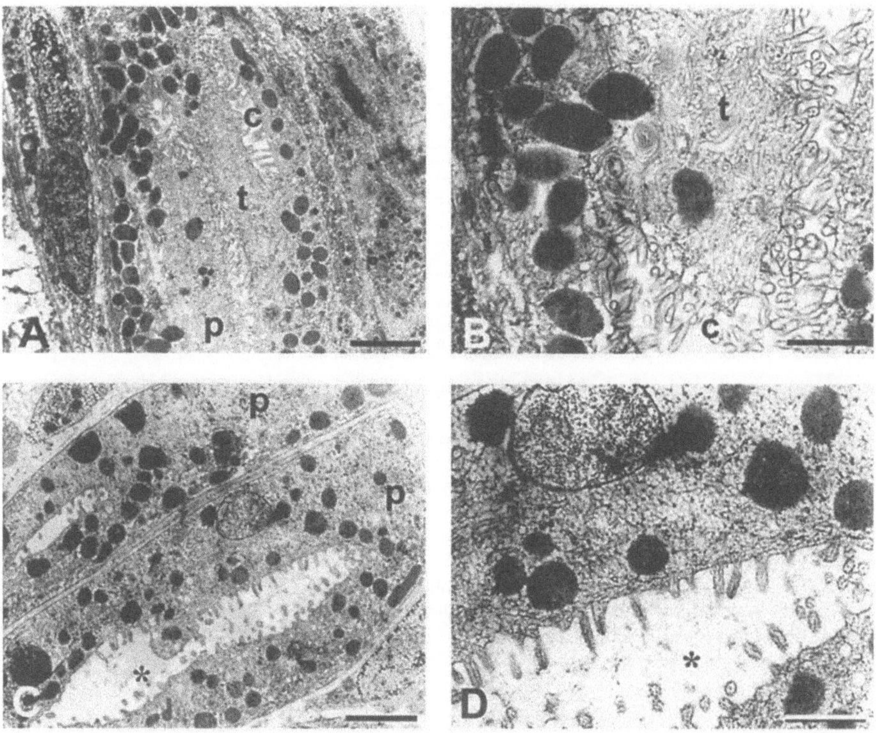


**Figure 2. Ultrastructure of parietal cells of 35-day-old H/Kβ–deficient mice.** Gastric mucosae of (A and B) wild-type and (C and D) H/Kβ–deficient mice were fixed by rapid freezing and cryosubstituted before imaging by transmission electron microscopy. In parietal cells (p) from wild-type animals, the tubular membranes (t) and secretory canaliculus (c) were present. In the parietal cells of H/Kβ–deficient mice an abnormal dilated canaliculus was present and tubulovesicular membranes were absent ( Bars in A and C=2.5 μm; bars in B and D=1 μm.)

locked in the apical canalicular form unable to be endocytosed. During ontogeny the secretory canaliculus appears prior to the tubular network (23). This also argues that the tubular membranes are derived from the apical canaliculus and does not form *de novo*.

One caveat to this interpretation was that the high levels of gastrin in the H/Kβ-deficient mice meant that the parietal cells were constitutively receiving a stimulatory signal that resulted in the conversion of all of the tubular membrane to the active canalicular form. To rule out this possibility we produced mice that lacked both functional *Atp4b* and *Gast* genes (H/Kβ/gastrin-deficient mice) (7). Mice deficient in gastrin are achlorhydric and furthermore parietal cells in these mice appear refractory to stimulation via cholinergic and histamine receptors (8). The parietal cell membranes of the H/Kβ/gastrin-deficient mice were very similar to that of the H/Kβ-deficient mice (7). Typical tubular membranes were absent and abnormal dilated canaliculi were apparent. This data indicates that the absence of the tubular membranes is not due to hyperstimulation of the parietal cells.

As discussed below the parietal cells of mice deficient in H/Kα have been reported to be very similar to those of the H/Kβ-deficient mice, in particular tubulovesicles were absent (29). This raises the possibility that both subunits and an intact $H^+/K^+$ ATPase heterodimer are required for tubular membrane formation. Although in heterologous expression systems a proportion of H/Kβ monomer can be transported to the cell surface it is not clear where the "orphaned" H/Kβ subunit is located in the parietal cells of H/Kα-deficient mice and it is possible that the putative endocytosis signal of the H/Kβ does not function as efficiently in a monomeric form.

## GASTRIC MUCOSAL DEVELOPMENT IN H/Kβ-DEFICIENT MICE AND THE ROLE OF HYPERGATRINEMIA

The majority of the gastric mucosa is a glandular epithelium. The glands or units are composed of three major differentiated cell types (12). Pit (surface mucous) cells occur in the upper pit zone of the unit and secrete protective mucus. Acid-secreting parietal cells are found in the middle and lower regions of the units. Zymogenic (chief) cells secrete pepsin, and in some species intrinsic factor, predominate towards the base of units. The cells of the gastric units are in continuous turnover. Mature epithelial cells develop from stem cells via a series of well-defined cellular intermediates (16). The cells of the gastric mucosa are in a continuous state of turnover. New cells are generated by the division of immature cell populations that include pluripotent stem cells and partially differentiated, committed precursors of pit, parietal and zymogenic cell lineages (15).

The gastric mucosa of H/Kβ deficient mice had significant abnormalities compared to wild type animals (7, 26). The mucosae are hypertrophic compared to normal animals from an early age (Fig 3.) We found a 2-fold increase in the total number of gastric mucosal cells and a significant perturbation in the proportions of epithelial cells found in the gastric units. In particular, we found that there was an accumulation of immature epithelial cell precursors, a depletion of zymogenic cells, an increase in apoptotic and dividing cells. The number of immature cells in the gastric mucosa of the H/Kβ-deficient mice was ~3-fold higher than in the units of wildtype animals. The increase in cell number was due to an increase in the rate of cell division in the mucosae of H/Kβ-deficient animals. Furthermore, the proliferating cells were no longer limited to the isthmal region of the gastric units as in normal gastric units but were also found distributed towards the base of the units. The increased production of immature cells and proliferation appeared to be largely the result of elevated levels of gastrin found in H/Kβ-deficient mice as H/Kβ/gastrin-deficient mice did not display these phenotypes (Fig 3.) (7). Gastrin is a hormone that does not appear to be essential for normal proliferation and development of the gastric mucosal lineages but in elevated quantities is able to stimulate hyperproliferation (8, 33).

The rate of production of gastric epithelial cells and their development into end-stage cells must be regulated in order to maintain cellular homeostasis. The total number of parietal cells and pit cells in *adult* H/Kβ-deficient gastric units was only slightly greater than normal (7). This is despite the enormously elevated number of potential precursors (immature cells). These data suggests the existence of a homeostatic mechanism that limits the number of immature cells that can progress to more mature cells, and that immature cells accumulate if production of immature cells exceeds this limit. In other words, production of end-stage cells may be determined not only by the number of immature cells present but also other factors that permit this further development. These factors could include the availability of growth factors or appropriate "niches" that can be occupied by mature cell types. This suggestion is supported by previous studies. Wang *et al.* (34) and Konda *et al.* (18) produced transgenic mice that overproduced gastrin. In both studies the gastric mucosa was hypertrophic and there appeared to be an overabundance of immature cells. This was substantiated in the latter work with a quantitative analysis that demonstrated that the "proliferative zone" of the gastric units was increased whereas the number of parietal cells was not. Mice with autoimmune gastritis also have elevated gastrin levels and mucosal hypertrophy due to an abundance of immature cells yet the number of end-stage cells is reduced (14).

We observed a ~35-fold reduction in the number of zymogenic cells in the gastric mucosa of the H/Kβ-deficient mice (7). A relationship between parietal and zymogenic cell lineages was first noted in mice that lack parietal cells due to genetic ablation (3, 20, 21). These mice also lack zymogenic cells, as do mice deficient in $Na^+/H^+$ exchanger isoform 2 (27), under autoimmune attack (14), treated chronically with anti-secretagogues (17) or infected with *Helicobacter* sp. (19). Interestingly, the H/Kα-deficient mice do not have a deficiency in zymogenic cells (29). Examination of the phenotype of these various situations does not reveal an obvious common trait that would explain the absence of zymogenic cells. Achlorhydria is not the cause as gastrin-deficient and H/Kα-deficient mice that lack stomach acid have normal numbers of zymogenic cells. The absence of parietal cells *per se* also does not appear to be a requirement as animals with normal numbers of parietal cells (H/Kβ and H/Kβ/gastrin-deficient, omeprazole-treated) are deficient in zymogenic cells. On the other hand, the parietal cells in these mice are structurally abnormal and perhaps unable to produce a factor that normally nurtures zymogenic cells. Many of these mice have highly elevated gastrin levels, but even those that do not (H/Kβ/gastrin-deficient mice) are zymogenic cell-deficient and further, some mice that are constitutively hypergastrinemic (gastrin-transgenic mice, H/Kα-deficient mice) have normal numbers of zymogenic cells. Many of these mice have abnormally high numbers of immature gastric epithelial cells that may be supplying a signal that inhibits zymogenic cell development (20). However, not all of them do (H/Kβ/gastrin-deficient). In short, the explanation for the failure of genesis of zymogenic cells does not appear to be as simple as the surfeit or deficit in another cell type.

## H/Kα-DEFICIENT MICE

Mice deficient in the H/Kα were produced by Gary Shull's group (29). These mice have a similar phenotype to the H/Kβ deficient mice although some differences were apparent. It should be noted, however, that H/Kβ and the H/Kα mice were analysed at different ages (35 days and 8-10 weeks, respectively). It is probable that this difference will influence the phenotype. Certainly, we have noted exacerbations of some abnormalities in the H/Kβ-deficient mice with increasing age.

The H/Kα-deficient mice like the H/Kβ-deficient mice appeared to be perfectly healthy and fertile and were born at the expected frequency indicating no role in normal development of the animal. The gastric mucosa of the H/Kα-deficient mice was hypertrophic but unlike the H/Kβ-deficient mice there was close to normal levels of zymogenic cells. Metaplastic change which included cystically dilated regions and ciliated cells were more obvious in the H/Kα-deficient mice. It is worth noting that with increasing age it is likely that gut flora normally kept at bay by gastric acid will be able to colonise the achlorhydric stomachs of H/K ATPase-deficient mice which would probably drastically affect the mucosa.

Importantly, the parietal cells of H/Kα mice were devoid of tubulovesicular membranes and contained a dilated canaliculus. The canaliculi were observed to open to the gastric lumen in the H/Kα-deficient mice whereas, as discussed above the, this was not found in the H/Kβ mice. Spicer et al. speculate that this could indicate that the H/Kβ is required for insertion of secretory membranes into the apical membranes. As the opening of the canaliculus to the lumen can be small compared to the diameter of parietal cells observation by electron microscopy of thin sections can be problematic in establishing lumenal connectivity. We feel it would be wise to experimentally address the issue of connectivity between the canaliculus and lumen in the H/K ATPase-deficient mice using tracers.

## PERSPECTIVES

Future work with mice deficient in the gastric H/K ATPase subunits will focus on defining the molecular mechanisms that underpin the various phenotypes observed. Of particular interest is the lack of tubulovesicular membranes in these mice and the associated structural cahnges in the apical canaliculus. It would seems likely that an absence of these molecules results in an alterations in the association of the apical membrane with various cytoskeletal components and other proteins such as elements of the endocytotic machinery.

We are also interested in the relationship between the membrane structure of the parietal cell membrane and the activity of the K and Cl ion channels that control acid secretory activity. (The activity H/K ATPase is not directly controlled but is rather a slave to regulated channels). While a functional link between parietal cell membrane transformations in co localising the K and Cl conductances and the H/K ATPase has been postulated there is as yet no experimental demonstration that this is a requirement for the control of acid secretory activity. Manipulation of the H/K ATPase genes has allowed us to in essence freeze the parietal cell membranes in the canalicular form. Thus, we will be able to examine the effect that this has on K and Cl conductance.

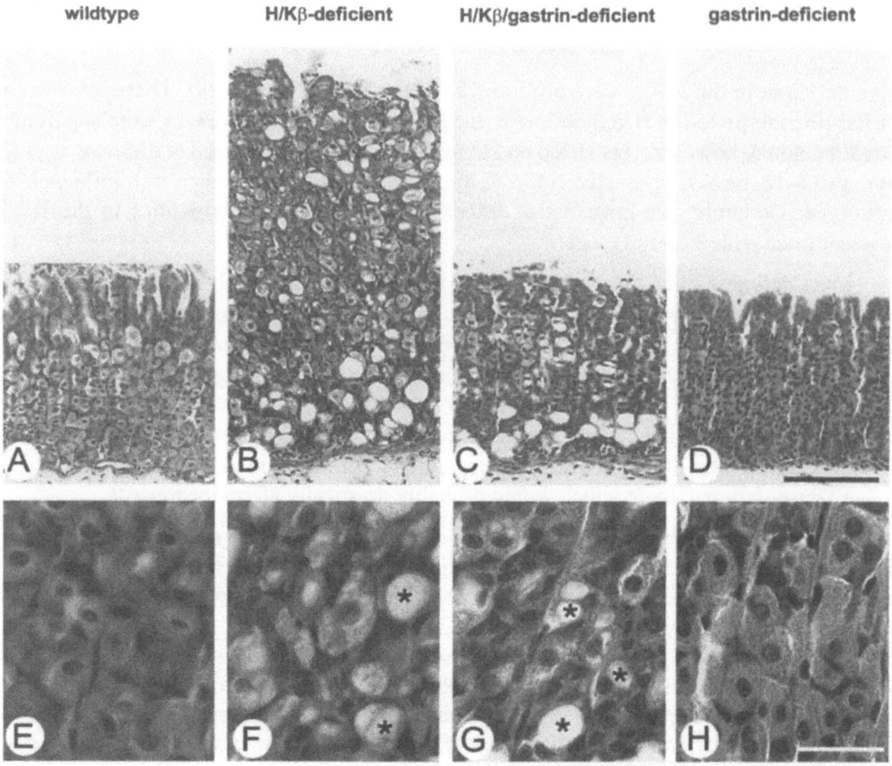


**Figure 3. The gastric mucosae of 35-day-old mice of varying genotypes.** Stomachs of mice of the genotypes indicated were removed fixed, sectioned, stained with haematoxylin and eosin and examined by light microscopy. The lumen of the stomach is at the top of each panel. Asterisks in F and G indicate dilated canaliculi in parietal cells. Bar in D applies to panels A-*D*, bar = 100 µm. Bar in H applies to panels E-H, bar = 25 µm.

## REFERENCES

1. **Besancon, M., J. Shin, F. Mercier, K. Munson, M. Miller, S. Hersey, and G. Sachs.** Membrane topology and omeprazole labeling of the gastric $H^+,K^+$-adenosine triphosphatase. *Biochemistry* 32: 2345-55, 1993.

2. **Callaghan, J. M., B. H. Toh, J. M. Pettitt, D. C. Humphris, and P. A. Gleeson.** Poly-N-acetyllactosamine-specifc tomato lectin interacts with gastric parietal cells: Identification of a tomato-lectin binding 60-90x10$^3$ Mr membrane glycoprotein of tubulovesicles. *J. Cell Sci.* 95: 563-576, 1990.

3. **Canfield, V., A. B. West, J. R. Goldenring, and R. Levenson.** Genetic ablation of parietal cells in transgenic mice: A new model for analyzing cell lineage relationships in the gastric mucosa. *Proc.Natl.Acad.Sci.USA* 93: 2431-2435, 1996.

4. **Courtois-Coutry, N., D. Roush, V. Rajendran, J. McCarthy, J. Geibel, M. Kashgarian, and M. Caplan.** A tyrosine-based signal targets H/K-ATPase to a regulated compartment and is required for the cessation of gastric acid secretion. *Cell* 90: 501-10, 1997.

5. **Forte, J. G., B. Ly, Q. Rong, S. Ogihara, M. Ramilo, B. Agnew, and X. Yao.** State of actin in gastric parietal cells. *Am. J. Physiology* 274: C97-104, 1998.

6. **Forte, J. G., and X. B. Yao.** The membrane-recruitment-and-recycling hypothesis of gastric HCl secretion. *Trends Cell Biol.* 6: 45-48, 1996.

7. **Franic, T. V., L. M. Judd, D. Robinson, S. P. Barrett, K. L. Scarff, P. A. Gleeson, L. C. Samuelson, and I. R. van Driel.** Regulation of gastric epithelial cell development revealed in $H^+/K^+$ ATPase β subunit- and gastrin-deficient mice. *American Journal of Physiology (Gastrointestinal and Liver)* 281: G1502-G1511, 2001.

8. **Friis-Hansen, L., F. Sundler, Y. Li, P. J. Gillespie, T. L. Saunders, J. K. Greenson, C. Owyang, J. F. Rehfeld, and L. C. Samuelson.** Impaired gastric acid secretion in gastrin-deficient mice. *Am. J. Physiol.* 274: G561-8, 1998.

9. **Gottardi, C. J., and M. J. Caplan.** An Ion-Transporting ATPase Encodes Multiple Apical Localization Signals. *J.Cell Biol.* 121: 283-293, 1993.

10. **Hanzel, D., H. Reggio, A. Bretscher, J. Forte, and P. Mangeat.** The secretion-stimulated 80K phosphoprotein of parietal cells is ezrin, and has properties of a membrane cytoskeletal linker in the induced apical microvilli. *EMBO J* 10: 2363-2373, 1991.

11. **Horisberger, J.-D., P. Jaunin, M. A. Reuben, L. S. Lasater, D. C. Chow, J. G. Forte, G. Sachs, B. C. Rossier, and K. Geering.** The H,K-ATPase β subunit can act as a surrogate for the α subunit of Na,K pumps. *J.Biol.Chem.* 266: 19131-19134, 1991.

12. **Ito, S.** *Functional Gastric Morphology.* New York: Raven Press, 1987.

13. **Jaunin, P., F. Jaisser, A. T. Beggah, K. Takeyasu, P. Mangeat, B. C. Rossier, J. D. Horisberger, and K. Geering.** Role of the Transmembrane and Extracytoplasmic Domain of β-Subunits in Subunit Assembly, Intracellular Transport, and Functional Expression of Na,K-Pumps. *J.Cell Biol.* 123: 1751-1759, 1993.

14. **Judd, L. M., P. A. Gleeson, B. H. Toh, and I. R. van Driel.** Autoimmune gastritis results in disruption of gastric epithelial cell development. *Am. J. Physiol. Gastointest. Liver Physiol.* 277: G209-G218, 1999.

15. **Karam, S., and C. Leblond.** Dynamics of epithelial cells in the corpus of the mouse stomach. I. Identification of proliferative cell types and pinpointing of the stem cell. *Anatomical Record* 236: 259-79, 1993.

16. **Karam, S. M.** New insights into the stem cells and the precursors of the gastric epithelium. *Nutrition.* 11: 607-613, 1995.

17. **Karam, S. M., and J. G. Forte.** Inhibiting gastric H(+)-K(+)-ATPase activity by omeprazole promotes degeneration and production of parietal cells. *Am J Physiol* 266: G745-58., 1994.

18. **Konda, Y., H. Kamimura, H. Yokota, N. Hayashi, K. Sugano, and T. Takeuchi.** Gastrin stimulates the growth of gastric pit with less-differentiated features. *Am. J. Physiol.* 277: G773-84, 1999.

19. **Lee, A., J. O'Rourke, M. C. De Ungria, B. Robertson, G. Daskalopoulos, and M. F. Dixon.** A standardized mouse model of Helicobacter pylori infection: introducing the Sydney strain. *Gastroenterology* 112: 1386-97., 1997.

20. **Li, Q., S. Karam, and J. Gordon.** Simian virus 40 T antigen-induced amplification of pre-parietal cells in transgenic mice. Effects on other gastric epithelial cell lineages and evidence for a p53-independent apoptotic mechanism that operates in a committed progenitor. *Journal of Biological Chemistry* 270: 15777-88, 1995.

21. **Li, Q. T., S. M. Karam, and J. I. Gordon.** Diphtheria toxin-mediated ablation of parietal cells in the stomach of transgenic mice. *J.Biol.Chem.* 271: 3671-3676, 1996.

22. **Pettitt, J. M., D. C. Humphris, S. P. Barrett, B. H. Toh, I. R. van Driel, and P. A. Gleeson.** Fast freeze-fixation/freeze-substitution reveals the secretory membranes of the gastric parietal cell as a network of helically coiled tubule - A new model for parietal cell transformation. *J.Cell Sci.* 108: 1127-1141, 1995.

23. **Pettitt, J. M., B. H. Toh, J. M. Callaghan, P. A. Gleeson, and I. R. van Driel.** Gastric Parietal Cell Development - Expression of the $H^+/K^+$ ATPase Subunits Coincides with the Biogenesis of the Secretory Membranes. *Immun.Cell Biol.* 71: 191-200, 1993.

24. **Pettitt, J. M., I. R. van Driel, B. H. Toh, and P. A. Gleeson.** From coiled tubules to a secretory canaliculus: A new model for membrane transformation and acid secretion by gastric parietal cells. *Tr.Cell Biol.* 6: 49-53, 1996.

25. **Sachs, G.** The gastric H,K ATPase - Regulation and structure/function of the acid pump of the stomach. In: *Physiology of the Gastrointestinal Tract* (3rd ed.), edited by L. R. Johnson. New York: Raven, 1994, p. 1119-1138.

26. **Scarff, K. L., L. M. Judd, B. H. Toh, P. A. Gleeson, and I. R. van Driel.** Gastric H+,K+-Adenosine triphosphatase β subunit is required for normal function, development, and membrane structure of mouse parietal cells. *Gastroenterology* 117: 605-618, 1999.

27. **Schultheis, P. J., L. L. Clarke, P. Meneton, M. Harline, G. P. Boivin, G. Stemmermann, J. J. Duffy, T. Doetschman, M. L. Miller, and G. E. Shull.** Targeted disruption of the murine $Na^+/H^+$ exchanger isoform 2 gene causes reduced viability of gastric parietal cells and loss of net acid secretion. *Journal of Clinical Investigation* 101: 1243-53, 1998.

28. **Scott, D., H. Helander, S. Hersey, and G. Sachs.** The site of acid secretion in the mammalian parietal cell. *Biochim Biophys Acta* 1146: 73-80, 1993.

29. **Spicer, Z., M. L. Miller, A. Andringa, T. M. Riddle, J. J. Duffy, T. Doetschman, and G. E. Shull.** Stomachs of mice lacking the gastric H,K-ATPase alpha-subunit have achlorhydria, abnormal parietal cells, and ciliated metaplasia. *Journal of Biological Chemistry* 275: 21555-21565, 2000.

30. **Toh, B. H., P. A. Gleeson, R. J. Simpson, R. L. Moritz, J. Callaghan, I. Goldkorn, C. M. Jones, T. M. Martinelli, F. T. Mu, D. C. Humphris, J. M. Pettitt, Y. Mori, T. Masuda, P. Sobieszczuk, J. Weinstock, T. Mantamadiotis, and G. S. Baldwin.** The 60-90 kDa parietal cell autoantigen associated with autoimmune gastritis is a β subunit of the gastric $H^+K^+$-ATPase (proton pump). *Proc. Natl. Acad. Sci. U.S.A.* 87: 6418-6422, 1990.

31. **Tyagarajan, K., R. R. Townsend, and J. G. Forte.** The beta-subunit of the rabbit H,K-ATPase: A glycoprotein with all terminal lactosamine units capped with alpha-linked galactose residues. *Biochemistry* 35: 3238-3246, 1996.

32. **van Driel, I. R., and J. M. Callaghan**. Proton and potassium transport by H+/K(+)-ATPases. *Clin Exp Pharmacol Physiol* 22: 952-60, 1995.

33. **Wang, T. C., and G. J. Dockray.** Lessons from genetically engineered animal models. I. Physiological studies with gastrin in transgenic mice. *Am J Physiol* 277: G6-11, 1999.

34. **Wang, T. C., T. J. Koh, A. Varro, R. J. Cahill, C. A. Dangler, J. G. Fox, and G. J. Dockray**. Processing and proliferative effects of human progastrin in transgenic mice. *J. Clin. Invest.* 98: 1918-29, 1996.

Chapter 20

# Genetically Engineered Animal Models: Physiological Studies with Gastrin in Transgenic Mice

G. J. Dockray[f], G. Bate, K.Hormi, T. Wang*, A. Varro, R. Dimaline

*Physiological Laboratory, Crown St., PO Box 147, University of Liverpool, Liverpool L69 3BX, UK., *Division of Gastroenterology, UMass Medical Center, 373 Plantation St, Worcester, MA01695-2377, USA*

## INTRODUCTION

The role of gastrin as a regulator of acid secretion has been appreciated for almost a century.(13) For many years it has been clear that while gastrin may act directly on parietal cells, it also acts on enterochromaffin-like (ECL) cells to control histamine release which in turn stimulates acid secretion.(33) It is presently thought that this effect is of primary physiological importance in the control of post-prandial acid secretion. In addition, gastrin has long been considered to be a regulator of gastric mucosal proliferation, although it remains controversial as to whether or not proliferating cells in the stomach normally express the gastrin-$CCK_B$ receptor, which is the main receptor mediating the effect of COOH-terminally amidated gastrins.

The gastrins are produced from G-cells in the pyloric antral part of the stomach. Multiple factors influence gastrin production at several different levels, including modulation of gene expression,(40) mRNA translation,(2) post-translational processing (see below), as well as secretion.(12) Recent studies in transgenic mice indicate that the biology of gastrin is more complex than formerly supposed.(36) Three issues are highlighted by these studies. First, biosynthetic precursors and intermediates in the production of the amidated gastrins appear to have their own characteristic spectrum of biological activities, raising the possibility that peptides derived from the *gastrin* gene influence more than just the acid-secreting epithelium. Second, gastrin increases the production of a variety of paracrine mediators that may account for at least some of the responses normally attributed to gastrin. Third, increased production of amidated gastrins is a feature in several genetically modified mice, which exhibit similar phenotypes characterised by gastric hyper-proliferation, foveolar hyerplasia and gland atrophy. Taken together these phenotypes raise other interesting questions regarding the biology of gastrin. The present account will review these issues.

## GASTRIN BIOSYNTHESIS

The gastrin gene encodes a precursor peptide of 101-104 residues depending on the species.(11) Removal of the signal peptide yields progastrin,(7;17) which is characterised by three putative cleavage sites (Arg-57,58; Lys-74,75; Arg-94,95), putative phosphorylation (Ser-96), sulphation (Tyr-87) and amidation (Phe-92) sites. The results of pulse-chase labelling studies in rat and human G-cells,(30-32) indicate that cleavage occurs after sequestration in secretory vesicles and is likely to be mediated by the prohormone convertases PC1/3 and PC2. It has been clear for many years that the events of post-translational processing may be functionally regulated. Aside from variations in expression of the prohormone convertases,(23) there are several factors may account for these differences. First, in cells that lack the regulated pathway of exocytosis eg non-endocrine cells, progastrin progresses directly from the *trans*-Golgi network (TGN) to the cell surface by the constitutive route of secretion. The main secretory product is therefore progastrin. In this case, control of secretion is exerted at the level of synthesis rather than exocytosis *per se*. Second, phosphorylation of Ser-96 inhibits the rate of subsequent cleavage at Arg-94,95.(3) Since the latter event occurs in secretory vesicles, reduced progastrin cleavage leads to release of precursor from the regulated secretory pathway. Third, secretory vesicle pH determines the cleavage of G34 to G17 and therefore the ratio of these two peptides secreted.(4) The $t_{1/2}$ of circulating G34 is five times greater than that of G17,(34) and so modulation of this cleavage directly influences the plasma concentrations of amidated gastrin. Secretory vesicle pH may be modulated by activity of vesicular monoamine transporters: these act as amine-proton exchangers with a stoichiometry of two protons for one amine.(4) They account for the ability of endocrine cells in the gut and elsewhere to co-store amines such as 5HT, dopamine, noradrenaline, histamine together with active peptides. In addition, secretory vesicle pH in G-cells may be increased by weak bases including dietary amines delivered via the lumen.(18)

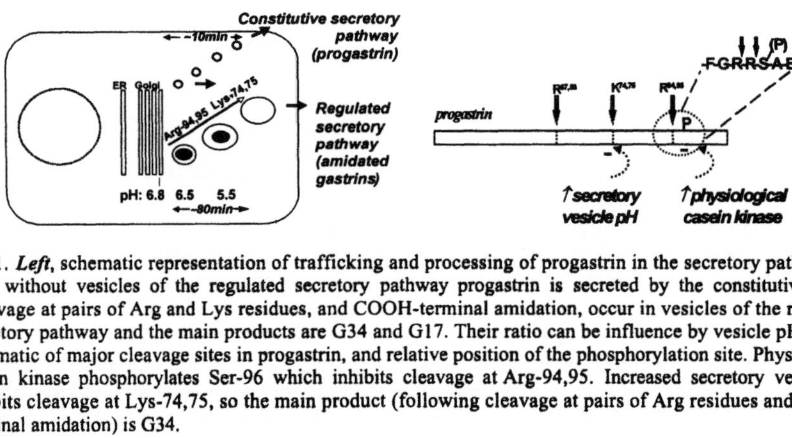


**Fig 1.** ***Left***, schematic representation of trafficking and processing of progastrin in the secretory pathway. In cells without vesicles of the regulated secretory pathway progastrin is secreted by the constitutive route. Cleavage at pairs of Arg and Lys residues, and COOH-terminal amidation, occur in vesicles of the regulated secretory pathway and the main products are G34 and G17. Their ratio can be influence by vesicle pH. ***Right***: schematic of major cleavage sites in progastrin, and relative position of the phosphorylation site. Physiological casein kinase phosphorylates Ser-96 which inhibits cleavage at Arg-94,95. Increased secretory vesicle pH inhibits cleavage at Lys-74,75, so the main product (following cleavage at pairs of Arg residues and COOH-terminal amidation) is G34.

## GENTICALLY MODIFIED ANIMAL MODELS

Transgenic mice over-expressing progastrin, Gly-gastrin and amidated gastrin have been described. In addition mice have been generated in which the *gastrin* or *gastrin-CCK*$_B$ receptor genes have been deleted by homologous recombination. In each case, new issues regarding the biology of gastrin have been raised. In addition there are many genetically modified mice strains with elevated plasma gastrin, mostly this can be attributed to decreased acid secretion, leading to reflex elevation of gastrin release from antral G-cells.(25;26)

***Progastrin*:** Mice in which the expression of the human *gastrin* gene has been targeted to the liver, have elevated plasma progastrin,(37) and to a much lesser extent Gly-extended gastrin. Plasma concentrations of amidated gastrin in these mice are normal. These animals exhibit increased proliferation in the colon and increased susceptibility to colon carcinogens.(29) They are of interest since colon tumours express the gastrin gene.(1;24) Moreover at least in colon cancer cell lines progastrin appears to be a mitogen.(28)

***Gly-gastrin*:** Mice expressing progastrin truncated at Gly-93 driven by the metallothionine promoter have moderately elevated plasma Gly-gastrin concentrations.(20) Their predominant phenotype is increased colon proliferation, and increased susceptibility to lung tumours. As with mice over-expressing progastrin these results provide experimental, *in vivo,* evidence which clarifies the interpretation of results in cell lines showing that Gly-gastrins are mitogens with actions clearly distinct from the amidated gastrins.(27;39)

***Amidated gastrin*.** Two groups have generated mice over-expressing amidated gastrins. Wang *et al* targeted expression of human preprogastrin to β-cells of the pancreatic islets (which are able to execute the full range of post-translational modifications) using the insulin promoter and the human preprogastrin coding sequence.(35) Konda *et al.*, expressed a transgene using the chicken β-actin promoter and a coding sequence consisting of human preprogastrin with two mutations: one conferred a furin cleavage site and the other terminated the coding sequence at Gly-93.(21) This transgene seems to be expressed in many tissues, although the major product is amidated gastrin. In mice expressing human progastrin in β-cells (Ins-Gas mice), there is modestly elevated plasma concentrations of amidated gastrin. These animals offer several advantages for selective release of peptide derived from the transgene (see below). They exhibit an increase in parietal cell number and acid secretion up to about 3 months of age, but thereafter there is loss of parietal cells and a progression to a hyper-proliferative epithelium exhibiting foveolar hyperplasia and loss of glands. The later is also described by Kondo *et al.* In the presence of *Helicobacter felis* the progression to atrophy in Ins-Gas mice is accelerated, and leads to gastric cancer by about 9 months of age. The similar atrophic phenotype in two quite different mice strains with elevated plasma gastrin concentrations suggests that in the mouse, gastrin exerts pro-atrophic actions. In man and in rats with hypergastrinaemia the most obvious change in the corpus mucosa is ECL cell hyperplasia, and in some cases the development of ECL cell carcinoid tumours.(5) The reasons for the different responses to hypergastrinaemia remain uncertain. It seems, however, that in both man and mouse, gastrin stimulates the production of paracrine mediators that might secondarily influence gland remodelling. Some of the paracrine mediators are briefly discussed below. It seems likely that many others are waiting to be discovered.

## PARACRINE MECHANISMS ACTIVATED BY GASTRIN

***ECL cell responses*.** The responses of the ECL cell to gastrin include increased proliferation, increased expression of genes required for ECL cell function, increased stability of certain mRNA species, increased synthesis of histamine, and of histamine release.(12) Earlier studies identified several gastrin-regulated genes expressed in ECL cells including histidine decarboxylase which converts histidine to histamine (HDC),(9;41) chromogranin A which is co-stored with histamine in secretory vesicles,(8;16) and vesicular monoamine transporter type 2 (VMAT2) which transports histamine into secretory vesicles.(10;38) In Gas-Ko mice, there is differential down regulation of the expression of these genes compared with wild type mice (Table 1). In particular, HDC mRNA is substantially depressed, VMAT2 by 50%, and CgA was not significantly changed. Presumably, other physiological factors, in addition to gastrin, influence expression of these three genes to varying degrees.

**Table 1.** *Relative abundance of mRNA for three genes expressed in ECL cells in wild type (Control) and Gas-Ko mice.* [#]

| mRNA | Control | Gas-Ko |
|---|---|---|
| HDC | 100 ± 4.3 | 9.9 ± 1.1** |
| VMAT2 | 100 ± 17.8 | 45.0 ± 8.7* |
| CgA | 100 ± 11.6 | 122.0 ± 15.5 |
| GAPDH | 100 ± 1.9 | 115.5 ± 7.5 |

[#]Data derived from Northern analysis of whole stomach extracts from fed *ad libitum* C57/Bl6 wild type mice (Control) and Gas-Ko mice. Probes were used as previously described,(8-10) signals were quantified using a phosphorimager and normalised as indicated to the relevant control value (=100). Values are means ± SE (n = 6 for each group). *p<0.05, **p<0.01

There are also differences in the extent to which these genes are up-regulated by increased plasma gastrin. In Ins-Gas mice, gastrin can be released from the islets of Langerhans by glucose. Thus in pancreatic islets obtained by collagenase digestion of pancreas from these mice, we have found that incubation in 16mM glucose for 10 min, leads to a 3-fold increase in gastrin release compared with incubation in 5mM glucose (there is no detectable release from islets of wild type mice). Similarly, administration of glucose leads to sustained increases in plasma gastrin in Ins Gas mice *in vivo*. For example after 5 days with *ad libitum* access to 20% glucose in the drinking water (which is highly palatable), plasma gastrin in Ins-GAS mice was 283 ± 50pm compared with 100 ± 17pM, in control Ins-GAS mice (p<0.05). There was no difference in plasma gastrin in wild type mice allowed access to

**Table 2.** *Relative abundance of mRNA and protein derived from three genes expressed in ECL cells in Ins-Gas mice either fed ad libitum, or fed ad libitum and given glucose.* [#]

| | mRNA | | Protein | |
|---|---|---|---|---|
| | Control | Glucose | Control | Glucose |
| HDC | 100 ± 2.6 | 163 ± 24.1* | 100 ± 1.8 | 313 ± 39* |
| VMAT2 | 100 ± 12.3 | 255 ± 28.2* | 100 ± 2.9 | 267 ± 28* |
| CgA | 100 ± 2.9 | 157 ± 8.6* | 100 ± 5.2 | 254 ± 38* |
| GAPDH | 100 ± 5.6 | 75 ± 14.3 | - | - |

[#]Data derived from Northern and Western analysis of whole stomach extracts from Ins-Gas mice, either fed *ad libitum* (control) or fed *ad libitum* with free access to 20% glucose in the drinking water for 5 days. Signals quantified using a phosphorimager and normalised as indicated to the relevant control value (=100). Comparable studies in FVB/N wild type mice showed no changes with glucose in either mRNA or protein abundance for the three ECL cell products. See Table 1 for further details.

glucose. The intragastric pH in fasted Ins-GAS mice (glucose treated: 2.85 ± 0.14; control: 3.0 ± 0.13) indicated that parietal cell function was normal at the time of these experiments. This model therefore provides a system to study ECL cell and other gastric responses to increased circulating gastrin in the presence of acid in the gastric lumen. For example, there was increased abundance of HDC, CgA and VMAT2 mRNA and protein after stimulating the release of gastrin from the pancrease of Ins Gas mice with oral glucose (Table 2). Unlike models previously used to study ECL cell responses to circulating gastrin *in vivo*, e.g omeprazole-treated animals, the Ins-Gas mouse allows study of ECL cell responses to endogenous gastrin without disruption of the normal processes of acid secretion, and consequent changes in endocrine reflexes e.g. D-cell function.

***Growth factors as paracrine mediators***. Studies in cell lines have established that gastrin increases production of HB-EGF, amphiregulin and TGFα.(35) These stimulate proliferation, and acutely inhibit acid secretion. In transgenic mice over-expressing TGFα the phenotype resembles the human condition of Menetrier's disease, characterised by foveolar hyperplasia, loss of gland cells and of acid secretion.(6) These changes resemble those in old InsGas mice,(35) The idea that prolonged gastrin stimulation may be responsible for the development of gastrin atrophy seems, at first sight, to be at odds with the effects in humans and rats with hypergastrinaemia. However, this response may reflect actions of gastrin that can be seen as protecting the epithelium from the aggressive effects of gastrin-stimulated acid secretion. During the normal production of gastric gland cells there is clearly a requirement to distinguish between the commitment of cells to mucus cell lineages compared with parietal, chief or endocrine cell lineages. The data raise the possibility that the tendency of gastrin to influence parietal and ECL cell abundance may be balanced by production of EGF-related growth factors which stimulate proliferation and production of mucus cells, thereby maintaining the overall distribution of different cell types in the gastric mucosa. Either way, it is likely that EGF-related growth factors are just one of several classes of gastrin-stimulated gastric growth factors and morphogens. Another candidate is Reg-1α (also known as lithostathine, pancreatic thread protein or pancreatic stone protein) which is found in ECL and chief cells (at least in man), as well as in pancreas. Gastrin increases Reg1α mRNA abundance in the stomach of rats and man, and in AR4-2J cells.(14;15) In both islets and gastric mucosal cells, it is thought that Reg1α is a mitogen; but in high concentrations it is reported to produce apoptosis. (14;19) Interestingly, inactivating mutations of Reg1α are associated with ECL cell carcinoid tumours, suggesting that it is also a tumour inhibitor, and compatible with the idea that locally high concentrations might lead to apoptoses.(15)

***Other paracrine mediators*** Studies in genetically modified mice are also revealing other paracrine mechanisms that may be regulated by gastrin. For example, by using mRNA display to identify genes that are differentially expressed in normal and GasKO mice, Khan *et al.* found multiple, novel, gastrin-dependent genes. One gastrin-regulated gene identified in this way was TTF1 (unpublished observations). The idea that gastrin stimulates TFF1 expression is interesting given that it is produced in cells that are not thought to express the gastrin-$CCK_B$ receptor. Presumably then, there is a paracrine link between parietal and ECL cells which express the gastrin-$CCK_B$ receptor, and surface epithelial cells which express TFF1. Mice in which the TFF1 gene has been deleted exhibit exhibited severe gastric hyperplasia and dysplasia and develop tumours in the antropyloric region.(22) Since TFF1 stimulates epithelial cell migration and wound healing, and inhibits proliferation, it seems that up-regulation by gastrin provides a protective mechanism counteracting both the aggressive effects of stimulation of acid secretion, and increased proliferation.

## OVERVIEW

Multiple active factors are generated from the gastrin gene. These include progastrin itself which is a stimulant of colon epithelial proliferation as well as the amidated gastrins which regulate gastric corpus function. The available transgenic models allow secretion of amidated gastrin to be regulated independently of G-cell function (and of changes in the gastric lumen). One of the themes to emerge from studies of the phenotype in transgenic mice with elevated plasma gastrin is that gastrin activates multiple paracrine mediators in the gastric corpus including histamine, EGF-family growth factors, Reg. Moreover gastrin-dependent targets extend to the control of non-classical regulatory molecules such as the trefoil factor TFF1. Elucidation of the function of these overlapping paracrine networks will be a major task for the future.

## REFERENCES

1. **Baldwin, G. S. and A. Shulkes.** Gastrin, gastrin receptors and colorectal carcinoma. *Gut* 42: 581-584, 1998.

2. **Bate, G. W., A. Varro, R. Dimaline, and G. J. Dockray.** Control of preprogastrin messenger RNA translation by gastric acid in the rat. *Gastroenterol.* 111: 1224-1229, 1996.

3. **Bishop, L., R. Dimaline, C. Blackmore, D. Deavall, G. J. Dockray, and A. Varro.** Modulation of the cleavage of the gastrin precursor by phosphorylation. *Gastroenterology* 115: 1154-1162, 1998.

4. **Blackmore, C. G., A. Varro, R. Dimaline, L. Bishop, D. V. Gallacher, and G. J. Dockray.** Measurement of secretory vesicle pH reveals intravesicular alkalinization by vesicular monoamine transporter type 2 resulting in inhibition of prohormone cleavage. *J Physiol* 531: 605-617, 2001.

5. **Bordi, C., B. Annibale, C. Azzoni, M. Marignani, G. Ferraro, G. Antonelli, T. D'Adda, G. D'Ambra, and G. Della-Fave.** Endocrine cell growths in atrophic body gastritis, critical evalualtion of a histological classification. *J.Pathol.* 182: 339-346, 1997.

6. **Dempsey, P. J., J. R. Goldenring, C. J. Soroka, I. M. Modlin, R. W. McClure, C. D. Lind, D. A. Ahlquist, M. R. Pittelkow, D. C. Lee, E. P. Sandgren, D. L. Page, and R. J. J. Coffey.** Possible role of transforming growth factor a in the pathogenesis of Menetrier's disease: Supportive evidence from humans and transgenic mice. *Gastroenterology* 103: 1950-1963, 1992.

7. **Desmond, H., S. Pauwels, A. Varro, H. Gregory, J. Young, and G. J. Dockray.** Isolation and characterization of the intact gastrin precursor from a gastrinoma. *FEBS Lett.* 210: 185-188, 1987.

8. **Dimaline, R., D. Evans, E. R. Forster, A. K. Sandvik, and G. J. Dockray**. Control of gastric corpus chromogranin A messenger RNA abundance in the rat. *Am.J.Physiol.* 264: G583-G588, 1993.

9. **Dimaline, R. and A. K. Sandvik**. Histidine decarboxylase gene expression in rat fundus is regulated by gastrin. *FEBS Lett.* 281: 20-22, 1991.

10. **Dimaline, R. and J. Struthers.** Expression and regulation of a vesicular monoamine transporter (VMAT2) in rat stomach: a putative histamine transporter. *J.Physiol.* 490: 249-256, 1996.

11. **Dockray, G. J.** Topical review. Gastrin and gastric epithelial physiology. *J Physiol (Lond)* 518 : 315-324, 1999.

12. **Dockray, G. J., Varro, A., Dimaline, R., and Wang, T. C.** The gastrins: their production and biological activities. Ann.Rev.Physiol. 63, 119-139. 2001. Ref Type: Generic

13. **Edkins, J. S.** On the chemical mechanism of gastric secretion. *Proc.Roy.Soc.Ser.B* 76: 376-376, 1905.

14. **Fukui, H., Y. Kinoshita, T. Maekawa, A. Okada, S. Waki, M. D. S. Hassan, H. Okamoto, and T. Chiba.** Regenerating gene protein may mediate gastric mucosal proliferation induced by hypergastrinemia in rats. *Gastroenterology* 115: 1483-1493, 1998.

15. **Higham, A. D., L. A. Bishop, R. Dimaline, C. G. Blackmore, A. C. Dobbins, A. Varro, D. G. Thompson, and G. J. Dockray.** Mutations of RegIalpha are associated with enterochromaffin-like cell tumor development in patients with hypergastrinemia. *Gastroenterology* 116: 1310-1318, 1999.

16. **Hocker, M., R. Raychowdhury, T. Plath, H. Wu, D. T. O'Connor, B. Wiedenmann, S. Rosewicz, and T. C. Wang.** Sp 1 and CREB mediate gastrin-dependent regulation of chromogranin A promoter activity in gastric carcinoma cells. *J.Biol.Chem.* 273: 34000-34007, 1998.

17. **Huebner, V. D., R. Jiang, T. D. Lee, K. Legesse, J. H. Walsh, J. E. Shively, P. Chew, T. Azuma, and J. R. Reeve.** Purification and structural characterization of progastrin-derived peptides from a human gastrinoma. *J.Biol.Chem.* 266: 12223-12227, 1991.

18. **Hussain, I., G. W. Bate, J. Henry, P. K. Djali, R. Dimaline, G. J. Dockray, and A. Varro.** Identification of vesicular monoamine transporter type 1 (VMAT1) in G-cells and its role in modulating cleavage of G34. *J.Physiol.* 517: 495-505, 1999.

19. **Kobayashi, S., T. Akiyama, K. Nata, M. Abe, M. Tajima, N. J. Shervani, M. Unno, S. Matsuno, H. Sasaki, S. Takasawa, and H. Okamoto.** Identification of a receptor for Reg (regenerating gene) protein, a pancreatic beta-cell regeneration factor. *J Biol.Chem.* 275: 10723-10726, 2000.

20. **Koh, T. J., G. J. Dockray, A. Varro, R. J. Cahill, C. A. Dangler, J. G. Fox, and T. C. Wang.** Overexpression of glycine-extended gastrin in transgenic mice results in increased colonic proliferation. *J.Clin.Invest.* 103: 1119-1126, 1999.

21. **Konda, Y., H. Kamimura, H. Yokota, N. Hayashi, K. Sugano, and T. Takeuchi.** Gastrin stimulates the growth of gastric pit with less-differentiated features. *Am.J Physiol* 277: G773-G784, 1999.

22. **Lefebvre, O., M. P. Chenard, R. Masson, J. Linares, A. Dierich, M. LeMeur, C. Wendling, C. Tomasetto, P. Chambon, and M. C. Rio.** Gastric mucosa abnormalities and tumorigenesis in mice lacking the pS2 trefoil protein. *Science* 274: 259-262, 1996.

23. **Macro, J. A., G. W. Bate, A. Varro, C. Vaillant, N. G. Seidah, R. Dimaline, and G. J. Dockray.** Regulation by gastric acid of the processing of progastrin- derived peptides in rat antral mucosa. *J.Physiol.* 502: 409-419, 1997.

24. **Nemeth, J., B. Taylor, S. Pauwels, A. Varro, and G. J. Dockray.** Identification of progastrin derived peptides in colorectal carcinoma extracts. *Gut* 34: 90-95, 1993.

25. **Scarff, K. L., L. M. Judd, B. H. Toh, P. A. Gleeson, and I. R. Van Driel.** Gastric H(+),K(+)-adenosine triphosphatase beta subunit is required for normal function, development, and membrane structure of mouse parietal cells. *Gastroenterology* 117: 605-618, 1999.

26. **Schultheis, P. J., L. L. Clarke, P. Meneton, M. Harline, G. P. Boivin, G. Stemmermann, J. J. Duffy, T. Doetschman, M. L. Miller, and G. E. Shull.** Targeted disruption of the murine Na+/H+ Exchanger isoform 2 gene causes reduced viability of gastric parietal cells and loss of net acid secretion. *J.Clin.Invest.* 101: 1243-1253, 1998.

27. **Seva, C., C. J. Dickinson, and T. Yamada.** Growth-promoting effects of glycine-extended progastrin. *Science* 265: 410-412, 1994.

28. **Singh, P., A. Owlia, A. Varro, B. Dai, S. Rajaraman, and T. Wood.** Gastrin gene expression is required for the proliferation and tumorigenicity of human colon cancer cells. *Cancer Research* 56: 4111-4115, 1996.

29. **Singh, P., M. Velasco, R. Given, M. Wargovich, A. Varro, and T. C. Wang.** Mice overexpressing progastrin are predisposed for developing aberrant colonic crypt foci in response to AOM. *Am.J Physiol Gastrointest.Liver Physiol* 278: G390-G399, 2000.

30. **Varro, A., G. J. Dockray, G. W. Bate, C. Vaillant, A. Higham, E. Armitage, and D. G. Thompson.** Gastrin biosynthesis in the antrum of patients with pernicious anemia. *Gastroenterology* 112: 733-741, 1997.

31. **Varro, A., J. Henry, C. Vaillant, and G. J. Dockray.** Discrimination between temperature- and brefeldin A-sensitive steps in the sulfation, phosphorylation, and cleavage of progastrin and its derivatives. *J.Biol.Chem.* 269: 20764-20770, 1994.

32. **Varro, A., S. Voronina, and G. J. Dockray.** Pathways of processing of the gastrin precursor in rat antral mucosa. *J.Clin.Invest.* 95: 1642-1649, 1995.

33. **Walsh, J. H.** Gastrin. In Walsh, J. H. and G. J. Dockray, eds. Gut Peptides. New York, Raven Press. 1994, 75-121.

34. **Walsh, J. H., J. I. Isenberg, J. Ansfield, and V. Maxwell.** Clearance and acid-stimulating action of human big and little gastrins in duodenal ulcer subjects. *J.Clin.Invest.* 57: 1125-1131, 1976.

35. **Wang, T. C., C. A. Dangler, D. Chen, J. R. Goldenring, T. J. Koh, R. Raychowdhury, R. J. Coffey, S. Ito, A. Varro, G. J. Dockray, and J. G. Fox.** Synergistic interaction between hypergastrinemia and *Helicobacter* infection in a mouse model of gastric cancer. *Gastroenterology* 118: 36-47, 2000.

36. **Wang, T. C. and G. J. Dockray.** Lessons from genetically engineered animal models. I. Physiological studies with gastrin in transgenic mice. *Am.J Physiol* 277: G6-11, 1999.

37. **Wang, T. C., T. J. Koh, A. Varro, R. J. Cahill, C. A. Dangler, J. G. Fox, and G. J. Dockray.** Processing and proliferative effects of human progastrin in transgenic mice. *J.Clin.Invest.* 98: 1918-1929, 1996.

38. **Watson, F., R. S. Kiernan, D. G. Deavall, A. Varro, and R. Dimaline.** Transcriptional activation of the rat vesicular monoamine transporter 2 promoter in gastric epithelial cells: Regulation by gastrin. *J Biol.Chem.* 276: 7661-7671, 2001.

39. **Watson, S. A., D. Michaeli, S. Grimes, T. M. Morris, G. Robinson, A. Varro, T. A. Justin, and J. D. Hardcastle.** Gastrimmune raises antibodies that neutralize amidated and glycine-exended gastrin-17 and inhibit the growth of colon cancer. *Cancer Research* 56: 880-885, 1996.

40. **Wu, S. V., A. Giraud, M. Mogard, K. Sunii, and J. H. Walsh.** Effects of inhibition of gastric secretion on antral gastrin and somatostatin gene expression in rats. *Am.J.Physiol.* 258: G788-G793, 1990.

41. **Zhang, Z., M. Hocker, T. J. Koh, and T. C. Wang.** The human histidine decarboxylase promoter is regulated by gastrin and phorbol 12-myristate 13-acetate through a downstream cis-acting element. *J.Biol.Chem.* 271: 14188-14197, 1996.


---

[f] *Tel*: 44 151 794 5324 *Fax*: 44 151 794 5325
*Email*: g.j.dockray@liverpool.ac.uk
The work described here was supported by the Medical Research Council (UK), The Wellcome Trust, and NIH.

Chapter 21

# Gastric Cytoprotection by Prostaglandin E2 - Relation to EP Receptor Subtypes -

KOJI TAKEUCHI, SHINICHI KATO, YUSAKU KOMOIKE
YOSHIHIRO OGAWA and MASANORI TAKEEDA
*Department of Pharmacology and Experimental Therapeutics*
*Kyoto Pharmaceutical University*
*Misasagi, Yamashina, Kyoto 607-8414, Japan*

## SUMMARY

This article overviews our recent studies on the relation between EP receptor subtypes and gastric cytoprotection afforded by endogenous and exogenous PGE2, using rats and EP-receptor knockout mice. Exogenous PGE2 dose-dependently reduced the severity of HCl/ethanol-induced gastric lesions. This action of PGE2 was mimicked by another E type prostanoids such as sulprostone (EP1/EP3) and 17-phenyl PGE2 (EP1) but not by butaprost (EP2), ONO-NT-012 (EP3) or 11-deoxy PGE1 (EP3/EP4), and significantly attenuated by ONO-AE-829 the selective EP1 antagonist. Similar results were obtained in EP-receptor knockout mice, and PGE2 inhibited these lesions in both wild type and EP3-receptor knockout mice but not in the animals lacking EP1-receptors. HCl/ethanol-induced gastric lesions were also prevented by mild irritants such 20 mM taurocholate (TC). This effect was accompanied by an increase of mucosal PGE2 production, attenuated by indomethacin as well as ONO-AE-829, and disappeared in EP1-receptor knockout mice. On the other hand, capsaicin also protected the stomach against HCl/ethanol, without increase of mucosal PGE2 production. This action was totally attenuated by chemical ablation of afferent neurons, partially mitigated by indomethacin but not by ONO-AE-829. Attenuation by indomethacin of the capsaicin protection was recoverd in the presence of butaprost but not other EP agonists. In addition, capsaicin exihibited gastric protection in knockout mice lacking EP1- and EP3-receptors but not in IP-receptor knockout mice. We conclude that 1) PGE2, either generated endogenously or administered exogenously, exhibit gastric cytoprotection directly through EP1 receptors, 2) this action may be functionally associated with inhibition of gastric motility but not with increase of gastric mucosal blood flow or mucus secretion, and 3) endogenous PGs also contribute to the mucosal protection induced by capsaicin by sensitizing sensory neurons, probably through EP2- and IP-receptors.

## INTRODUCTION

Prostaglandins (PGs), produced from arachidonic acid by two isoforms of cyclooxygenase (COX), are present throughout the gastrointestinal tract and are known to bring about a wide variety of actions in the gut, including control of acid secretion, bicarbonate secretion, mucus production, mucosal blood flow and maintenance of mucosal integrity (14). Indeed, PGs administered exogenously protect the gastrointestinal mucosa against various ulcerogenic stimuli such as stress, necrotizing agents and nonsteroidal antiinflammatory drugs (NSAIDs). The pioneering study by Robert et al (16) in 1979 was the first to demonstrate that PGs administered exogenously protect the stomach against a variety of necrotizing agents, a phenomenon called "gastric cytoprotection". PGE2 is one of the major PGs most effective in this action. It is also known that when the stomach is pre-exposed to a mild irritant such as taurocholate (TC), the resistance of the mucosa to subsequently applied necrotizing agents increases, a phenomenon called "adaptive cytoprotection" (14,17). Since this effect disappears in the presence of indomethacin, a COX inhibitor, it is assumed to be mediated through an enhanced production of endogenous PGs. Likewise, endogenous PGs are also involved in gastric protection induced by capsaicin, the selective stimulator of capsaicin-sensitive afferent neurons. Capsaicin provides gastric protection, essentially by stimulating afferent neurons, yet this effect is also in part dependent on endogenous PGs (8,22).

Recent pharmacological studies have classified PGE2 receptors into four specific G protein-coupled subtypes, EP1 to EP4 (4). The distribution of these receptors is considered to explain the multiple effects of PGE2 in various tissues including the gastrointestinal tract. In addition, mice lacking various receptors for prostanoids have been established (18), and by using these "knockout mice" the roles of specific PG receptors in various biological actions of PGs have been demonstrated (25). We have recently performed a series of experiments to determine the EP receptor subtypes mediating the gastrointestinal protection afforded by PGE2, using various models in both rats and EP-receptor knockout mice (1,23,24). In these studies, we also used prostanoids, subtype-specific EP receptor agonists and an antagonist, as a tool to characterize the EP receptor subtypes involved in gastrointestinal protection.

In this article, we review our recent publications on the relation between EP receptor subtypes and the gastric cytoprotection afforded by exogenous PGE2, mild irritant and capsaicin, those obtained in both rats and EP-receptor knockout mice, and discuss possible functional alterations responsible for the protective action of PGE2.

## MATERIALS AND METHODS

### Animals

Male Sprague Dawley rats (200-220 g) and C57BL/6 mice (25-30 g) were used. Mice lacking the EP1- or EP3-receptors were generated as described previously (18,25). In brief, the mouse genes encoding the EP1- and EP3-receptors were individually disrupted, and chimaeric mice were generated. These animals were then back crossed with C57BL/6 mice, and the resulting heterozygous litter mates were bred to produce homozygous EP1-, EP3- or IP-knockout mice. Distribution of the EP1, EP3 or IP receptor genes was verified by northern blot hybridization, which failed to detect messenger RNAs encoding the respective receptors in EP1 (-/-), EP3 (-/-) and IP (-/-) mice. These knockout mice and rats were deprived of food but allowed free access to tap water for 18 hr before the experiments. All studies were performed using 4-8 animals per group under unanesthetized conditions.

### Induction of Gastric Lesions

The rats were given 1 ml of HCl/ethanol (60% in 150 mM HCl) p.o. through esophageal intubation, and killed 1 hr later under deep ether anesthesia (10). The stomachs were removed, inflated by injecting 10 ml of 1% formalin for 10 min to fix the tissue walls, and opened along the greater curvature. The area ($mm^2$) of hemorrhagic lesions developed in the stomach was measured under a dissecting microscope with a square grid (x10). Using this model, we examined the effects of exogenous PGE, mild irritant and capsaicin.

**A) Various EP agonists**

The following prostanoids were administered i.v. as a single injection from the tail vein 10 min before HCl/ethanol treatment; PGE2 (EP1/ EP2/EP3/EP4 agonist: 3-30 μg/kg), sulprostone (EP1/ EP3 agonist: 3-30 μg/kg), butaprost (EP2 agonist: 1 and 3 mg/kg), 17-phenyl- ω -trinor-PGE2 (17-phenyl PGE2: EP1 agonist: 3-30 μg/kg), ONO-NT-012 (EP3 agonist: 0.1-1 mg/kg) and 11-

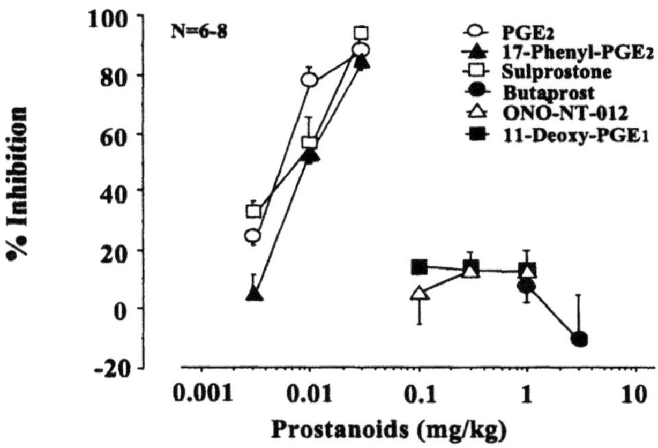


**Figure 1.** Dose-response relationships for various EP agonists in protecting the rat stomach against HCl/ethanol. The animals were administered with 1 ml of HCl/ethanol (60% in 150 mM HCl), and killed 1 hr later. Each agent was administered i.v. as a single injection 10 min before HCl/ethanol. Data are indicated as % inhibition from control and represent the means±SE from 6~8 rats.

deoxy-PGE1 (EP3/EP4 agonist: 0.1-1 mg/kg). These prostanoids are considered to be the EP receptor subtype specific agonists. In some animals, the effect of ONO-AE-829 (AE-829: 10 mg/kg), the EP1 receptor antagonist (1), on the gastric protection afforded by PGE2 (30 µg/kg, i.v.) was examined by administering this agent s.c. 30 min before PGE2.

**B) Mild irritant**

We used sodium taurocholate (TC) as a mild irritant. TC (5-20 mM) was given p.o. as a mild irritant 30 min before administration of HCl/ethanol. In some cases, indomethacin (5 mg/kg) or AE-829 (10 mg/kg) was given s.c. 1 hr before administration of TC. In addition, the protective effect of TC against HCl/ethanol was also examined in rats with chemical ablation of capsaicin-sensitive sensory neurons (chemical deafferentation). Chemical deafferentation was induced by s.c. injections of capsaicin once daily for three consecutive days (total dose: 100 mg/kg) 2 weeks before the experiment (22). All capsaicin injections were performed under ether anesthesia, and the rats were pretreated with terbutaline (0.1 mg/kg, i.m.) and aminophylline (10 mg/kg, i.m.) to counteract the respiratory impairment associated with capsaicin injection. To check for the effectiveness of the treatment, a drop of capsaicin solution (0.1 mg/ml) was instilled into one eye of each rat, and the wiping movements were counted as previously reported (8).

**C) Capsaicin**

Capsaicin (1-10 mg/kg) was given p.o. 30 min before administration of HCl/ethanol. In some cases, indomethacin (5 mg/kg) or AE-829 (10 mg/kg) was given s.c. 1 hr before administration of capsaicin. In addition, the protective effects of PGE2 (30 µg/kg, i.v.) and capsaicin (10 mg/kg, p.o.) on HCl/ethanol were also examined in rats with chemical deafferentation.

**D) Knockout mice**

In separate experiments, wild-type (WT) mice and EP1-, EP3- or IP-receptor knockout mice were given HCl/ethanol orally in a volume of 0.3 ml, and killed 1 hr later (1). Then, the stomach was removed, treated with formalin, and the mucosa was examined for hemorrhagic lesions under a dissecting microscope, as described previously. PGE2 (0.3 mg/kg, s.c.), 20 mM TC (p.o.) or capsaicin (10 mg/kg, p.o.) was given 30 min before administration of HCl/ethanol. In some cases, indomethacin (5 mg/kg) or AE-829 (10 mg/kg) was given s.c. 1 hr before administration of TC in WT mice.

**Measurement of Mucosal Prostaglandin E2 Levels**

PGE2 levels in the gastric mucosa were measured 30 min after p.o. administration of 20 mM TC or capsaicin (10 mg/kg) in rats. In some cases, indomethacin (5 mg/kg) or AE-829 (10 mg/kg) was given s.c. 1 hr before administration of TC. Under ether anesthesia, the stomachs were quickly removed, opened along the greater curvature, and rinsed with ice-cold saline. To separate the mucosal layer, the corpus mucosa was placed between two glass slides squeezed with rubber band, and placed in hexane-frozen Dry Ice and acetone. These glasses were separated, the mucosa was collected, weighed, and put in a tube containing 100% ethanol plus 0.1 M indomethacin (7). Then, the samples were homogenized, and centrifuged for 10 min at 12000 r.p.m. at 4℃. The supernatant of each sample was used for determination of PGE2 by EIA using PGE2-kit (Cayman Chemical Co., Ann Arbor, MI, USA).

**Preparation of Drugs**

Drugs used were PGE2, 17-phenyl- $\omega$ -trinor PGE2, (17-phenyl PGE2),11-deoxy PGE1 (Funakoshi, Tokyo, Japan), ONO-AE-829 (AE-829), butaprost, sulprostone, ONO-NT-012 (Ono, Osaka, Japan), terbutaline (Bricanyl[R], Fujisawa, Osaka, Japan), taurocholate Na (Difco Lab., Detroit, Michigan, USA), capsaicin (Nakarai Tesque, Kyoto, Japan), aminophylline (Neophyllin[R], Eisai, Tokyo, Japan) and indomethacin (Sigma Chemicals, St. Louis, Mo., USA). Capsaicin was dissolved in Tween 80-ethanol solution (10% ethanol, 10% Tween 80, 80% saline, w/w) for s.c. injection, while indomethacin was suspended in saline with a drop of Tween 80 (Wako, Osaka, Japan). AE-829 were dissolved in saline. PGE2 and other EP receptor ligands were first dissolved in absolute ethanol and then diluted with saline to desired concentrations. Each agent was prepared immediately before use and given in a volume of 0.5 ml/100 g body weight (rat) or 0.1 ml/10 g body weight (mouse) for p.o. and s.c. administration, respectively, and given i.v. in a volume of 0.1 ml/100 g body weight (rat). Control animals received saline in place of the active agent.

**Statistics**

Data are presented as the means±SE from 4-8 animals per group. Statistical analyses were performed using a two-tailed Dunnett's multiple comparison test, and values of $P<0.05$ were regarded as significant.

## RESULTS

**Effects of Various EP Agonists on HCl/Ethanol-Induced Gastric Lesions**

Oral administration of ethanol (60% in 150 mM HCl) produced hemorrhagic band-like lesions in the glandular stomach, the lesion score being 108.6±7.2 $mm^2$. These lesions were significantly prevented by prior i.v. administration of PGE2 (3~30 μg/kg) in a dose-dependent manner, the degree of inhibition being 91.2% at 30 μg/kg (Figure 1). The mucosal ulcerogenic response to HCl/ethanol was also dose-dependently inhibited by i.v. administration of both sulprostone (3-30 μg/kg) and 17-phenyl PGE2 (3-30 μg/kg); the potency of these prostanoids in preventing HCl/ethanol-induced gastric lesions was almost similar to that of PGE2, the $ED_{50}$ being about 10 μg/kg. However, neither butaprost (1 and 3 mg/kg), ONO-NT-012 (0.1-1 mg/kg) nor 11-deoxy PGE1 (0.1-1 mg/kg) had any effect on the development of gastric lesions in response to HCl/ethanol. As expected, the protective action of PGE2 was significantly attenuated by prior administration of AE-829 (10 mg/kg) the EP1 antagonist, but not by sensory deafferentation (Figure 2).

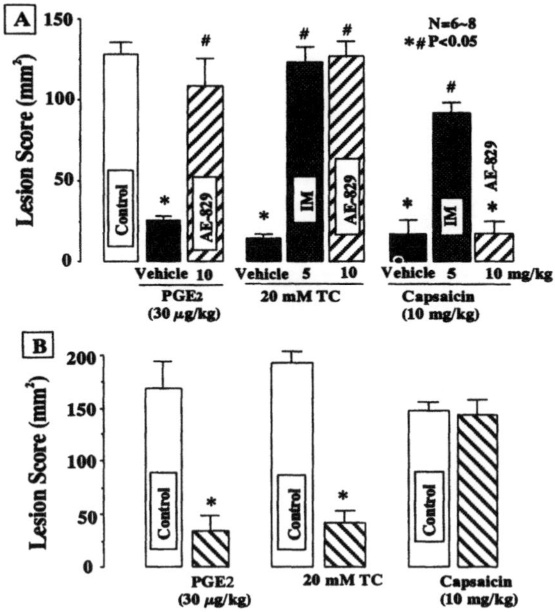


**Figure 2.** **A**: Effects of PGE2, TC and capsaicin on HCl/ethanol-induced gastric lesions in intact rats. The animals were given 1 ml of HCl/ethanol (60% ethanol in 150 mM HCl) p.o. and killed 1 hr later. PGE2 (30 μg/kg), TC (1 ml: 20 mM) or capsaicin (10 mg/kg) was given p.o. 30 min before HCl/ethanol. indomethacin (IM: 5 mg/kg) or AE-829 (10 mg/kg) was given s.c. 1 hr before the above agents. Data are presented as the means±SE from 6-8 rats. Significant difference at $P<0.05$; ∗ from control; # from vehicle. **B**: Effects of PGE2, TC and capsaicin on HCl/ethanol-induced gastric lesions in sensory deafferented rats. Data are presented as the means±SE from 6~8 rats. ∗ Significant difference from the corresponding control, at $P<0.05$.

To futher examine the relationship of EP receptors and gastric cytoprotection, we compared the protective action of PGE2 against HCl/ethanol between WT and knockout mice, lacking EP1- or EP3-receptors. PGE2, given s.c. at 0.3 mg/kg, significantly inhibited gastric lesions in response to HCl/ethanol, in both WT and EP3-receptor "knockout" mice (Figure 3). However, this agent at the same dose did not afford any protection against gastric damage in the knockout mice, lacking EP1-receptors.

**Protection by TC Against HCl/Ethanol-Induced Gastric Lesions**

The severity of HCl/ethanol-induced gastric lesions was dose-dependently reduced when the animals were given TC (5-20 mM) p.o. before challenge with HCl/ethanol, and a significant effect was obtained at over 10 mM, the inhibition at 20 mM TC being 92.4% (not shown). The protective action of TC (20 mM) was almost totally attenuated by prior administration of AE-829 (10 mg/kg) as well as indomethacin (5 mg/kg), but not by chemical deafferentation (Figure 2). As shown in Figure 3, 20 mM TC also significantly reduced the gastric ulcerogenic response to HCl/ethanol in both WT and EP3- or IP-receptor knockout mice, but the effect totally disappeared in the animals lacking EP1-receptors. The protective action of TC against HCl/ethanol

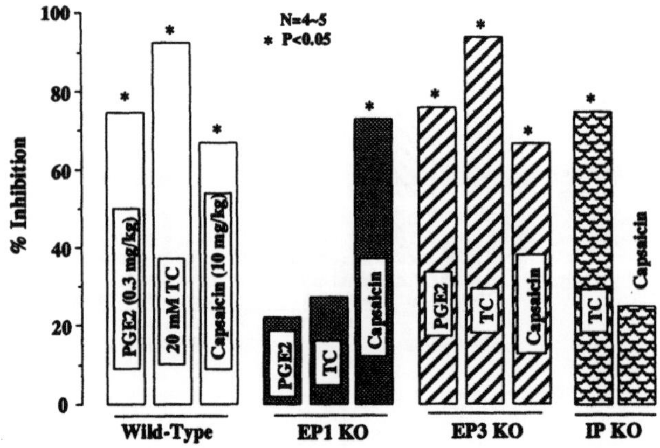


**Figure 3.** Effects of PGE2, TC and capsaicin on HCl/ethanol-induced gastric lesions in wild-type mice and those lacking EP1-, EP3- or IP-receptors. The animals were given 0.3 ml of HCl/ethanol (60% ethanol in 150 mM HCl) p.o. and killed 1 hr later. PGE2 (0.3 mg/kg, s.c.), TC (0.3 ml: 20 mM, p.o.) or capsaicin (10 mg/kg, p.o.) was given 30 min before HCl/ethanol. Data are presented as % inhibition of the lesion score observed in the corresponding controls and represent the mean from 4-5 mice. * Significant protection from control at $P<0.05$.

in WT mice was significantly attenuated by pretreatment with either indomethacin or AE-829 (not shown).

**Protection by Capsaicin Against HCl/Ethanol-Induced Gastric Lesions**

The severity of HCl/ethanol-induced gastric lesions was dose-dependently reduced in the animals pretreated with capsaicin (1-10 mg/kg) p.o. before challenge with HCl/ethanol, and a significant effect was obtained at over 3 mg/kg (not shown). The protective action of capsaicin (10 mg/kg) was totally attenuated by chemical ablation of sensory neurons as well as prior administration of indomethacin (5 mg/kg) but not by AE-829 (10 mg/kg)(Figure 2). However, when the animals were given various EP agonists i.v. 20 min after indomethacin, the protective action of capsaicin was again observed in the rats pretreated with butaprost (EP2 agonist). Neither ONO-NT-012 nor 11-deoxy PGE1 rescued the capsaicin action against HCl/ethanol in the presence of indomethacin (Figure 4). Furthermore, when butaprost (3 mg/kg) was given i.v. prior to capsaicin p.o. (1-10 mg/kg), the protective action of capsaicin was significantly enhanced at 1 and 3 mg/kg (not shown).

Since a recent study showed a role for PGI2 in the release of calcitonin gene-related peptide (CGRP) in the stomach following capsaicin stimulation (2), the protective effect of capsaicin was also examined in IP-receptor knockout mice as well as those lacking EP1- or EP3-receptors. HCl/ethanol caused gastric lesions in both EP1-, EP3- and IP-receptor knockout mice, similar to WT mice, and the severity of these lesions was about the same among these groups. Capsaicin (10 mg/kg) significantly reduced the severity of these lesions in both WT mice and those lacking either EP1- or EP3-receptors (Figure 3). However, this agent failed to protect the stomach against HCl/ethanol in IP-receptor knockout mice; the lesion score in these animals was $18.1\pm3.2$ $mm^2$, which was not significantly different from that observed in the IP-receptor knockout animals treated with saline.

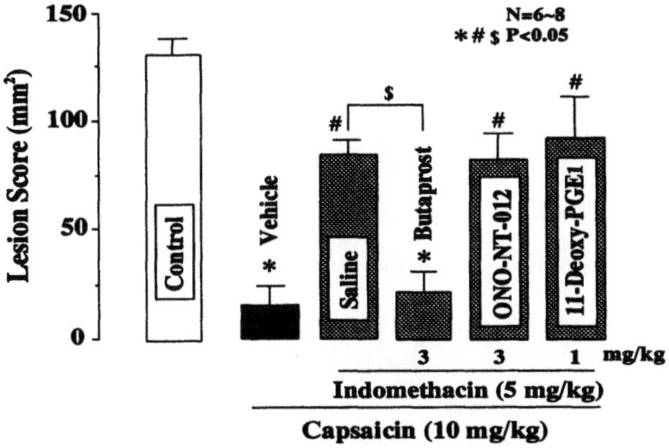


Figure 4. Effects of various EP agonists on the mucosal protective action of capsaicin against HCl/ethanol-induced gastric lesions in the rat stomach in the presence of indomethacin. Animals were given 1 ml of HCl/ethanol (60% in 150 mM HCl) and killed 1 hr later. Capsaicin (10 mg/kg) was given p.o. 30 min before HCl/ethanol. Indomethacin (5mg/kg) was given s.c. 30 min before capsaicin. Various EP agonists were given i.v. 10 min before capsaicin. Data are presented as means±SE from 6~8 rats. Significant difference at $P<0.05$: * from control; # from vehicle; $ from saline.

**Effects of TC and Capsaicin on Mucosal PGE2 Contents**

Levels of PGE2 in the normal rat gastric mucosa were 41.31±7.6 ng/g tissue. Oral administration of 20 mM TC significantly increased the PGE2 levels in the stomach when determined at 0.5 hr after the administration, the levels being 230.6±27.3 ng/g tissue (Figure 5). Prior administration of indomethacin (5 mg/kg, s.c.) markedly reduced the enhanced PG generation in the gastric mucosa following TC treatment, while AE-829 (10 mg/kg) had any effect on the increased PGE2 response caused by TC. The enhanced PG generation by TC was similarly observed in chemically deafferented rats following capsaicin pretreatment. The increase of mucosal PGE2 response to TC in these animals was also totally blocked in the presence of indomethacin. On the other hand, the mucosal PGE2 levels in the stomach was not affected by oral administration of capsaicin (10 mg/kg).

## DISCUSSION

PGs, either endogenous or exogenous derivatives, act on multiple receptors. These receptors that are preferentially activated by PGE over other PGs have been described, and synthetic agonists exhibiting preference for one or more of these receptors have been found. In the present study, we determined the EP receptor subtypes responsible for gastric cytoprotection afforded by exogenous and endogenous PGE2 and discussed which functional alteration is most closely associated with this action. In addition, we also investigated the role of endogenous PGs in gastroprotective action of capsaicin.

We confirmed that PGE2 exhibited a potent cytoprotective action against HCl/ethanol-induced gastric lesions (1). This action was observed dose-dependently with the $ED_{50}$ of about 10 µg/kg. Likewise, gastric cytoprotection against HCl/ethanol was observed by 17-phenyl PGE2 and sulprostone, and the potency of these prostanoids was almost same as that of PGE2.

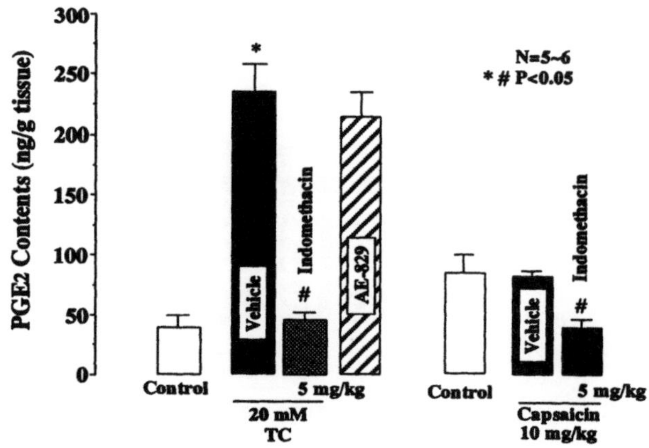


Figure 5. PGE2 generation in the gastric mucosa after oral administration of TC or capsaicin in rats. The animals were given 20 mM TC or capsaicin (10 mg/kg) p.o., killed 0.5 hr later, and the mucosal PGE2 contents were determined by EIA. Indomethacin (5 mg/kg) or ONO-AE-829 (10 mg/kg) was given s.c. 1 hr before TC treatment. Data are presented as the means±SE from 5~6 rats per group. Significant difference at $P<0.05$; * from control in the corresponding group; # from vehicle in the corresponding group.

However, neither butaprost, ONO-NT-012 nor 11-deoxy $PGE_1$ was effective in reducing the severity of HCl/ethanol-induced gastric lesions, indicating that the activation of the EP2, EP3, and EP4 receptors does not lead to gastric cytoprotection. These results strongly suggest that $PGE_2$-induced gastric cytoprotection is brought about by activation of EP1 receptor. This contention is supported by the fact that AE-829, a selective EP1 antagonist, significantly attenuated the protective action of $PGE_2$ against HCl/ethanol. Furthermore, we also confirmed these findings in EP-receptor knockout mice and showed that $PGE_2$ did not affect HCl/ethanol-induced lesions in EP1-receptor knockout mice, although the agent showed a potent gastric cytoprotection in both WT- and EP3-receptor knockout mice.

It is known that endogenous PGs also mediate adaptive cytoprotection observed in the stomach pre-exposed to mild irritants (14,17). In the present study, we used 20 mM TC as a mild irritant and showed that this agent increased $PGE_2$ generation in the stomach and inhitibited the occurrence of damage in response to HCl/ethanol (24). These changes caused by TC totally disappeared in the presence of indomethacin, confirming the involvement of endogneous $PGE_2$ in this phenomenon. Of interest, the protective effect of TC was also attenuated by AE-829 the EP1 antagonist, similar to that afforded by exogenous $PGE_2$. Since this agent did not affect the increased formation of $PGE_2$ in the stomach after TC treatment, it is considered that the reversal by this agent of TC-induced gastric protection occurs through antagonism at the EP1-receptors. Indeed, we observed in EP-receptor knockout mice that TC failed to show gastric protection against HCl/ethanol in mice lacking EP1-receptors, although this agent exhibited a potent protection against these lesions in both WT- and EP3-receptor knockout mice. Since TC increased $PGE_2$ generation in the stomach by the same degree, irrespective of whether EP1- or EP3-receptors were knocked out (not shown), it is likely that the absence of TC-induced gastric protection is due to a lack of EP1-receptors in these animals. These data strongly suggest that mild irritants induce adaptive gastric protection, mediated by endogenous $PGE_2$ through activa-

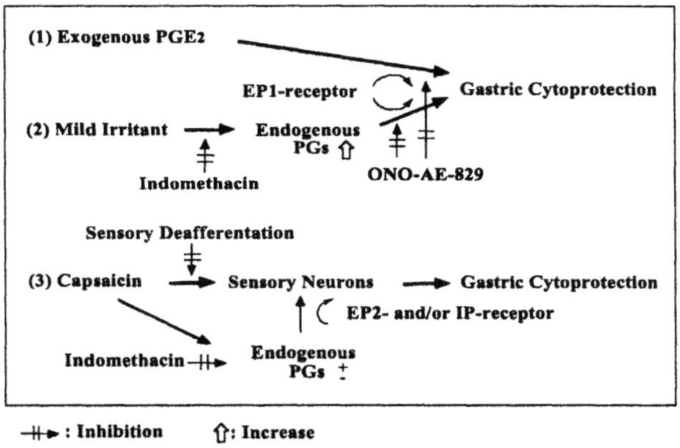


Figure 6. EP receptor subtypes responsible for gastric cytoprotection induced by exogenous PGE2, a mild irritant and capsaicin. 1) Exogenous PGE2 provides direct gastric cytoprotection mediated by activation of EP1 receptors, and this effect is totally blocked by the EP1 antagonist, ONO-AE-829. 2) A mild irritant increases endogenous PGE2 production in the stomach and offers adaptive gastric cytoprotection. This action is prevented by indomethacin as well as the EP1 antagonist. 3) Capsaicin provides gastric cytoprotection, essentially mediated by capsaicin-sensitive afferent neurons. Although capsaicin does not increase PGE2 production in the gastric mucosa, this protective action is facilitated by endogenous PGs through EP2 and IP

tion of EP1-receptors.

Number of studies have been conducted to clarify the mechanisms involved in adaptive protection in the stomach following mild irritants (10,11,20), yet the exact mechanism remains unknown. Since this phenomenon is mediated by endogenous PGE2, the mechanisms should be related to the actions reproduced by exogenous PGE2. Endogenous PGs play a role in the regulation of various gastric functions, such as acid secretion, mucus/bicarbonate secretion, mucosal blood flow and motility, that may contribute to gastric cytoprotection. According to previous studies including our own (1,19,23,26), PGE2 inhibits acid secretion through EP3 receptors and increases mucus and bicarbonate secretion through EP4 and EP1 receptors, respectively. We also observed that gastric mucosal blood flow was increased by EP2, EP3 and EP4 agonists but not EP1 agonists (1). Of interest, prostanoids exhibiting a preference for only EP1 receptors affected gastric motility and provided mucosal protection against gastric lesions induced by HCl/ethanol (1). Certainly, these effects were both antagonized by AE-892, the EP1 antagonist, suggesting that the motility effect of PGE2 is paralleled by a reduction in gastric mucosal damage. We have reported that a variety of compounds afforded gastric cytoprotection at doses that inhibit gastric motility (21). The inhibition of gastric motility (relaxation of circular smooth muscle) may lead to a flattening of the mucosal foldings and a decrease in mucosal vulnerability to irritants, resulting in prevention of the fold-related band-like lesions, as observed following administration of HCl/ethanol (12). Since neither butaprost, ONO-NT-012 nor 11-deoxy PGE1 provided any gastric protection against HCl/ethanol, despite causing an increase in gastric mucosal blood flow, it is unlikely that the gastric cytoprotection afforded by PGE2 is functionally associated with an increase of gastric mucosal blood flow. Certainly, because inhibition of gastric motility may lead to attenuation of microvascular disturbances due to stomach contraction, it is possible that prostanoids through EP1 receptors help to maintain mucosal blood flow during exposure to noxious agents.

The mechanism by which PGE2 inhibits gastric motility through EP1 receptors remains unknown. Milenov and Golenhofen (13) reported that PGE2 relaxed the circular muscle but contracted the longitudinal muscle of the canine stomach. Narumiya and his group reported the localization of mRNAs of the EP receptors along the gastrointestinal tract (5,15). They showed that strong signals for EP1 transcripts occurred in the smooth muscle cells in the muscularis mucosa throughout the tract. Since EP1 receptors are coupled to phosphatidyl inositol turnover (4), it is assumed that contraction of longitudinal smooth muscle by PGE2 is associated with an increase of cytosolic calcium. At present, the mechanism by which PGE2 relaxes circular smooth muscle through activation of EP1 receptors is unknown.

On the other hand, endogenous PGs play a role in the gastric cytoprotection induced by capsaicin and some antiulcer drugs. Capsaicin in particular is unique in that it causes a selective stimulation of capsaicin-sensitive afferent neurons through interaction with vaniloid type 1 receptors (3). The protective action of capsaicin was totally blocked by chemical ablation of these afferent neurons and significantly attenuated by the antagonist of CGRP as well as nitric oxide (NO) synthase inhibitors (8,9). Interestingly, the protective action of capsaicin was also significantly mitigated in the presence of indomethacin, suggesting an involvement of endogenous PGs, similar to the case of adaptive cytoprotection induced by a mild irritant (22). However, this effect of capsaicin was not affected by AE-829, the selective EP1 antagonist, in contrast to that of taurocholate as a mild irritant (24). It should also be noted that neither stimulation of sensory neurons by capsaicin nor sensory deafferentation affected mucosal PGE2 levels in the stomach. These results suggest that although endogenous PGs are involved in the gastric protection induced by both mild irritants and capsaicin, the mode of action seems to be different in these two cases. It is assumed that the stimulation of afferent neurons by capsaicin does not increase the generation of PG in the stomach, yet it exerts a gastroprotective action partly dependent on endogenous PGs. We found that the protective action of capsaicin was significantly restored even in the presence of indomethacin by prior administration of butaprost, the EP2 agonist, but not EP3 or EP4 agonist. Since the capsaicin-induced gastric protection was not affected by the EP1 antagonist, it is unlikely that EP1 receptors are involved in the facilitation by endogenous PGs of this action. Indeed, significant protection by capsaicin was observed even in the knockout mice lacking EP1 and EP3 receptors, confirming that the capsaicin-induced gastric protection has nothing to do with the EP1 and EP3 receptors. However, we found that capsaicin did not provide gastric cytoprotection against HCl/ethanol in IP-receptor knockout animals. These findings in knockout mice suggest that IP receptors are also involved in the protective action of capsaicin in the stomach, in addition to EP2 receptors. At present, the exact mechanism by which endogenous PGs contribute to the protective action of capsaicin remains unknown. Boku et al. (2) recently reported a lack of release of CGRP in response to mild injury in the stomach of IP-receptor knockout mice. Thus, it is assumed that endogenous PGI2 plays a supportive role in the mechanism of capsaicin-induced gastric cytoprotection, probably by sensitizing capsaicin-sensitive afferent neurons.

It should be noted in the present study that the protective effect of exogenous PGE2 against HCl/ethanol was not affected by chemical ablation of capsaicin-sensitive sensory neurons. Likewise, TC-induced gastric protection was not also affected by chemical deafferentation. These observations are controversial with previous findings by others that sensory deafferentation significantly attenuated the protective effect of PGE2 or mild irritants against ethanol (6,11). Although the reason for this controversy remains unknown, it may be due to different experimental conditions, including the period of fasting, the strain of rats or the environmental factors such as breeding, housing and chow. The effect of chemical deafferentation might appear differently, depending on the tonus of the sensory neurons; if it were maintained in "tonic conditions", then the chemical deafferentation would negatively affect the protective action of mild irritants.

Taken together, it is concluded that PGE2, either generated endogenously or administered exogenously, exhibits gastric protection by activation of EP1-receptors and that endogenous PGs also contribute to the mucosal protection induced by capsaicin by sensitizing sensory neurons, probably through EP2- and IP-receptors (Figure 6). In addition, the protective action of PGE2 may be functionally related with inhibition of gastric motility (relaxation of circular smooth muscle) but not increase of mucosal blood flow or mucus secretion.

## REFERENCES


1. **Araki H, Yagi K, Suzuki K, Furukawa O and Takeuchi K.** Roles of prostaglandin E receptor subtypes in cytoprotective action of prostaglandin E2 in rat stomachs. *Alimental Pharmacol Ther* 2000; 14 Suppl 1: 18-25.
2. **Boku K, Ohno T, Saeki T, Hayashi H, Hayashi I, Katori M, Murata T, Narumiya S, Saigenji K, Majima M.** Adaptive cytoprotection mediated by prostaglandin I2 is attributable to sensitization of CGRP-containing sensory nerves. *Gastroenterology* 2001; 120: 134-43.
3. **Caterina MJ, Schumacher MA, Tominaga M. et al.**: The capsaicin receptor: a heat-activated ion channel in the pain pathway. *Nature* 1997; 389: 316-324.
4. **Coleman RA, Smith WL and Narumiya S.** Classification of prostanoid receptors: Properties, distribution, and structure of the receptors and their subtypes. *Pharmacol Rev* 1994; 46: 205-229.
5. **Ding M, Kinoshita Y, Kishi K, Nakata H, Hassan S, Kawanami C, Sugimoto Y, Katsuyama M, Negishi M, Narumiya S, Ichikawa A, Chiba T:** Distribution of prostaglandin E receptors in the rat gastrointestinal tract. *Prostaglandins* 1997; 53: 199-216.
6. **Esplugues JV, Whittle BJ and Moncada S.** Modulation by opioids and by afferent sensory neurones of prostanoid protection of the rat gastric mucosa. *Br J Pharmacol* 1992; 106: 846-852.
7. **Futaki N, Takahashi S, Yokoyama M, Arai I, Higuchi S and Otomo S.** NS-398, a new antiinflammatory agent, selectively inhibits prostaglandin G/H synthase/cyclo-oxygenase (COX-2) activity in vitro. *Prostaglandins* 1994; 47: 55-59.
8. **Holzer P and Sametz W.** Gastric mucosal protection against ulcerogenic factors in the rat mediated by capsaicin-sensitive afferent neurons. Gastroenterology 1986; 91: 975-981.
9. **Holzer P.,** Neural emergency system in the stomach. *Gastroenterology* 1998; 114: 823-839.
10. **Matsumoto J, Ueshima K, Takeuchi K and Okabe S.** Capsaicin-sensitive afferent neurons in adaptive response of the rat stomach induced by a mild irritant. *Jpn J Pharmacol* 1991; 55: 181-185.
11. **Mercer DW, Cross JM, Castaneda AA and Gunter JA.** Gastroprotective actions of bombesin, L-DOPA, and mild irritants: roles of prostaglandins and sensory neurons. *Surgery* 1988; 124: 864-870.
12. **Mersereau WA and Hinchey EJ:** Role of gastric mucosal folds in formation of focal ulcers in the rat. *Surgery* 1982; 91: 150-155.
13. **Milenov K and Golenhofen K:** Contractile responses of longitudinal and circular smooth muscle of the canine stomach to prostaglandin E2 and F2 $\alpha$. *Prostaglandins Leukotrines Med* 1982; 8: 287-300.
14. **Miller TA.** Protective effects of prostaglandins against gastric mucosal damage: Current

knowledge and proposed mechanisms. *Am J Physiol* 1983; 245: G601-G623.

15. **Morimoto K, Sugimoto Y, Katsuyama M, Oida H, Tsuboi K, Kishi K, Kinoshita Y, Negishi M, Chiba T, Narumiya S, Ichikawa A:** Cellular localization of mRNAs for prostaglandin E receptor subtypes in mouse gastrointestinal tract. *Am J Physiol* 1997; 272: G681-G687.
16. **Robert A, Nezamis JE, Lancaster C and Hanchar AJ:** Cytoprotection by prostaglandins in rats. Prevention of gastric necrosis produced by alcohol, HCl, NaOH, hypertonic NaCl, and thermal injury. *Gastroenterology* 1979; 77: 433-443.
17. **Robert A, Nezamis JE, Lancaster C, Davis JP, Field SO and Hanchar AJ.** Mild irritants prevent gastric necrosis through "adaptive cytoprotection" mediated by prostaglandins. *Am J Physiol* 1983; 245: G113-G121.
18. **Sugimoto Y. Namba T, Honda A, Hayashi Y, Negishi M, Ichikawa A and Narumiya S.** Cloning and expression of a cDNA for mouse prostaglandin E receptor EP3 subtype. *J Biol Chem* 1992; 267: 6463-6466.
19. **Takahashi S, Takeuchi K and Okabe S:** EP4 receptor mediation of prostaglandin E2-stimulated mucus secretion by rabbit gastric epithelial cells. *Biochem Pharmacol* 1999; 58: 1997-2002.
20. **Takeuchi K, Ohno T and Okabe S.** Irritative and protective activity of mild irritants in the rat stomach. *Dig Dis Sci* 1987; 32: 889-896.
21. **Takeuchi K, Matsumoto J, Ueshima K. and Okabe S:** Gastric motility changes in the cytoprotective action of N-ethyl-maleimide and capsaicin in the rat stomach. in Tsuchiya M (ed); Frontier of Mucosal Immunology, Elsevier Science Publisher (Biomedical Division), 1991, Vol. 2, pp 581-584.
22. **Takeuchi K, Niida H, Matsumoto J, Ueshima K, Okabe S.** Gastric motility changes in capsaicin-induced cytoprotection in the rat stomach. *Jpn J Pharmacol* 1991; 55: 147-155.
23. **Takeuchi K, Yagi K, Kato S and Ukawa H.** Roles of prostaglandin E-receptor subtypes in gastric and duodenal bicarbonate secretion in rats. *Gastroenterology* 1997; 113: 1553-1559.
24. **Takeuchi K, Araki H, Umeda M, Komoike Y, Suzuki K.** Adaptive gastric cytoprotection is mediated by prostaglandin EP1 receptors: A study using rats and knockout mice. *J Pharmacol Exp Ther,* 2001; 297: 1160-1165.
25. **Ushikubi F, Segi E, Sugimoto Y, Murata T, Matsuoka T, Kobayashi T, Hizaki H, Tuboi K, Katsuyama M, Ichikawa A, Tanaka T, Yoshida N and Narumiya S.** Impaired febrile response in mice lacking the prostaglandin E receptor subtype 3. *Nature* 1998; 395: 281-284.
26. **Yokotani K, Okuma Y and Osumi Y:** Inhibition of vagally mediated gastric acid secretion by activation of central prostanoid EP3 receptors in urethane-anesthetized rats. *Br J Pharmacol* 1996; 117: 653-656.

Chapter 22

# Gastric Mucus Accumulation is Reduced in *Helicobacter pylori* Infected Transgenic Mice Expressing the Human α-1,3/4-Fucosyltransferase in Surface Epithelial Cells

JOHANNA HENRIKSNÄS, MIA PHILLIPSON, LARS ENGSTRAND*, PER FALK‡, LENA HOLM

*Dep. of Medical Cell Biology, Division of Integrative Physiology, Uppsala University, Uppsala, Sweden, Swedish Institute for Infectious Disease Control *, Stockholm, Sweden, AstraZeneca‡, Mölndal, Sweden*

## INTRODUCTION

*Helicobacter pylori* has been shown to be the major cause of gastritis and duodenal ulcers. The mechanisms behind development of injury are, however, not yet fully known. The human Lewis$^b$ blood group antigen (Le$^b$) functions as a receptor for adhesins on *H. pylori* and mediates the attachment of the bacteria to gastric pit and surface mucous cells. A transgenic mouse expressing the human α-1,3/4-fucosyltransferase will produce Le$^b$ epitopes and can thus be infected with clinical isolates of *H. pylori* (3). We have used these mice and developed an experimental in vivo model in which we can study gastric mucus thickness and accumulation in chronically infected mice.

## MATERIALS AND METHODS

FVB/N mice expressing the human α-1,3/4-fucosyltransferase were inoculated twice with *H. pylori* 1 ($10^7$ CFU/mouse) when 5 weeks of age. The inoculation was done at least 8 weeks prior to the experiment. The mice were anaesthetized by inhalation of isoflurane. Body temperature was maintained at 37 C and a cannula was placed in the left carotic artery to monitor blood pressure. The jugular vein was cannulated for continuous infusion of a modified Ringer solution at a rate of 0.35 ml $h^{-1}$. The stomach was exteriorized through a midline abdominal incision, and the forestomach was opened along the greater curvature. The mouse was placed on its left side on a Lucite microscope stage, and the stomach was everted and loosely draped over a truncated cone with the luminal surface facing upwards. A mucosal chamber with a hole in the bottom was placed over the mucosa exposing 0.3cm$^2$ of the corpus. The chamber was filled with unbuffered 0.9% saline, kept at 37°C by warm water perfusing the mucosal chamber.

*Mucus measurements:* Mucus gel thickness was measured by inserting a microelectrode from the luminal solution into the gel at an angle of 34-43° to the surface using a micromanipulator. A "digimatic indicator" was connected to the micromanipulator for measurement of the distance covered by the microelectrode.

*Experimental protocol:* Total mucus thickness was measured, whereafter as much as possible of the mucus was removed by gentle suction. Mucus thickness was measured immediately after removal and after 15 and 30 minutes accumulation. After the measurement at 30 minutes the mucus was removed a second time and the thickness of the firmly adherent mucus measured. Blood pressure and body temperature were monitored during the entire experiment.
All experimental procedures in this study have previously been approved by the Swedish Laboratory Animal Ethical Committee in Uppsala and were conducted in accordance with guidelines of the Swedish National Board for Laboratory Animals.

## RESULTS

As previously shown in rats, the mucus layer could be divided into two layers; a loosely adherent layer that could be removed by gentle suction or mechanically removed with a cotton tip and a firmly adherent layer not possible to remove.

Total mucus thickness did not differ between the infected (n=5) and the non-infected (n=5) mice. The thickness of the firmly adherent mucus layer was however significantly thinner in the infected mice than in the control mice (figure 1). In addition, the mucus accumulation seen in the control group was abolished in the infected group (figure 2).

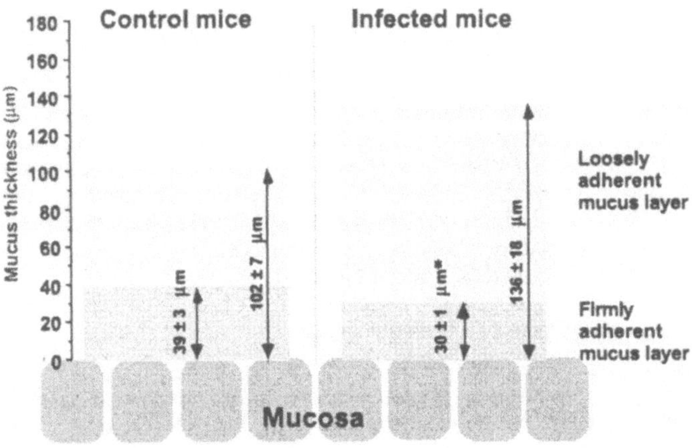


*Figure 1, Thickness of the mucus layers in control and infected mice. The values represent the total mucus thickness at the beginning of the experiment and the thickness of the firmly adherent mucus layer after the first removal of the loosely adherent layer. *P<0.05 compared with control animals.*

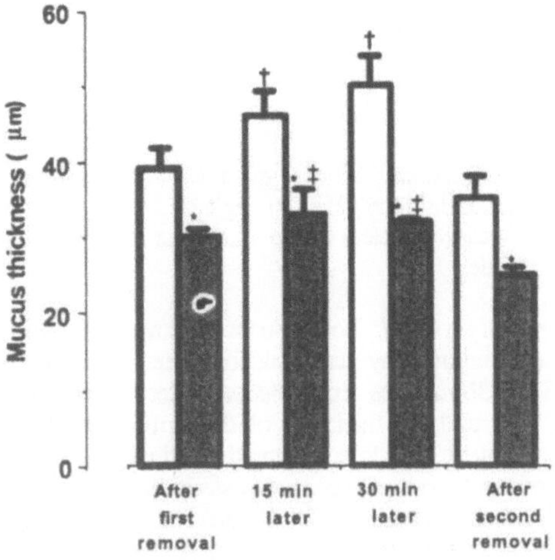


*Figure 2, Mucus thickness in control mice (open bars) and in infected animals (filled bars). Fifteen and thirty minutes after the first removal, the mucus thickness was measured again revealing the rate of mucus accumulation. Finally, the loosely adherent layer was removed again, and the thickness of the firmly adherent layer was measured (second removal). *p<0.05 compared with the control animals, †p<0.05 compared with the first removal, ‡p<0.05 compared with the second removal*

## DISCUSSION

The thinner firmly adherent mucus layer and the abolished mucus accumulation in *H. pylori* infected mice indicate a reduced mucus secretion and/or increased mucus degradation. This could be due to decreased mucin expression, alterations in the types of mucins expressed and/or break down of the mucin gel structure, which has been shown in vitro and in humans infected with *H. pylori* (1,2,4).

Studies in our laboratory in rats show that the firmly adherent layer is adequate for stabilizing a pH gradient with a neutral value at the cell surface while the pH in the lumen is 1 (5). However, the ability to maintain the gradient when the thickness of the firmly adherent layer is reduced, as in the H. pylori infected mice, remains to be investigated.

## CONCLUSION

We have developed an *in vivo* mouse model in which we can measure mucus thickness and accumulation rate in mice chronically infected with *H. pylori*.

The thickness of the firmly adherent mucus layer is reduced and the accumulation of mucus abolished in mice infected with *H. pylori.*

## REFERENCES

1. **Byrd JC, Yan P, Sternberg L, Yunker CK, Scheiman JM, and Bresalier RS.** Aberrant expression of glandtype gastric mucin in the surface epithelium of Helicobacter pylori-infected patients. Gastroenterology, 113, 455-464, 1997

2. **Byrd JC, Yunker CK, Xu Q, Sternberg LR, Bresalier RS**. Inhibition of gastric mucin synthesis by Helicobacter pylori. Gastroenterology, 118, 1072-1079, 2000

3. **Falk PG, Bry L, Holgersson J, and Gordon JI.** Expression of a human α-1,3/4-fucosyltransferase in the pit cell lineage of FVB/N mouse stomach results in production of Leb-containing glycoconjugates: A potential transgenic mouse model for studying Helicobacter pylori infection. Proc. Natl. Acad. Sci. USA 92, 1515-1519, 1995

4. **Newton JL, Jordan N, Oliver L, Strugala V, Pearson J, James FW and Allan A.** Helicobacter pylori in vivo causes structural changes in the adherent gastric mucus layer but barrier thickness is not compromised. Gut, 43, 470-475, 1998

5. **Phillipson M, Atuma C, Henriksnäs J, Holm L.** The importance of mucus layers and bicarbonate transport in preservation of gastric juxtamucosal pH. Submitted

Chapter 23

# New Insights Into Second Messenger Regulation Of Parietal Cell Function By Novel Downstream Signaling Proteins

C.S. CHEW, X. CHEN, H-Y, QIN AND T. STOMING
*Institute of Molecular Medicine and Genetics, Medical College of GA, Augusta, GA*

## INTRODUCTION

HCl secretion by the gastric parietal cell is increased by independent activation of both the cAMP and calcium-dependent signaling pathways. The dramatic internal membrane rearrangements that accompany HCl secretion appear to involve an exocytotic process in which the $H^+,K^+$-ATPase, or proton pump, is translocated from cytoplasmic tubulovesicles to the canalicular membrane, whereupon it becomes active (16). It is well established that histamine binding to H2-type receptors activates adenylyl cyclase, increases [cAMP]i and activates cAMP-dependent protein kinase. In contrast, cholinergic agonists such as carbachol bind to muscarinic M3-type receptors and increase intracellular concentrations of inositol 1,3,5 trisphosphate, diaclyglycerol and $[Ca^{2+}]$i (21, 52). Although the downstream protein kinases in the cholinergic signaling pathway have not been unequivocally defined, most evidence supports the involvement of both calcium-dependent protein kinase(s) and protein kinase C (21, 52). The activation of multiple protein kinases in the various signaling pathways predicts the involvement of multiple phosphoprotein substrates. To date, two regulated phosphoproteins have been identified in the cAMP signaling pathway, ezrin and lasp-1 (11, 20). Two other proteins, CSPP28 and coroninse, have been identified as calcium and protein kinase C-dependent phosphoproteins, respectively (38, 39). The identification of these new signaling proteins has generated many more questions. Which cellular activities are modified by changes in protein phosphorylation? Are other phosphoproteins also regulated within these signaling pathways? Do the pathways intersect? If so, where? Our goal has been to address these questions systematically by 1) identifying the regulated phosphoproteins and 2) applying a wide range of molecular and physiological approaches to characterize these proteins.

## MATERIALS AND METHODS

The following methods have been described in detail in previous publications: isolation of parietal cells and gastric glands from male New Zealand White rabbits (8, 9), in situ $^{32}P$ labeling of cellular proteins, two dimensional (2D) electrophoresis of labeled proteins, phosphoprotein purification, in situ tryptic digestion, PCR cloning, EGFP-tagged-constructs, western blot analysis and enhanced chemiluminescence (ECL) detection, immunofluorescence and confocal microscopic analysis (10, 11, 38, 39).

Antibodies included previously generated monoclonals: anti-lasp-1 (10); anti-CSPP28 (38), anti-coroninse (39) and those from commercial sources: anti-rac-1 and anti-Cdc42 (mouse monoclonals, Upstate Biotechnology (UBI), Lake Placid, NY), anti-Arp 3 (goat polyclonal IgG, Santa Cruz, CA). Activated Cdc42 was identified using an assay that detects Cdc42-GTP by affinity precipitation with the p21-binding domain of p21-activated kinase 1 (PAK-1) bound to agarose beads (UBI kit).

Actin polymerization was assessed by low speed centrifugation, which differentially pellets polymerized (F) actin but not monomeric (G) actin, and by a pyrene-based fluorescent assay (24).

In this latter assay, pyrene-labeled non-muscle actin was mixed with unlabeled non-muscle actin (Cytoskeleton, Inc., Denver, CO) at a ratio of 1:10 and a concentration of 0.34 µg/µl. His-tagged lasp-1 was added at varying concentrations and actin polymerized using a standard actin polymerization buffer (Cytoskeleton). Fluorescence was monitored with a PTI Deltascan spectrofluorimeter (excitation and emission λ' s: 365 and 407 nm, respectively). Actin bundling was assessed by the method of Korneeva et al. (23).

2D-DIGE was performed as follows: Extracts from control and forskolin-stimulated (10 μM, 15 min.) parietal cells (8) were differentially labeled with cyanine(Cy)-3 and Cy-5 fluorescent dyes according to the standard Amersham Pharmacia protocol. Equal amounts of differentially labeled proteins were pooled and separated according to pI on Immobiline strips (pH 3-10 NL) with an IPGPhor (Amersham Pharmacia) and according to Mr by SDS-PAGE in the second dimension. Fluorescently tagged proteins were analyzed by excitation at appropriate wavelengths and computer-based scanning. Changing proteins were identified based on differences in the merged images of Cy-3 vs Cy-5 labeling patterns. In some experiments, proteins in the gels were subsequently labeled with Sypro Ruby Red fluorescent stain (Molecular Probes, Eugene, OR) for automatic "spot" picking and tryptic digestion and mass spec analyses.

## RESULTS AND DISCUSSION

### *cAMP-dependent phosphoproteins*

To date, the best characterized phosphoprotein is the ERM (ezrin-radixin-moesin) family member, ezrin, which is a microvillar component of a number polarized epithelial cell types and has been localized most prominently within the F-actin rich apically-directed canalicular membrane region of the parietal cell (19). In this highly differentiated cell type, ezrin preferentially associates with β-actin in the canalicular region (57). Work with other cell types and cell lines suggests that ezrin crosslinks actin to cell membranes by binding to specific membrane proteins such as EBP50 and the close homologue, E3KARP, which confers cAMP-dependent protein kinase (PKA)-dependent regulation to the sodium-hydrogen exchanger, NHE3 (2, 29). Ezrin also binds to type II PKA extracted from parietal cells (15). Many studies have focused on the role ezrin and other ERM proteins play in regulating actin association with the cortical cell membrane with reference to changes in cell shape, motility and adhesion. For example, ezrin is a candidate for the direct control of elongation of actin microfilaments (29) and, like other ERM proteins, is phosphorylated on a conserved threonine residue upon activation of both Rho kinase and PKC θ (2, 29). Phosphorylation also alters the conformation of ezrin such that it binds to F-actin. Once phosphorylated, ezrin can also disassociate Rho-GDI, a GDP dissociation inhibitor, from Rho thereby activating the Rho-dependent signaling pathway. Rho family members, Rho, Rac and Cdc42, have well known effects on the actin cytoskeleton, particularly in the formation of cell processes (43). These proteins have also been localized to vesicular compartments and are likely to play an important role in regulating vesicular trafficking (42). Recent research suggests that ezrin and other ERM proteins may also be more directly involved in the regulation of endocytic trafficking (3), possibly by blocking the endocytosis of specific transporters within the apical membrane (2). Ezrin might also serve to regulate de novo actin assembly on membrane surfaces thereby providing a "track" for organelles to move toward actin nucleating sources and aggregate prior to fusion (14). Such functions have not yet been specifically ascribed to the parietal cell nor have the upstream protein kinase(s) responsible for cAMP-dependent phosphorylation of ezrin in this cell type been identified. However, interesting parallels can be drawn with respect to microfilament elongation and translocation of the $H^+,K^+$-ATPase to these structures.

Compared to ezrin, lasp-1 (LIM and SH3 domain-containing protein) is much less well characterized. Lasp-1 was first identified as a ~40 kDa cAMP-responsive parietal cell phosphoprotein in 1987, but the sequence was not defined until recently (11). Lasp-1 not only contains an amino terminal LIM domain and a carboxyl terminal SH3 domain but also two nebulin-like actin binding repeats. Recently, lasp-1 was also shown to associate with actin *in vitro* (44). Blast-initiated GenBank searches indicate that lasp-1 shares the highest degree of domain-based homology with previously identified members of the nebulin family: nebulin, nebulette and N-RAP. Lasp-1 appears to be a composite of other nebulin family members in that it contains an N-terminal LIM domain like N-RAP and a carboxy-terminal SH3 domain like nebulin and nebulette (Fig. 1).

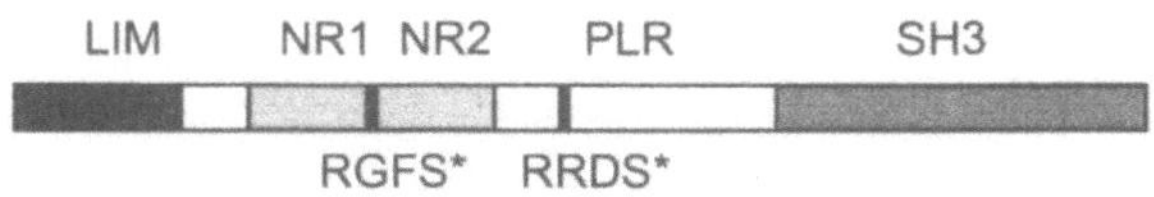


Fig. 1. Diagrammatic representation of various domains present in lasp-1 beginning with the carboxyl terminal LIM domain. NR, nebulin repeat; PLR, proline-rich linker region; RGFS*/RRDS*, cAMP-dependent protein kinase consensus phosphorylation sites.

Independent analyses of the two nebulin repeats in lasp-1 indicate that the first nebulin repeat is most homologous with a highly conserved repeat labeled module #8 in nebulette, whereas the second nebulin repeat is most homologous with another highly conserved module (#21) in the same protein (31). Phosphobase analyses (25) detect a number of PKA consensus sites in human adult nebulin (36) and nebulette (13) as well as in murine N-RAP (11). As with the two putative PKA consensus sites in human and rabbit lasp-1 (RGF<u>S</u>$^{99}$ and RRD<u>S</u>$^{146}$), the majority of these sites fall within the nebulin repeat regions of these proteins (not shown). Such detailed domain and phosphorylation site analyses in conjunction with previous findings that lasp-1 is not expressed in skeletal, cardiac or non-vascular smooth muscle (10), have led us to propose lasp-1 as an epithelial representative of the nebulin repeat family of proteins.

Like other nebulin family members, the physiological functions of lasp-1 are unknown. The primary family member, nebulin, is a large (~600-800 kDa) skeletal muscle protein, that may be involved in specifying the length of thin filaments (26). In vitro, nebulin fragments promote actin polymerization and induce the formation of tightly packed actin bundles (6,17). Nebulin also appears to be phosphorylated *in vivo* on serine/threonine residues in response to isoproterenol, which elevates skeletal muscle [cAMP]i (47). Nebulette is exclusively expressed in cardiac muscle where it has been proposed to have a function similar to that of nebulin in skeletal muscle (31). N-RAP is expressed in both skeletal and cardiac muscle and has been suggested to play a role in anchoring terminal actin filaments to the myotendinous junction (27).

In the face of this limited information, we have initiated several experimental approaches to begin to define the functions of lasp-1 including sequence analyses across organisms, cellular and subcellular localization studies, protein interaction profiles, and phosphorylation site analyses. Extensive Blast searches of complete, or near complete, genomic sequences of yeast, bacteria, worms and flies have not identified a single protein with clearly defined overall homology to lasp-1. However, homologous SH3 and LIM domains are present in two different putative yeast (*C. elegans*) proteins, YLZ3 (P34416, 81% positives) and YLZ4 (P34417, 75% positives). This latter protein also contains two nebulin repeats that are most homologous with the second nebulin repeat of lasp-1 (86 and 78% positives). In flies (*Drosophila*), no significant LIM domain-based homologies were identified but the putative CG3849 gene product (AAF4926.1) is most homologous with the lasp-1 SH3 domain (78% positives) and also contains two nebulin repeats that are most homologous to the second nebulin repeat in lasp-1 (70 and 63% positives). These findings are generally in keeping with a recent report by Venter and colleagues (53) who noted a striking expansion in nebulin repeat modules in humans (148) as compared to fly (2), worm (1), yeast (0) and plant (0). Thus, like other nebulin family members, lasp-1 appears to be a recently evolved mammalian protein and, as such, expected to have specialized function(s) in epithelia. This hypothesis is reinforced by observations that lasp-1 is highly expressed in certain ion transporting epithelial cells including, for example, the gastric parietal cell, intestinal enterocytes, pancreatic, salivary and hepatic duct cells as well as certain cells in the distal tubule/collecting duct region of the kidney (10). Interestingly, within the SH3 domain region, lasp-1 also shares a relatively high degree of homology with several actin binding proteins including cortactin (NP005222.1), yeast ABP1 (NP10012.1) and human drebin F (AF077353.1) as well as with proteins linked to vesicle trafficking including several synaptic dynamin-associated protein (syndapin) II isoforms (as AF139493.1, for example). Thus, it is possible that lasp-1, like ezrin, is not only an actin-binding protein but is also involved in vesicular trafficking.

Lasp-1 has been immunolocalized mainly to the subcortical membrane region in polarized parietal cells within gastric glands (10) and to the tips of membrane extensions and podosomes in cultured cells and/or fibroblasts (10, 44). However, a more detailed analysis suggests that in the "resting" parietal cell lasp-1 is also present within small F-actin-rich "patches" that appear form a between the base of the intracellular canaliculi and F-actin present in the cortical membrane regione (Fig. 2 a,b; arrowheads). There is also a less prominent lasp-1 signal within the apically-directed regions of the canaliculi (Fig. 2a, arrows). After stimulation with forskolin (which induces an elevation in [cAMP]i and a dramatic microvillar expansion of the intracellular canaliculi), the lasp-1 signal becomes more prominent in this region (Fig. 2 c; arrows) as previously shown (10). Interestingly, lasp-1 is also associated with thin, F-actin containing fibers that appear connect the cell cortex to the canaliculi (Fig. 2 c,d arrowheads).

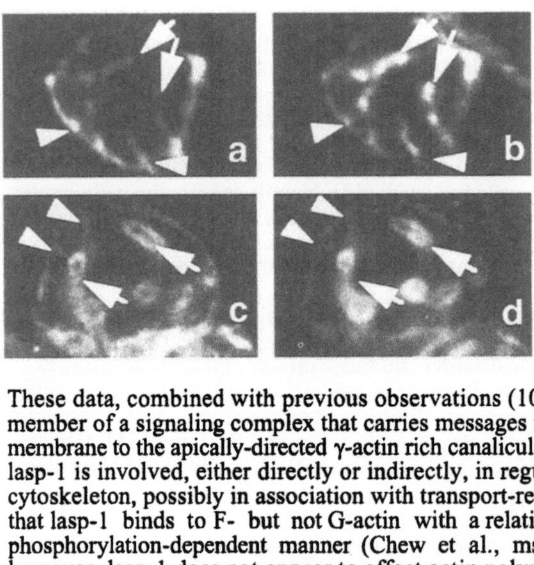


Fig.2. Immunofluorescent localization of lasp-1 (a,c) and F-actin (b,d) in polarized parietal cells within gastric glands. In unstimulated cells (a,b), lasp-1 is prominent at the cell cortex and in actin rich "patches" (arrowheads) which appear to be associated with the base of the intracellular canaliculi (arrows). After forskolin stimulation (c,d; 10 μM, 30 min), both lasp-1 and F-actin signals are prominent in the canalicular region (arrows). F-actin and lasp-1 can also be detected along "strings" that appear tether the canaliculi to the basolateral plasma membrane (arrowheads).

These data, combined with previous observations (10), suggest that lasp-1 may serve as a member of a signaling complex that carries messages from the γ-actin rich basolateral plasma membrane to the apically-directed γ-actin rich canalicular membrane. Another possibility is that lasp-1 is involved, either directly or indirectly, in regulating the plasticity of the actin-based cytoskeleton, possibly in association with transport-related functions. Recent studies indicate that lasp-1 binds to F- but not G-actin with a relatively low affinity (Kd ~ 3 μM) in a phosphorylation-dependent manner (Chew et al., ms submitted). In contrast to nebulin, however, lasp-1 does not appear to affect actin polymerization over a wide range of lasp-1 concentrations (Fig. 3). Lasp-1 also does not appear to modulate actin bundling (not shown).

The strong similarities between the SH3 domain of lasp-1 and cortactin/syndapin II led us to analyze the known functions of these proteins for clues that might lead to a better understanding of lasp-1. Both cortactin and syndapin II bind to the large GTPase, dynamin II (30, 40, 55). Dynamin isoforms control the fission reaction in vesicular endocytosis (46) and may also be functionally linked to the actin cytoskeleton (34). Recently, lasp-1 has also been found to associate with dynamin II in vitro and to partially co-localize with this protein in the parietal cell (35). Syndapins not only interact with dynamin but also with several other proteins implicated in membrane trafficking and expression of the SH3 domains of both brain-specific syndapin I and the more ubiquitously expressed syndapin II, blocks receptor-mediated endocytosis (40). Syndapin II also binds neural Wiskott-Aldrich syndrome protein (N-WASP) (40), a widely expressed protein that promotes actin filament nucleation by interacting with the Arp2/3 complex (28). Interestingly, overexpression of either full length syndapin II or lasp-1 induces filopodial formation (35,40). Moreover, both proteins have been localized to filopodial tips and lamellipodia (10,40). Thus, lasp-1 and syndapin II might also play similar and/or complimentary roles in linking membrane trafficking activities and actin cytoskeleton dynamics (35,40).

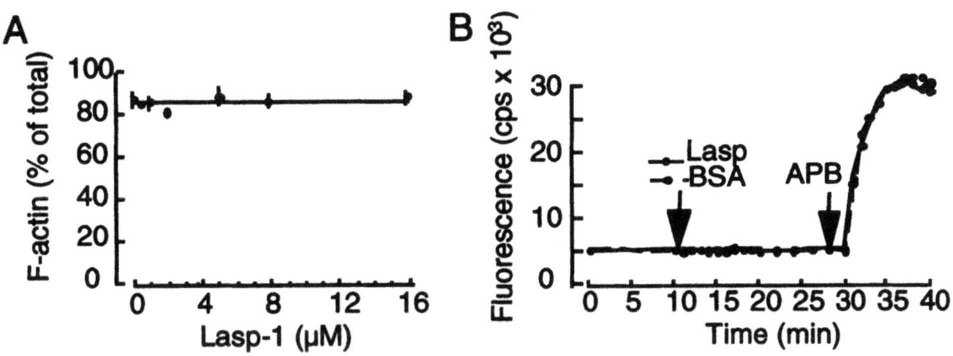


Fig. 3. Lasp-1 has no detectable effect on actin polymerization.

A. Results from low speed pelleting assays (means ±SEMs, N=5) showing constant level of F-actin over a wide range of lasp-1 concentrations.

B. Superimposed results from two independent experiments showing that neither lasp-1 (8 μM) nor BSA (8 μM) induce actin polymerization or affect the rate of actin polymerization induced upon addition of actin polymerization buffer (APB). The same results were obtained with lasp-1 at concentrations ranging from 0.5 to 8 μM (not shown).

The similarities between cortactin and lasp-1 are even more striking. In addition to associating with dynamin, the cortactin/amplaxin EMS1 gene, like the lasp-1 MLN 50 gene, is amplified in certain cancers including breast carcinomas (45, 50). The binding of both lasp-1 and cortactin to F-actin is modified by phosphorylation (22). Both proteins also undergo phosphorylation-dependent translocation in cells and mobility shifts on western blots (10, 11, 37) and have been localized to actin-rich cell surface projections (10, 44, 56). Recent findings that cortactin stimulates the actin-nucleating activity of the Arp2/3 complex (51, 54) are particularly intriguing. Previously, only members of the Wiskott-Aldrich-syndrome protein (WASP) family and the ActA protein from *Listeria* were known to have such an effect. Thus, cortactin may function cooperatively with WASP family members to maintain the active state of the Arp2/3 complex (59). Further work is required to determine if lasp-1 also modulates actin polymerization by a similar indirect action. However, as shown in figure 4, several of the necessary players are readily detectable in the parietal cell including Cdc42, Rac-1 and Arp3.

*PKC and calcium-dependent phosphoproteins*

Coroninse was first identified as a 66 kDa protein (pp66) that underwent increased phosphorylation on serine residues following stimulation with the cholinergic agonist, carbachol, or the PKC activator, phorbol myristate acetate (4). Chelation of both intracellular and extracellular calcium has no effect on pp66 phosphorylation. Thus, it appears that this protein is regulated by novel or calcium-independent, phorbol ester-dependent protein kinase(s). Once pp66 was isolated, partially sequenced and cloned, it was found to be a new coronin family member. It was named coroninse because it was found to be highly expressed in secretory epithelia (39). There are now six known mammalian coronin-like protein family members, coronins 1A-C, 2A, clipin C and IR10 (32, 36, 39, 48, 58). Rabbit coroninse is most homologous with coronin 1B, which has also been labeled coronin-2 in rat.

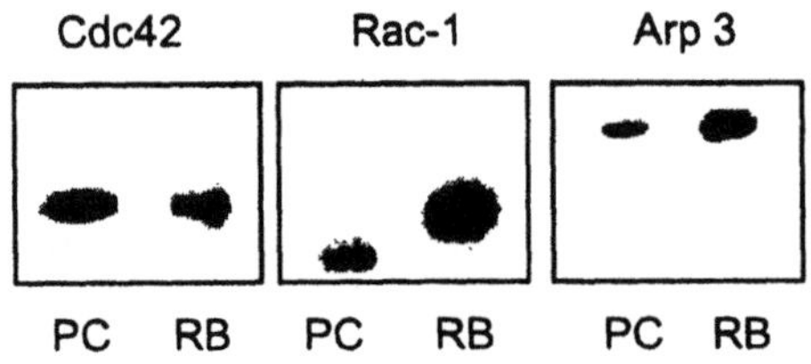


Fig. 4. Representative western blots comparing Cdc42, Rac-1 and Arp 3 immunoreactivity in extracts from rabbit parietal cells and rat brain. In each panel, the signal on the left is from parietal cells (PC) and that on the right is from brain (RB). The apparent molecular weight of Rac-1 was consistently lower in rabbit vs rat (17 and 21 kDa, respectively). In contrast, in both species Arp3 migrated at ~ 50 kDa and Cdc42 migrated at ~21 kDa.

As shown in the diagram in figure 5, coroninse contains eleven PKC phosphorylation motifs, five of which are serine residues. Depending on the stringency of the analysis, there are 5-7 WD40 (Gβ) repeats plus several endocytosis motifs. There is also a potential transmembrane region near the N-terminus and a coiled-coiled region that may be involved in actin binding at the carboxyl terminus. Both the WD40 repeats and the predicted coiled-coiled secondary structures are conserved in the known coronin isoforms. However, the coiled-coil region is the most variable among the isoforms in terms of amino acid sequence. In *Dictyostilium*, the original coronin has been implicated in remodeling of the cortical actin cytoskeleton as related to phagocytosis and macropinocytosis and has been specifically associated with a late phase endocytotic compartment (41). In mammalian neutrophils, coronin 1A (p57) is also associated with the phagocytic apparatus (13); and in *Listeria*, coronin has been localized to actin comet tails along with Rac, the Arp2/3 complex and other proteins known to be critical for cellular actin rearrangements (12). Within the parietal cell, EGFP-tagged coroninse localizes to vesicular structures in the subcanalicular region (39). Thus, the possibility that coroninse plays a role in the endocytic retrieval of the $H^+,K^+$-ATPase is attractive, particularly given that the major effect of phorbol esters appears to be inhibition of HCl secretion and the transient nature of the cholinergic acid secretory response (21, 52).

CSPP28 (calcium-sensitive phoshoprotein) is a 28 kDa protein that undergoes increased phosphorylation in parietal and chief cells in response to cholinergic agonists and calcium ionophores (4). When isolated and cloned, CSPP28 was found to share >95% homology with the predicted human gene product, D52 (38). D52 is a member of a small gene family and is overexpressed in some mammary and lung cancers (5, 7, 33). CSPP28 appears to be identical to cholinergically-regulated phosphoproteins detected in rat colon and pancreas (18, 38). Recent evidence suggests that the pancreatic protein (CRHSP28) is associated with vesicles, perhaps by binding to annexin VI in a calcium-dependent manner (49). In gastric mucosal fractions, CSPP28 is also enriched in a light membrane fraction (Fig. 6). Interestingly, although both lasp-1 and coroninse are somewhat enriched in these same fractions, these two proteins are also prominent in low speed particulate fractions which suggests an association with other cellular compartments, most likely cytoskeletal in nature.

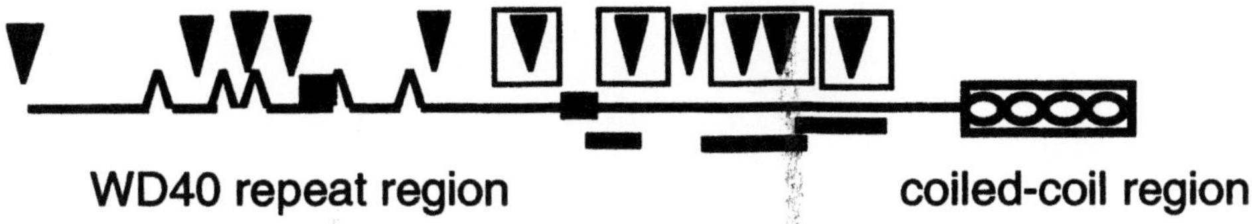


Fig. 5. Diagrammatic representation of coroninse showing location of WD40 repeats, coiled-coil actin binding region, endocytosis-associated motifs and PKC consensus phosphorylation sites. Arrowheads, threonine residues; boxed arrowheads, serine residues. Small black boxes, dileucine motifs; Solid lines, regions with homology to YGR clathrin adaptor complex, NSF-vesicle fusion region in AAA protein family, and LRL dynamin family (39).

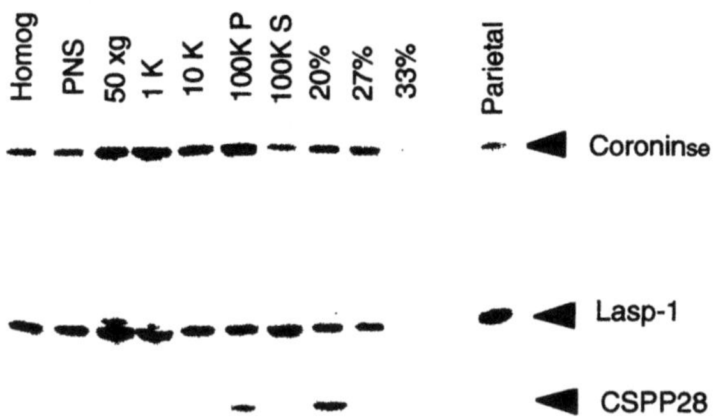


Fig. 6. Western blot analyses of the distribution of coroninse, lasp-1, and CSPP28 in various fractions from a gastric mucosal homogenate and from a parietal cell extract. The same blots were probed for all three proteins as described in Methods. Homog, whole homogenate; PNS, post nuclear supernatant; P, particulate, S, supernatant; 20-33%, vesicular fractions from sucrose gradient.

### *Future Directions*

Although much work remains to be done in defining the specific cellular functions that are regulated by the known signaling phosphoproteins described above, all evidence to date points in the direction of the actin cytoskeleton and/or vesicular trafficking associated with the acid secretory process. This linkage with actin is particularly exciting in light of recent findings with cultured parietal cells in which the formation of cell extensions was found to be significantly more sensitive to actin cytoskeleton disrupters such as latrunculin A and cytochlalasin D, as compared to the intracellular canalicular region. Thus, in addition to differences in isoform distribution patterns (57), there appear to be at least two pools of differentially regulated actin in the parietal cell (1). As shown in figure 7, results with 2D DIGE further suggest that a number of other phosphoproteins may be involved in the regulation of parietal cell functions. With the rapid advances in the field of proteomics, these proteins should be identified much more rapidly than their prototypes. However, such data make it clear that we have only just begun the process of understanding how so many different proteins function in a unified fashion to control cellular activities!

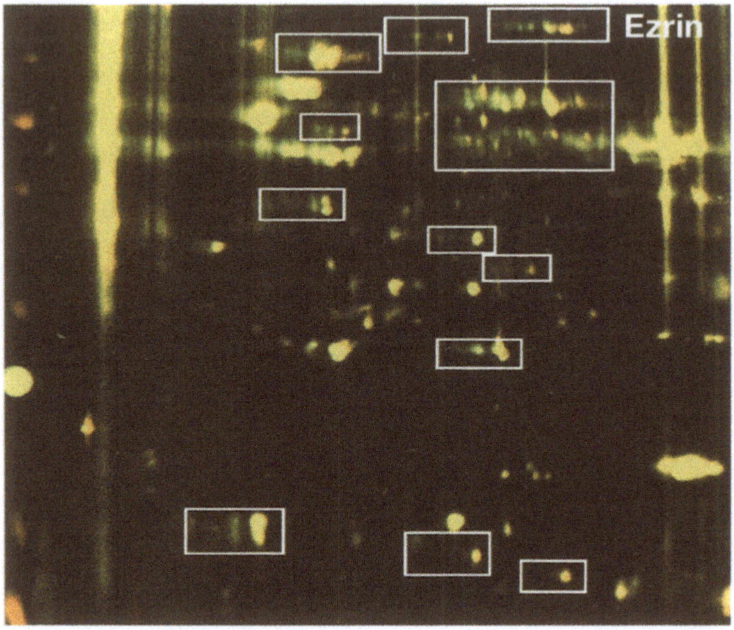


Fig. 7. Representative 2D-DIGE gel showing the location of differentially labeled proteins from control (Cy5, red) and forskolin-stimulated (Cy3, green) extracts of parietal cells. Acidic end of the gel on the left. Red staining proteins on the far left are BioRad Precision Mr standards (75-15 kDa). Large yellow "dot" is an internal marker. White boxes highlight proteins that did not co-migrate. An acidic shift was observed with all of these proteins which most likely indicates an increase in negative charge resulting from the addition of phosphate groups to the protein.

## REFERENCES

1. **Ammar, D. A., P. N. Nguyen, and J. G. Forte.** Functionally distinct pools of actin in secretory cells. *Am J Physiol Cell Physiol* 281: C407-17, 2001.
2. **Bretscher, A.** Regulation of cortical structure by the ezrin-radixin-moesin protein family. *Curr Opin Cell Biol* 11: 109-16, 1999.
3. **Bretscher, A., D. Chambers, R. Nguyen, and D. Reczek.** ERM-Merlin and EBP50 protein families in plasma membrane organization and function. *Annu Rev Cell Dev Biol* 16: 113-43, 2000.
4. **Brown, M. R., and C. S. Chew.** Carbachol-induced protein phosphorylation in parietal cells: regulation by [Ca2+]i. *Am J Physiol* 257: G99-110, 1989.
5. **Byrne, J. A., M. G. Mattei, and P. Basset.** Definition of the tumor protein D52 (TPD52) gene family through cloning of D52 homologues in human (hD53) and mouse (mD52). *Genomics* 35: 523-32, 1996.
6. **Chen, M. J., C. L. Shih, and K. Wang.** Nebulin as an actin zipper. A two-module nebulin fragment promotes actin nucleation and stabilizes actin filaments. *J Biol Chem* 268: 20327-34, 1993.

7. **Chen, S. L., X. K. Zhang, D. O. Halverson, M. K. Byeon, C. W. Schweinfest, D. K. Ferris, and N. K. Bhat.** Characterization of human N8 protein. *Oncogene* 15: 2577-88, 1997.
8. **Chew, C. S.** cAMP technologies, functional correlates in gastric parietal cells. *Methods Enzymol* 191: 640-61, 1990.
9. **Chew, C. S., K. Nakamura, and A. C. Petropoulos.** Multiple actions of epidermal growth factor and TGF-alpha on rabbit gastric parietal cell function. *Am J Physiol* 267: G818-26, 1994.
10. **Chew, C. S., J. A. Parente, Jr., X. Chen, C. Chaponnier, and R. S. Cameron.** The LIM and SH3 domain-containing protein, lasp-1, may link the cAMP signaling pathway with dynamic membrane restructuring activities in ion transporting epithelia. *J Cell Sci* 113: 2035-45, 2000.
11. **Chew, C. S., J. A. Parente, Jr., C. Zhou, E. Baranco, and X. Chen.** Lasp-1 is a regulated phosphoprotein within the cAMP signaling pathway in the gastric parietal cell. *Am J Physiol* 275: C56-67, 1998.
12. **David, V., E. Gouin, M. V. Troys, A. Grogan, A. W. Segal, C. Ampe, and P. Cossart.** Identification of cofilin, coronin, Rac and capZ in actin tails using a Listeria affinity approach. *J Cell Sci* 111: 2877-84, 1998.
13. **de Hostos, E. L.** The coronin family of actin-associated proteins. *Trends Cell Biol* 9: 345-50, 1999.
14. **Defacque, H., M. Egeberg, A. Habermann, M. Diakonova, C. Roy, P. Mangeat, W. Voelter, G. Marriott, J. Pfannstiel, H. Faulstich, and G. Griffiths.** Involvement of ezrin/moesin in de novo actin assembly on phagosomal membranes. *Embo J* 19: 199-212, 2000.
15. **Dransfield, D. T., A. J. Bradford, J. Smith, M. Martin, C. Roy, P. H. Mangeat, and J. R. Goldenring.** Ezrin is a cyclic AMP-dependent protein kinase anchoring protein. *Embo J* 16: 35-43, 1997.
16. **Forte, J. G., D. K. Hanzel, C. Okamoto, D. Chow, and T. Urushidani.** Membrane and protein recycling associated with gastric HCl secretion. *J Intern Med Suppl* 732: 17-26, 1990.
17. **Gonsior, S. M., M. Gautel, and H. Hinssen.** A six-module human nebulin fragment bundles actin filaments and induces actin polymerization. *J Muscle Res Cell Motil* 19: 225-35, 1998.
18. **Groblewski, G. E., M. J. Wishart, M. Yoshida, and J. A. Williams.** Purification and identification of a 28-kDa calcium-regulated heat- stable protein. A novel secretagogue-regulated phosphoprotein in exocrine pancreas. *J Biol Chem* 271: 31502-7, 1996.
19. **Hanzel, D., H. Reggio, A. Bretscher, J. G. Forte, and P. Mangeat.** The secretion-stimulated 80K phosphoprotein of parietal cells is ezrin, and has properties of a membrane cytoskeletal linker in the induced apical microvilli. *Embo J* 10: 2363-73, 1991.
20. **Hanzel, D., H. Reggio, A. Bretscher, J. G. Forte, and P. Mangeat.** The secretion-stimulated 80K phosphoprotein of parietal cells is ezrin, and has properties of a membrane cytoskeletal linker in the induced apical microvilli [published erratum appears in EMBO J 1991 Dec;10(12):3978-81]. *EMBO J.* 10: 2363-73, 1991.
21. **Hersey, S. J., and G. Sachs.** Gastric acid secretion. *Physiol. Revs.* 75: 155-89, 1995.
22. **Huang, C., Y. Ni, T. Wang, Y. Gao, C. C. Haudenschild, and X. Zhan.** Down-regulation of the filamentous actin cross-linking activity of cortactin by Src-mediated tyrosine phosphorylation. *J Biol Chem* 272: 13911-5, 1997.
23. **Korneeva, N. L., and B. M. Jockusch.** Light microscopic analysis of ligand-induced actin filament suprastructures. *Eur J Cell Biol* 71: 351-5, 1996.
24. **Kouyama, T., and K. Mihashi.** Fluorimetry study of N-(1-pyrenyl)iodoacetamide-labelled F-actin. Local structural change of actin protomer both on polymerization and on binding of heavy meromyosin. *Eur J Biochem* 114: 33-8, 1981.

25. **Kreegipuu, A., N. Blom, and S. Brunak.** PhosphoBase, a database of phosphorylation sites: release 2.0. *Nucleic Acids Res* 27: 237-9, 1999.
26. **Littlefield, R., and V. M. Fowler.** Defining actin filament length in striated muscle: rulers and caps or dynamic stability? *Annu Rev Cell Dev Biol* 14: 487-525, 1998.
27. **Luo, G., J. Q. Zhang, T. P. Nguyen, A. H. Herrera, B. Paterson, and R. Horowits.** Complete cDNA sequence and tissue localization of N-RAP, a novel nebulin-related protein of striated muscle. *Cell Motil Cytoskeleton* 38: 75-90, 1997.
28. **Machesky, L. M., and K. L. Gould.** The Arp2/3 complex: a multifunctional actin organizer. *Curr Opin Cell Biol* 11: 117-21, 1999.
29. **Mangeat, P., C. Roy, and M. Martin.** ERM proteins in cell adhesion and membrane dynamics. *Trends Cell Biol* 9: 187-92, 1999.
30. **McNiven, M. A., L. Kim, E. W. Krueger, J. D. Orth, H. Cao, and T. W. Wong.** Regulated interactions between dynamin and the actin-binding protein cortactin modulate cell shape. *J Cell Biol* 151: 187-98, 2000.
31. **Moncman, C. L., and K. Wang.** Architecture of the thin filament-Z-line junction: lessons from nebulette and nebulin homologies. *J Muscle Res Cell Motil* 21: 153-69, 2000.
32. **Nakamura, T., K. Takeuchi, S. Muraoka, H. Takezoe, N. Takahashi, and N. Mori.** A neurally enriched coronin-like protein, ClipinC, is a novel candidate for an actin cytoskeleton-cortical membrane-linking protein. *J Biol Chem* 274: 13322-7, 1999.
33. **Nourse, C. R., M. G. Mattei, P. Gunning, and J. A. Byrne.** Cloning of a third member of the D52 gene family indicates alternative coding sequence usage in D52-like transcripts. *Biochim Biophys Acta* 1443: 155-68, 1998.
34. **Ochoa, G. C., V. I. Slepnev, L. Neff, N. Ringstad, K. Takei, L. Daniell, W. Kim, H. Cao, M. McNiven, R. Baron, and P. De Camilli.** A functional link between dynamin and the actin cytoskeleton at podosomes. *J Cell Biol* 150: 377-89, 2000.
35. **Okamoto, C. T., C.S. Chew, and R. Li.** Characterization of lasp-1-dynamin II interactions in vitro and in vivo. *FASEB J.* 15: A499, 2001.
36. **Okumura, M., C. Kung, S. Wong, M. Rodgers, and M. L. Thomas.** Definition of family of coronin-related proteins conserved between humans and mice: close genetic linkage between coronin-2 and CD45- associated protein. *DNA Cell Biol* 17: 779-87, 1998.
37. **Ozawa, K., K. Kashiwada, M. Takahashi, and K. Sobue.** Translocation of cortactin (p80/85) to the actin-based cytoskeleton during thrombin receptor-mediated platelet activation. *Exp Cell Res* 221: 197-204, 1995.
38. **Parente, J. A., J. R. Goldenring, A. C. Petropoulos, U. Hellman, and C. S. Chew.** Purification, cloning, and expression of a novel, endogenous, calcium-sensitive, 28-kDa phosphoprotein. *J Biol Chem* 271: 20096-101, 1996.
39. **Parente, J. A., Jr., X. Chen, C. Zhou, A. C. Petropoulos, and C. S. Chew.** Isolation, cloning, and characterization of a new mammalian coronin family member, coroninse, which is regulated within the protein kinase C signaling pathway. *J Biol Chem* 274: 3017-25, 1999.
40. **Qualmann, B., and R. B. Kelly.** Syndapin isoforms participate in receptor-mediated endocytosis and actin organization. *J Cell Biol* 148: 1047-62, 2000.
41. **Rauchenberger, R., U. Hacker, J. Murphy, J. Niewohner, and M. Maniak.** Coronin and vacuolin identify consecutive stages of a late, actin- coated endocytic compartment in Dictyostelium. *Curr Biol* 7: 215-8, 1997.
42. **Ridley, A. J.** Rho proteins: linking signaling with membrane trafficking. *Traffic* 2: 303-10, 2001.
43. **Schmidt, A., and M. N. Hall.** Signaling to the actin cytoskeleton. *Annu Rev Cell Dev Biol* 14: 305-38, 1998.

44. **Schreiber, V., C. Moog-Lutz, C. H. Regnier, M. P. Chenard, H. Boeuf, J. L. Vonesch, C. Tomasetto, and M. C. Rio.** Lasp-1, a novel type of actin-binding protein accumulating in cell membrane extensions. *Mol Med* 4: 675-87, 1998.
45. **Schuuring, E.** The involvement of the chromosome 11q13 region in human malignancies: cyclin D1 and EMS1 are two new candidate oncogenes--a review. *Gene* 159: 83-96, 1995.
46. **Sever, S., H. Damke, and S. L. Schmid.** Garrotes, springs, ratchets, and whips: putting dynamin models to the test. *Traffic* 1: 385-92, 2000.
47. **Somerville, L. L., and K. Wang.** In vivo phosphorylation of titin and nebulin in frog skeletal muscle. *Biochem Biophys Res Commun* 147: 986-92, 1987.
48. **Suzuki, K., J. Nishihata, Y. Arai, N. Honma, K. Yamamoto, T. Irimura, and S. Toyoshima.** Molecular cloning of a novel actin-binding protein, p57, with a WD repeat and a leucine zipper motif. *FEBS Lett* 364: 283-8, 1995.
49. **Thomas, D. D., W. B. Taft, K. M. Kaspar, and G. E. Groblewski.** CRHSP-28 regulates $Ca^{2+}$-stimulated secretion in permeabilized acinar cells. *J Biol Chem* 276: 28866-72, 2001.
50. **Tomasetto, C., C. Moog-Lutz, C. H. Regnier, V. Schreiber, P. Basset, and M. C. Rio.** Lasp-1 (MLN 50) defines a new LIM protein subfamily characterized by the association of LIM and SH3 domains. *FEBS Lett* 373: 245-9, 1995.
51. **Uruno, T., J. Liu, P. Zhang, Y. Fan, C. Egile, R. Li, S. C. Mueller, and X. Zhan.** Activation of Arp2/3 complex-mediated actin polymerization by cortactin. *Nat Cell Biol* 3: 259-66, 2001.
52. **Urushidani, T., and J. G. Forte.** Signal transduction and activation of acid secretion in the parietal cell. *J Membr Biol* 159: 99-111, 1997.
53. **Venter, J. C., M. D. Adams, E. W. Myers, P. W. Li, R. J. Mural, et al.** The sequence of the human genome. *Science* 291: 1304-51, 2001.
54. **Weaver, A. M., A. V. Karginov, A. W. Kinley, S. A. Weed, Y. Li, J. T. Parsons, and J. A. Cooper.** Cortactin promotes and stabilizes Arp2/3-induced actin filament network formation. *Curr Biol* 11: 370-4, 2001.
55. **Wigge, P., K. Kohler, Y. Vallis, C. A. Doyle, D. Owen, S. P. Hunt, and H. T. McMahon.** Amphiphysin heterodimers: potential role in clathrin-mediated endocytosis. *Mol Biol Cell* 8: 2003-15, 1997.
56. **Wu, H., and J. T. Parsons.** Cortactin, an 80/85-kilodalton pp60src substrate, is a filamentous actin-binding protein enriched in the cell cortex. *J Cell Biol* 120: 1417-26, 1993.
57. **Yao, X., C. Chaponnier, G. Gabbiani, and J. G. Forte.** Polarized distribution of actin isoforms in gastric parietal cells. *Mol Biol Cell* 6: 541-57, 1995.
58. **Zaphiropoulos, P. G., and R. Toftgard.** cDNA cloning of a novel WD repeat protein mapping to the 9q22.3 chromosomal region. *DNA Cell Biol* 15: 1049-56, 1996.
59. **Zettl, M., and M. Way.** New tricks for an old dog? *Nat Cell Biol* 3: E74-5, 2001.

Chapter 24

# Functional Role of Protein KinaseB/Akt in Gastric Acid Secretion

ANDREA TODISCO, NONTHALEE PAUSAWASDI, SARAVANAN RAMAMOORTHY, JOHN DEL VALLE, REBECCA W. VAN DYKE, FREDERICK K. ASKARI

*Department of Internal Medicine, University of Michigan, Ann Arbor, MI, USA*

## INTRODUCTION

Epidermal growth factor (EGF) is the prototypic member of a large family of peptide growth factors that are known to exert numerous biological actions in the gut (30).

In the stomach, EGF affects acid secretion in a divergent fashion. Under acute conditions, EGF has a long recognized inhibitory effect on acid secretion, whereas prolonged administration of EGF increases both basal and maximal acid secretion in vivo (12,16,17,36). Similar results have been observed in vitro, using preparations of highly purified gastric parietal cells in primary culture which have been shown to express specific receptors for EGF/TGFα (4,12,16,24). In this system, prolonged exposure of the parietal cells to EGF (>16 h) has been shown to enhance the stimulatory effect of carbachol, gastrin and histamine, on gastric acid secretion (12,16,24). The mechanisms responsible for this phenomenon are currently only partially understood.

One hypothesis, based on in vivo observations, is that EGF stimulates the proliferation of progenitor neck cells present in the gastric glands, leading to an overall increase in the number of acid producing parietal cells present in the gastric mucosa (17). Alternatively, recent studies performed both in isolated, cultured, parietal cells and in AGS human gastric adenocarcinoma cells, have indicated that EGF is able to induce both the expression and the transcription of the $H^+,K^+$-ATP-ase α-subunit gene (22,24), suggesting that EGF might have a direct stimulatory effect on the expression of the gene responsible for the process of gastric acid production.

Little is known about the intracellular signal transduction mechanisms that mediate the stimulatory effects of EGF on gastric acid production and on $H^+,K^+$-ATP-ase α-subunit gene expression. Some studies, performed in isolated gastric parietal cells in primary culture have suggested that the stimulatory actions of EGF on gastric acid secretion, could be mediated by activation of the MAPK signal transduction pathway since they can be fully reversed either by genistein, a protein tyrosine kinase inhibitor, or by the MEK1 inhibitor, PD98059 (1,16,34).

One signal transduction pathway which has been recently shown to play an important role in the mediation of some of the physiological actions of EGF, is that involving phosphoinositide 3-kinase (PI3K) and protein kinase B/Akt (8,10,29,37). Activation of Akt is known to induce cellular growth and survival and to promote the expression and the maintenance of highly differentiated cellular phenotypes (8,10,23,25). In addition, Akt has also been shown to regulate the translocation of GLUT-4 to the plasma membrane (10,25) and to contribute to the regulation of vescicular trafficking and endocytosis (10). Akt is homologous to the PKA and PKC families of protein kinases (2,10). In vivo, the activity of Akt is regulated by growth factors and serum through the induction of PI3K (2,10,20). Phosphorylation of Akt appears to be critical for Akt activation (2,10,20). The major phosphorylation sites required for activation of Akt 1 have been identified as threonine 308 and serine 473 which are the target of the phosphoinositide-dependent kinases I (PDK1) and II (PDK2), respectively (2,10). Similarly, Akt 2 and Akt 3 are also phosphorylated on serine and threonine residues (Thr-309 and Ser-474 in Akt 2 and Thr-305 and Ser-472 in Akt 3) by PDK 1 and PDK2 (10,29). Recent investigations have suggested that growth factors induce the activation of PI3K which in turn, leads to the production of PtdIns-3,4-$P_2$ at the cell membrane

(8,10). Akt binds to this lipid, dimerizes and it is stabilized in a partially active state (8,10). The location at the membrane and the dimerization appear to enhance the ability of Akt to be phosphorylated and fully activated (8,10).

Currently, little is known about the regulation and function of protein kinase Akt in the stomach. Thus, we have undertaken these studies to examine the role of Akt in the process of gastric acid production using highly purified gastric parietal cells in primary culture. Since EGF has a direct stimulatory effect on $H^+,K^+$-ATP-ase α subunit gene expression, we explored the possibility that this action of EGF could be mediated by the activation of Akt.

## MATERIAL AND METHODS

**Adenoviral vectors** - The replication defective adenoviral vectors expressing either dominant negative, hemagglutinin- (HA) tagged Akt 1(with serine 473 and threonine 308 mutated to alanine) (Ad.dom.neg.Akt) or HA-tagged, myristoylated, constitutively active Akt 1 (Ad.Myr-Akt), under the control of the CMV promoter, were gifts of K. Walsh (Tufts University, Boston, MA) (21). The adenoviral vector expressing the β-galactosidase enzyme under the control of the CMV promoter (Ad.CMV-β-gal) was previously described (3).
**Primary parietal cell preparation and culture** - For preparation of primary parietal cells we utilized a modification of the method of Soll et al (7,11,15,16,24,26,32,34,35). The isolated parietal cells ($2 \times 10^6$ cells/well) are cultured according to the method of Chew et al. with some modifications (11,26,28,35). The parietal cells were transduced, when indicated, with 100 multiplicity of infection (moi) of the adenoviral vectors expressing dominant negative Akt, constitutively active Akt, or β-galactosidase, for different time periods.
**Northern blot analysis** - The parietal cells were lysed with TRIzol (Gibco BRL, Grand Island, NY) according to the manufacturer's instructions. Northern blot hybridization assays were performed as previously described (28).
**Immunoprecipitations, Western Blots and kinase assays** - Immunoprecipitations, western blots and kinase assays were performed according to previously described techniques (28,31,35).
**Amplification and purification of adenoviral vectors** - The recombinant adenoviruses were amplified as previously described using 293 cells (3).
**Immunohistochemistry** - The parietal cells were transduced with the adenoviral vectors expressing HA-tagged dominant negative Akt, HA-tagged Myr-Akt or with those expressing the β-galactosidase enzyme and cultured on slides for 16-18 h. At the end of the incubation period the cells were fixed in 4% formalin-PBS and stained with a mouse monoclonal anti-HA antibody (1:500) (Babco, CA) and a FITC-conjugated donkey anti-mouse igG secondary antibody (Jackson Immunoresearch Laboratories, West Grove, PA). After a wash with PBS, the cells were mounted in Vectashield (Vector Laboratories, Burlingame, CA) and visualized by confocal fluorescence microscopy.
**Aminopyrine uptake** - Gastric acid secretion was measured according to previously described methods (24,34).

## RESULTS

We examined the effect of EGF on Akt phosphorylation using western blots with a specific anti-phospho-Akt antibody directed against phospho-Ser 473. As shown in fig. 1, EGF stimulated Akt phosphorylation, with a maximal effect detected after 5 min of incubation at doses ranging between $10^{-8}$ M and $10^{-6}$ M. Akt activation returned to baseline after prolonged exposure to EGF [30-60 min, (fig. 1) and 24 h (data not shown)].

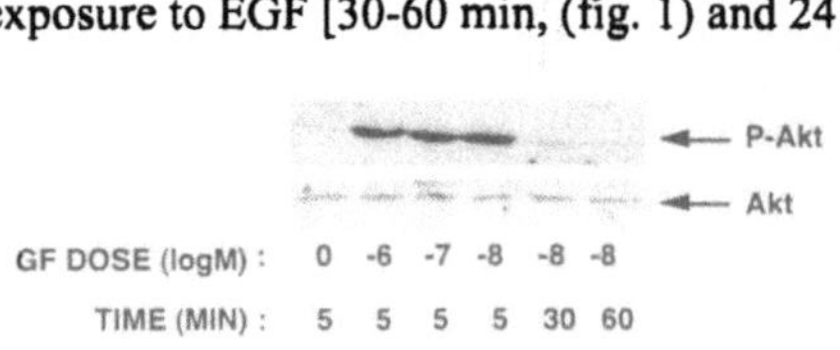


**FIG.1**: EGF stimulates Akt phosphorylation.

By contrast, the cholinergic agonist carbachol, ($10^{-4}$ M) failed to have any effect (fig. 2A). Total Akt levels were not changed by any of the treatments. In order to confirm that phosphorylation of Akt correlated with Akt activation, we examined the effect of EGF ($10^{-8}$ M) on Akt activity measured by an in-vitro kinase assay using glycogen synthase kinase 3α (GSK3α) as substrate. GSK3α phosphorylation was quantitated by western blots with a specific anti-phospho-GSK3α/β antibody. EGF exposure for 5 minutes induced Akt activation but it did not alter the amount of total Akt immunoprecipitated from the cell lysates (data not shown). Since activation of PI3K is though to be an important step in Akt activation, we examined whether EGF induction of Akt phosphorylation, required PI3K activity. EGF-stimulated Akt phosphorylation was inhibited by $10^{-5}$ M LY294002 (fig. 2B), a specific and well characterized inhibitor of PI3K, indicating that in the gastric parietal cells, EGF targets Akt through PI3K dependent signaling pathways. In contrast, both the MAPK inhibitor PD98059 ($5x10^{-5}$ M) and the PKC inhibitor GF109203X ($3.5x10^{-6}$ M) failed to have any effect on EGF-stimulated Akt phosphorylation (fig. 2C).

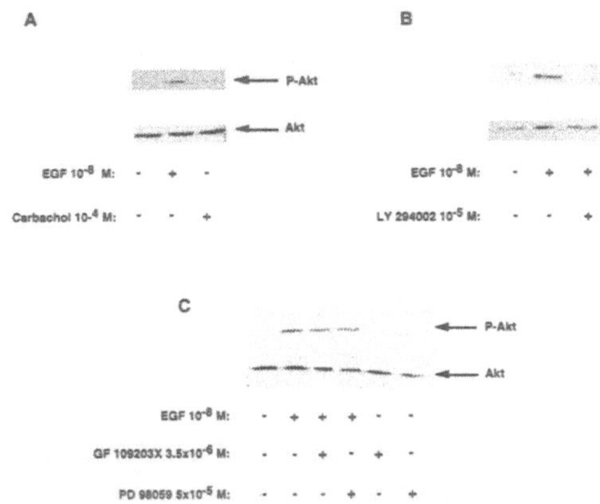


**FIG.2**: Signal transduction pathways regulating Akt phosphorylation.

In order to examine the role of Akt on $H^+/K^+$-ATP-ase gene expression and on gastric acid production, we transduced the gastric parietal cells for 18 h with 100 moi of the adenoviral vectors expressing either a hemagglutinin-tagged myristoylated, constitutively active form of Akt or a hemagglutinin-tagged dominant negative Akt gene. Control experiments were performed with the adenoviral vector Ad.CMV-β-gal which expresses β-galactosidase. Non-infected cells did not stain with X-gal (fig. 3A). In contrast, as shown in fig. 3B, histochemical staining for β-galactosidase demonstrated that more than 80% of the parietal cells were transduced with the adenoviral vector expressing β-galactosidase. In order to document expression of both dominant negative and of constitutively active Akt, parietal cells transduced with either Ad.dom.neg.Akt or Ad.Myr-Akt, were stained with an anti-hemagglutinin primary antibody and a FITC-conjugated secondary antibody. Eighteen to twenty, 1 μm thick z-sections of parietal cells from three different fields of one slide of transduced cells, were analyzed by confocal microscopy in order to examine the localization of the virally expressed kinases. As indicated by the representative images shown in fig. 3, hemagglutinin-tagged dominant negative Akt (fig. 3D) and hemagglutinin-tagged myristoylated Akt (fig. 3F), were expressed in the parietal cells. Dominant negative Akt- and myristoylated Akt-transduced parietal cells, exhibited similar morphology. Ad.CMV-β-gal-transduced parietal cells stained with the anti-hemagglutinin primary antibody and the FITC-conjugated secondary antibody, showed no significant cell membrane staining (fig. 3H). The percentage of parietal cells transduced with the adenoviral vectors expressing dominant negative and myristoylated Akt

varied between 50 and 80%, depending on the preparation. The data represented in fig. 3D and F indicate that more then 50% of the cells were successfully transduced. Corresponding differential interference contrast images (DIC) are shown in C, E and G.
Time course studies showed that expression of HA-tagged dom.neg.Akt was detectable as early as 1 h after infection (data not shown).

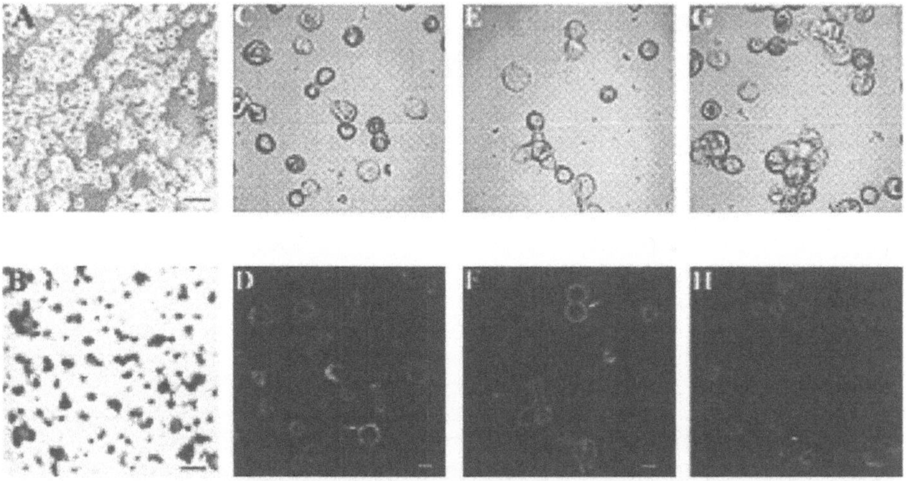


**FIG.3**: Transduction of cultured gastric parietal cells with adenoviral vectors. Images are representative 1 µm confocal fluorescence sections. Arrows indicate specific staining of the surface plasma membrane with the anti-HA antibody. Magnification 20X in A and B, 63X in C,D,E,F,G and H. Bars 50 µm in A and B, 10 µm in D, F and H.

$H^+/K^+$-ATP-ase mRNA was isolated from the parietal cells and quantitated by Northern blot analysis using a canine $H^+/K^+$-ATP-ase α-subunit cDNA probe. The autoradiograms were controlled for RNA quantity by hybridization of the RNA with a GAPDH cDNA probe. In agreement with previously published observations (24), stimulation of the parietal cells with EGF ($10^{-8}$ M) for either 7 h or 16 h, induced canine $H^+/K^+$-ATP-ase α-subunit gene expression in these gastric parietal cells (data not shown). Transduction of the parietal cells with the adenoviral vector expressing constitutively active Akt for 18 h, induced $H^+/K^+$-ATP-ase α-subunit gene expression three-fold, while no effect was observed in the presence of the control vector expressing β-galactosidase (fig. 4A and B). Identical results were observed when the parietal cells were transduced with Ad.Myr-Akt for seven hours (data not shown). In order to confirm the specificity of this observation, we examined the effect of Akt on the expression of the highly inducible gene cyclooxygenase-2 (COX-2). Transduction of the parietal cells for 18 h with the adenoviral vector expressing constitutively active Akt, failed to stimulate COX-2 gene expression which, in contrast, was potently induced by stimulation of parietal cells from the same dog preparation, with carbachol ($10^{-4}$ M), for two hours. In addition, stimulation with EGF ($10^{-8}$ M) for 30 min, 2 h and 24 h, failed to induce COX-2 gene expression in the gastric parietal cells (data not shown). These results indicate that Akt exerts specific stimulatory effects on $H^+/K^+$-ATP-ase gene expression in the gastric parietal cells (fig. 4 C and D).

**FIG.4** : Akt regulates $H^+/K^+$-ATP-ase α subunit gene expression.

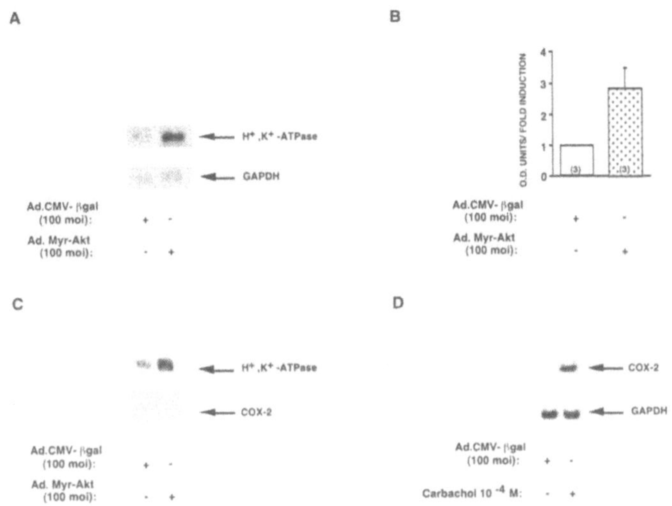


To examine the functional significance of Akt in gastric acid secretion, we tested the effect of Akt on $^{14}$C-aminopyrine uptake. As depicted in fig. 5, transduction of the parietal cells for 18 h with the adenoviral vector expressing constitutively active Akt led to a four-fold induction in basal aminopyrine uptake and to an enhancement of the stimulatory effect of carbachol. In contrast, the control adenoviral vector expressing β-galactosidase did not alter the function of the parietal cells since carbachol stimulated uptake of $^{14}$C-aminopyrine was similar in both infected and non infected cells. Thus, activation of Akt appears to be important for stimulation of gastric acid production.

**FIG.5**: Effect of adenoviral vectors on $^{14}$C aminopyrine uptake.

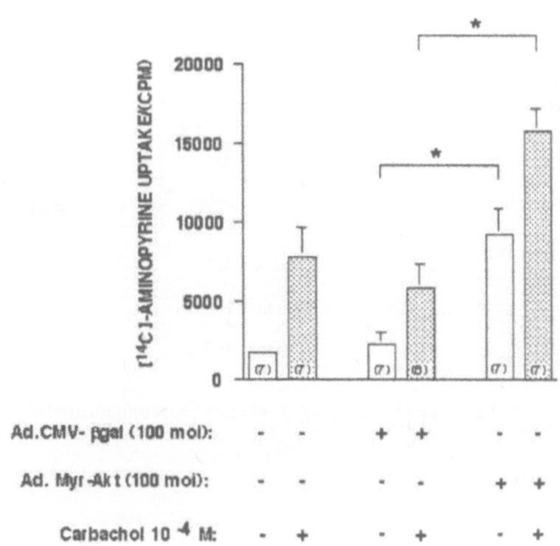

In order to confirm the involvement of Akt in the stimulatory action of EGF on gastric acid secretion, we conducted studies in which we transduced the gastric parietal cells with the adenoviral vector expressing dominant negative Akt. We examined the effect of dominant negative Akt on gastric acid secretion. Transduction of gastric the parietal cells with the adenoviral vector expressing dominant negative Akt, reversed the stimulatory action of EGF on carbachol induced increase in $^{14}C$ aminopyrine uptake (fig. 6), confirming the notion that activation of Akt plays an important role in the mediation of the stimulatory effect of EGF on gastric acid production.

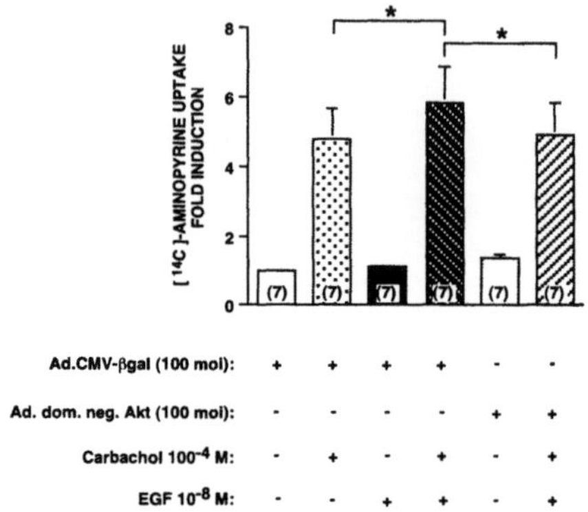


**FIG.6**: Effect of Ad.dom.neg.Akt on EGF enhancement of gastric acid secretion.

Since in the gastric parietal cells, EGF appears to target Akt through PI3K dependent signaling pathways, we examined the effect of LY294002 on the stimulatory action of EGF on gastric acid secretion. As shown in fig. 7, preincubation of the parietal cells with LY294002, completely reversed the stimulatory effect of EGF on AP uptake, suggesting that EGF requires the sequential activation of PI3K and Akt kinase in order to regulate gastric acid production.

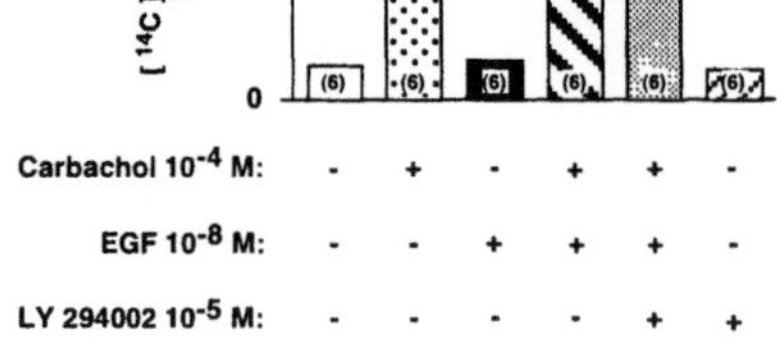


**FIG.7**: LY294002 inhibits EGF enhancement of gastric acid secretion.

To further demonstrate that EGF stimulation of gastric acid secretion is mediated by increased production of the $H^+/K^+$-ATP-ase through an Akt-dependent mechanism, we performed western blots with a specific anti-$H^+/K^+$-ATP-ase $\alpha$-subunit antibody. As shown in fig. 8, EGF stimulated the production of the $H^+/K^+$-ATP-ase $\alpha$-subunit, while dominant negative Akt inhibited this effect. Thus, EGF induces Akt to regulate $H^+/K^+$-ATP-ase production and gastric acid secretion in the gastric parietal cells.

**FIG.8**: EGF induces the production of the $H^+/K^+$-ATP-ase.

# DISCUSSION

In this study, we explored the functional role of Akt in the gastric parietal cells. Akt is known to exert numerous important physiological functions and to mediate some of the intracellular effects of EGF (37,29).

Gastric acid secretion is a complex biological process that involves the coordinated interaction of numerous hormones and neurotransmitters (16). The importance of carbachol, gastrin, and histamine as stimulants of gastric acid secretion has been confirmed in studies with isolated parietal cells (16). These agents are known to trigger a complex cascade of intracellular events that leads to the activation of specific genes including that encoding the key enzyme involved in $H^+$ production, the $H^+$,$K^+$-ATP-ase (7,16).

The resting parietal cells are characterized by the presence of a collapsed canalicular system and of cytoplasmic tubulovesicles containing the gastric proton pump, $H^+$,$K^+$-ATP-ase. Upon stimulation with the gastric acid secretagogues, the cells rapidly develop an intracellular canalicular system bulging with microvilli with concomitant fusion of the tubulovesicles with the canaliculi (16,19,33). This fusion translocates the $H^+$,$K^+$-ATP-ase from the vesicular membranes to the apical plasma membrane, where it actively pumps $H^+$, ions in exchange for $K^+$ (16).

In addition to carbachol, gastrin, and histamine, numerous other agents are known to regulate the secretory activity of the parietal cells (16). EGF, in particular, is known to interact with specific EGF receptors to modulate gastric acid production (4,16). Several studies, conducted in isolated and cultured parietal cells have indicated that prolonged incubation of the cells (>16 h) with EGF leads to enhancement of the stimulatory effect of carbachol, gastrin and histamine on gastric acid secretion (12,16,17,24,34). Additional investigations have suggested that EGF-enhancement of secretagogue-stimulated gastric acid secretion, could reflect the ability of EGF to induce the expression of the $H^+$,$K^+$-ATP-ase gene (24). A study performed in the canine parietal cells, has indicated that EGF is able to induce the transcription of the $\alpha$-subunit of the $H^+$,$K^+$-ATP-ase gene through a novel EGF-response element (ERE) located between bases -162 and -156 (5'-GACATGG-3') relative to the cap site (24). This ERE was found to be homologous to the 3'half-site of the c-fos serum response element (SRE). The specific nuclear proteins that regulate the transcriptional activation of this element are currently unknown.

We reported that EGF induces Akt through PKC- and MAPK-independent signaling pathways since neither the PKC inhibitor GF109203X nor the MEK1 inhibitor PD98059 had any effect on EGF stimulation of Akt activation. These observations are in agreement with other studies that have indicated that MAPK is not involved in the activation of Akt (13). Although the role of PKC in Akt activation has been poorly characterized (18), our results suggest that GF109203X-sensitive PKC isoforms do not play any role in the regulation of Akt in the parietal cells. In contrast, we confirmed the involvement of PI3K-dependent signal transduction pathways, since the PI3K inhibitor LY294002 completely blocked EGF stimulation of Akt activation.

Interestingly, we also observed that carbachol does not induce Akt in the gastric parietal cells, while it potently activates this kinase in the PC-12 cells (27). Accordingly, there appear to be considerable cell type specific differences regarding the ability of muscarinic receptors to induce the Akt signal transduction pathway.

In order to analyze the functional significance of Akt activation in the parietal cells, we took advantage of adenoviral vectors expressing either constitutively active or dominant negative forms of Akt. Using immunocytochemistry and confocal microscopy, we confirmed that these molecules could be effectively delivered to the parietal cells. As expected, the constitutively active Akt protein, which contains a myristoylation sequence, was expressed at the level of both the surface membrane and of intracellular canalicular structures. Although it lacked a specific membrane localization signal, the dominant negative Akt protein was also predominantly localized at the level of the plasma membrane. We hypothesize that this observation might be the result of the high intracellular level of expression of dominant negative Akt achieved by the CMV promoter.

Transduction of the parietal cells with either Ad.Myr-Akt or ad.dom.neg.Akt demonstrated that Akt is involved in the regulation of $H^+,K^+$-ATP-ase gene expression and in the mediation of the stimulatory effect of EGF on gastric acid production. On the basis of these observations we postulate that activation of the parietal cell EGF receptor initiates a cascade of intracellular reactions that leads to the activation of Akt and to the stimulation of $H^+,K^+$-ATP-ase gene expression. These data suggest a novel role for Akt in the regulation of the complex process of gastric acid secretion.

Akt has been shown to regulate the transcription of the early response genes c-myc and c-fos (10). In addition, Akt can phosphorylate and regulate the transcriptional activity of CREB and of forkhead/winged-helix transcription factors (6,10). Thus, it is conceivable that in the parietal cells, Akt might exert similar regulatory actions on nuclear proteins that direct the transcription of the $H^+,K^+$-ATP-ase gene.

We have previously reported that activation of the parietal cell EGF receptor induces MAPK and that incubation of the parietal cells with the MEK1 inhibitor PD98059 leads to a complete reversal of EGF enhancement of carbachol-stimulated gastric acid secretion (34). Accordingly, it appears that multiple signal transduction pathways could be involved in the stimulatory effect of EGF on gastric acid production. Although a direct link between induction of MAPK and stimulation of the $H^+,K^+$-ATP-ase gene has not been demonstrated, it is possible that this gene might receive input from signaling pathways involving both Akt and MAPK. It is clear that additional studies are necessary to examine in more details, the relative contribution of MAPK and Akt in the mediation of the stimulatory action of EGF on $H^+,K^+$-ATP-ase gene expression.

Akt is known to induce the expression of highly specialized cellular phenotypes (10,23,25). In particular, Akt has been shown to play an important role in the complex process of myogenesis and muscle differentiation (23). Similarly, expression of constitutively active forms of Akt can induce the differentiation of 3T3-L1 fibroblasts into adipocytes (25). Another important function of Akt, is to promote cell survival (10) through its ability to phosphorylate proteins such as bad (10,13,14), caspase-9 (9) and forkhead/winged-helix transcription factors (6,10) that appear to play an important role in the induction of apoptosis. Accordingly, the ability of Akt to induce the expression of the $H^+,K^+$-ATP-ase gene might also represent a more generalized effect of this kinase on parietal cell differentiation and survival. It is possible that induction of Akt might be an important signal transduction pathway activated by the parietal cell EGF receptor to promote cellular differentiation and to inhibit apoptosis.

In conclusion, EGF induces a cascade of events in the parietal cells that results in the activation of Akt. The functional role of Akt appears to be stimulation of gastric acid secretion and of $H^+/K^+$-ATP-ase gene expression. These findings suggest a novel role for Akt in the regulation of the secretory functions of the gastric parietal cells.

## REFERENCES


1. **Alessi DR, Cuenda A, Cohen P, Dudley DT, and Saltiel A.** PD 098059 is a specific inhibitor of the activation of mitogen-activated protein kinase kinase in Vitro and in Vivo. *J Biol Chem.* 270: 27489-27494, 1995.

2. **Alessi DR, Andjelkovic M, Caudwell B, Cron P, Morrice N, Cohen P, and Hemmings BA.** Mechanism of activation of protein kinase B by insulin and IGF-1. *EMBO J.* 15: 6541-6551, 1996.

3. **Askari FK, Hitomi Y, Mao M, and Wilson JM.** Complete correction of hyperbilirubinemia in the Gunn rat model of Crigler-Najjar syndrome type I following transient in vivo adenovirus-mediated expression of human bilirubin UDP-glucuronosyltransferase. *Gene Ther.* 3: 381-388, 1996.

4. **Beauchamp RD, Barnard JA, McCutchen CM, Cherner JA, and Coffey RJ Jr.** Localization of transforming growth factor alpha and its receptor in gastric mucosal cells. Implications for a regulatory role in acid secretion and mucosal renewal. *J Clin Invest.* 84: 1017-1023, 1989.

5. **Bradford MM.** A rapid and sensitive method for the quantitation of microgram quantities of protein utilizing the principle of protein-dye binding. *Anal Biochem.* 72: 248-254, 1983.

6. **Brunet A, Bonni A, Zigmond MJ, Lin MZ, Juo P, Hu LS, Anderson MJ, Arden KC, Blenis J, and Greenberg ME.** Akt promotes cell survival by phosphorylating and inhibiting a Forkhead transcription factor. *Cell.* 96: 857-868, 1999.

7. **Campbell V, and Yamada T.** Acid secretagogue-induced stimulation of gastric parietal cell gene expression. *J Biol Chem.* 264: 11381-11386,1989.

8. **Cantley LC, and Neel BG.** New insights into tumor suppression. PTEN suppresses tumor formation by restraining the phosphoinositide 3-kinase/AKT pathway. *Proc Natl Acad Sci. USA* 96: 4240-4245, 1999.

9. **Cardone MH, Roy N, Stennicke HR, Salvesen GS, Franke TF, Stanbridge E, Frisch S, and Reed JC.** Regulation of cell death protease caspase-9 by phosphorylation. *Science.* 282: 1318-1321, 1998.

10. **Chan TO, Rittenhouse SE, and Tsichlis PN.** AKT/PKB and other D3 phosphoinositide-regulated kinases: kinases activation by phosphoinositide-dependent phosphorylation. *Annu Rev Biochem.* 68: 965-1014, 1999.

11. **Chew CS, Ljungstroem M, Smolka A, and Brown MR.** Primary culture of secretagogue-responsive parietal cells from rabbit gastric mucosa. *Am J Physiol.* 256: G254-263, 1989.

12. **Chew CS, Nakamura K, and Petropoulos AC.** Multiple actions of epidermal growth factor and TGF-α on rabbit gastric parietal cell function. *Am J Physiol.* 267: G818-G826, 1994.

13. **Datta SR, Dudek H, Tao X, Masters S, Fu H, Gotoh Y, and Greenberg ME.** Akt phosphorylation of BAD couples survival signals to the cell-intrinsic death machinery. *Cell.* 91: 231-41, 1997.

14. **del Peso L, Gonzalez-Garcia M, Page C, Herrera R, Nunez G.** Interleukin-3-induced phosphorylation of BAD through the protein kinase Akt. Science. 278: 687-689, 1997.

15. **Del Valle, J, Tsunoda Y, Williams JA, and Yamada T.** Regulation of $[Ca^{2+}]i$ by secretagogue stimulation of canine gastric parietal cells. *Am J Physiol.* 262: G420-G426, 1992.

16. **Del Valle J, and Todisco A.** Gastric secretion. In: Textbook of Gastroenterology (3nd. ed.), edited by T. Yamada. Philadelphia, PA: J.B. Lippincott Company, 1999, pp. 278-319.

17. **Dembinski AB, and Johnson LR.** Effect of epidermal growth factor on the development of rat gastric mucosa. *Endocrinology.* 116: 90-4, 1985.

18. **Doornbos RP, Theelen M, van der Hoeven PCJ, van Blitterswijk WJ, Verkleij AJ, and Van Bergen en Henegouwen PMP.** Protein kinase Cζ is a negative regulator of protein kinase B activity. *J Biol Chem.* 274: 8589-8596, 1999.

19. **Forte JG, Ly B, Rong Q, Ogihara S, Ramilo M, Agnew B, and Yao X.** State of actin in gastric parietal cells. *Am J Physiol.* 274: C97–C104, 1998.

20. **Franke TF, Yang SI, Chan TO, Datta K, Kazlauskas A, Morrison DK, Kaplan DR, and Tsichlis PN.** The protein kinase encoded by the Akt proto-oncogene is a target of the PDGF-activated phosphatidylinositol 3-kinase. *Cell.* 81: 727-36,1995.

21. **Fujio Y, and Walsh K.** Akt mediates cytoprotection of endothelial cells by vascular endothelial growth factor in an anchorage-dependent manner. *J Biol Chem.* 274: 16349-16354, 1999.

22. **Gooz M, Hammond CE, Larsen K, Mukhin YV, and Smolka AJ.** Inhibition of human gastric H(+)-K(+)-ATPase α-subunit gene expression by Helicobacter pylori. *Am J Physiol. Gastrointest Liver Physiol.* 278: G981-991, 2000.

23. **Jiang BH, Aoki M, Zheng JZ, Li J, and Vogt PK.** Myogenic signaling of phosphoinositol 3-kinase requires the serine-threonine kinase Akt/protein kinase B. *Proc Natl Acad Sci. USA.* 96: 2077-2081, 1999.

24. **Kaise M, Muraoka A, Yamada J, and Yamada T.** Epidermal growth factor induces $H^+$, $K^+$-ATPase α subunit gene expression through an element homologous to the 3' half-site of the c-*fos* serum response element. *J Biol Chem.* 270: 18637-18642, 1995.

25. **Kohn AD, Summers SA, Birnbaum MJ, and Roth RA.** Expression of a constitutively active Akt Ser/Thr kinase in 3T3-L1 adipocytes stimulates glucose uptake and glucose transporter 4 translocation. *J Biol Chem.* 271: 31372-8, 1996.

26. **Muraoka A, Kaise M, Guo Y, Yamada J, Song I, DelValle J, Todisco A, and Yamada T.** Canine $H^+$-$K^+$-ATPase α-subunit gene promoter: studies with canine parietal cells in primary culture. *Am. J. Physiol.* 271: 1104-1113, 1996.

27. **Murga C, Laguinge L, Wetzker R, Cuadrado A, and Gutkind JS.** Activation of Akt/protein kinase B by G protein-coupled receptors. A role for α and βγ subunits of heterotrimeric G proteins acting through phosphatidylinositol-3-OH kinase γ. *J Biol Chem.* 273: 19080-5, 1998.

28. **Nagahara A, Wang L, Del Valle J, and Todisco A.** Regulation of c-Jun NH2-terminal kinases in isolated canine gastric parietal cells. *Am J Physiol.* 275: G740-748, 1998.

29. **Okano J, Gaslightwala I, Birnbaum MJ, Rustgi AK, and Nakagawa H.** Akt/protein B isoforms are differentially regulated by epidermal growth factor stimulation. *J Biol Chem.* 275: 30934-30942, 2000.

30. **Podolsky DK.** Peptide growth factors in the gastrointestinal tract. In: Physiology of the Gastrointestinal Tract. (3nd. ed.), edited by LR. Johnson. New York, NY: Raven Press, New York, 1994, pp. 129-167.

31. **Smolka, AJ, Larsen KA, Schweinfest CV, and Hammond CE.** *Biochem J.* 340, 601-611, 1999.

32. **Soll AH.** The actions of secretagogues on oxygen uptake by isolated mammalian parietal cells. *J Clin Invest.* 61: 370-380, 1978.

33. **Soroka CJ, Chew CS, Hanzel DK, Smolka A, Modlin IM, and Goldenring JR.** Characterization of membrane and cytoskeletal compartments in cultured parietal cells: immunofluorescence and confocal microscopic examination. *Eur J Cell Biol.* 60: 76-87, 1993.

34. **Takeuchi Y, Yamada J, Yamada T, and Todisco A.** Functional role of extracellular signal-related kinases in gastric acid secretion. *Am J Physiol.* 273: G1263–G1272, 1997.

35. **Takeuchi Y, Pausawasdi N, and Todisco A.** Carbachol activates ERK2 in isolated gastric parietal cells via multiple signaling pathways. *Am J Physiol.* 276: G1484–G1492, 1999.

36. **Wang L, Wilson EJ, Osburn J, and Del Valle J.** Epidermal growth factor inhibits carbachol-stimulated canine parietal cell function via protein kinase C. *Gastroenterology* 110: 469-477, 1996.

37. **Wang X, McCullough KD, Franke TF, and Holbrook NJ.** Epidermal growth factor receptor-dependent Akt activation by oxidative stress enhances cell survival. *J Biol Chem.* 275: 14624-14631, 2000.

Chapter 25

# Permeabilized Gastric Gland Models - A New Strategy For Analyzing Parietal Cell Signal Transduction

TETSURO URUSHIDANI

*Laboratory of Pharmacology & Toxicology, Graduate School of Pharmaceutical Sciences, The University of Tokyo, Tokyo 113-0033, Japan.*

## INTRODUCTION

In various cell types, intracellular signal transduction has been clarified by newly developed techniques due to progress in molecular and cellular biology (21). Although several candidate components have been suggested in acid secreting parietal cells, there are a few crucial disadvantages in using this cell type. One is a limitation in the use of pharmacological probes. We previously reported (18) that a large number of pharmacological probes, frequently used to analyze intracellular signal transduction, showed non-specific effects other than the expected effects on parietal cell function (Fig. 1). The most misleading cases are when a probe has the property of protonophore, e.g., various kinase inhibitors (H89, H7, ML7, ML9, KN93) and all the available calmodulin antagonists. In other cell types, this effect is not so serious unless mitochondrial uncoupling effects emerge. However, in acid secretion by parietal cell, protonophore directly cancels the proton gradient formed and concomitantly inhibitory effects become outstanding. Among these, calmodulin antagonists (W-7, calmidazolium, chlorpromazine, and trifluoperazine) also show potent inhibitory activity on H,K-ATPase in a similar concentration range (17, 18). Furthermore, a widely used anion-exchanger inhibitor, DIDS, also inhibits H,K-ATPase at μM range, at a much lower concentration than is usually used (18). We also reported that the representative PLC inhibitor, U-73122, did not inhibit but potentiated aminopyrine accumulation in rabbit glands (14). Taking these together, it should be concluded that analysis of intracellular signal transduction using cell permeable probes has its limitations, at least in the acid secretory process. It is thus desirable to develop a model system that enables the use of more specific probes, e.g., functional peptide fragments, functional protein itself, or antibody against the protein.

## β-ESCIN-PERMEABLIZED GASTRIC GLAND MODEL (1)

Table 1 summarizes the features of three representative models of permeabilized rabbit gastric glands. Hersey and Steiner (6) first introduced digitonin-permeabilized glands. However, their model lost its ability to respond to any stimulus after permeabilization, thus limiting its application in the study of signal transduction. Thibodeau et al. (19) developed the α-toxin-permeabilized cell. This model has a big advantage in that it retains an ability to respond to receptor-mediated stimuli (12, 19). However, the permeability of the membrane is restricted to molecules smaller than 1,000 M.W. This limited its utilization in that it worked only when small but impermeable molecules, e.g., cAMP, labeled ATP (19) or GTPγS (12)

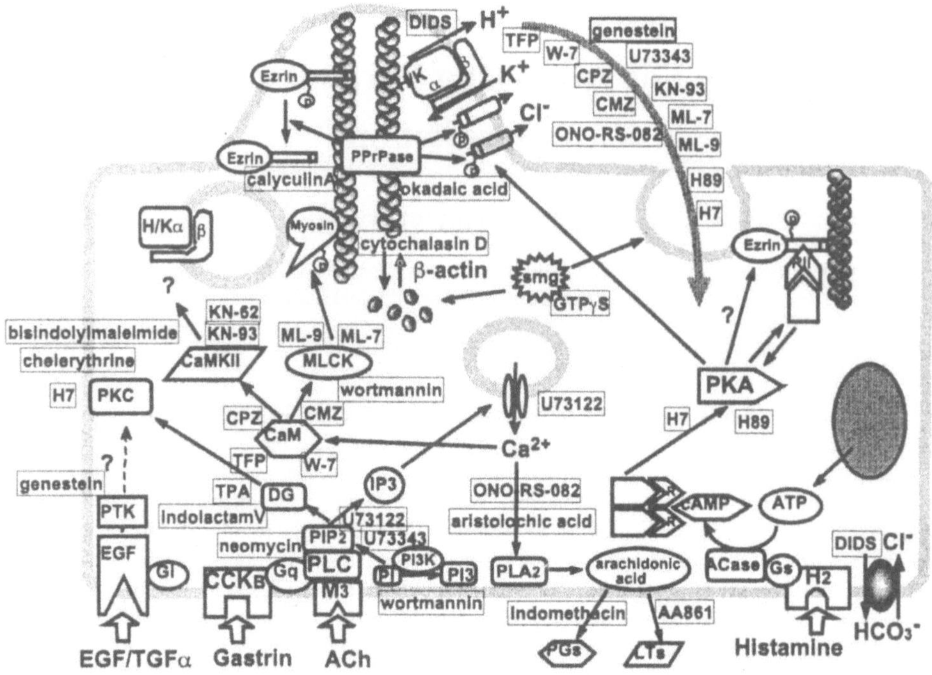


Fig 1. Schematic presentation of parietal cell signal transduction and the sites of action of various pharmacological probes commonly used for the analysis. Drug names are surrounded by lectangles and placed on the site of action. The large arrow at the top indicates the backflux of secreted protons by protonophores alighed on both sides. The drugs on the left side of the arrow have inhibitory affect on the H,K-ATPase in addition to protonophoretic effect. As for the explanation for the intracellular signaltransduction, please see the reference (21).

Table 1. Comparison of three permeabilized gland models.

| Permeabilized glands | Responsiveness to secretagogues | | | | $Ca^{2+}$ sensitivity | Redistribution of H, K-ATPase | Pore size (kDa) |
|---|---|---|---|---|---|---|---|
| | carbachol | histamine | forskolin | cAMP | | | |
| digitonin | - | - | - | - | - | - | >150 |
| β-escin | - | + | + | + | - | + | ~10 |
| α-toxin | + | + | + | + | - | + | ~1 |

were introduced in the cell. It was thus necessary to develop another system where molecules larger than 5,000 M.W. could be introduced into the cell (corresponding to peptides with ca. 25 amino acids).

We then developed a β-escin-permeabilized system following the case of smooth muscle cells (1). It was reported that the saponin ester, β-escin, rendered the cell membrane permeable to relatively large molecules, e.g., calmodulin (M.W. 14,000), and preserved cell responsiveness via receptor activation (9). We showed that β-escin-permeabilized isolated rabbit gastric glands retain their ability to respond not only to the second messenger, cAMP, but also to forskolin and histamine (Table 1). Using this model, we reported the effects of specific peptide fragments to elucidate the role of various putative components of signal transduction in gastric parietal cells. Aminopyrine accumulation stimulated by 100 μM cAMP was dose-dependently inhibited by cAMP-dependent protein kinase (PKA) inhibitory peptide (M.W. 2222.4) and was almost abolished at 20 μM or more. This peptide also inhibited acid secretion stimulated by forskolin and by histamine. As the peptide was ineffective on aminopyrine accumulation stimulated by dibutyryl cAMP in intact glands, this result demonstrated the usefulness of the model for introducing functional peptides into the cell. There seems to be a consensus that activation of PKA is involved in the acid secretory process, but there had been no direct demonstration that PKA is involved. However, there exists a pathway without activation of PKA downstream of cAMP, e.g., small GTP binding proteins (20). One reliable piece of evidence for this is the inhibitory effect of H89, a "specific" inhibitor of cAMP-dependent protein kinase, on acid secretion. However, this drug was probed and determined to be a protonophore as described above (2). In contrast to previous studies with this pharmacological agent of questionable specificity, these findings provide strong evidence that PKA actually plays a central role in regulation of acid secretion, from using the much more specific probe of this permeabilized cell model.

In the next series of experiments, the effect of peptides related to other protein kinases on cAMP-stimulated aminopyrine accumulation were examined. The following peptides were without effect: the pseudo-substrate peptide [19-31] of protein kinase C (the specific inhibitor of protein kinase C), pseudo-substrate [281-301] for calmodulin dependent protein kinase II (the specific inhibitor of calmodulin dependent protein kinase II), and the calmodulin binding domain [290-309] of calmodulin kinase II (the specific inhibitor of calmodulin). In contrast, the pseudo substrate peptide [480-501] of myosin light chain kinase (MLCK) (the specific inhibitor of MLCK) showed potent inhibitory activity at 30 μM. These data, except that of MLCK, are reasonable since we observed no potentiating interaction between stimulation by cAMP and $Ca^{2+}$ (Table 1), so we consider the observed activation in the present model to be a part of the secretory process, i.e., PKA pathway. We do not know why the obvious crosstalk between cAMP and $Ca^{2+}$ observed in the intact cell (15) disappeared in permeabilized glands, but a similar phenomenon was also observed in α-toxin-permeabilized glands (12).

If the involvement of $Ca^{2+}$/calmodulin were excluded from this model, this would be contradicted by the observation that the specific inhibitory peptide of MLCK inhibits cAMP-stimulated secretion. We pointed out the possible involvement of this enzyme in previous work (22) and our present result supports the hypothesis that the MLCK-like enzyme plays a key role in the activation of parietal cell. As there is no room for $Ca^{2+}$/calmodulin in this model and MLCK is a $Ca^{2+}$/calmodulin-dependent enzyme, it should be concluded that either a new enzyme or a new property of a known enzyme is involved in acid secretion.

In various secretory cells, it is known that GTPγS, a non-hydrolyzable GTP analog, accelerates the secretory response. Experiments using specific peptide fragments have supported the involvement of the secretory granule-associated Rab protein, a member of the small GTP-binding protein family, e.g., Rab4 in insulin-mediated translocation of the glucose transporter in rat adipocytes, and Rab3 in amylase secretion from pancreatic acinar cells. In contrast to these cells, GTPγS inhibits acid secretion in α-toxin- (12) or β-escin-(1) permeabilized glands. It has been revealed that Rab11 co-localizes in the tubulovesicles as a parietal cell-specific small GTP-binding protein (3). Following on the assumption that the target of GTPγS is one of the small GTP-binding proteins, the effects of several functional fragments of small GTP-binding proteins were examined. The fragments employed were: the inhibitory peptide [2-17] of Arf1 (Arf1p), Rab3 effector domain (Rab3AL), the N-terminal peptides of Rab11 (Rab11NT) and Rab25 (Rab25NT), the fragments that the sequence of the N-terminal peptide of Rab11 was replaced with AL to strengthen the effector (Rab11AL), and the C-terminal hyper variable region of Rab11. None of the peptides showed stimulatory effects by themselves or potentiating effects on 10 μM cAMP (submaximal concentration)-stimulated secretion. Moreover, all of the Rab-related peptides up to 100 μM failed to show any inhibitory effect on aminopyrine accumulation stimulated by 100 μM cAMP in β-escin-permeabilized glands. Instead, Arf1p partially inhibited cAMP-stimulated aminopyrine accumulation in this model. It is an interesting future project to elucidate whether Arf participates in activation of parietal cell stimulation.

## DIGITONIN-PERMEABILIZED GASTRIC GLAND MODEL (2)

In the previous section, it was shown that β-escin permeabilized gastric gland was an excellent model which is useful in introducing specific peptide fragments into parietal cell in order to analyze signal transduction of the cell. However, it is often more desirable to use proteins, e.g., functional proteins, specific antibodies, etc. To do this, pores with a larger size have to be created in the plasma membrane using a detergent like digitonin. As shown in Table 1, however, it has been shown that digitonin-permeabilized glands completely lose their ability to secrete acid in response to any secretagogues. This is not because the secretory membranes are disturbed by detergent but possibly because some essential component(s) or structure(s) for the activation process of acid secretion are lost during permeabilization. Resting or stimulated glands are "fixed" in their state after digitonin-permeabilization and they are able to create proton gradient in the presence (resting) or absence (stimulated) of the potassium ionophore, valinomycin (6). This feature could be turned into an advantage. If we could reconstitute acid secretion by digitonin-permeabilized glands, it should be a useful tool to identify the essential factors for acid secretion. In fact, there are model systems responding to larger molecules, such as digitonin-permeabilized chromaffin cell, and several factors essential for the secretory process have been identified using these models (13,23).

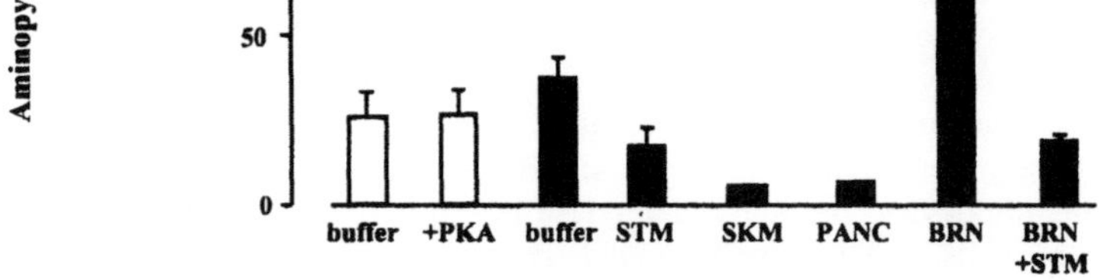


**Figure 2.** Effects of cytosol from various tissues on digitonin-permeabilized isolated gastric glands. Cytosolic fractions (1.5 mg/ml final) were prepared from stomach (STM), brain (BRN), skeletal muscle (SKM), and pancreas (PNC) of rabbit, and added to digitonin-permeabilized glands without (control; open column) or with (filled column) 100 μM cAMP. Catalytic subunit of cAMP-dependent protein kinase (PKA; 1000 units/ml) was also tested. The aminopyrine ratio was measured in duplicate and Mean ± SEM of 3 - 5 independent experiments are shown (except skeletal muscle and pancreas which were both two experiments apiece).

We first tried to reconstitute acid secretion of digitonin-permeabilized glands by a supplement of various proteins. As shown in Fig. 2, digitonin-permeabilized glands do not respond to cAMP. This is not due to the simple loss of PKA, because addition of PKA failed to increase the aminopyrine ratio. The cytosol prepared from gastric mucosa also failed to stimulate acid secretion even in the presence of cAMP. This does not make sense if the non-responsiveness is simply due to the loss of cytosol. We then tried cytosol from other tissues including skeletal muscle and pancreas but with no effect. We finally found that brain cytosol strongly stimulated aminopyrine accumulation by itself. More interestingly, when gastric cytosol was added together with brain cytosol, stimulation was abolished.

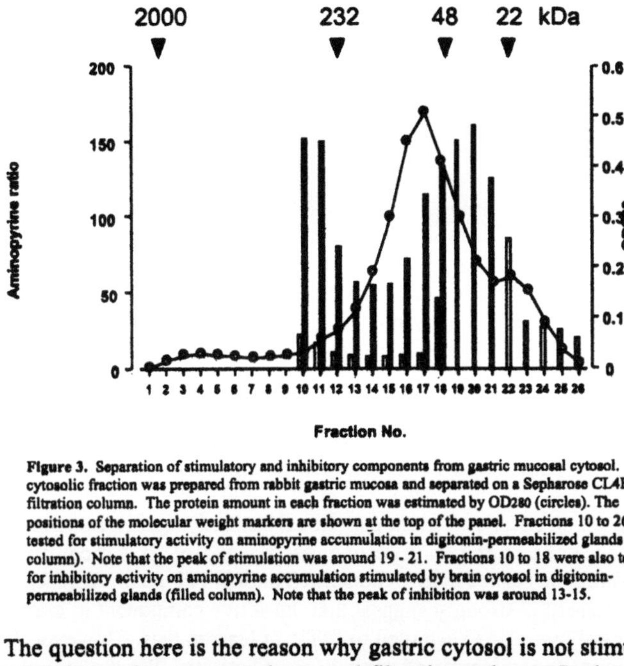


**Figure 3.** Separation of stimulatory and inhibitory components from gastric mucosal cytosol. The cytosolic fraction was prepared from rabbit gastric mucosa and separated on a Sepharose CL4B gel filtration column. The protein amount in each fraction was estimated by OD280 (circles). The positions of the molecular weight markers are shown at the top of the panel. Fractions 10 to 26 were tested for stimulatory activity on aminopyrine accumulation in digitonin-permeabilized glands (open column). Note that the peak of stimulation was around 19 - 21. Fractions 10 to 18 were also tested for inhibitory activity on aminopyrine accumulation stimulated by brain cytosol in digitonin-permeabilized glands (filled column). Note that the peak of inhibition was around 13-15.

The question here is the reason why gastric cytosol is not stimulatory but inhibitory. Gastric cytosol was then separated on a gel filtration column, Sepharose CL-4B, and each fraction was assayed using digitonin-permeabilized glands (Fig. 3). It was found that the fraction around 30 kDa stimulated permeabilized glands by itself, whereas the 200 kDa fraction potently inhibited brain cytosol-stimulated acid secretion. When brain cytosol was separated on the same column, the stimulatory component eluted in the same position, i.e, around 30 kDa but no inhibitory activity was detected. The overall effect of gastric cytosol is considered to be inhibitory because the effect of the high molecular weight component is dominant. We then tried to purify this stimulatory factor from rabbit brain cytosol.

When the 30 kDa stimulatory factor was further separated on a Superdex 200, the total activity drastically decreased and the activity split into two dull peaks, peak 1 (app. M.W. ca. 20 kDa) and peak 2 (app. M.W. ca. 2 kDa). We thought then that these were multiple components working synergistically. We then repeated the assay in the presence of threshold concentration of peak 2 and found that the activity of peak 1 was markedly potentiated by peak 2. We decided to purify peak 1, and changed the source to bovine brain to increase the purification scale.

**Table 2. Purification of peak 1 from bovine brain.** Each fraction was assayed for aminopyrine accumulation in digitonin-permeabilized gastric glands. In order to quantify the activity, the maximal aminopyrine ratio stimulated by brain cytosol (1.5mg/ml) was arbitrarily designated as 100 units and the activity of the other fractions was expressed by this unit. Assay for steps 2 to 5 contained 0.1 ml of peak 2 fraction from rabbit brain per 1ml to attain the maximal stimulation.

| step | protein (mg) | total activity (units) | specific activity (units/mg) | purification (fold) | yield (% of total) |
|---|---|---|---|---|---|
| 1. cytosol | 6070 | 400620 | 66.7 | 1 | 100 |
| 2. DEAE-Sepharose | 1370 | 177278 | 130 | 2 | 44 |
| 3. Red-agarose | 50 | 6700 | 134 | 2 | 1.7 |
| 4. Superdex 200 | 0.42 | 590 | 1404 | 21 | 0.147 |
| 5. Diol-150 | 0.03 | 84 | 2800 | 42 | 0.021 |

Table 2 shows the purification steps of peak 1 from bovine brain. Bioassay was performed with digitonin-permeabilized glands in the presence of threshold concentration of peak 2 purified from rabbit brain. It was obvious from the purification rate that the stimulatory activity came from a multiple source, since the total activity recovered in the active fraction was quite small. This means that the active fraction obtained by this procedure only explains part of the stimulatory activity in bovine brain. The final step of the purification by Diol-150 (gel filtration) HPLC column gave a sharp active peak (apparent molecular weight about 10 kDa) containing a single band around 35 kDa on SDS-polyacrylamide gel. We cut out this band from the gel, trypsinized, and separated on a reverse-phase HPLC and analyzed on a peptide sequencer. Two fragments gave reliable sequences (EYQVCQLYSVAE and KAEEDPAK) and both of them showed striking homology to phosphatidylinositol transfer protein (PITP) whose molecular weight is 35 kDa (24). In order to identify peak 1 as PITP, we made recombinant βPITP and applied it to digitonin-permeabilized glands in the presence of peak 2. PITP, but not control protein GST, stimulated acid secretion. We also raised rat polyclonal antibody against recombinant βPITP and tested for the presence of this protein in brain or gastric cytosol. It was confirmed that the active fractions from brain or gastric cytosol contained anti-βPITP positive 35 kDa bands by Western blotting. These results strongly suggest that the active component in brain cytosol as well as in gastric mucosal cytosol is attributed to PITP, at least in part. However, PITP alone is not active, but demands the unidentified peak 2 for its activity. We thought possibly that peak 2 was phosphatidylinositol, but a bioassay using phosphatidylinositol instead of peak 2 excluded this possibility. In fact, it has been shown that the amount of phosphatidylinositol is not rate limiting within the cell (24). Further study is necessary for identifying the low molecular weight cofactor, peak 2.

As for the inhibitory component in gastric mucosal cytosol, we kept trying to purify it with little success. Any chromatography other than Sepharose CL-4B obscured the peak of inhibitory activity, suggesting that the inhibitory factor consists of multiple components. In order to clarify the feature of this activity further, we tried to quantify both the inhibitory and stimulatory activities in cytosol prepared from stimulated and resting gastric glands. Rabbit gastric glands were isolated and separated into two aliquotes, one of which was stimulated with 0.1 mM histamine + 50 μM isobutylmethylxanthine for 30min, and the other was kept resting with 0.1 mM cimetidine at 37 °C. The glands were then homogenized and the cytosol was prepared. The cytosol of each was separated on a Sepharose CL-4B from which both stimulatory and inhibitory fractions were harvested, and their activity was assayed on digitonin-permeabilized glands using brain cytosol as agonist to estimate inhibitory activity.

The results were entirely unexpected. The stimulatory activity was unchanged by stimulation. Moreover, the inhibitory activity was drastically increased, not decreased, with stimulation. This result has brought a new understanding of the mechanism of parietal cell activation, i.e., the activation of parietal cell is now thought to be triggered by the release of putative inhibitory factor(s) from the membrane where the secretory machinery is suppressed. It appears that the stimulatory factor(s), including PITP found in the present study, are not the switch but are constitutively active in permeabilized glands. This phenomenon, together with the loss of responsiveness to calcium ions, are limitations of the permeabilized gland models which should be improved in the future.

## POTASSIUM PERMEABILITY OF STIMULATED APICAL MEMBRANE OF THE PARIETAL CELL REQUIRES PHOSPHOINOSITIDES (16)

The discovery in the previous section that PITP was an essential factor in gastric acid secretion was not surprising for us because we had identified phosphoinositides as the essential component for $K^+$ permeability in the apical membrane of stimulated parietal cell by using experiments performed in parallel.

H,K-ATPase in gastric microsomes purified from resting stomach requires ionophore, e.g., valinomycin, for its full activity (4), since the membrane has limited $K^+$ permeability and the activation site for $K^+$ of the enzyme faces the luminal side. The activation mechanism of the enzyme was clarified by Wolosin and Forte (25) who showed that vesicles purified from acid secreting stomach display $K^+$ permeability manifested by valinomycin-independent proton pumping of H,K-ATPase. Subsequently, Im et al. (10) found that the apparent $K^+$ permeability was attenuated by treatment of the rat membrane with 5 mM $Mg^{2+}$ and that this effect of $Mg^{2+}$ was prevented by pyrophosphate, a non-specific phosphatase inhibitor. They concluded that putative potassium channels were phosphorylated when parietal cell was activated, and their conductance was reduced by a putative membrane-bound $Mg^{2+}$-dependent phosphoprotein phosphatase. This is strong evidence that activation of kinase(s) directly leads to the final activation step of acid secretion, i.e., the opening of the potassium channel which supplies potassium ion for the extracellular potassium site of H,K-ATPase. In those days, however, no appropriate inhibitors for phosphoprotein phosphatases were available.

We re-examined their experiments using rabbit membrane and confirmed their reproducibility. However, we found that any inhibitors of phosphoprotein phosphatase I, $II_A$, and $II_B$ failed to prevent the loss of $K^+$ permeability in membrane by treatment with $Mg^{2+}$. Although we excluded the possible involvement of protein phosphatases, we kept searching for the involvement of metabolic events including other phosphatases, since the loss of $K^+$ permeability needed about 10 min incubation with $Mg^{2+}$ at 37 °C, and pyrophosphate was still an effective inhibitor for the $Mg^{2+}$ effect.

We continued to test various drugs and finally found that 100 μM neomycin surrogated the effect of $Mg^{2+}$. Although neomycin is known to be a phospholipase C inhibitor, another phospholipase C inhibitor, U73122, did not show such an effect, suggesting that neomycin acted on other than phosphopolipase C. As for the mechanism by which neomycin inhibits phospholipase C, it has been suggested (5) that the compound traps its substrate, phosphatidylinositol-4,5-bisphosphate ($PIP_2$). We thus attribute the effect of neomycin to some metabolic events of phosphoinositides. In order to verify our hypothesis, experiments were conducted as shown in Fig. 4.

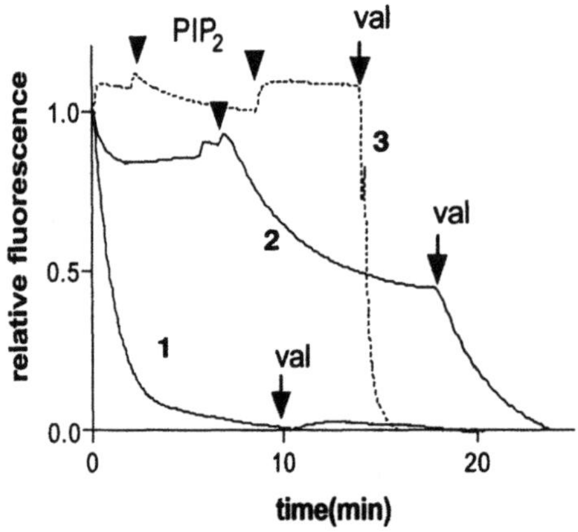


Fig 4. Induction of $K^+$ permeability by Phosphatidylinositol-4,5-bisphosphate ($PIP_2$) in the stimulated, but not resting, gastric vesicles. Gastric vesicles purified from acid-secreting rabbit were pre-incubated with or without 5 mM $Mg^{2+}$ for 10 min at 37° C. The treated vesicles were added to a reaction mixture (at time 0) and acridine orange quenching was continuously monitored by a spectrofluorometer (CAF 110) Trace 1:control apical membrane. Note that a rapid quenching occurred and the addition of valinomycin (10 μM; val) did not accelerate. Trace 2: The membrane in trace 1 was pre-incubated with 5 mM $Mg^{2+}$ at 37° C for 10min to reduce the $K^+$ permeability. At the point indicated by arrowhead, 5mg/ml $PIP_2$ was added to the cuvétte. Note that $PIP_2$ surrogated the effect of valinomycin in some extent. Trace 3: Tubulovesicle-rich fraction purified from resting rabbit stomach. Note that $PIP_2$, added twice at the points of arrowhead, did not, but valinomycin caused rapid quenching.

Trace 1 is a control, showing the valinomycin-independent quenching, or high $K^+$ permeability of the membrane. Trace 2 is the membrane pretreated with 5 mM $Mg^{2+}$ for 10 min showing lowered quenching due to reduced $K^+$ permeability. When $PIP_2$ was added, acceleration of the quenching occurred, indicating that the attenuated $K^+$ permeability was restored by $PIP_2$. Although not shown here, similar results were obtained using neomycin-treated membrane, and furthermore either phosphatidylinositol (PI) bound to PITP or phosphatidylinositol-4,5,6-trisphosphate ($PIP_3$) surrogated the effect of $PIP_2$. These results strongly suggested that the effects of $Mg^{2+}$ and neomycin were related to the phosphoinositides. In order to confirm our hypothesis, we labeled the phospholipids with $^{32}P$ and measured the contents of $PIP_2$ and $PIP_3$ in the membrane by thin-layer chromatography. We observed that treatment of the membrane with $Mg^{2+}$ actually reduced the contents of both $PIP_2$ and $PIP_3$. In contrast to $Mg^{2+}$, neomycin did not reduce, but rather increased the amount of $PIP_2$ and $PIP_3$, which suggested that neomycin trapped $PIP_2$ to isolate it from the metabolic pathway.

Interestingly, $PIP_2$ did not cause acridine orange quenching in the tubulovesicle-rich fraction purified from resting rabbit gastric mucosa, in which acridine orange quenching was entirely valinomycin-dependent (Fig. 4). This clearly means that $PIP_2$ cannot induce de novo $K^+$ permeability in the resting tubulovesicle membrane, but recovers $K^+$ permeability in the stimulated apical membrane that had been lost by treatment with $Mg^{2+}$ or neomycin. As PI itself was not effective on this system, it was suggested that PITP was necessary for appropriate incorporation of PI into the membrane, and when PI was supplied, it was converted to $PIP_2$ or $PIP_3$ by enzymes present in the apical membrane. These results also suggest two possible activation mechanisms. One is that the putative $K^+$ channel or transporter sensitive to phosphoinositides is present in the membrane in a location other than the tubulovesicles and is recruited to the apical membrane. The other is that the putative $K^+$ channel is changed into a structure sensitive to phosphoinositides when acid secretion is activated.

In the above experiments, the $K^+$ permeability was estimated by an indirect measure, i.e., the valinomycin dependency of H,K-ATPase-driven proton accumulation in the vesicle. It was still possible that the phenomenon observed above was due to effects not related to the $K^+$ permeability itself but to other factors, like H,K-ATPase activity. Therefore, the $K^+$ permeability was assessed by other techniques, i.e., the passive proton diffusion in the vesicle monitored by acridine orange quenching and the $^{86}Rb$ flux rate. It was confirmed using these that $Mg^{2+}$ or neomycin-treatment actually reduced $K^+$ permeability of the apical membrane from stimulated parietal cell.

Taking these results together, $PIP_2$ and/or $PIP_3$ could be essential for $K^+$ permeability in the apical membrane of secreting parietal cell. Within the membrane, the phosphorylation/desphosphorylation cycle of phosphoinositides appears to take place, since $Mg^{2+}$ ion and neomycin affect their contents. We postulate that $Mg^{2+}$ either activates kinases or phosphatases for $PIP_2$ and/or $PIP_3$, and subsequently reduces the contents of these phosphoinositides in the membrane. Considering the fact that $PIP_3$ surrogated $PIP_2$, the metabolite(s) of $PIP_2$ would be the key molecule(s). Alternatively, $PIP_3$ might have been metabolized into $PIP_2$ within the membrane. Of the phosphoinositides, $PIP_2$ has been reported to modulate the activity of various channels or transporters including the inward rectifier $K^+$ channels (8, 11), the $Na^+$-gated nonselective cation channel (26), and Na/Ca exchanger (7). This information should be quite useful to identify the molecular entity of the putative $K^+$ channel or transporter in future experiments. More work is necessary to identify the molecule responsible for $K^+$ permeability that is modified by phosphoinositides in gastric membrane.

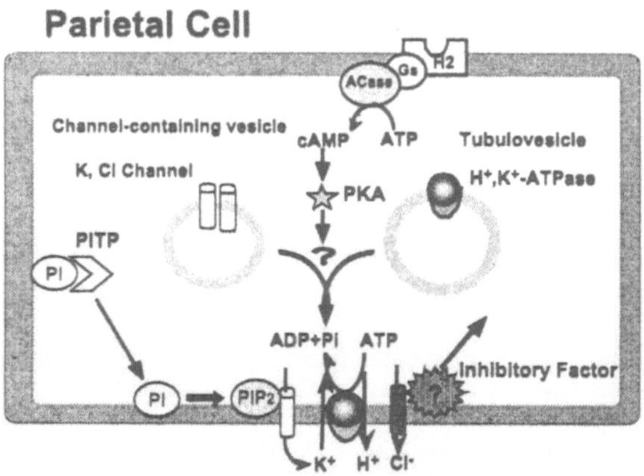


**Fig. 5. Schematic presentation of putative signal transduction mechanism in the parietal cell based on the data presently described.**
At least in the rabbit parietal cell, the main route of activation is triggered by the elevation of intracellular cAMP, which is created by adenylate cyclase (ACase) activated by Gs liberated from histamine H2 receptor. Activation of PKA by cAMP leads to fuse H,K-ATPase containing tubulovesicles and channel containing vesicles with the apical membrane to construct a complete secretory machinery. Alternatively, the channels are the residents and activated there. At the same time of fusion events, unidentified inhibitory factor(s) would be release to activate the machinery. The K channel, essential for operating proton pump, requires PIP2 (or PIP3) for its full activity and PIP2 is synthesized from PI which is supplied by PITP.

## CONCLUSION

Based on the data and discussion described above, a new scheme for parietal cell signal transduction could be presented as shown in Fig. 5. At least in rabbit parietal cell employed in the present work, the main stimulatory pathway involves PKA activation via histamine $H_2$ receptor/Gs. The direct targets of PKA have not yet been identified, but they should trigger at least two activation steps, i.e., the fusion of vesicles containing H,K-ATPase and $K^+$, $Cl^-$ channels to the apical membrane, and also the release of inhibitory factor from the membrane. Neither the identity of these channels nor the place where they exist in the resting state is known. However, on the stimulated apical membrane the $K^+$ channel requires $PIP_2$ for its full activity, and $PIP_2$ is metabolically produced from PI, and the PI is continuously supplied by PITP. This new concept would contribute not only to the understanding of parietal cell physiology, but also constitute a new strategy in the development of drugs controlling the secretory process.

## ACKNOWLEDGMENTS

This article was based on the works mainly performed by the graduate students, Yuko Muto, Keiko Akagi, Yuko Sugita, and Noriaki Omi. These works was supported in part by Japanese Ministry of Education, Science, Sports, and Culture, and published as No. 1, 2, 14, 15, 16, 18, and 22, in the references.


## REFERENCES


1. **Akagi K, Nagao T, and Urushidani T.** Responsiveness of β-escin-permeabilized rabbit gastric gland model —Effects of functional peptide fragments. *Am. J. Physiol. Gastrointest. & Liver Physiol.* 277: G736-745, 1999.

2. **Akagi K, Nagao T, and Urushidani T.** Reconstitution of acid secretion in digitonin-permeabilized rabbit gastric glands—identification of cytosolic regulatory factors. *J. Biol. Chem.* 276:28171-28178, 2001.

3. **Calhoun BC, Lapierre LA, Chew CS, and Goldenring JR.** Rab11a redistributes to apical secretory canaliculus during stimulation of gastric parietal cells. *Am. J. Physiol. Cell Physiol.* 275:C163-170, 1998.

4. **Forte JG, Ganser A, Beesley R, and Forte TM.** Unique enzymes of purified microsomes from pig fundic mucosa. $K^+$-stimulated adenosine triphosphatase and $K^+$-stimulated pNPPase.*Gastroenterol.* **69:** 175-189, 1975.

5. **Gabev E, Kasianowicz J, Abbott T, and McLaughlin, S.** Binding of neomycin to phosphatidylinositol 4,5-bisphosphate (PIP2). *Biochim. Biophys. Acta* **979,** 105-112, 1989.

6. **Hersey SJ and Steiner L.** Acid formation by permeable gastric glands: enhancement by prestimulation. *Am. J. Physiol. Gastrointest. & Liver Physiol.* 248:G561-G568, 1985.

7. **Hilgemann DW, and Ball, R.** Regulation of cardiac $Na^+$,$Ca^{2+}$ exchange and $K_{ATP}$ potassium channels by PIP2. *Science* **273,** 956-959, 1996.

8. **Huang CL, Feng S, and Hilgemann DW.** Direct activation of inward rectifier potassium channels by PIP2 and its stabilization by Gβγ. *Nature* **391,** 803-806, 1998.

9. **Iizuka K, Dobashi, K, Yoshii A, Horie T, Suzuki H, Nakazawa T, and Mori M.** Receptor-dependent G protein-mediated $Ca^{2+}$ sensitization in canine airway smooth muscle. *Cell Calcium.* 22:21-30, 1997.

10. **Im WB, Blakeman DP, Bleasdale JE, and Davis JP.** A protein phosphatase associated with rat heavy gastric membranes enriched with ($H^+$-$K^+$)-ATPase influences membrane $K^+$ transport activity. *J. Biol. Chem.* **262,** 9865-9871, 1987.

11. **Liou HH, Zhou SS, and Huang CL.** Regulation of ROMK1 channel by protein kinase A via a phosphatidylinositol 4,5-bisphosphate-dependent mechanism. *Proc. Natl. Acad. Sci.*

*USA* **96,** 5820-5825, 1999.

12. **Miller MD and Hersey SJ.** Differential effects of GTP gamma S on acid and pepsinogen secretion by permeable gastric glands. *Am. J. Physiol. Gastrointest. & Liver Physiol.* 270:G962-968, 1996.

13. **Morgan A, and Burgoyne RD.** Exo1 and Exo2 proteins stimulate calcium-dependent exocytosis in permeabilized adrenal chromaffin cells. *Nature 355*, 833-836, 1992.

14. **Muto Y, Nagao, T and Urushidani T.** A putative phospholipase C inhibitor, U73122, and its negative control, U73343, both elicit unexpected effects on the rabbit parietal cell. *J. Pharm. Exp. Ther.* 282: 1379-1388, 1997.

15. **Muto Y, Nagao T, Yamada M, Mikoshiba K, and Urushidani T.** A proposed mechanism for the potentiation of cAMP-mediated acid secretion by carbachol. *Am. J. Physiol. Cell Physiol.* 280: C155-C165, 2001.

16. **Omi N, Nagao T, and Urushidani T.** Phosphatidylinositol is essential determinant for $K^+$ permeability involved in gastric proton pumping. *Am. J. Physiol. Gastrointest. & Liver Physiol.* 281:G786-G797, 2001.

17. **Sewing K-F and Beil W.** Experimental studies with proton pump inhibitors. in *Advances in Drug Therapy of Gastrointestinal Ulceration.* Edited by A. Garner and B. J. R. Whittle. John Wiley & Sons, p.53-66, 1989.

18. **Sugita Y, Nagao T, and Urushidani T.** Nonspecific effects of the pharmacological probes commonly used to analyze signal transduction in rabbit parietal cells. *Eur. J. Pharmacol.* 365, 77-89, 1999.

19. **Thibodeau A, Yao X, and Forte JG.** Acid secretion in alpha-toxin-permeabilized gastric glands. *Biochem. Cell Biol.* 72:26-35, 1994.

20. **Tsygankova OM. Kupperman E. Wen W. Meinkoth JL.** Cyclic AMP activates Ras. *Oncogene* 19:3609-15, 2000.

21. **Urushidani T and Forte JG.** Signal transduction and activation of acid secretion in the parietal cell. *J. Memb. Biol.* 159: 99-111, 1997.

22. **Urushidani T, Muto Y, Nagao T, Yao X, and Forte JG.** ME3407, a new antiulcer agent, inhibits acid secretion by interfering with intracellular redistribution of the gastric proton pump. *Am. J. Physiol. Gastrointest. & Liver Physiol.* 272, G1122-1134, 1997.

23. **Walent JH, Porter BW, and Martin TFJ.** A novel 145 kd brain cytosolic protein reconstitutes $Ca^{2+}$-regulated secretion in permeable neuroendocrine cells. *Cell 70*, 765-775. 1992.

24. **Wiedemann C, and Cockcroft S.** The role of phosphatidylinositol transfer proteins (PITPs) in intracellular signaling. *Trends Endocrinol. Metab. 9*, 324-328, 1998.

25. **Wolosin JM and Forte JG.** Changes in the membrane environment of the ($K^+ + H^+$)-ATPase following stimulation of the gastric oxyntic cell. *J. Biol. Chem.* **256,** 3149-3152, 1981.

26. **Zhainazarov AB and Ache BW.** Effects of phosphatidylinositol 4,5-bisphosphate and phosphatidylinositol 4-phosphate on a $Na^+$-gated nonselective cation channel. *J. Neurosci.* **19,** 2929-2937, 1999.

Chapter 26

# Parietal Cell Volume Regulation During Acid Secretion

URSULA SEIDLER, OLIVER BACHMANN, THORSTEN SONNENTAG, ALEXANDER HEINZMANN, WOLF-KRISTIAN SIEGEL, *ANDREAS MACK, AND MICHAEL GREGOR

*1. Medizinische Klinik und *Anatomisches Institut der Eberhard-Karls Universität Tübingen Tübingen, Germany*

## SUMMARY

We have previously shown that concomitant $Na^+/H^+$ and $Cl^-/HCO_3^-$ exchange activation occurs during stimulation of acid secretion in cultured rabbit parietal cells, possibly related to a necessity for volume regulation during the secretory process. We now investigated whether cytoplasmic volume changes occur during secretagogue stimulation of cultured rabbit parietal cells, and the activation of which ion transport mechanisms is responsible for these changes. Cells were loaded with the fluorescent dye calcein and the calcein concentration within a defined cytoplasmic volume was recorded by confocal microscopy. $10^{-5}$M forskolin, $10^{-4}$M carbachol and hyperosmolarity (400 mosmol) resulted in a rapid increase in the cytoplasmic dye concentration, indicative of cell shrinkage, followed by recovery to baseline within several minutes, indicative of regulatory volume increase (RVI). We hypothesized that the initial shrinkage, because it was more rapid than fluid accumulation in the secretory vesicles, may be due to the opening of agonist-dependent basolateral or apical $K^+$ or anion channels, and tested the effect of a variety of $K^+$ and $Cl^-$ channel inhibitors. The results can be summarized as follows: $Ba^{2+}$-sensitive $K^+$-channels, likely in the basolateral membrane, are constitutively open and their blockade results in rapid cell swelling and the prevention of forskolin- or carbachol-induced cytoplasmic shrinkage. The inhibition of agonist-activated Hoe 293B- and ChTX-sensitive and NPPB-sensitive anion channels, which are possibly apically located, strongly decreases agonist-activated shrinkage. $Na^+/H^+$ exchange inhibitors slightly reduced the initial cell shrinkage and significantly slowed the RVI, whereas 100µM bumetanide had no significant effect on either parameter. In conclusion, acid secretagoguges induce a rapid loss of parietal cell cytoplasmic volume, secondary to the opening of agonist-induced $K^+$ and anion channels. Initial shrinkage is followed by RVI, which is predominantly mediated by $Na^+/H^+$ and $Cl^-/HCO_3^-$ exchange.

## INTRODUCTION

The stomach secretes about 2 l gastric juice per day, a fluid mainly consisting of $H^+$, $Cl^-$, $K^+$ and water. Transport pathways for $H^+$, $Cl^-$, and $K^+$ have been identified in the apical membrane of the secreting parietal cell, but not for water. Quite the contrary, isolated apical membrane vesicles from secreting gastric mucosa, socalled "stimulation-associated" vesicles, were

found to have a very low water permeability (29). Therefore, it has been assumed that the water permeates through paracellular pathways.
However, we and others observed a strong immunoreactivity of parietal cell basolateral membranes with antibodies against aquaporin 4 (25;39), and it has recently been shown that the aquaporin 4 knockout mouse has a reduced pentagastrin-stimulated acid secretion (19), suggesting that water uptake into the parietal cell is necessary for acid secretion. Moreover, we had observed that acid secretagogues activate parietal cell $Na^+/H^+$ exchange, but that in the presence of $CO_2/HCO_3^-$, this activation is not associated with an increase in $pH_i$ or basolateral proton efflux, suggesting 1. that a $Cl^-/HCO_3^-$ exchanger is activated concomitantly and 2. that the physiological role for $Na^+/H^+$ exchange activation during stimulation of acid secretion is not to increase parietal cell $pH_i$ (1). Since the concomitant activation of $Na^+/H^+$ and $Cl^-/HCO_3^-$ exchange is one of the known mechanisms that cells use to increase their volume (20), we speculated that the secretagogue-induced $Na^+/H^+$ exchange activation may occur in the context of parietal cell volume regulation. Therefore, we investigated whether 1. the cytoplasmic volume of parietal cells in primary culture changes during stimulation of acid secretion and 2. What principal ion transport mechanisms mediate the volume changes in these cells during stimulation of acid formation.

## MATERIAL AND METHODS

**Materials:** The materials used for cell preparation and culture were the same as previously described . Dimethyl-amiloride, forskolin and carbachol were from Sigma (Deisenhofen, Germany), 2′-7′biscarboxyethyl-5(6)-carboxyfluorescein (BCECF/am) and calcein-acetoxymethylester (calcein/am) were purchased from Molecular probes (Leiden, Netherlands). 4-isopropyl-3-methylsulphonylbenzoyl-guanidine methanesulphonate (Hoe 642) was a generous gift from Hoechst (Frankfurt, Germany).

**Cell isolation and culture:** Cell isolation and culture was carried out as previously described (1;31). Parietal cells for volume measurements were used during the first three days after plating, when they have still retained a relatively round shape.

**Cytoplasmic volume measurements:** The method used for cytoplasmic volume measurements has been described in detail (36). In brief, parietal cells cultured on coverslips were incubated with 3 μM calcein/am for 30 min at 37°C in perfusion buffer (in mM: 118 NaCl, 22 $NaHCO_3$, 14 HEPES, 7 TRIS, 3 $KH_2PO_4$, 2 $K_2HPO_4$, 1,2 $CaCl_2$, 1,2 $MgSO_4$, 20 glucose, pH 7.4, gassed with 5% $CO_2$ /95% $O_2$), then washed twice and incubated a further 30 min in buffer without dye. The coverslips were then mounted in a custom-made heated perfusion chamber which allowed extremely fast buffer change. Measurements were performed using a Zeiss LSM 410 inverted laser scanning microscope (Carl Zeiss, Oberkochen, Germany). The light from a single line argon laser (488 nm) was directed through the objective (Plan Neofluar, 40x/1.3 oil). Emitted fluorescence was directed through a long pass filter (LP 515, Zeiss), resulting in a 1000fold reduction of the light intensitiy, and a pinhole of 20 (which defines the accuracy of one point in the xy-plane, not relevant for z-scans) before it was detected by a photomultiplier. A time series of z-scans (z-step height of 1 μm) was made through the middle of a parietal cell. The thickness of optic sectioning in the z-plane was approx. 500 nm. The duration of each z-scan was approx. 1.5 sec, the interval between scans was 15 sec in the first 5 min of the experiment and 30 sec thereafter. An area within the cytoplasm was marked, fluorescence intensity measured, and the voxels in each 1 $\mu m^3$ within this area in all z-scans of the series were calculated. This procedure was used to maximize the scanning speed and to minimize the light exposure. A series of z-scans were made to assess dye bleaching/dye leakage over time and to construct a bleaching curve for subtraction.

**Statistics**

To assess the effect of inhibitors on volume regulation, we compared the slopes of the individual time courses for calcein concentration change of the control and the inhibitor-treated cells in the linear phase of volume recovery using the students-T test or the Spearman-rank test for paired samples. Identical time periods of the scans were compared between control and inhibitor-treated cells. n = number of separate experiments.

## RESULTS

**Changes in parietal cell morphology during stimulation of acid formation**

Figure 1 shows a z-scan through three parietal cells on the second day of culture, before and 15 min after stimulation with $10^{-5}$M forskolin. Albeit the chosen picture is an extreme example of variability, it is evident that cellular morphology may vary considerably within individual cells. While one cell has no discernible intracellular vacuoles and does not develop any during stimulation, one has large vacuoles which further enlarge during stimulation, and a third has small vacuoles that enlarges during stimulation. Experiments with a conventional fluorescent microscope and the fluorescent dye 9 aminoacridine has shown that the pH in these vacuoles is acidic, indicative of acid formation ongoing in cultured parietal cells in the resting state, and becomes more acidic when stimulated. All cells stain for $H^+/K^+$ ATPase and also show other morphological features of parietal cells (1).

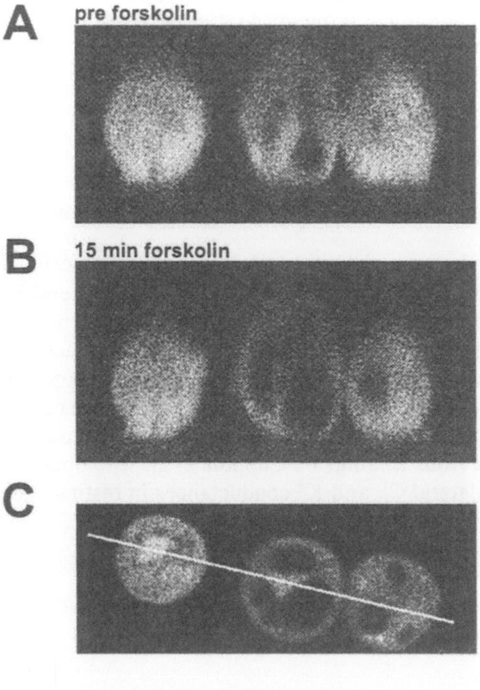


**Figure 1:**
Z-scan through three parietal cells before (Figure 1a) and 15 min after (Figure 1b) stimulation with $10^{-5}$M forskolin. Figure 1c shows the corresponding xy-scan and the line indicates the plane of the z-scan. All three cells had the typical characteristics of parietal cells which can be well distinguished from other cells by light microscopy and an experienced eye. One clearly sees the enlargement of intracellular vacuoles during stimulation. The marked difference of individual parietal cells in both the time course and the intensity of the response to a stimulus has been noted before (5).

**Effect of forskolin and carbachol on parietal cell cytoplasmic volume**

The application of $10^{-5}$ M forskolin, which in this concentration elicits a rapid stimulation of acid formation in cultured parietal cells, resulted in a rapid increase in cytoplasmic dye concentration with a maximum of 21±6% after 120 sec, indicative of cell shrinkage, followed by recovery to baseline within the next 6 min (Figure 2a). $10^{-4}$M carbachol is a weaker agonist of acid formation in rabbit parietal cells, resulting in approx 50% of the $^{14}$C-aminopyrin uptake rates compared to $10^{-5}$M forskolin (1). Interestingly, carbachol also resulted in a rapid cell shrinkage, but only by 9±4%, with recovery within 5 min (Figure 2b). This indicates that the magnitude of cell shrinkage correlates with the acid stimulatory capacity of the secretagogue.

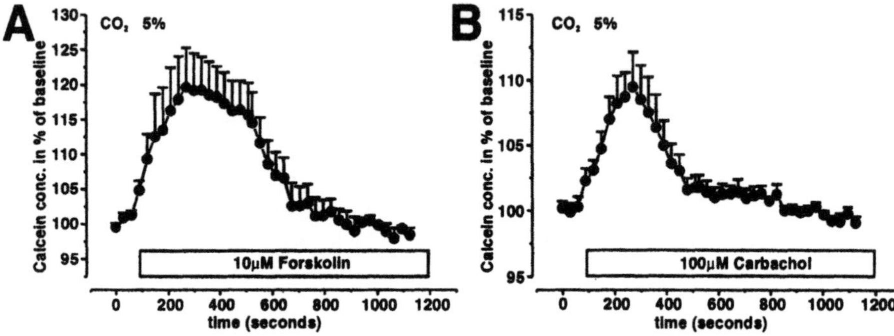


**Figure 2:**
Figure 2A: $10^{-5}$M forskolin results in rapid cytoplasmic shrinkage, indicated by an increase in the calcein concentration, followed by recovery to baseline values. Figure 2B: $10^{-4}$M carbachol also caused very rapid shrinkage, albeit less pronounced, followed by similarly rapid recovery. Since carbachol is a weaker secretagogue than forskolin in cultured rabbit parietal cells, the degree of shrinkage is correlated with the degree of acid formation stimulated by the secretagogue. n=5

**Effect of $K^+$ and $Cl^-$ channel blockade on secretagogue-induced parietal cell shrinkage**

The rapidity of shrinkage, which was much faster than the enlargement of intracellular vesicles, suggested to us that volume loss occurred secondary to opening of basolateral $K^+$ channels in addition to apical $Cl^-$ and $K^+$ channels. We therefore tested whether barium, a nonspecific inhibitor of $K^+$ channels, would inhibit the shrinkage. 5 mM barium actually resulted in cell swelling, followed by volume recovery (Figure 3a) and prevented forskolin-or carbachol-induced cell shrinkage (Figure 3b). These results show that $K^+$ channels are constitutively open in the absence of secretagogues, and their blockade results in reversible parietal cell swelling. This is in agreement with previous findings by Coppola et al, who obtained findings supportive of the presence of basolateral $Cl^-$ and $K^+$ conductances in resting state bullfrog oxyntic cells (9).

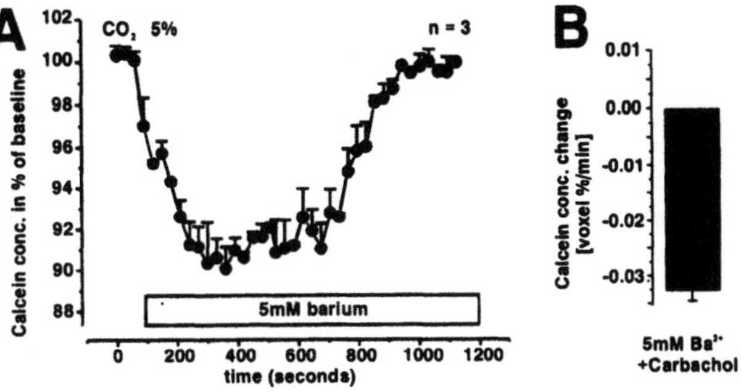


**Figure 3:**

Figure 3A: Calcein concentration change after application of 5 mM barium. This high barium concentration will block most $K^+$ channels and depolarize the cell, resulting in cell swelling. Indeed, we observed a rapid decline in calcein concentration, followed by recovery. These results suggest that $Ba^{2+}$-sensitive $K^+$ channels are constitutively open in resting parietal cells. Figure 3B.. After $Ba^{2+}$-mediated $K^+$-channel blockade, no secretagogue was able to induce shrinkage. n = 4-6, * $p<0.05$

We then tested the effect of more selective $K^+$ channel inhibitors. Recently, a KCNQ1 $K^+$ channels has been found to be expressed in parietal cells, possibly in an apical location (12;14). This KCNQ $K^+$-channel family is sensitive to chromanols (3), and has been implicated in cAMP-dependent $Cl^-$ secretion in the rat colon (15). We tested the effect of the chromanol Hoe 293b on forskolin- and carbachol-induced acid formation in cultured rabbit parietal cells, and found this substance to strongly inhibit $^{14}C$-AP accumulation stimulated by either agent, in the same concentration range that KCNQ1 channel inhibition has been described (3). When we tested the effect of Hoe 293b on parietal cell calcein concentration, we found virtually no change in the presence of Hoe 293b alone, but a very strong reduction in forskolin-evoked shrinkage (Figure 4A), and a somewhat lesser but still pronounced effect on carbachol-induced shrinkage (Figure 4B). This suggests that cAMP and $Ca^{2+}$-mediated stimulation activates 293b-$K^+$ channels which are largely quiescent in resting parietal cells.

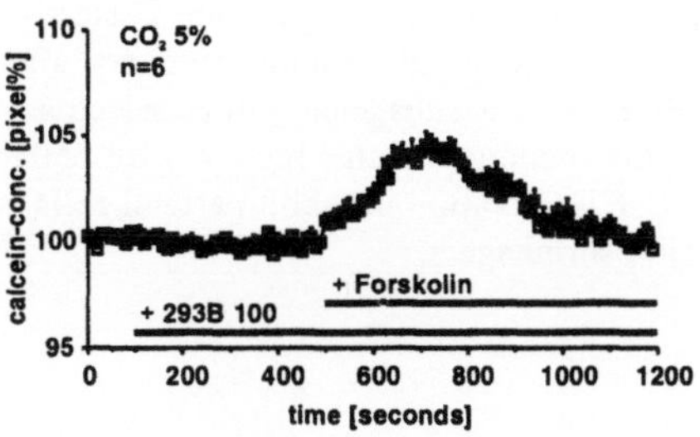


**Figure 4:**

Effect of the $K^+$-channel inhibitor Hoe 293b on parietal cell volume. In contrast to $Ba^{2+}$, Hoe 293b did not cause a marked swelling (even in the very high concentration used), but very markedly slowed subsequent secretagogue-associated shrinkage. These results suggest that the activation of a 293b-sensitive $K^+$-channel is one of the mechanisms of the initial cell shrinkage.

Charybdotoxin (ChTX) is an inhibitor of a subgroup of $Ca^{2+}$-sensitive $K^+$ channels (2;10). We tested the effect of ChTX on forskolin- and carbachol-induced acid formation in cultured rabbit parietal cells, and found this substance caused a partial inhibition of carbachol-induced $^{14}$C-AP accumulation and had no effect on forkolin-stimulated $^{14}$C-AP uptake. When parietal cell cytoplasmic volume changes were measured in the presence of ChTX, we found no significant volume increase, but a strong reduction of carbachol-induced shrinkage (whereas forskolin-associated volume decrease was not affected). This indicates that at least two $Ca^{2+}$-sensitive $K^+$-channels are activated during cholinergic stimulation of acid secretion.
NPPB is a rather nonselective inhibitor of $Cl^-$ channels and strongly inhibits acid secretion. Since it is likely that initial $K^+$ and $Cl^-$ loss needs to occur simultaneously, we tested the effect of NPPB on secretagogue-induced shrinkage and found it to be almost completely inhibited. Thus, the inhibition of agonist-activated $K^+$ and $Cl^-$ channels causes the initial cellular water loss after stimulation.

**Effect of $Na^+/H^+$ exchange and $Na^+K^+2Cl^-$ cotransport inhibition on secretagogue-induced cytoplasmic volume changes.**

$Na^+/H^+$ exchangers have been shown to be involved in RVI in several cell types (17;20;28). We have previously demonstrated that forskolin activates $Na^+/H^+$ exchange in cultured parietal cells, and that the involved isoforms are NHE1 and NHE4 (1). Since high DMA concentrations are needed to inhibit NHE4, we tested the effect of 500 µM DMA on forskolin-induced parietal cell volume changes. DMA reduced forskolin-induced cell shrinkage by approx. 25% (Figure 5B) and inhibited the speed of volume recovery by 60% and prolonged the time to complete recovery from approx. 6 min to approx. 16 min. A low concentration of DMA ($10^{-5}$M) did not result in a significant reduction in the speed of volume recovery, suggesting that the effect of DMA is primarily due to $Na^+/H^+$ exchange inhibition and not due to inhibition of an amiloride-sensitive $Na^+$ -channel. In many cell types, the bumetanide-sensitive $Na^+K^+2Cl^-$ cotransporter which is also expressed in the stomach (16) is involved in RVI. We therefore investigated if inhibition of $Na^+K^+2Cl^-$ cotransport by 100 µM bumetanide had an effect on forskolin-induced parietal cell volume changes. Bumetanide had no significant effect on forskolin-induced shrinkage or recovery (Figure 5C), and, when added together with DMA, did not result in a stronger inhibition of recovery than DMA alone. Therefore, the NKCC does not appear to regulate volume in acid-secreting parietal cells.
In a previous study, carbachol was found to primarily activate NHE1 (1). We speculated that if carbachol-induced $Na^+/H^+$ exchange activation was related to volume regulation, a selective inhibition of NHE1 would have a significant effect on cytoplasmic volume recovery after stimulation-associated shrinkage. Indeed, 1 µM HOE 642, a substance which selectively inhibits NHE1 at this low concentration, inhibited the speed of volume recovery by >50% (Figure 6). This indicates that carbachol-induced NHE1 activation in rabbit parietal cells is likely related to NHE1-mediated volume recovery after shrinkage.

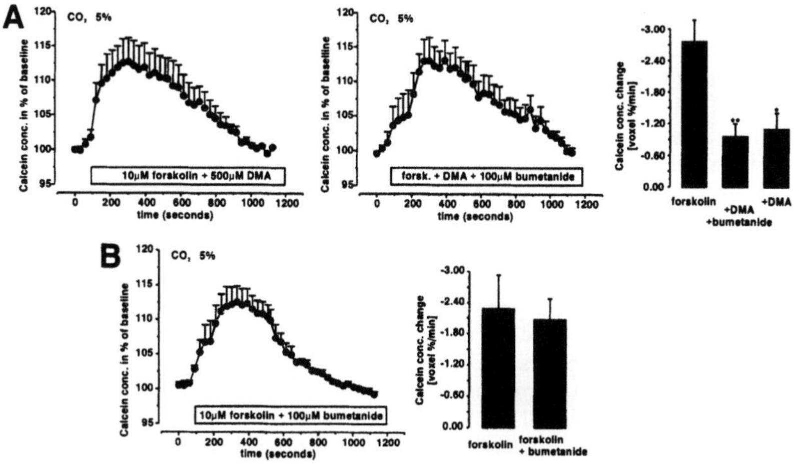


**Figure 5:**

Figure 5A: 500 µM DMA, which inhibits all forskolin-activated $Na^+/H^+$ exchanger isoforms in rabbit parietal cells (1), results in a significant slowing of volume recovery after forskolin application, both in the absence and additional presence of 100 µM bumetanide. As observed with hyperosmolarity, 500 µM DMA also reduces the initial shrinkage from 21% to 14%. n = 5-7, * p<0.05, **p<0.01. Figure 5B: 100 µM bumetanide, on the other hand, did not have a significant effect on the slope of the calcein concentration change. Therefore, the $Na^+ K^+ 2Cl^-$ cotransporter does not take part in the RVI after forskolin-associated parietal cell shrinkage. n = 3.

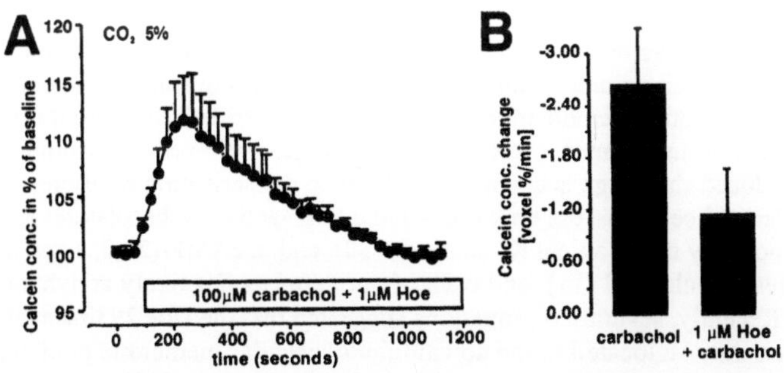


**Figure 6:**

Carbachol predominantly activates NHE1 in rabbit parietal cells (1). 1 µM Hoe 642, which very specifically inhibits NHE1, markedly slows volume recovery after carbachol-induced cytoplasmic shrinkage (Figure 5b,c). Unlike 500 µM DMA, 1 µM Hoe 642 does not reduce carbachol-induced shrinkage. This demonstrates that the carbachol-induced NHE1 activation in rabbit parietal cells is indeed related to shrinkage-associated volume recovery. n = 5, *p<0.05

## DISCUSSION

In this study, we have shown that 1.) isolated rabbit parietal cells in primary culture are able to regulate their cell volume after hyperosmolarity-induced cell shrinkage and depolarization-induced swelling 2.) secretagogue activation causes cytoplasmic shrinkage due at least in part to the opending of secretagogue-induced $K^+$ and $Cl^-$ channel activation, and 3.) $Na^+/H^+$ exchange is one of the mechanisms of rabbit parietal cell RVI after secretagogue activation, whereas $Na^+K^+2Cl^-$ cotransport appears not to be involved during RVI in this cell type.

Despite considerable scientific interest in the parietal cell pathways for ion and fluid movement, nothing is known about the ability of gastric parietal cells for volume regulation and the involved ionic mechanisms (20). A few years ago, Chew and collegues developed a method for culture of parietal cells for a period of several days and reported an increased sensitivity of these cultured cells to secretagogues (6). It was further shown that these cells develop a polarized distribution of membrane proteins, that they secrete acid into intracellular secretory vesicles whose membrane contains the apical proteins involved in HCl secretion and who enlarge during stimulation (23;37). Although this latter feature suggests that they do not fully resemble native parietal cells in situ, probably due to the lack of specialized cellular contacts, they proved useful for the study of signal transduction (7;8), cellular trafficking events (4) and gene transcription (27). We adapted this technique for the fluorometric study of ion transport processes and found that in contrast to the situation in freshly isolated parietal cells, secretagogues strongly stimulated $Na^+/H^+$ exchange in cultured parietal cells (1;34). Interestingly, a concomitant $pH_i$ increase and proton efflux was only observed in the absence, not in the presence of $CO_2/HCO_3^-$. Since the latter is the physiological situation, this suggested to us 1. that the physiological significance of secretagogue-induced $Na^+/H^+$ exchanger activation does not lie in an increase in $pH_i$ and that 2. $Cl^-/HCO_3^-$ exchange activation is likely occurring concomitantly with, and not sequentially to and caused by $Na^+/H^+$ exchange activation, as has been previously suggested (26). Since the concomitant activation of $Na^+/H^+$ and $Cl^-/HCO_3^-$ exchange is one mechanism for RVI, we speculated that parietal cells may shrink during stimulation of acid secretion, and that the observed $Na^+/H^+$ exchange activation was related to a parietal cell volume regulatory phenomenon during stimulation of acid secretion.

The present investigation demonstrates that secretagogue stimulation of cultured parietal cells does indeed result in a rapid cytoplasmic shrinkage, followed by volume recovery. Secretagogue-induced shrinkage is at least in part due to the activation of secretagogue-sensitive $K^+$ and $Cl^-$. Parietal cells possess both $Ca^{2+}$- and cAMP-activated basolateral $K^+$ channels (38), at least one, possibly more, apical $K^+$ channels (30), and a cAMP-(22;35) and possibly a $Ca^{2+}$-activated apical $Cl^-$ channel (11), and each of our agonists has likely activated a different set of channels. Recently, several investigations suggested that the Hoe 293b-sensitive $K^+$ channel in the stomach may be located in the apical/tubulovesicular membrane pool and therefore the long-searched for apical $K^+$ channel in the parietal cell (12;14). However, recent work by Cuppoletti et al (see this volume) suggest that members of the Kir $K^+$ channel family are apically located in parietal cells. By Western blotting of highly purified tubulovesicles, secretory and basolateral membranes, wee are presently trying to localize the Hoe 293b- and ChTX-sensitive $K^+$ channels within the parietal cell.

500 µM DMA caused an approx. 25% reduction of both hyperosmolarity- and forskolin induced cell shrinkage, probably due to a direct or indirect – via a $pH_i$ decrease – effect on $K^+$ channels. McLeod et al have found that in isolated enterocytes, cellular volume decrease after

nutrient-induced mild swelling requires $Na^+/H^+$ exchange activation, and that the $Na^+/H^+$ exchanger-induced alkalinization was essential for the activation of a $K^+$ channel (21). In cultured parietal cells, an alkalinization does not appear essential for cell shrinkage, since none occurs in the presence of 500 µM DMA, whereas the cells shrink still to 75% of the control value.

Secretagogue-induced $Na^+/H^+$ exchange activation occurs concomitantly with the onset of cellular volume loss (1). Rabbit parietal cells express NHE1, NHE2 and NH4 (33), and we have found that the forskolin-induced $Na^+/H^+$ exchange activation is mediated largely by NHE4 . The high DMA concentrations necessary to inhibit NHE4 reduced the speed of recovery by >60%, consistent with an important role of $Na^+/H^+$ exchangers in parietal cell volume regulation during acid secretion. However, this DMA concentration would likely also inhibit an epithelial sodium-channel which has been described to mediate a part of volume recovery in hepatocytes (40), and furthermore exert secondary effects due to the strong $pH_i$-decrease during DMA treatment. Therefore, we also studied carbachol-induced volume changes in cultured rabbit parietal cells. We have previously shown that carbachol predominantly activates the NHE1 isoform in cultured rabbit parietal cells, and that 1 µM Hoe 642 is able to inhibit this activation almost completely and very specific (1). Thus, we speculated that if the carbachol-induced NHE1 activation was related to a volume-regulatory phenomenon, we may inhibit the RVI after carbachol-induced shrinkage by selectively inhibiting NHE1. This was indeed the case. Therefore, the results of this study strongly support the concept that in rabbit parietal cells, the NHE isoforms activated during stimulation of acid secretion play an important role in the volume regulation necessary during acid secretion. The fact that the activation of $Na^+/H^+$ exchange is not accompanied by a $pH_i$ increase demonstrates that a $Cl^-/HCO_3^-$ exchange process, likely the AE2 isoform which is strongly expressed in rabbit parietal cells (32), is activated concomitantly.

However, volume recovery was not completely prevented by $Na^+/H^+$ exchange inhibition, indicative that other mechanisms may exist for a RVI in rabbit parietal cells. Inhibition of the bumetanide-sensitive $Na^+ K^+ 2Cl^-$ cotransporter NKCC did not result in an additional inhibition of volume recovery, suggesting that in rabbit parietal cells, this transporter is not involved in secretagogue-associated volume regulation. We also observe no effect of bumetanide on forskolin-stimulated acid formation in cultured rabbit parietal cells; and although bumetanide causes a marked reduction in short circuit current in isolated mouse gastric mucosa in the Ussing-chamber, it has no effect on forskolin-stimulated acid secretion (data not shown). This suggests that while NKCC is clearly involved in gastric $Cl^-$ transport, it is not a $Cl^-$ supply mechanism in acid secreting parietal cells, as has been speculated (18). This is further supported by the normal low gastric pH of NKCC knockout mice (13). Recent immunohistochemical data show that NKCC1 antibodies stain only parietal cells in the base of the glands, and that these parietal cells do not stain with AE2 antibodies, suggesting that they may not mediate HCl but NaCl or KCl secretion (24). Since we have chosen parietal cells which responded to acid secretagogues with an increase in the size in their intracellular vesicles, we may have selected parietal cells that secrete acid, as was our aim, and thus not studied those parietal cells that express an NKCC.

Additional putative mechanisms for the DMA-insensitive volume increase are the $Na^+HCO_3^-$ cotransporter NBC1, which is expressed in these cells and is activated by a low $pH_i$ (31), but whose role in volume regulation is as yet speculative, unspecific cation channels, the $H^+/K^+$ATPase which reabsorbs $K^+$ ions from the secretory vesicular space, or the fact that after

stimulation- and low $pH_i$- associated depolarization, cells gradually reswell because the equilibrium potential for $Cl^-$ is reversed.
In summary, this study demonstrates that cultured parietal cells are a useful model to study parietal cells volume regulatory mechanisms. Agonists of acid secretion cause a rapid cytoplasmic shrinkage, secondary at least in part to secretagogue-induced activation of K and Cl channels, followed by volume recovery which is predominantly mediated by $Na^+/H^+$ and $Cl/HCO_3^-$ exchange.

## ACKNOWLEDGEMENTS

We are sincerely grateful for the advice of Markus Ritter, Ewald Wöll and Florian Lang for the helpful advice and discussion concerning the volume measurements.
This work was supported by grants from the Deutsche Forschungsgemeinschaft Se 460/2-5 and Se 460/9-2 and Se 460/9-4 and a grant from the Federal Ministry of Education, Science, Research and Technology and the Interdisciplinary Center for Clinical Research (IZKF) of the University of Tübingen (Fö. 01KS9602). Thorsten Sonnentag was a stipend of the Graduiertenkolleg „Zellbiologie in der Medizin" of the Deutsche Forschungsgemeinschaft.
This work contains experiments performed by Thorsten Sonnentag and Andreas Heinzmann towards fulfilment of the requirements for their doctoral theses.

## REFERENCES

1. **Bachmann, O., T. Sonnentag, W. K. Siegel, G. Lamprecht, A. Weichert, M. Gregor, and U. Seidler.** Different acid secretagogues activate different $Na^+/H^+$ exchanger isoforms in rabbit parietal cells. *Am J. Physiol.* 275: G1085-G1093, 1998.
2. **Bleich, M., N. Riedemann, R. Warth, D. Kerstan, J. Leipziger, M. Hor, W. V. Driessche, and R. Greger.** $Ca^{2+}$ regulated $K^+$ and non-selective cation channels in the basolateral membrane of rat colonic crypt base cells. *Pflugers Arch.* 432: 1011-1022, 1996.
3. **Bleich, M. and R. Warth.** The very small-conductance $K^+$ channel KvLQT1 and epithelial function. *Pflugers Arch.* 440: 202-206, 2000.
4. **Calhoun, B. C., L. A. Lapierre, C. S. Chew, and J. R. Goldenring.** Rab11a redistributes to apical secretory canaliculus during stimulation of gastric parietal cells. *Am.J.Physiol.* 275: C163-C170, 1998.
5. **Chew, C. S. and M. Ljungstrom.** HCl secretion and $[Ca^{2+}]i$ in cultured parietal cells. *J.Intern.Med.Suppl.* 732: 9-15, 1990.
6. **Chew, C. S., M. Ljungstrom, A. Smolka, and M. R. Brown.** Primary culture of secretagogue-responsive parietal cells from rabbit gastric mucosa. *Am.J.Physiol.* 256: G254-63, 1989.
7. **Chew, C. S., K. Nakamura, and A. C. Petropoulos.** Multiple actions of epidermal growth factor and TGF-alpha on rabbit gastric parietal cell function. *Am.J.Physiol.* 267: G818-26, 1994.
8. **Chew, C. S., C. J. Zhou, and J. Parente-JA.** $Ca^{2+}$-independent protein kinase C isoforms may modulate parietal cell HCl secretion. *Am.J.Physiol.* 272: G246-G256, 1997.
9. **Coppola, S., R. Caroppo, and E. Fromter.** Electrophysiological investigation of microdissected gastric glands of bullfrog. II. Basolateral membrane properties in the presence of histamine. *Pflugers Arch.* 429: 203-209, 1994.

10. **Cuthbert, A. W., M. E. Hickman, P. Thorn, and L. J. MacVinish.** Activation of Ca(2+)- and cAMP-sensitive $K^+$ channels in murine colonic epithelia by 1-ethyl-2-benzimidazolone. *Am.J.Physiol* 277: C111-C120, 1999.
11. **Debellis, L., R. Caroppo, E. Fromter, and S. Curci.** Alkaline secretion by frog gastric glands measured with pH microelectrodes in the gland lumen. *J. Physiol. (Lond).* 513: 235-241, 1998.
12. **Dedek, K. and S. Waldegger.** Colocalization of KCNQ1/KCNE channel subunits in the mouse gastrointestinal tract. *Pflugers Arch.* 442: 896-902, 2001.
13. **Flagella, M., L. L. Clarke, M. L. Miller, L. C. Erway, R. A. Giannella, A. Andringa, L. R. Gawenis, J. Kramer, J. J. Duffy, T. Doetschman, J. N. Lorenz, E. N. Yamoah, E. L. Cardell, and G. E. Shull.** Mice lacking the basolateral Na-K-2Cl cotransporter have impaired epitheial chloride secretion and are profoundly deaf. *Nat.Genet.* 22:192-195, 1999.
14. **Grahammer, F., A. W. Herling, H. J. Lang, A. Schmitt-Graff, O. H. Wittekindt, R. Nitschke, M. Bleich, J. Barhanin, and R. Warth. The cardiac $K^+$ channel KCNQ1 is essential for gastric acid secretion.** *Gastroenterology* 120: 1363-1371, 2001.
15. **Greger, R.** Role of CFTR in the colon. *Annu.Rev.Physiol* 62: 467-491, 2000.
16. **Hallonquist, H., R. R. Cima, M. E. Klingensmith, M. J. Purdy, E. Delpire, M. J. Zinner, and D. I. Soybel.** Selective increase in gastric mucosal mRNA encoding basolateral Na-K-2C1 cotransporter following ileostomy in the rat. *J.Gastrointest.Surg.* 2: 238-243, 1998.
17. **Hoffmann, E. K. and P. B. Dunham.** Membrane mechanisms and intracellular signalling in cell volume regulation. *Int.Rev.Cytol.* 161: 173-262, 1995.
18. **Klingensmith, M. E., H. Hallonquist, B. P. McCoy, R. R. Cima, E. Delpire, and D. I. Soybel.** Pentagastrin selectively modulates levels of mRNAs encoding apical H/K adenosine triphosphatase and basolateral Na-K-Cl cotransporter in rat gastric fundic mucosa. *Surgery* 120: 242-246, 1996.
19. **Komar, A. R., Ma, T, Wang, K. S., McLeroy, J., Cross, I. V., Verkman, A. S., and Bastidas, J. A.** Decreased gastric acid secretion in aquaporin 4 knockout mice. Gastroenterology 114, S0126. 1998.
20. **Lang, F., G. L. Busch, and H. Volkl.** The diversity of volume regulatory mechanisms. *Cell Physiol.Biochem.* 8: 1-45, 1998.
21. **MacLeod, R. J. and J. R. Hamilton.** Activation of $Na^+/H^+$ exchange is required for regulatory volume decrease after modest "physiological" volume increases in jejunal villus epithelial cells. *J.Biol.Chem.* 271: 23138-23145, 1996.
22. **Malinowska, D. H., E. Y. Kupert, A. Bahinski, A. M. Sherry, and J. Cuppoletti.** Cloning, functional expression, and characterization of a PKA-activated gastric $Cl^-$ channel. *Am.J.Physiol.* 268: C191-C200, 1995.
23. **Mangeat, P., T. Gusdinar, A. Sahuquet, D. K. Hanzel, J. G. Forte, and R. Magous.** Acid secretion and membrane reorganization in single gastric parietal cell in primary culture. *Biol.Cell* 69: 223-231, 1990.
24. **McDaniel, N. and C. Lytle.** Parietal cells express high levels of Na-K-2Cl cotransporter on migrating into the gastric gland neck. *Am. J. Physiol.* 276: G1273-G1278, 1999.
25. **Misaka, T., K. Abe, K. Iwabuchi, Y. Kusakabe, M. Ichinose, K. Miki, Y. Emori, and S. Arai.** A water channel closely related to rat brain aquaporin 4 is expressed in acid- and pepsinogen-secretory cells of human stomach. *FEBS Letters.* 381: 208-212, 1996.

26. **Muallem, S., D. Blissard, E. J. Cragoe, Jr., and G. Sachs.** Activation of the $Na^+/H^+$ and Cl-/HCO3- exchange by stimulation of acid secretion in the parietal cell. *J.Biol.Chem.* 263: 14703-14711, 1988.
27. **Muraoka, A., M. Kaise, Y. J. Guo, J. Yamada, I. Song, J. DelValle, A. Todisco, and T. Yamada.** Canine H(+)-K(+)-ATPase alpha-subunit gene promoter: studies with canine parietal cells in primary culture. *Am.J.Physiol.* 271: G1104-G1113, 1996.
28. **Pedersen, S. F., B. Kramhoft, N. K. Jorgensen, and E. K. Hoffmann.** Shrinkage-induced activation of the $Na^+/H^+$ exchanger in Ehrlich ascites tumor cells: mechanisms involved in the activation and a role for the exchanger in cell volume regulation. *J.Membr.Biol.* 149: 141-159, 1996.
29. **Priver, N. A., E. C. Rabon, and M. L. Zeidel.** Apical membrane of the gastric parietal cell: water, proton, and nonelectrolyte permeabilities. *Biochemistry* 32: 2459-2468, 1993.
30. **Reenstra, W. W. and J. G. Forte.** Characterization of $K^+$ and $Cl^-$ conductances in apical membrane vesicles from stimulated rabbit oxyntic cells. *Am.J.Physiol.* 259: G850-G858, 1990.
31. **Rossmann, H., O. Bachmann, W.-K. Siegel, G. Lamprecht, M. Gregor, and U. Seidler.** $Na^+HCO_3^-$ cotransport and expression of NBC1 and NBC2 in cultured rabbit parietal and surface cells. *Gastroenterology* 116: 1389-1398, 1999.
32. **Rossmann, H., O. Bachmann, Z. Wang, G. E. Shull, B. Obermaier, A. Stuart-Tilley, S. L. Alper, and U. Seidler.** Differential expression and regulation of AE2 anion exchanger subtypes in rabbit parietal and mucous cells. *J Physiol* 534: 837-848, 2001.
33. **Rossmann, H., T. Sonnentag, A. Heinzmann, B. Seidler, O. Bachmann, D. Vieillard-Baron, M. Gregor, and U. Seidler.** Differential expression and regulation of Na(+)/H(+) exchanger isoforms in rabbit parietal and mucous cells. *Am J Physiol.* 281: G447-G458, 2001.
34. **Seidler, U., S. Roithmaier, M. Classen, and W. Silen.** Influence of acid secretory state on Cl(-)-base and Na(+)-H+ exchange and pHi in isolated rabbit parietal cells. *Am.J.Physiol.* 262: G81-G91, 1992.
35. **Sherry, A. M., K. Stroffekova, L. M. Knapp, E. Y. Kupert, J. Cuppoletti, and D. H. Malinowska.** Characterization of the human pH- and PKA-activated ClC-2G(2 alpha) $Cl^-$ channel. *Am.J.Physiol.* 273: C384-C393, 1997.
36. **Sonnentag, T., W. K. Siegel, O. Bachmann, H. Rossmann, A. Mack, H. J. Wagner, M. Gregor, and U. Seidler.** Agonist-induced cytoplasmic volume changes in cultured rabbit parietal cells. *Am J Physiol.* 279: G40-G48, 2000.
37. **Soroka, C. J., C. S. Chew, D. K. Hanzel, A. Smolka, I. M. Modlin, and J. R. Goldenring.** Characterization of membrane and cytoskeletal compartments in cultured parietal cells: immunofluorescence and confocal microscopic examination. *Eur.J.Cell Biol.* 60: 76-87, 1993.
38. **Ueda, S., D. D. Loo, and G. Sachs.** Regulation of $K^+$ channels in the basolateral membrane of Necturus oxyntic cells. *J.Membr.Biol.* 97: 31-41, 1987.
39. **Valenti, G., J. M. Verbavatz, I. Sabolic, D. A. Ausiello, A. S. Verkman, and D. Brown.** A basolateral CHIP28/MIP26-related protein (BLIP) in kidney principal cells and gastric parietal cells. *Am.J.Physiol.* 267: C812-C820, 1994.
40. **Wehner, F. and H. Tinel.** Role of $Na^+$ conductance, $Na^+$-$H^+$ exchange, and $Na^+$-$K^+$-$2Cl^-$ symport in the regulatory volume increase of rat hepatocytes. *J. Physiol. (Lond).* 506: 127-142, 1998.

Chapter 27

# Characterization, Regulation and Localization of the Gastric $Cl^-$ Channel Associated with Gastric Acid Secretion

DANUTA H. MALINOWSKA, ANN M. SHERRY, KIRTI P. TEWARI AND JOHN CUPPOLETTI

*University of Cincinnati College of Medicine, Department of Molecular and Cellular Physiology, Cincinnati, OH 45267-0576.*

## ABSTRACT

The $Cl^-$ channel of the apical membrane of the gastric parietal cell was characterized in native rabbit H/K ATPase-containing membrane vesicles in bilayer reconstitution studies. It is a linear $Cl^-$ channel which exhibits activation by reduction of extracellular pH and PKA. The ClC-2 $Cl^-$ channel cloned from rabbit gastric mucosa and expressed in *Xenopus laevis* oocytes, exhibited similar electrophysiological characteristics and regulation as the native gastric $Cl^-$ channel. Potential PKA phosphorylation sites in rabbit ClC-2 appear to underlie PKA activation. The pH sensor on the extracellular surface of ClC-2 was identified as E419 and amidation resulted in activation of the channel. Recombinant ClC-2 stably expressed in HEK-293 cells showed $Cl^-$ currents that were activated by forskolin and IBMX, low extracellular pH, amidation, arachidonic acid and acid-activated omeprazole. Using immunofluorescent confocal microscopy, ClC-2 was localized to the secretory canalicular membrane of stimulated parietal cells and to intracellular structures (tubulovesicles) in resting parietal cells, which was confirmed by immunogold electron microscopy. Distribution of the gastric H/K ATPase and ClC-2 were similar. These findings strongly support the view that ClC-2 is the $Cl^-$ channel in the apical secretory membrane of the gastric parietal cell which, together and in concert with the H/K ATPase and a $K^+$ channel, result in HCl secretion.

## INTRODUCTION

The $K^+$ and $Cl^-$ conductances in the apical/secretory membrane of the gastric parietal cell have been the focus of research in this laboratory for many years (6-9,13-15,23-26). These two conductances allow $K^+$ into the lumen to be recycled by the H/K ATPase in exchange for $H^+$ and $Cl^-$ into the lumen to produce HCl (**Fig. 1**). Gastric membrane vesicles containing the H/K ATPase isolated from the gastric mucosae of rabbits either stimulated to secrete HCl with histamine and diphenhydramine (stimulated vesicles) or treated with cimetidine to render the stomach non-secreting (resting vesicles) have been highly characterized with respect to ATPase activity and $H^+$ transport, but only indirect evidence suggested that KCl movement occurred through conductances (7,17,28). This paper focuses on the $Cl^-$ channel. Characterization, identification, regulation and localization of the gastric $Cl^-$ channel associated with HCl secretion is described.

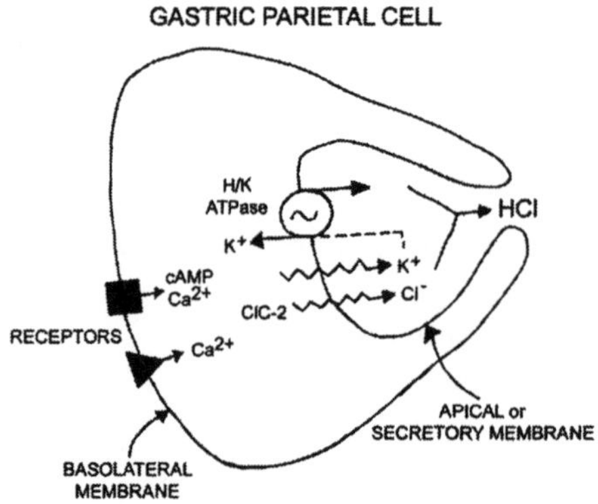


**Fig. 1.** Model of the gastric parietal cell showing apical secretory membrane containing H/K ATPase and $K^+$ and $Cl^-$ channels.

## MATERIALS and METHODS

Gastric vesicles were isolated as previously described (7). Expression of ClC-2 in *Xenopus laevis* oocytes was as described in (15,24,25). Lipid bilayer reconstitution methods were as described in (6,15,24,25). Stable transfection of wild type and mutant channels into HEK-293 cells and whole cell patch clamp were as described in (8,26). Molecular biology, cloning and site-directed mutagenesis were as described (15,25). *In vitro* translation was carried out according to the manufacturer's instructions using the Promega TNT® T7 Coupled Reticulocyte Lysate System. Immunofluorescent confocal microscopy was carried as described in (23).

## RESULTS

**$Cl^-$ channels are present in rabbit gastric membrane vesicles.** To obtain direct evidence for the presence of $Cl^-$ channels, stimulated gastric H/K ATPase-containing vesicles were fused to planar lipid bilayers and channel currents were measured. **Fig. 2A** (from (6)) shows channel current recordings obtained with 800 mM CsCl pH 7.4 on both *cis* (intracellular) and *trans* (extracellular) sides of the channel. Channel currents were also evident when the pH of the *trans* (extracellular) side was lowered to 3.0 (a physiologically relevant condition that mimics the pH of the primary secretion of the gastric parietal cell) (**Fig. 2B**). The I/V relationship of the channel was linear (**Fig. 2C**), the slope conductance was 28±1 pS at 800 mM CsCl and 7 pS at 150 mM CsCl, and the reversal potential was +22 mV under a 5-fold *cis:trans* gradient indicating an anion/cation discrimination ratio of 6:1 (**Fig. 2D**). Shown in **Fig. 3A** (from (6)), a channel of the same conductance and reversal potential was also observed when vesicles from cimetidine-treated (non-secreting or resting) rabbits were fused to the bilayer. Unlike channels from stimulated preparations, the resting preparations gave channels that were mainly closed, even at the long times used for recording of the channel in resting membranes. The histogram in **Fig. 3B** compares open probabilities (Po) for the channel from resting and stimulated preparations. Channels in resting vesicle preparations were mainly closed, while channels in stimulated vesicle preparations had a high Po of 0.8 (**Figs. 3 and 4** from (6)).

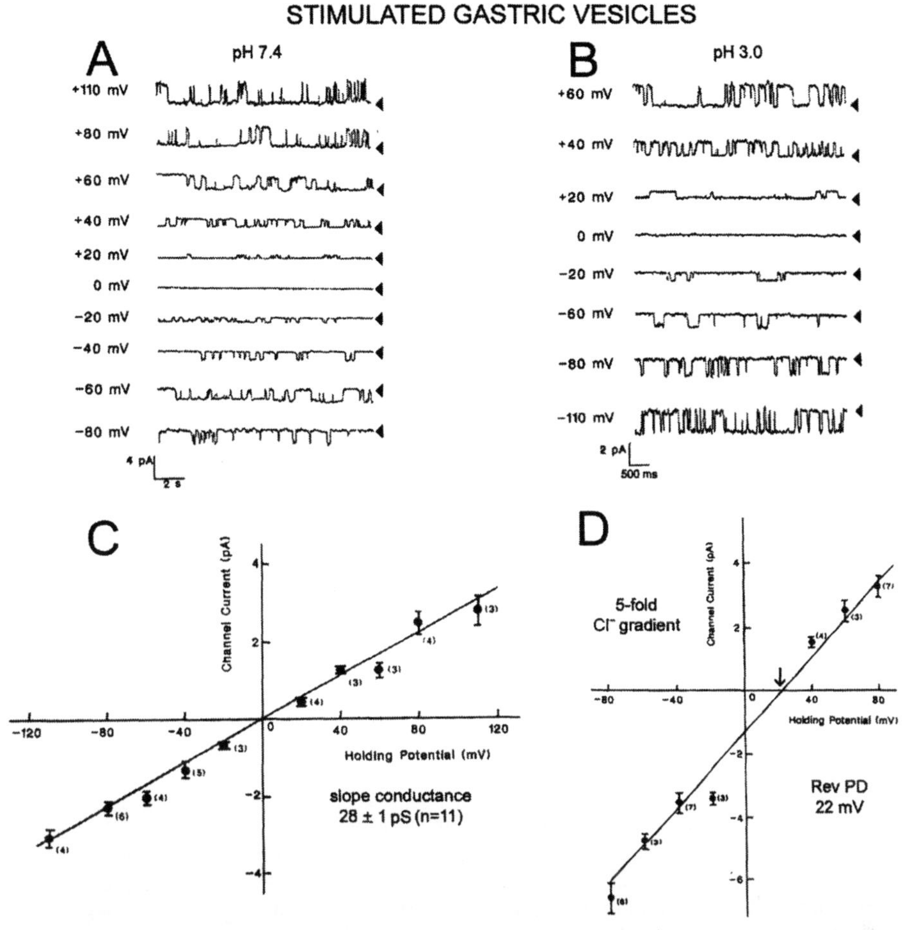


**Fig. 2.** Electrophysiological properties of stimulated H/K ATPase-containing membrane vesicle preparations. Data in C & D are plotted as the mean ± SEM. For details see (1).

**Cloning the gastric Cl⁻ channel and its functional characteristics.** The molecular nature of the gastric Cl⁻ channel was investigated by molecular cloning (15) using a combination of expression cloning and homology cloning. In expression cloning, mRNA for the channel was size fractionated, injected into oocytes, and the membranes from each fraction tested for Cl⁻ channel function at low pH. Simultaneous homology cloning was carried out using a DNA probe made from the known sequence of ClC-1, a member of the ClC Cl⁻ channel family. A rabbit gastric library was screened and a cDNA was isolated which was present in the same mRNA fraction that resulted in expression of pH stable Cl⁻ channel activity. The cDNA encoded an 898 amino acid (98 kDa) protein with 12 trans-

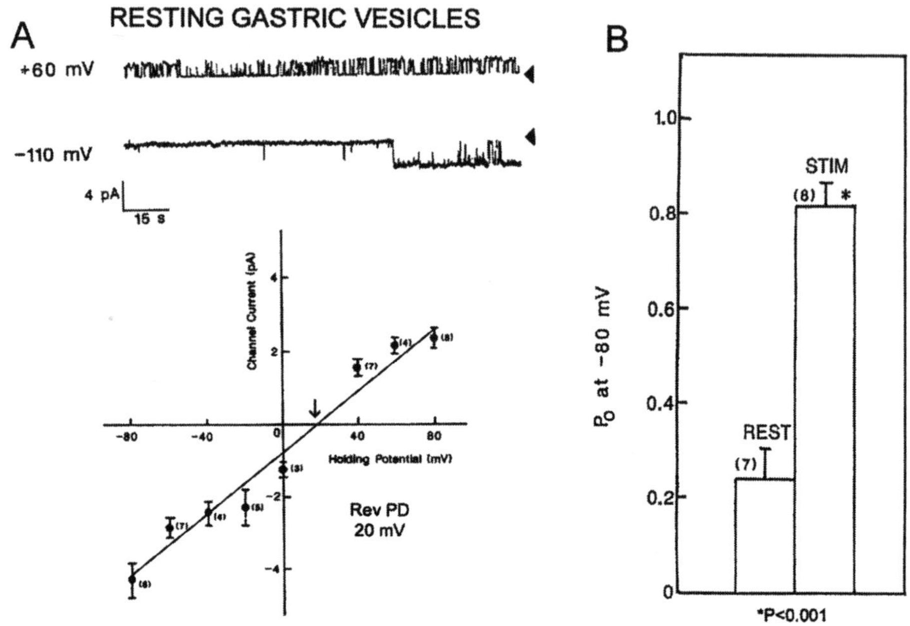


**Fig. 3.** Electrophysiological properties of resting H/K ATPase-containing membrane vesicle preparations. Data are plotted as the mean ± SEM. For details see (1).

membrane domains with overall 93% similarity to rat brain ClC-2, which dropped to 74% similarity in a 151 amino acid stretch of the C-terminus. To investigate ClC-2 function, cRNA was made and injected into *Xenopus* oocytes, and after 4-5 days, oocyte plasma membranes were prepared and fused to planar lipid bilayers (15,24,25). The characteristics of recombinant ClC-2 were indistinguishable from those of native gastric vesicles: channel currents evident at $pH_{trans}$ 7.4 and 3.0, linear I/V curve, 29 pS conductance at 800 mM CsCl, reversal potential of about 25 mV (15,24,25).

**PKA activation of the rabbit and human form of ClC-2.** As shown in **Fig. 4** (data from (6,15)), when PKA catalytic subunit and ATP-Mg were added to preparations of native resting gastric vesicles, channel Po increased from about 0.1 to about 0.7, the latter value reflecting the Po of channels in stimulated vesicle preparations. The important feature of the C-terminus of the cloned rabbit (15) and human (2,24) ClC-2 is that it contains 2 potential PKA phosphorylation sites (RRAT and RGET in the rabbit (15); RRQS and KRKS in the human (2,24) that were not present in the rat form of ClC-2 (27). Rat ClC-2 is not stimulated by PKA (11). The unique sites in rabbit and human ClC-2 could be responsible for the PKA activation observed with the native channel. As also shown in **Fig. 4** (data in (15)), in the absence of PKA catalytic subunit, recombinant ClC-2 was virtually closed, (Po~0.1), but upon PKA catalytic subunit and ATP-Mg addition, ClC-2 opened with Po~0.75.

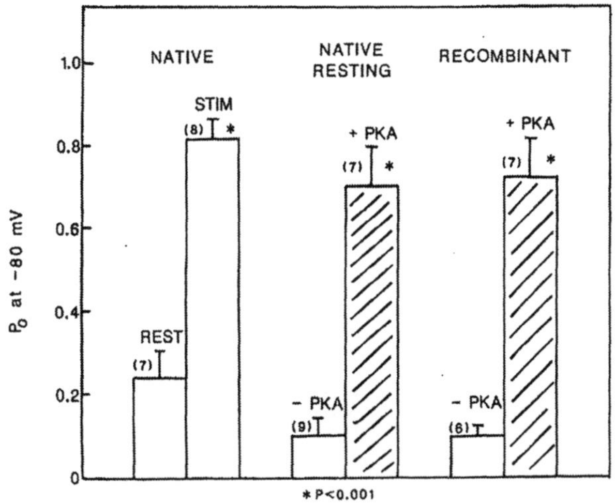


**Fig. 4.** Effect of PKA on Cl⁻ channel Po of native resting gastric membrane preparations and recombinant ClC-2. For comparison, channel Po values for resting and stimulated gastric membrane preparations are also plotted. Data are plotted as the mean ± SEM. For details see (6,15).

**Effect of low external pH: Identification of the pH sensor.** In addition to being PKA-activated, Cl⁻ channel activity in native gastric vesicles was found to be activated by low pH (6). The recombinant rabbit and human ClC-2 are also pH activated (8,15,22,24-26). It is important to recognize that ClC-2 is only pH activated when phosphorylated ((26) and Fig. 7). Using recombinant ClC-2, the site(s) of pH activation was investigated. In these studies, the effect of lowering extracellular pH over the range 7.4 to 3.0 was first investigated. Reduced pH on the extracytoplasmic face of the channel progressively increased channel Po (25). Half-maximal stimulation was found at pH 4.5, suggesting involvement of histidine, glutamate or aspartate at the extracellular face of the channel protein. The effect of amidation with glycine methyl ester (GME) catalyzed by the water-soluble EDC, was therefore investigated next. Amidation catalyzed by EDC acts upon carboxyl groups, but does not react with histidine. If neutralization of the negative charge of carboxyl groups by protonation was responsible for the change in Cl⁻ channel activity, then neutralization by amidation should also activate the channel. **Fig. 5** (from (25)) shows that amidation significantly increased ClC-2 channel Po. These reagents are large, water-soluble compounds and are incapable of entering the membrane, indicating that externally facing carboxylic acids (not histidines) are involved in channel activation at low extracellular pH. Glutamates on the extracellular face of the channel (12,19) between transmembrane segments D8 and D9 (EELE419) were candidates to serve proton sensor function. To examine whether these residues were involved in pH activation and stimulation by amidation, site directed mutagenesis was used to change negatively charged glutamic acids to neutral glutamine. As shown in Fig. 6 (from (25)) $pH_{trans}$ 3.0 had no effect on the Po of the EELQ mutant (**Fig. 6C**), but increased Po of both WT ClC-2 (**Fig. 6A**) and the QQLE mutant (**Fig. 6B**) similarly, indicating that E419 plays an important role in pH sensor function.

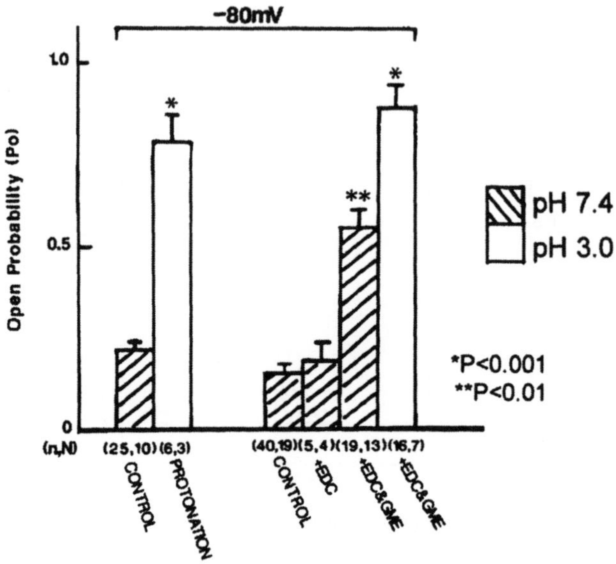


**Fig. 5.** Effect of protonation (extracytoplasmic pH 3.0) and amidation (with EDC and GME) on recombinant ClC-2 Po at –80 mV holding potential. For details see (25).

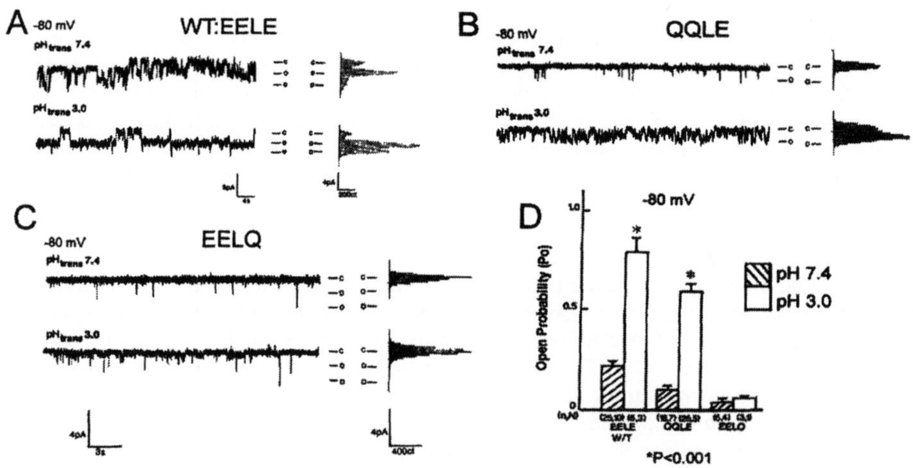


**Fig. 6.** Effect of low extracellular pH on channel activity of **(A)** WT recombinant ClC-2 (EELE419) and ClC-2 mutants QQLE **(B)** and EELQ **(C)**. **(D)** Summary of the Po data for these channels is plotted as the mean ± SEM. For details see (25).

**Characteristics of ClC-2 stably transfected into HEK-293 cells.** Whole cell current measurements by patch clamp provide a number of advantages over single channel studies in bilayers. Such studies facilitate determination of the effects of intracellular mediators and drugs. For these studies, ClC-2 was stably transfected into HEK-293 cells, which have low endogenous $Cl^-$ currents and have been widely used for studies of $Cl^-$ channels (for details, see (8)). As shown in **Fig. 7** (from (26)) both forskolin/IBMX and low external pH of 6.5 increased $Cl^-$ currents and activation by forskolin/IBMX was inhibited by the cell-permeant myristoylated PKI, indicating that they are mediated by PKA (26). Non-transfected HEK-293 cells showed no effects of forskolin/IBMX or external pH of 6.5 or a variety of other effectors of ClC-2 (8,26). The properties of human ClC-2 in HEK-293 cells were similar to those of the recombinant channel and native gastric ClC-2 channel as studied in bilayers.

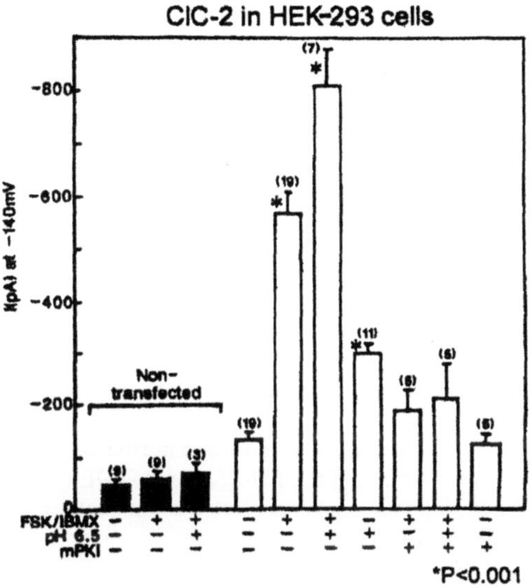


**Fig. 7.** Whole cell $Cl^-$ currents at −140 mV holding potential in non-transfected HEK-293 cells and in HEK-293 cells stably transfected with human recombinant ClC-2: effect of forskolin (5 μm)/IBMX (20 μm), pH 6.5 and myristoylated PKI (0.4 μm). Data are plotted as the mean ± SEM.

**Effect of arachidonic acid, fatty acids, and COX inhibitors on ClC-2 $Cl^-$ channel currents.** Arachidonic acid has been shown to affect other ion channels (16,26), therefore the effect on ClC-2 whole cell currents was examined. **Fig. 8** shows the effect of arachidonic acid (1 μM) on whole cell currents. Arachidonic acid increased the currents in ClC-2 transfected HEK-293 cells with no effect in non-transfected cells. Similar activation was seen with COX and lipoxygenase inhibitors, suggesting that ClC-2 might be a target for the action of COX inhibitors (3,26).

**Effect of amidation and acid-activated omeprazole on ClC-2 $Cl^-$ channel function.** As expected from bilayer studies (25), whole cell current activation was observed with amidation shown in **Fig 8** from (26). Since ClC-2 has one or more externally facing

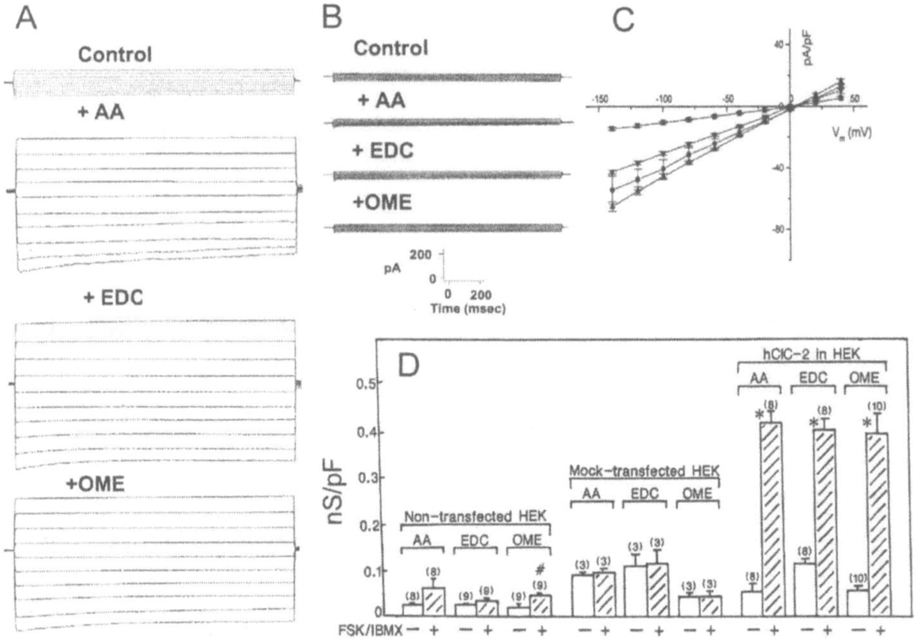


**Fig. 8.** Effects of arachidonic acid (AA), amidation (EDC) and acid-activated omeprazole (OME) on whole cell $Cl^-$ currents in HEK-293 cells stably expressing human recombinant ClC-2. Representative current scans in transfected cells **(A)** and non-transfected cells **(B)** are shown. **(C)** I/V curves for $Cl^-$ currents normalized to cell capacitance before (■) and after treatment with AA(●), EDC (▲) and OME (▼). **(D)** shows summarized data for normalized slope conductance plotted as the mean ± SEM. #P<0.01; *P<0.001. For further details see (8).

cysteines (8) the effect of acid-activated omeprazole was examined and also found to be a potent ClC-2 channel activator. Both of these approaches have promise in activation of $Cl^-$ channels in disease (4,5).

**ClC-2 localization in the gastric mucosa.** It was essential to address the cellular and subcellular localization of the ClC-2 $Cl^-$ channel protein and mRNA in the gastric mucosa. As shown in **Fig 9A** (from (23)) ClC-2 was used first to probe a Western blot of stimulated gastric vesicles. A protein of about 93 kDa was detected by the antibody. An antigenic peptide was used to establish specificity of the antibody. A protein of the same size is made when recombinant ClC-2 cDNA is translated *in vitro* as shown in **Fig. 9B**.

Using immunofluorescent confocal microscopy, ClC-2 was localized in rabbit gastric glands. **Fig. 10A** (from (23)) shows that in stimulated gastric glands ClC-2 (red fluorescence) localizes to parietal cells and to what seems to be the apical secretory canaliculus within the parietal cells. ClC-2 is absent from the basolateral membrane. The location of ClC-2 was then compared to that of the H/K ATPase (**Fig. 10B**) using resting

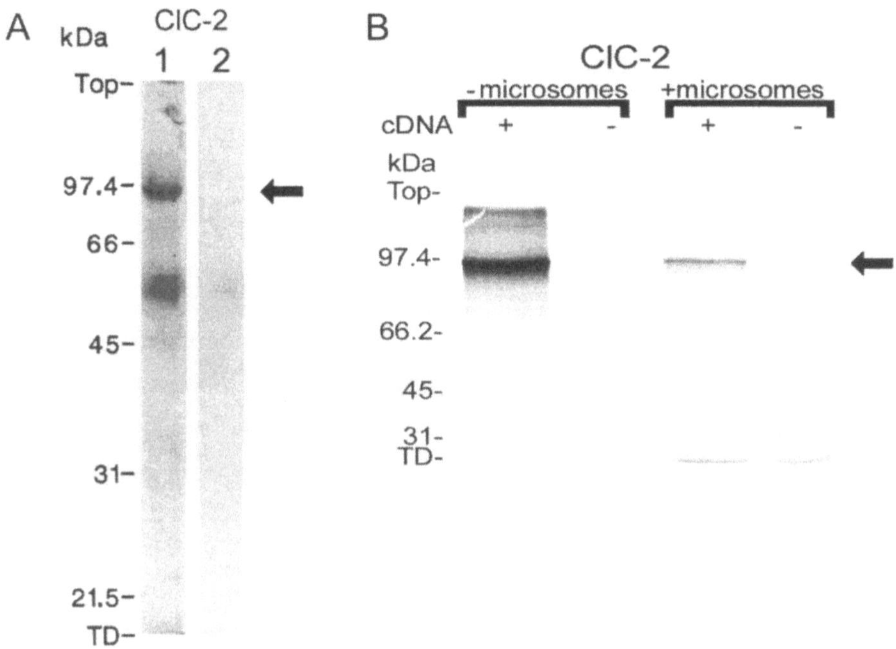


**Fig. 9.** **(A)** Western blot of native rabbit gastric vesicles probed with anti-ClC-2 antibody. **(B)** *In vitro* translation of recombinant ClC-2. Arrows indicate the ClC-2 protein of about 93 kDa in both (A) and (B).

gastric glands. Phalloidin was used to stain actin, which outlines the parietal cells and the glandular lumen. ClC-2 and H/K ATPase co-localized in the parietal cell. Immunogold electronmicroscopy showed that ClC-2 is found in the tubulovesicles in resting parietal cells and in the secretory canaliculus of stimulated parietal cells (see Fig. 5 of (23)).

## DISCUSSION

The objectives of the present studies were to identify the molecules and mechanisms involved in regulated gastric HCl secretion across the apical secretory membrane of the parietal cell. The gastric parietal cell apical membrane, as reflected in H/K ATPase-containing membrane vesicles from resting (cimetidine-treated) and stimulated (histamine and diphenhydramine-treated) rabbit gastric mucosae, contain a 28 pS $Cl^-$ channel that was stable and activated when bathed at pH 3.0 on the extracytosolic surface of the channel (6). Channels in resting preparations had low Po, and channels in stimulated preparations had a much higher Po, as expected for a channel responsible for regulated HCl secretion (6). Expression and homology cloning of $Cl^-$ channels from rabbit gastric mucosa yielded a form of ClC-2 that differed from rat brain ClC-2 (27) in that it contained unique potential PKA

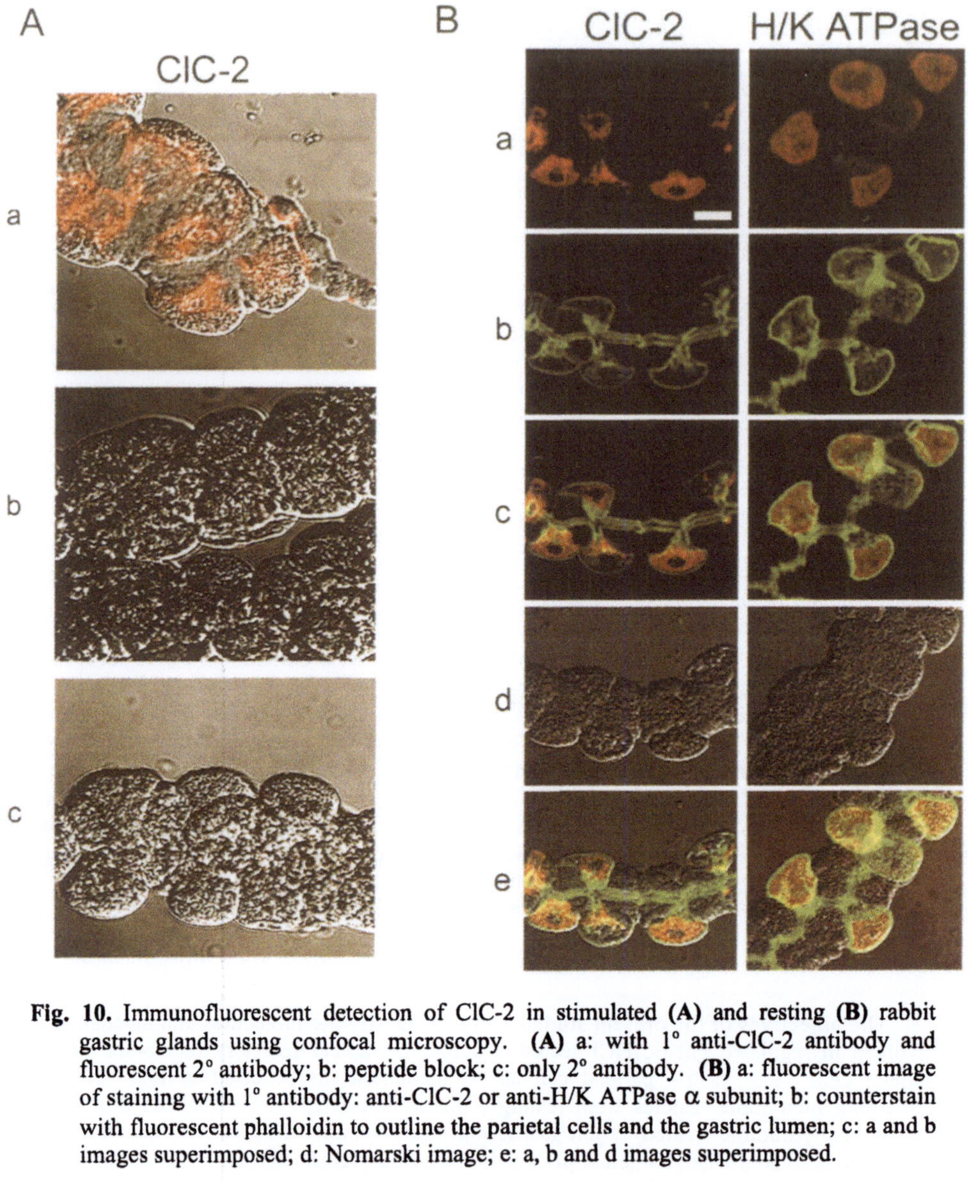


**Fig. 10.** Immunofluorescent detection of ClC-2 in stimulated **(A)** and resting **(B)** rabbit gastric glands using confocal microscopy. **(A)** a: with 1° anti-ClC-2 antibody and fluorescent 2° antibody; b: peptide block; c: only 2° antibody. **(B)** a: fluorescent image of staining with 1° antibody: anti-ClC-2 or anti-H/K ATPase α subunit; b: counterstain with fluorescent phalloidin to outline the parietal cells and the gastric lumen; c: a and b images superimposed; d: Nomarski image; e: a, b and d images superimposed.

phosphorylation sites, RRQS and KRKS (15). Similar unique sites were found in the human form (2,24). The unit conductance, linearity of I/V curves, reversal potential, stability and activation with low pH bathing the extracellular face of the cloned ClC-2 channels were indistinguishable from the native channel (6,15,24). PKA activated both native rabbit gastric

$Cl^-$ channels (15) and rabbit (15) and human (24) recombinant ClC-2 $Cl^-$ channels. ClC-2 was localized exclusively to the gastric parietal cell in gastric glands (23). Immunoelectron microscopy showed the channel to be present in tubulovesicles of resting parietal cells and in the canalicular membrane of stimulated parietal cells (23).

All of the above is consistent with a previously published model of regulated HCl secretion (6), that was initially developed on the basis of studies of native gastric vesicles with potential sensitive dyes (7,14), $Cl^-$ channel inhibitors (13), and studies with ionophores (17,28). Please see model in the accompanying paper in this series on $K^+$ channels of the gastric parietal cell by the same authors (9). This model states that $Cl^-$ transport is mediated by an anion channel. The electrophysiological and molecular (6,15,24) studies and earlier studies (7,13,14,17,28) helped to establish that $Cl^-$ transport across the apical membrane of the gastric parietal cell is electrogenic and mediated via ClC-2. It should also be noted that patch clamp studies of parietal cells (18) are consistent with bilayer insertion studies of native gastric $Cl^-$ channels and recombinant ClC-2 (6,15,24).

The model for regulation of the gastric $Cl^-$ channel involves activation by voltage, pH, and covalent modification by PKA of the $Cl^-$ (and $K^+$) channels when HCl secretion is activated. Although rat ClC-2 appears to be voltage-activated under some conditions, this does not seem to be a consistent feature of native rabbit gastric $Cl^-$ channels or recombinant rat, rabbit or human ClC-2 (10), and will not be further considered here. Evidence for pH activation of the native rabbit gastric $Cl^-$ channel was first demonstrated in bilayer insertion studies (6), and this has been confirmed using recombinant rabbit ClC-2 (15,25), and recombinant human ClC-2 (8,24,26) in a variety of electrophysiological studies. The molecular basis for pH activation has been demonstrated using chemical modification and site-directed mutagenesis (25). The studies of regulation using vesicles from the gastric mucosae of rabbits stimulated by histamine to secrete HCl and treated with cimetidine to inhibit HCl secretion (6,7,14,15,17,28), and regulation by PKA of recombinant rabbit (15,25) and human (8,24,26) ClC-2 demonstrate the physiologically relevant role of PKA phosphorylation of ClC-2 in gastric HCl secretion. Studies are currently in progress to determine the role of protein phosphatase action in regulation of gastric acid secretion. Recent studies suggest that pH activation requires activation by PKA (26). Future studies will be directed at determination of whether the order of pH and PKA treatment affects the level of ClC-2 $Cl^-$ channel activity in the parietal cell and in recombinant systems. These studies could help lead to an understanding of the sequence of cellular events that leads to PKA-activated HCl secretion.

A very recent report from studies of a ClC-2 knockout mouse (1) suggests that histamine-stimulated HCl secretion in the mouse is independent of ClC-2 since the small histamine induced drop in pH of the stomach contents measured as described (21) was similar in WT and KO mouse. It has been previously shown (20) that there is a very large variance in the acidity of mouse gastric contents that preclude physiologically relevant measurement of gastric HCl secretion without measurement of rates of secretion in perfusion studies. It is also well established that the mouse is essentially insensitive to stimulation by histamine alone, requiring instead in addition luminal perfusion of carbachol to achieve physiologically relevant HCl secretion (27). Studies with the ClC-2 knockout mouse under physiologically relevant conditions, with perfusion with carbachol and measurement of rates of HCl secretion, must be carried out to determine whether ClC-2 knockout mice have altered HCl secretion. Importantly, the lack of histamine stimulation of HCl secretion in the mouse correlates with the lack of the unique PKA sites that are present in the rabbit and human ClC-2, mammals in which HCl secretion is strongly stimulated by histamine.

In summary the results of these studies with the native rabbit gastric $Cl^-$ channel from the gastric mucosa of cimetidine- and histamine-treated rabbits and recombinant ClC-2 help to elucidate the molecules and mechanisms involved in regulated HCl secretion.

Acknowledgements: Supported by NIH DK43816, NIH HL58399 and CFF R457.

## REFERENCES

1. **Bosl MR, Stein V, Hubner C, Zdebik AA, Jordt SE, Mukhopadhyay AK, Davidoff MS, Holstein AF, and Jentsch TJ.** Male germ cells and photoreceptors, both dependent on close cell-cell interactions, degenerate upon ClC-2 $Cl^-$ channel disruption. *EMBO J* 20: 1289-1299, 2001.
2. **Cid LP, Montrose-Rafizadeh C, Smith DI, Guiggino WB, and Cutting GR.** Cloning of a putative human voltage-gated chloride channel (ClC-2) cDNA widely expressed in human tissues. *Hum Mol Genet* 4: 407-413, 1995.
3. **Cuppoletti J.** Possible role of ion channels in the action of nonsteroidal anti-inflammatory drugs. *NIPS* 15: 106-107, 2000.
4. **Cuppoletti J,** inventor, University of Cincinnati, assignee. Chemical modification of chloride channels as a treatment for cystic fibrosis and other diseases. US Patent 6015828, January 18, 2000.
5. **Cuppoletti J,** inventor, University of Cincinnati, assignee. Activation of chloride channels for correction of defective chloride transport. US Patent 6159968, December 12, 2000.
6. **Cuppoletti J, Baker AM, and Malinowska DH.** $Cl^-$ channels of the gastric parietal cell that are active at low pH. *Am J Physiol Cell Physiol* 264: C1609-1618, 1993.
7. **Cuppoletti J and Sachs G.** Regulation of gastric acid secretion via modulation of a chloride conductance. *J Biol Chem* 259: 14952-14959, 1984.
8. **Cuppoletti J, Tewari KP, Sherry AM, Kupert EY, and Malinowska DH.** ClC-2 $Cl^-$ channels in human lung epithelia: activation by arachidonic acid, amidation, and acid-activated omeprazole. *Am J Physiol Cell Physiol* 281: C46-54, 2001.
9. **Cuppoletti J, Tewari KP, Sherry AM, and Malinowska DH.** Kir2.1 $K^+$ Channels of the Gastric Parietal Cell. Present Volume, p.255-264, 2002.
10. **Fahlke C.** Ion permeation and selectivity in ClC-type chloride channels. *Am J Physiol Renal Physiol* 280: F748-757, 2001.
11. **Jordt SE and Jentsch TJ.** Molecular dissection of gating in the ClC-2 chloride channel. *EMBO J* 16: 1582-1592, 1997.
12. **Kieferle S, Fong P, Bens M, Vandewalle A, and Jentsh TJ.** Two highly homologous members of the ClC chloride channel family in both rat and human kidney. *Proc Natl Acad Sci USA* 91: 6943-6947, 1994.
13. **Malinowska DH.** $Cl^-$ channel blockers inhibit acid secretion in rabbit parietal cells. *Am J Physiol Gastrointest Liver Physiol* 259: G536-G543, 1990.
14. **Malinowska DH, Cuppoletti J, and Sachs G.** The $Cl^-$ requirement of acid secretion in isolated gastric glands. *Am J Physiol Gastrointest Liver Physiol* 245: G573-G581, 1983.

15. **Malinowska DH, Kupert EY, Bahinski A, Sherry AM, and Cuppoletti J.** Cloning, functional expression, and characterization of a PKA-activated gastric $Cl^-$ channel. *Am J Physiol Cell Physiol* 268: C191-200, 1995.
16. **Meves H.** Modulation of ion channels by arachidonic acid. *Prog Neurobiol* 43: 175-186, 1994.
17. **Reenstra WW and Forte JG.** Characterization of $K^+$ and $Cl^-$ conductances in apical membrane vesicles from stimulated rabbit oxyntic cells. *Am J Physiol Gastrointest Liver Physiol* 259: G850-G858, 1990.
18. **Saccomani G, Psarras CG, Smith PR, Kirk KL, and Shoemaker RL.** Histamine-induced chloride channels in apical membrane of isolated rabbit parietal cells. *Am J Physiol Cell Physiol* 260: C1000-1011, 1991.
19. **Schmidt-Rose T and Jentsch TJ.** Transmembrane topology of a CLC chloride channel. *Proc Natl Acad Sci USA* 94: 7633-7638, 1997.
20. **Schofield GC, Ito S, and Bolender RP.** Changes in membrane surface areas in mouse parietal cells in relation to high levels of acid secretion. *J Anat* 128: 669-692, 1979.
21. **Schultheis PJ, Clarke LL, Meneton P, Harline M, Boivin GP, Stemmermann G, Duffy JJ, Doetschman T, Miller ML, and Shull GE.** Targeted disruption of the murine $Na^+/H^+$ exchanger isoform 2 gene causes reduced viability of gastric parietal cells and loss of net acid secretion. *J Clin Invest* 101: 1243-1253, 1998.
22. **Schwiebert EM, Cid-Soto LP, Stafford D, Carter M, Blaisdell CJ, Zeitlin PL, Guggino WB, and Cutting GR.** Analysis of ClC-2 channels as an alternative pathway for chloride conduction in cystic fibrosis airway cells. *Proc Natl Acad Sci USA* 95: 3879-3884, 1998.
23. **Sherry AM, Malinowska DH, Morris RE, Ciraolo M, and Cuppoletti J.** Localization of ClC-2 $Cl^-$ channels in rabbit gastric mucosa. *Am J Physiol Cell Physiol* 280: C1599-1606, 2001.
24. **Sherry AM, Stroffekova K, Knapp LM, Kupert EY, Cuppoletti J, and Malinowska DH.** Characterization of the human pH- and PKA-activated ClC-2G(2$\alpha$) $Cl^-$ channel. *Am J Physiol Cell Physiol* 273: C384-393, 1997.
25. **Stroffekova K, Kupert EY, Malinowska DH, and Cuppoletti J.** Identification of the pH sensor and activation by chemical modification of the ClC-2G $Cl^-$ channel. *Am J Physiol Cell Physiol* 275: C1113-1123, 1998.
26. **Tewari KP, Malinowska DH, Sherry AM, and Cuppoletti J.** PKA and arachidonic acid activation of human recombinant ClC-2 chloride channels. *Am J Physiol Cell Physiol* 279: C40-50, 2000.
27. **Thieman A, Grunder S, Pusch M, and Jentsch TJ.** A chloride channel widely expressed in epithelial and non-epithelial cells. Nature Lond 356: 57-60, 1992.
28. **Wolosin JM and Forte JG.** Stimulation of oxyntic cell triggers $K^+$ and $Cl^-$ conductances in apical $H^+$-$K^+$-ATPase membrane. *Am J Physiol Cell Physiol* 246: C537-C545, 1984.

Chapter 28

# IDENTIFICATION OF POTASSIUM CHANNELS ASSOCIATED WITH $H^+/K^+$-ATPASE IN THE PARIETAL CELL

GRAHAMMER F [1], HERLING AW [3], VON HAHN T [1], HEITZMANN D [1], LANG HJ [3], SCHMITT-GRÄFF A [2], NITSCHKE R [1], WITTEKINDT O [1], BLEICH M [3], BARHANIN J [4], WARTH R [4]

*[1] Institute of Physiology, Albert-Ludwigs-University, Hermann-Herder-Strasse 7, D-79104 Freiburg; [2] Institute of Pathology, Albert-Ludwigs-University, Albertstrasse 19, D-79104 Freiburg; [3] Aventis Pharma GmbH, D-65926 Frankfurt am Main; [4] IPMC, CNRS, 660 route des Lucioles, F-06560 Valbonne Sophia Antipolis*

## SUMMARY

Gastric $H^+$ secretion via the $H^+/K^+$-ATPase is coupled to the uptake of $K^+$. However, the molecular identity of luminal $K^+$ channels enabling $K^+$ recycling in parietal cells was unknown. Cuppoletti and coworkers presented at this meeting data on Kir2.1 as a $K^+$ channel in the luminal membrane. We identified KCNQ1, which is mutated in cardiac long QT syndrome, as another $K^+$ channel located in tubulo-vesicles and apical membrane compartment of parietal cells, where it co-localized with $H^+/K^+$-ATPase in immunofluorescence experiments. Blockade of KCNQ1 current by the chromanol 293B led to complete inhibition of acid secretion in mouse, rat, and dogs *in-vivo* and in isolated rabbit gastric glands *in-vitro*. The putative KCNQ1 subunits, KCNE2 and KCNE3, were abundant; KCNE1, however, was absent in human stomach. Co-expression of KCNE3/KCNQ1 in COS cells led to a cAMP- and $IP_3/Ca^{2+}$-activated and acid-insensitive $K^+$ current. In contrast, cAMP, IP3/Ca2+, and low extracellular pH activated KCNE2/KCNQ1. In conclusion, we identified KCNQ1 as a luminal $K^+$ channel in parietal cells. 293B almost completely abolished acid secretion in several species, although Kir2.1, the other putative luminal $K^+$ channel, is unaffected by 293B. Moreover, in KCNQ1 knockout mice gastric acid secretion is impaired (14). These data speak in favor for the crucial role of KCNQ1 during gastric proton secretion. The putative KCNQ1 β-subunits KCNE3 and KCNE2 are expressed in human stomach. One or both are candidates to co-assemble with KCNQ1 in parietal cells to form a luminal $K^+$ channel. Stomach- and subunit-specific inhibitors of KCNQ1 might offer new therapeutical perspectives for peptic ulcer disease.

## INTRODUCTION

The precise mechanisms by which HCl is secreted in parietal cells have been a matter of debate since decades. The $H^+/K^+$-ATPase couples $H^+$ exit to $K^+$ uptake (11). Therefore, luminal $K^+$ recycling and $Cl^-$ exit are essential for HCl secretion (26). A splice variant of the ubiquitous CLC-2 $Cl^-$-channel (16;22) and parchorin (18), a CLC-related protein, were identified as luminal $Cl^-$ channels. There is also good evidence for $K^+$ channels in the luminal membrane (26). Cuppuletti and coworkers presented at this meeting convincing data that

Kir2.1 is such a luminal $K^+$ channel. However, the functional role of Kir2.1 during acid secretion *in vivo* remains to be elucidated.

The $K^+$ channel gene *KCNQ1* (also called *KvLQT1*) is mutated in the hereditary "Long QT Syndrome Type 1" (25). In heart and inner ear KCNQ1 co-assembles with its β-subunit KCNE1 (IsK, minK (2;17;20)) to form the native $K^+$ channels (Fig. 1).
Besides its expression in cardiac muscle and inner ear KCNQ1 is expressed in a variety of epithelia, i.e. kidney, pancreas, small and large intestine and stomach (4;6;28). In colonic mucosa the 293B-sensitive KCNQ1 is a prerequisite for cAMP-activated $Cl^-$ secretion (15). Here it co-assembles with KCNE3 (MiRP2) in the basolateral membrane to form the native channel (8;21). KCNE2 (MiRP1) was also demonstrated to co-assemble with KCNQ1 (8;23). Very recently, the KCNQ1 knockout mouse was reported to have an impaired gastric acid secretion, hypergastrinemia, hyperplasia of gastric glands and vacuolated parietal cells(14).
We examined the distribution of KCNQ1 in gastric mucosa and characterized its function during acid secretion. Moreover, we found the putative KCNQ1 subunits KCNE2 and KCNE3 abundant in human stomach.

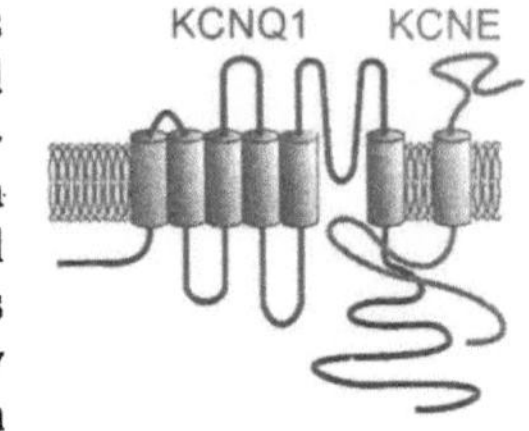


***Fig. 1:** The KCNQ1 $K^+$ channel protein has 6 transmembrane domains and co-assembles with small KCNE proteins to form native channels.*

## MATERIAL AND METHODS

**Immunofluorescence**
Paraffin embedded sections of human oxyntic gastric biopsies (archive material) were used. After removal of paraffin the sections were boiled for 3 min in citrate buffer (10 mM, pH 6.0). The KCNQ1 antibodies were obtained by immunizing rabbits with a peptide (CPADLGPRPRVSLDPRVSIY) of the cytosolic N-terminus (Davids Biotechnologie, Regensburg, Germany). The KCNQ1 antibodies were affinity purified and characterized using transfected cells and Westernblot (12). A monoclonal antibody was used for $H^+/K^+$-ATPase staining (Dianova, Hamburg, Germany (7)). The antibodies (1:500 and 1:1000) were dissolved in antibody dilution solution (Dako, Carpinteria, USA). Alexa488 goat anti-rabbit (Molecular Probes, Leiden, The Netherlands) and cy3 goat anti-mouse (Dianova) were used as secondary antibodies, HOE 33342 (2 µmol/l, Molecular Probes, Leiden, The Netherlands) was used for staining of the nuclei. The sections were examined with a confocal microscope using sequential excitation to avoid emission cross talk (LSM 510, Zeiss, Jena, Germany; Alexa488: excitation 488 nm, emission 505-550 nm; Cy3: excitation 543 nm, emission >560 nm).

**$^{14}C$-aminopyrine accumulation in isolated gastric glands**
Rabbits (2-3 kg BW) were killed by cervical dislocation during anesthesia. Gastric glands were isolated as described elsewhere (3). The glands were diluted to a concentration of 2-4 g dry weight/l and samples of 1 ml gland suspension were equilibrated for 20 min at 37°C in 1 ml medium containing 0.1 mCi/l $^{14}C$-aminopyrine and 293B (Aventis Pharma GmbH, Frankfurt, Germany) or vehicle, respectively. Next either carbachol (CCH, 100 µM, Sigma, Deisenhofen, Germany) or dbcAMP (300 µM, Merck, Darmstadt, Germany) was added for 30 min and 45 min, respectively. Radioactivity of aliquots of both the digested gland pellet and the supernatant were determined in triplicate in a liquid scintillation counter. The $^{14}C$-

aminopyrine accumulation as a measure of acid secretion was calculated as the ratio between $^{14}C$-aminopyrine in the intraglandular water and in the incubation medium (19).

**Stomach-lumen perfused mice**

KCNE1-knockout mice were fasted 2-4 hours prior to the experiment with free access to 5 % Glucose solution. Mice were anaesthetized with ketamine (100 mg/kg bw i.p.)/xylazine (4 mg/kg bw i.p.). A modified saline solution (0.9% NaCl, 5% Glucose and 0.1% bovine serum albumin, pH 7.4) was infused i.v. at a rate of 0.15 µl/min/g. The abdominal cavity was opened and pylorus was ligated. After preparation of the vagus nerve the esophagus was ligated, too. A double lumen perfusion cannula was fixed in the forestomach. The stomach was perfused continuously with saline (37°C; pH 7.0; 200 µl/min). The perfusate was collected every 10 min and re-titrated to pH 7.0.

**Stomach-lumen perfused rats**

Male Sprague-Dawley rats (300-350g) were anaesthetized with 30% (w/v) urethane (5 ml/kg i.m.) and tracheotomized. Abdominal surgical procedures for rats were similar to those described for mice. The stomach was perfused at a rate of 1 ml/min and the perfusate was collected every 15 min. The corresponding acid output was measured by re-titration to pH 7.

**Heidenhain-pouch in dogs**

Male Beagle dogs were prepared with Heidenhain-pouches (9;13). After surgery they were allowed to recover for at least 6 weeks and were trained to stay in a Pawlov stand. Food was withdrawn 18 hours prior to the experiment but tap water was given ad libitum. During the experiments gastric juice was collected from the pouches every 30 min, acid output was measured by re-titration to pH 7.0.

**Patch-clamp experiments on transfected cells**

Human KCNE2 and KCNE3 were cloned in the pIRES-CD8 mammalian expression vector (10). COS cells were transiently co-transfected with pCI-hKCNQ1 and either KCNE2 or KCNE3 (each 0.5 µg per dish) using the FuGENE transfection reagent (Roche, Mannheim, Germany). Standard whole cell recordings were performed on transfected cells, as visualized by CD8 expression. The patch pipette solution was (in mmol/l) 110 K-gluconate, 30 KCl, 4.8 $Na_2HPO_4$, 1.2 $NaH_2PO_4$, 5 Glucose, 2.38 $MgCl_2$, 0.73 $CaCl_2$, 1 EGTA, 3 ATP, pH 7.2. The extracellular solution was (in mmol/l): 150 NaCl, 5 KCl, 2 $MgCl_2$, 1 $CaCl_2$, 5 HEPES, pH 7.4 or 5.5.

## RESULTS AND DISCUSSION

In the present study we examined the localization of KCNQ1 in human gastric mucosa. A weak KCNQ1-specific staining was observed at the basolateral membrane of cells in the foveolar region. However, one cell type in gastric glands was strongly stained and could be identified as parietal cells by co-staining with a $H^+/K^+$-ATPase antibody (Fig. 2A). The subcellular distribution pattern was very similar to that of the $H^+/K^+$-ATPase but not identical speaking in favor for a functional coupling of the two proteins rather than a direct protein-protein interaction with a fixed stoichiometry (Fig. 2B). Meanwhile, the distribution of KCNQ1 in gastric mucosa was confirmed by another group using a different KCNQ1 antibody (8).

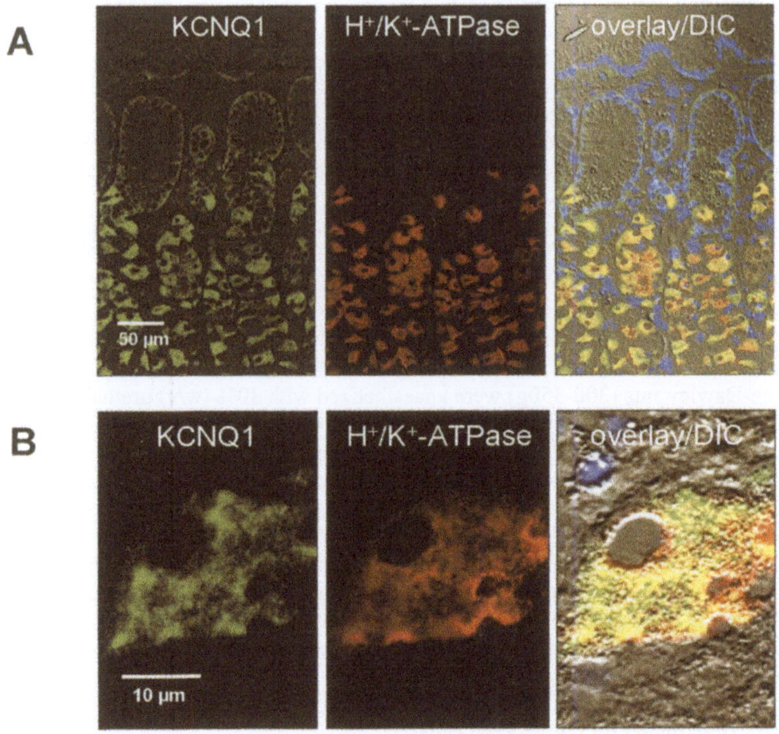


***Fig. 2: Immunolocalisation of KCNQ1 in human gastric mucosa***
*(A) The KCNQ1-specific antibody stained preferentially one cell type of gastric gland (green), which could be co-stained with a monoclonal $H^+/K^+$-ATPase antibody (red). The overlay of both fluorescence images (yellow for perfect match) with the differential interference contrast image (grey scale) and a HOE33342 nucleus staining (blue) is depicted at the right hand side. (B) shows a single parietal cell at a high magnification suggesting a localisation of KCNQ1 in $H^+/K^+$-ATPase-containing tubulo-vesicles*

To examine the physiological role of KCNQ1 in parietal cells we performed measurements of gastric acid secretion *in-vivo* using the KCNQ1 inhibitors 293B, 377B, and 389B as pharmacological tools. 293B reversibly blocked acid secretion in mouse (Fig. 3) and rat perfused stomach and Heidenhain-pouch dogs. 377B exerted a similar effect; 389B was less potent.

At concentrations higher than 20 µM 293B also blocks Kv4.3, Kv1.5 (27), $I_{To}$ channels (5) and CFTR (1). However, gastric $H^+$ secretion was much more sensitive for inhibition by 293B: The 293B serum concentration of approximately 8 µmol/l was sufficient to inhibit acid secretion in mice by some 85% (12). These data speak in favor for 293B inhibition of KCNQ1 associated to a subunit that increased 293B sensitivity as described for KCNE2/KCNQ1 (23) and KCNE3/KCNQ1 (21). However, unspecific effects of 293B via inhibition of other ion channels cannot be completely ruled out, although such effects are not likely at the concentrations used in this study. Moreover, a protonophoretic action of 293B as described for other inhibitors of gastric acid secretion (24) could be excluded (12).

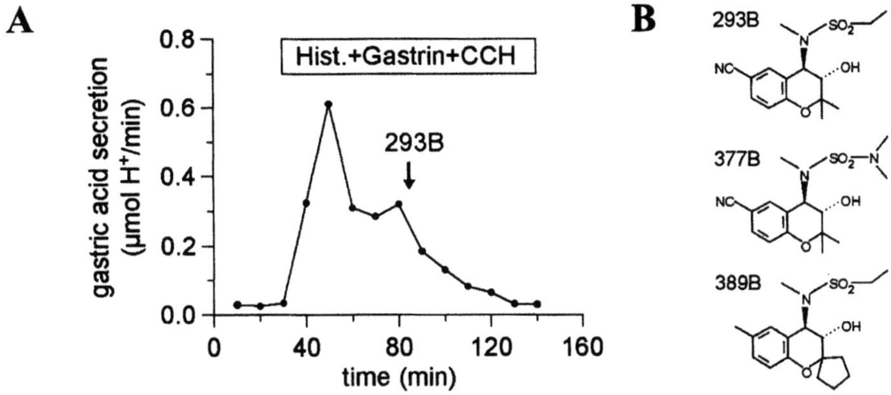


*Fig. 3: Effect of chromanols on gastric acid secretion in perfused mouse stomach*
*(A) Acid secretion was stimulated using a cocktail (pentagastrin 44 µg/kg/h i.v., carbachol 3 µg/kg/h i.v., histamine 128 µg/kg/h i.v.). 293B (5 mg/kg bolus i.p.) inhibited proton secretion almost completely. (B) Chemical structure of the chromanols 293B, 377B and 389B. The effect of 377B was similar to 293B; 389B was less potent.*

In order to exclude an indirect inhibition of acid secretion, we performed *in-vitro* experiments with isolated rabbit gastric glands. Using the $^{14}$C-aminopyrine accumulation method we observed activation of secretion in the presence of CCH, histamine, and dbcAMP. 293B blocked $H^+$ secretion irrespectively of the hormone by which secretion was activated with $IC_{50}$ values in the range of 3-20 µmol/l. These data suggest a direct way of action of 293B on parietal cells. Moreover, an indirect effect on $H^+$ secretion via 239B blockade of hormone receptors or cAMP production could be ruled out by 293B inhibition of dbcAMP-induced secretion.

In heart and inner ear KCNQ1 co-assembles with KNCE1 (IsK or minK) to form the native channel complex (2;17). We found KCNE2 and KCNE3 but not KCNE1 expression in human stomach. In a very recent study, in-situ hybridization revealed a strong expression of KCNE2 in rodent parietal cells (8). Interestingly, we observed in patch-clamp experiments on KCNE2/KCNQ1 co-transfected COS-cells a 293B-sensitive $K^+$ current which was activated by low extracellular pH (12), by ATP (purinergic P2Y receptor stimulation activates the IP3/$Ca^{2+}$ cascade) and via cAMP (unpublished data). Therefore, KCNE2 is a very likely candidate to co-assemble with KCNQ1 in gastric parietal cells (Fig. 4).

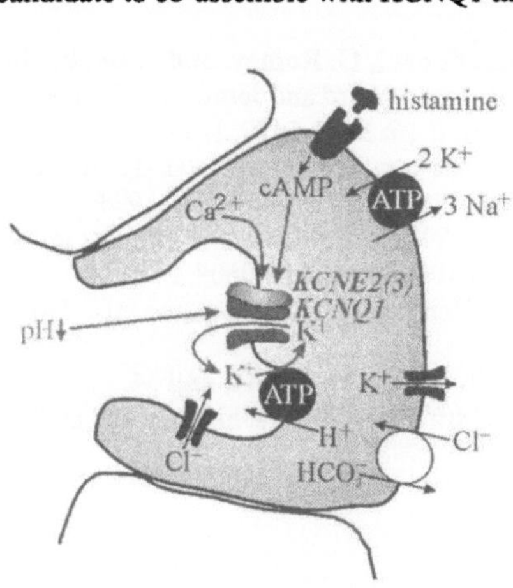


***Fig. 4: KCNQ1 channels recycle $K^+$ over the luminal membrane of parietal cells***
*KCNQ1 probably associated with KCNE2 is located in the luminal membrane compartment and enables $K^+$ recycling whenever $K^+$ is taken up by $H^+/K^+$-ATPase. The channel complex is activated by low extracellular pH, stimulation of the $IP_3/Ca^{2+}$ cascade and cAMP. Inhibition of KCNQ1 channels by 293B almost completely abolishes proton secretion.*

Cuppoletti and coworkers presented convincing data that the 293B-insensitive Kir2.1 is another $K^+$ located in the luminal membrane of parietal cells. However, the almost complete inhibition of acid secretion by the KCNQ1 inhibitor 293B together with an abolished acid secretion in KCNQ1 knockout mice (14) suggest a crucial role of KCNQ1 channels for acid secretion at least in mice. Further studies are required to assess acid secretion and plasma gastrin levels in patients homozygous for KCNQ1 mutations (Jervell-Lange-Nielsen syndrome).

In conclusion, our data suggest that KCNQ1 is the missing luminal $K^+$ channel in gastric parietal cells of several species including humans. Most likely, KCNE2 (and/or KCNE3) co-assembles with KCNQ1 in human parietal cells. The tissue-specific subunit composition of KCNQ1 and the peculiar pharmacokinetics of the stomach could offer new therapeutical perspectives for the treatment of peptic ulcer disease via inhibition of the KCNQ1 channel complex.

**Acknowledgements:** The authors gratefully acknowledge the expert assistance by S. Haxelmanns, M.M. Larroque, W. Rohm and B. Weinhold. The study was supported by the Forschungsförderung des Landes Baden-Württemberg; RW was supported by an EMBO fellowship.

## REFERENCES

1. **Bachmann, A., U. Quast, and U. Russ.** Chromanol 293B, a blocker of the slow delayed rectifier K+ current (IKs), inhibits the CFTR Cl- current. *Naunyn Schmiedebergs Arch Pharmacol.* 363: 590-596, 2001.
2. **Barhanin, J., F. Lesage, E. Guillemare, M. Fink, M. Lazdunski, and G. Romey.** $K_vLQT1$ and IsK (minK) proteins associate to form the $I_{Ks}$ cardiac potassium current. *Nature* 384: 78-80, 1996.
3. **Berglindh, T. and K. J. Obrink.** A method for preparing isolated glands from the rabbit gastric mucosa. *Acta Physiol Scand.* 96: 150-159, 1976.
4. **Bleich, M. and R. Warth.** The very small conductance $K^+$ channel $K_vLQT1$ and epithelial function. *Pflügers Arch* 440: 202-206, 2000.
5. **Bosch, R. F., R. Gaspo, A. E. Busch, H. J. Lang, G. R. Li, and S. Nattel.** Effects of the chromanol 293B, a selective blocker of the slow, component of the delayed rectifier $K^+$ current, on repolarization in human and guinea pig ventricular myocytes. *Cardiovasc.Res.* 38: 441-450, 1998.
6. **Chouabe, C., N. Neyroud, P. Guicheney, M. Lazdunski, G. Romey, and J. Barhanin.** Properties of KvLQT1 $K^+$ channel mutations in Romano-Ward and Jervell and Lange-Nielsen inherited cardiac arrhythmias. *EMBO Journal* 16: 5472-5479, 1997.
7. **Chow, D. C. and J. G. Forte.** Characterization of the beta-subunit of the $H^+$-$K^+$-ATPase using an inhibitory monoclonal antibody. *Am J Physiol* 265: C1562-C1570, 1993.
8. **Dedek, K. and S. Waldegger.** Colocalization of KCNQ1/KCNE channel subunits in the mouse gastrointestinal tract. *Pflügers Arch* DOI 10.1007/s004240100609, 2001.
9. **DeVito, R. V. and H. N. Harkins.** Techniques in Heidenhain pouch experiments. *J Appl Physiol* 14: 138-139, 1959.
10. **Fink, M., F. Lesage, F. Duprat, C. Heurteaux, R. Reyes, M. Fosset, and M. Lazdunski.** A neuronal two P domain K+ channel stimulated by arachidonic acid and polyunsaturated fatty acids. *EMBO J* 17: 3297-3308, 1998.

11. **Forte, J. G., G. M. Forte, and P. Saltman.** $K^+$-stimulated phosphatase of microsomes from gastric mucosa. *J.Cell Physiol* 69: 293-304, 1967.
12. **Grahammer, F., A. W. Herling, H. J. Lang, A. Schmitt-Graff, O. H. Wittekindt, R. Nitschke, M. Bleich, J. Barhanin, and R. Warth.** The cardiac $K^+$ channel KCNQ1 is essential for gastric acid secretion. *Gastroenterology* 120: 1363-1371, 2001.
13. **Herling, A. W., T. Scholl, M. Bickel, H. J. Lang, K. H. Scheunemann, K. Weidmann, and R. Rippel.** Gastric acid inhibitory profile of saviprazole (HOE 731) compared to omeprazole. *Pharmacology* 43: 293-303, 1991.
14. **Lee, M. P., J. D. Ravenel, R. J. Hu, L. R. Lustig, G. Tomaselli, R. D. Berger, S. A. Brandenburg, T. J. Litzi, T. E. Bunton, C. Limb, H. Francis, M. Gorelikow, H. Gu, K. Washington, P. Argani, J. R. Goldenring, R. J. Coffey, and A. P. Feinberg.** Targeted disruption of the Kvlqt1 gene causes deafness and gastric hyperplasia in mice. *J Clin.Invest* 106: 1447-1455, 2000.
15. **Lohrmann, E., I. Burhoff, R. B. Nitschke, H. J. Lang, D. Mania, H. C. Englert, M. Hropot, R. Warth, W. Rohm, M. Bleich, and R. Greger.** A new class of inhibitors of cAMP-mediated $Cl^-$ secretion in rabbit colon, acting by the reduction of cAMP-activated $K^+$ conductance. *Pflügers Arch* 429: 517-530, 1995.
16. **Malinowska, D. H., E. Y. Kupert, A. Bahinski, A. M. Sherry, and J. Cuppoletti.** Cloning, functional expression, and characterization of a PKA-activated gastric $Cl^-$ channel. *Am J Physiol* 268: C191-C200, 1995.
17. **Nicolas, M., D. Dememes, A. Martin, S. Kupershmidt, and J. Barhanin.** KCNQ1/KCNE1 potassium channels in mammalian vestibular dark cells. *Hear.Res.* 153: 132-145, 2001.
18. **Nishizawa, T., T. Nagao, T. Iwatsubo, J. G. Forte, and T. Urushidani.** Molecular cloning and characterization of a novel chloride intracellular channel-related protein, parchorin, expressed in water-secreting cells. *J.Biol.Chem.* 275: 11164-11173, 2000.
19. **Sack, J. and J. G. Spenney.** Aminopyrine accumulation by mammalian gastric glands: an analysis of the technique. *Am.J.Physiol* 243: G313-G319, 1982.
20. **Sanguinetti, M. C., M. E. Curran, A. Zou, J. Shen, P. S. Spector, D. L. Atkinson, and M. T. Keating.** Coassembly of $K_vLQT1$ and minK (IsK) proteins to form cardiac $I_{Ks}$ potassium channel. *Nature* 384: 80-83, 1996.
21. **Schroeder, B. C., S. Waldegger, S. Fehr, M. Bleich, R. Warth, R. Greger, and T. J. Jentsch.** A constitutively open potassium channel formed by KCNQ1 and KCNE3. *Nature* 403: 196-199, 2000.
22. **Sherry, A. M., D. H. Malinowska, R. E. Morris, G. M. Ciraolo, and J. Cuppoletti.** Localization of ClC-2 Cl(-) channels in rabbit gastric mucosa. *Am J Physiol Cell Physiol* 280: C1599-C1606, 2001.
23. **Tinel, N., S. Diochot, M. Borsotto, M. Lazdunski, and J. Barhanin.** KCNE2 confers background current characteristics to the cardiac KCNQ1 potassium channel. *EMBO J.* 19: 6326-6330, 2000.
24. **Urushidani, T. and J. G. Forte.** Signal transduction and activation of acid secretion in the parietal cell. *J Membr Biol* 159: 99-111, 1997.
25. **Wang, Q., M. E. Curran, I. Splawski, T. C. Burn, J. M. Millholland, T. J. VanRaay, J. Shen, K. W. Timothy, G. M. Vincent, T. de Jager, P. J. Schwartz, J. A. Toubin, A. J. Moss, D. L. Atkinson, G. M. Landes, T. D. Connors, and M. T. Keating.** Positional cloning of a novel potassium channel gene: KVLQT1 mutations cause cardiac arrhythmias. *Nature Genet.* 12: 17-23, 1996.
26. **Wolosin, J. M. and J. G. Forte.** Stimulation of oxyntic cell triggers $K^+$ and $Cl^-$ conductances in apical $H^+$-$K^+$-ATPase membrane. *Am J Physiol* 246: C537-C545, 1984.

27. **Yang, I. C., M. W. Scherz, A. Bahinski, P. B. Bennett, and K. T. Murray.** Stereoselective interactions of the enantiomers of chromanol 293B with human voltage-gated potassium channels. *J.Pharmacol.Exp.Ther.* 294: 955-962, 2000.
28. **Yang, W.-P., P. C. Levesque, W. A. Little, M. L. Conder, F. Y. Shalaby, and M. A. Blanar.** KvLQT1, a voltage-gated potassium channel responsible for human cardiac arrhythmias. *Proc Natl Acad Sci U S A* 94: 4017-4021, 1997.

Chapter 29

# Kir2.1 $K^+$ Channels of the Gastric Parietal Cell

JOHN CUPPOLETTI, KIRTI P. TEWARI, ANN M. SHERRY, AND DANUTA H. MALINOWSKA
*University of Cincinnati College of Medicine, Department of Molecular and Cellular Physiology, Cincinnati, OH 45267-0576*

## ABSTRACT

The purpose of this article is to review the literature regarding the characteristics of the gastric parietal cell apical membrane $K^+$ channel and provide new evidence that suggests that the inward rectifying $K^+$ channel, Kir2.1, may be involved in $K^+$ recycling at the apical membrane of the gastric parietal cell. It has been previously reported that gastric $H^+/K^+$ ATPase-containing vesicles exhibit a 10 pS $K^+$ channel when measured in planar lipid bilayers. The native gastric $K^+$ channel in these vesicles is insensitive to ATP, is active in the absence of $Ca^{2+}$, is stable at an extracellular pH of 3.0, and is activated by PKA. A literature search suggested that Kir2.1 has many of the properties of this channel. This is the first report to show that recombinant rabbit Kir2.1 $K^+$ channels are active in the presence of ATP, in the absence of $Ca^{2+}$, and are stable at an extracellular pH of 3.0. Preliminary results demonstrate that Kir2.1 is also PKA activated. Kir2.1 also exhibits single channel currents that are of similar magnitude as the native channel under these conditions. In an accompanying paper, Grahammer *et al* suggest from 293B inhibition of acid secretion that KCNQ1 is the $K^+$ channel involved in $K^+$ recycling at the apical membrane. Kir2.1 is not inhibited by 293B. Acid accumulation by $H^+/K^+$ ATPase-containing vesicles from stimulated rabbits is insensitive to 293B. This finding strongly suggests that KCNQ1 is not involved in $K^+$ recycling at the apical membrane of the gastric parietal cell. Kir2.1 is a candidate for this process.

## INTRODUCTION

This article briefly reviews published data and provides new information on the nature of $K^+$ channels (including Kir2.1) that may be candidates for the apical membrane $K^+$ channel involved in regulated acid secretion by the gastric parietal cell.

Acid secretion requires the gastric $H^+/K^+$ ATPase and a pathway for the recycling of $K^+$ and provision of equivalents of $Cl^-$ for HCl production. It is now generally agreed that $K^+$ and $Cl^-$ transport is accomplished by $K^+$ channels and $Cl^-$ channels in the secretory membrane. Vesicle preparations from the gastric mucosa of histamine-stimulated rabbits provide useful *ex vivo* models for the study of the molecules and mechanisms of HCl accumulation (5,6,13,18), since they contain all of the proteins required for acid secretion. These vesicle preparations have been used in bilayer reconstitution studies to study the $K^+$ and $Cl^-$ channels involved in acid secretion (4,5,11,12). Successes using these vesicles include studies of the single channel characteristics and regulation of the gastric $Cl^-$ channel (5), which led to the cloning and identification of ClC-2 as the gastric $Cl^-$ channel (12).

Reports of $K^+$ channels that are stable to extremes of acidic extracellular pH (pH 3.0) in stimulated gastric vesicle preparations have appeared (4,11).

The parietal cell contains a variety of $K^+$ channels, including KCNQ1, KCNE2 and KCNE3 as reported by Grahammer *et al* (7,8), as well as Kir2.1 (11), Kir4.1 and Kir7.1 (9,11). There may be combinations of $K^+$ channel subunits and other subunits that have not yet been identified which are involved in acid secretion. In this regard, the $K^+$ channel diversity of the parietal cell may be similar to that found in other tissues such as the pituitary, where there exist 21 different Kv channel alpha subunits, 9 different Kir alpha subunits and 6 auxiliary subunits (19), which could not be distinguished by electrophysiology and pharmacological approaches. The presence of multiple $K^+$ channel subunits of different families greatly complicates identification of the $K^+$ channel(s) involved in $K^+$ recycling.

In the present article, known characteristics of the native gastric apical membrane $K^+$ channel are reviewed and the properties of Kir2.1 are compared and contrasted with the properties of the gastric $K^+$ channel. The effect of 293B on acid secretion by stimulated gastric vesicles and upon Kir2.1 channel activity is examined in view of the study by Grahammer *et al* (7,8) that concluded that KCNQ1 is involved in $K^+$ recycling at the secretory membrane.

## METHODS

*Membrane vesicle preparations.* $H^+/K^+$ ATPase-containing membrane vesicles were purified by differential and sucrose gradient fractionation of vesicles from the gastric mucosae of secreting (histamine + diphenhydramine injected) and non-secreting (cimetidine injected) New Zealand white rabbits as previously described (6) and are referred to as stimulated and resting membrane vesicles respectively.

*Measurement of cation channels in planar bilayers.* The procedures and equipment were as described in detail previously (4,5,12). All of the solutions contained 400 mM $K_2SO_4$, 2 mM $MgSO_4$, 1 mM K-ATP (pH 7.4), 10 mM potassium ethylene-glycol-bis(beta-aminoethyl ether)-N,N, N',N'-tetraacetic acid (EGTA) pH 7.4, and were buffered with either 10 mM potassium N-2-hydroxyethylpiperazine-N'-2-ethanesulfonic acid (HEPES), pH 7.4 or 10 mM potassium citrate, pH 3. A fire-polished glass capillary tube was used to apply lipids, and also to apply vesicles from a suspension (1-2 mg/ml protein) to the planar lipid bilayer. Reconstitutions of the channels lasted from 10 min to several hours. When specified, PKA catalytic subunit (50 U/ml) was added to the *cis* side of the bilayer, which corresponds to the cytosolic side of the channel. $pH_{cis}$ was 7.4 in all experiments.

*Acid accumulation by gastric vesicles.* Acid accumulation was measured as previously described (6). Briefly, acid accumulation was measured by following the time course of acridine orange fluorescence quench (excitation 492 nm, emission 530 nm), using a Perkin Elmer LS5B fluorimeter. Approximately 10 µg membrane vesicles were added to medium containing 200 mM sucrose, 50 mM KCl, 10 µM acridine orange and 8 mM NaHEPES pH 7.5 in a 3.0 ml reaction volume. The reaction was initiated by the addition of 1 mM ATP-Mg. 16 µg/ml nigericin was added at the end of the experiment to dissipate acid gradients. Where indicated, 293B was dissolved in ethanol and added at a final concentration of 1% ethanol plus 100 µM 293B. 293B was the kind gift of Dr. H. J. Lang of Aventis Corporation.

*Patch clamp measurements.* Patch clamp data was carried out essentially as described previously (2,3,16), using a bath solution contain 140 mM N-methyl D-gluconate, 5.4 mM KCl, 10 mM glucose plus 10 mM HEPES, pH 7.4, and a pipette solution of 140 mM K-gluconate, 5 mM $MgCl_2$, 10 mM HEPES, 1 mM Mg-ATP, pH 7.4.

## RESULTS

**Fig. 1A** shows a typical recording of a gastric $K^+$ channel from the $H^+/K^+$ ATPase-containing vesicles at pH 7.4 and **Fig. 1B** shows a recording at pH 3.0 (from (4)). Unit conductance of the channel was approximately 11 pS. The vesicles used for channel reconstitution are largely oriented with the cytoplasmic face facing the bathing media as measured by detergent-sensitive ATPase activity, and insert with the cytosolic face of the membrane proteins facing the side of the bilayer on which they are applied (*cis* side). The single channel currents were similar whether measured under symmetric conditions of pH 7.4 **(A)** or asymmetric conditions of pH, with the channel bathed at pH 3.0 on the extracytosolic face (*trans* side of the bilayer), and pH 7.4 on the intracellular face of the channel **(B).** The channel is stable under these conditions. The channels recorded in **Fig. 1** were obtained from a stimulated gastric vesicle preparation (6). Similar channels have been recorded in preparations from the stomachs of rabbits treated with cimetidine.

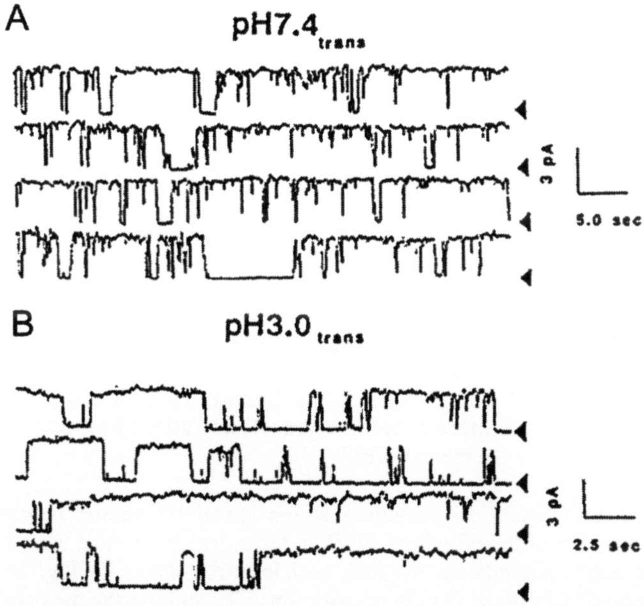


**Fig. 1. The gastric apical membrane $K^+$ channel at pH 7.4 (A) and pH 3.0 (B).** Single channel current recording with solutions on the *trans* (extracellular) face of the channel at pH 7.4 **(A)** or pH 3.0 **(B)**. In both recordings, solution on the *cis* side was at pH 7.4 and the holding potential was −40 mV. Arrowheads, closed state of the channel. From (4).

An extensive survey of the biophysical properties of various cloned $K^+$ channels, their distribution, and their regulatory properties suggested that Kir2.1 (and possibly other Kir channels) had properties in common with the native gastric $K^+$ channel. Several of these candidate $K^+$ channels were shown to be present at the level of mRNA in the parietal cell. Some of these were expressed. Kir2.1 was active under the same conditions as used to study the native $K^+$ channel (11): in the presence of ATP, in the absence of $Ca^{2+}$, and with pH 3.0 bathing the extracellular face of the channel.

Kir2.1, when expressed in CHO cells, exhibits low basal currents that are increased by 100 μM 293B (**Fig. 2A**). The I/V curve in **Fig. 2B** shows the summarized data for 3 separate determinations. The 293B-activated channel shows the inward rectification that is characteristic of this channel. Activation of Kir2.1 by 293B suggests either that Kir2.1 is not involved in acid secretion (since Grahammer *et al* (7,8) postulated from 293B inhibition of acid secretion *in vivo* and in gastric glands, that the 293B-sensitive $K^+$ channel KCNQ1 was the $K^+$ channel involved in apical membrane recycling of $K^+$), or that 293B sensitivity is not a hallmark of the $K^+$ channel involved in $K^+$ recycling by the apical membrane of the gastric parietal cell.

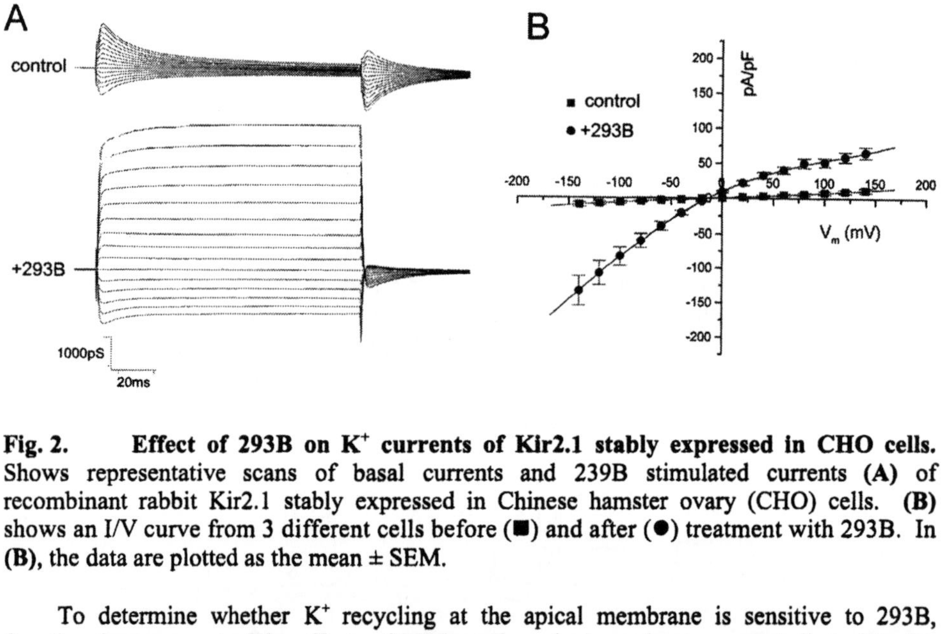


**Fig. 2. Effect of 293B on $K^+$ currents of Kir2.1 stably expressed in CHO cells.** Shows representative scans of basal currents and 239B stimulated currents **(A)** of recombinant rabbit Kir2.1 stably expressed in Chinese hamster ovary (CHO) cells. **(B)** shows an I/V curve from 3 different cells before (■) and after (●) treatment with 293B. In **(B)**, the data are plotted as the mean ± SEM.

To determine whether $K^+$ recycling at the apical membrane is sensitive to 293B, functional assessment of the effects of 293B on the apical membrane were studied. **Fig. 3A** shows control acid accumulation by such vesicles. As shown in **Fig. 3B**, 293B did not inhibit acid secretion in these vesicle preparations. To summarize the results from several trials, there were no statistically significant differences between the normalized extents (or rates) of fluorescence change (1.0 ± 0.2 in the control, and 1.1 ± 0.2 in the presence of 100 μM 293B, n=3). In data not shown, there was also no effect of 293B upon the rate or extent of acidification in the presence of valinomycin, demonstrating that 293B does not inhibit the $H^+/K^+$ ATPase, and that 293B is not a protonophore.

**Fig. 4A** shows a model of the gastric parietal cell showing the $H^+/K^+$ ATPase, a $K^+$ channel and a $Cl^-$ channel in the apical membrane and **Fig. 4B** shows a scheme for regulated acid secretion (4,5) wherein a non-secreting parietal cell contains an inactive $Cl^-$ channel, ClC-2

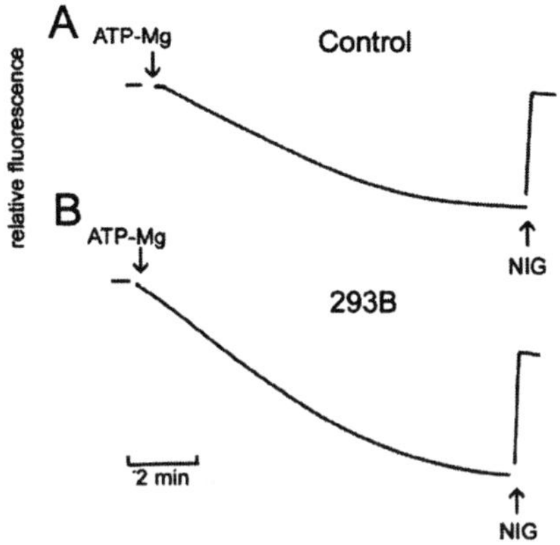


**Fig. 3.** Effect of 293B on acidification of gastric vesicles. Shows representative fluorescence-based acidification assays of rabbit gastric vesicles in the absence **(A)** and presence **(B)** of 100 μM 293B, respectively. The reaction was initiated by the addition of 1 mM ATP-Mg, and the acid gradient was dissipated by addition of 16 μg/ml nigericin (NIG). The summarized data for 3 experiments showed no effect of 293B on acid accumulation.

(12,15), an inactive $K^+$ channel, Kir2.1 (11), and the $H^+/K^+$ ATPase. Upon histamine stimulation, re-organization of the gastric parietal cell tubulovesicular membranes containing the pump and channels and covalent modification of the channels by PKA occur. The PKA-modified channels, when in the canalicular membrane, are exposed to a different membrane potential and low pH. All of these changes lead to increased activity of the ion channels needed for acid secretion. This model does not address the effect of protein phosphatases that are also expected to play important roles in the regulation acid secretion.

## DISCUSSION

Gastric vesicles derived from the stomachs of histamine plus diphenydramine treated rabbits contain all of the elements that are essential for accumulation of ATP-driven HCl accumulation (6,18). When reconstituted into planar lipid bilayers, ion channel activity is present in solutions of $K_2SO_4$. Substitution studies showed that the channel is selective for $K^+$ over $SO_4^{2-}$ (4). The channel is stable when bathed at pH 3.0 on the extracytosolic face of the channel, consistent with a role of the channel in acid secretion at the apical membrane of the gastric parietal cell. The channel was active in the presence of ATP and in the absence of $Ca^{2+}$. The native $K^+$ channel is activated by PKA. The properties of this channel are consistent with the properties of Kir2.1, and inconsistent with the properties of a large variety of other $K^+$ channels. Kir2.1 is present in a variety of epithelial tissues, including the stomach (11). The properties of recombinant Kir2.1 reconstituted into planar lipid bilayers are similar to the native gastric $K^+$ channel in that it is active with pH 3.0 bathing the extracytosolic surface of the channel, with ATP present in the bath, and with chelation of $Ca^{2+}$. It has a similar unit conductance as the native gastric $K^+$ channel studied under the same conditions. These data suggest that Kir2.1 may be the parietal cell apical membrane $K^+$ channel.

Grahammer *et al* (7,8) showed that the chromanols 293B, 377B, and 389B are potent inhibitors of gastric acid secretion in a variety of animal models and in rabbit gastric glands. They reported that 8 μM 293B was sufficient to inhibit acid secretion in mice by 85%, with an $IC_{50}$ between 3-20 μM in isolated rabbit gastric glands. 10 μM 293B inhibited $K^+$ currents in COS cells expressing the recombinant channel and KCNQ1 was shown to be present in the gastric parietal cell. Based on these findings, Grahammer *et al* (7,8) suggested that KCNQ1 is the "missing" $K^+$ channel responsible for $K^+$ recycling at the parietal cell apical membrane during acid secretion.

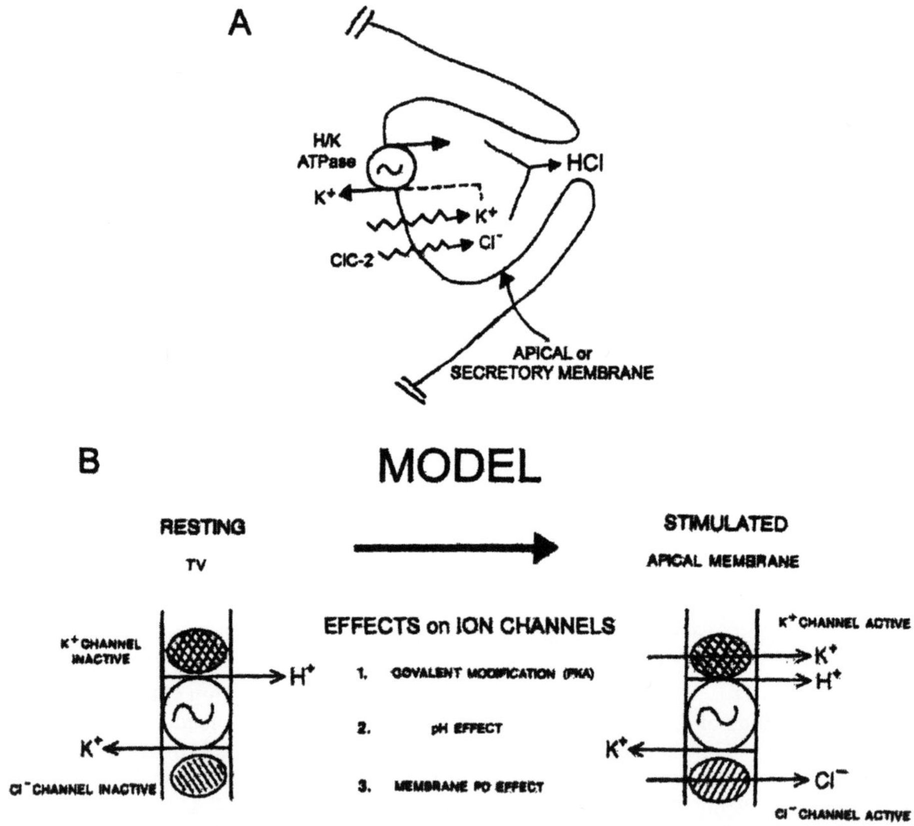


**Fig. 4. Proposed model for regulated HCl secretion by the gastric parietal cell.** Cross hatched area, candidate $K^+$ channel(s) including Kir2.1; hatched area, $Cl^-$ channel (ClC-2); circle, $H^+/K^+$ ATPase. TV, tubulovesicles. Modified from (7).

Recombinant Kir2.1 expressed in CHO cells was stimulated by 100 μM 293B. This suggests that Kir2.1 is not involved in the same processes that led to inhibition of acid secretion in the studies by Grahammer *et al* (7,8). Those authors suggested (on the basis of inhibition of acid secretion in intact animals and gastric glands and immunofluorescence showing the presence of protein for KCNQ1 in the parietal cell), that this channel was in the apical membrane of the parietal cell.

The present study tested this hypothesis by examination of the effect of 293B on acid accumulation by $H^+/K^+$ ATPase-containing membrane vesicles from stimulated rabbit gastric mucosa. These functional studies showed a lack of effect of 293B on acid accumulation. The lack of effect of 293B on HCl accumulation in a gastric vesicle model system suggests that the $K^+$ channel that is responsible for $K^+$ recycling in the parietal cell apical membrane is not KCNQ1. 293B inhibits ion channels other than KCNQ1, including CFTR (1) at concentrations as low as 20 μM. It is possible that the effects of 293B on acid secretion are due to inhibition of other ion channels or other processes. It is also possible that KCNQ1 plays another essential role in acid secretion distinct from $K^+$ recycling at the apical membrane.

The apical membrane of the mammalian gastric parietal cell is not easily accessible to patch clamp. This largely precludes measurement of ion channel activity in the context of the apical membrane of the intact parietal cell. However, heroic efforts in the past have demonstrated that ClC-2-like currents were present in the apical membrane of the gastric parietal cell (14). The properties of these currents appear to be identical to currents measured in $H^+/K^+$ ATPase-containing vesicles and recombinant ClC-2 (4,5,12). Studies with gastric vesicles and the successful identification of the molecules that are involved in apical membrane acid secretion provide a means of study of the regulatory properties of these ion channels that can be subsequently related to the properties of the channels in the physiologically relevant context of the intact organ.

The molecular identification of Kir2.1 as a candidate for $K^+$ recycling at the parietal cell apical membrane raises a number of new opportunities for understanding regulated acid secretion by the gastric parietal cell as outlined in **Fig. 4.** This model suggests that regulation of acid secretion depends in part upon regulation of the apical membrane $K^+$ and $Cl^-$ channels. It will now be possible to address whether this channel shows stimulus-dependent recruitment from $H^+/K^+$ ATPase-containing tubulovesiclular membranes to the canalicular membrane. Such studies can be carried out with immunoblots of Kir2.1 using $H^+/K^+$ ATPase-containing vesicles or immunopurification of vesicle preparations from secreting and non-secreting animals. Confocal immunoelectron microscopy (15) could also be used. Such studies would help to address whether the $H^+/K^+$ ATPase and Kir2.1 reside in the same membrane populations when the parietal cell is in the non-secreting and resting state, a long-standing debate in the parietal cell literature (10). Detailed studies of the relationship between acid activation of the channel(s) and PKA activation of the channel(s) is now be possible. Such studies will help to elucidate the relationship between acid activation and secretagogue activation. Studies of the ClC-2 $Cl^-$ channel shows that pH activation of ClC-2 is dependent upon prior modification of the ClC-2 channel by PKA (2,3,16). This suggests that PKA modification of ClC-2 is required for acid accumulation, and that acid produced by the secretory apparatus. If this same effect holds for the $K^+$ channel, the sequence of events for secretagogue (histamine) stimulation of acid secretion would be PKA activation followed by pH activation. If the assumption that the activity of the ion channels of the apical membrane serves to control the rate of acid secretion (6) is correct, the identification of the

$K^+$ channel involved in apical membrane $K^+$ recycling (Kir2.1), can also help to address the control points in acid secretion.

Membrane potential differences that are experienced by the ion channels can also lead to changes in their activity. The extent of movement of $K^+$ is dependent upon the membrane potential and the activity of the channel. The tubulovesiclular membrane is not expected to have a membrane potential, whereas the canalicular membrane is exposd to a membrane potential giving rise to movement of $K^+$ when the channel is activated by PKA. It will be of interest to determine whether the extent of rectification of the $K^+$ channel changes upon stimulation of $K^+$ currents with secretagogues and PKA.

It will also be important to examine whether Kir2.1 is present in $H^+/K^+$ ATPase-containing membranes of both secreting and non-secreting animals to help address whether the channel is always present in the same membranes that contain the $H^+/K^+$ ATPase, an important question in understanding the mechanisms of acid secretion. With the exception of Im *et al* (10) all of the evidence accumulated to date suggests that it is always associated with the $H^+/K^+$ ATPase-containing membranes, as suggested in the model shown in Fig. 4. Preliminary studies suggest that it is not in the basolateral membrane. Knockout models of Kir2.1 could be important to helping resolve the issue of whether Kir2.1 is important to acid secretion, but would not be useful to help understand whether Kir2.1 is involved in $K^+$ recycling.

It is important to note that the gastric parietal cell contains additional $K^+$ channels (including Kir4.1 and 7.1 (11)) that may also play a role in acid secretion. The present data does not preclude a role for these channels in acid secretion.

In summary, the results of the present study suggest that Kir2.1 is a candidate $K^+$ channel involved in $K^+$ recycling at the apical membrane. The lack of inhibition of acid secretion in $H^+/K^+$ ATPase containing vesicles from stimulated rabbits by 293B challenges the view that KCNQ1 may be involved in $K^+$ recycling at the apical membrane. Further studies will be required to definitively address whether Kir2.1 acts alone or in concert with other $K^+$ channels and/or ancillary subunits at the apical membrane.

Acknowledgements: Supported by NIH DK43816, NIH HL58399 and CFF R457.

## REFERENCES

1. **Bachmann A, Quast U, and Russ U.** Chromanol 293B, a blocker of the slow delayed rectifier $K^+$ current (IKs), inhibits the CFTR $Cl^-$ current. Naunyn Schmiedebergs Arch Pharmacol 363: 590-596, 2001.

2. **Cuppoletti J, Tewari KP, Sherry AM, Kupert EY, and Malinowska DH.** ClC-2 $Cl^-$ channels in human lung epithelia: activation by arachidonic acid, amidation, and acid-activated omeprazole. Am J Physiol Cell Physiol 281: C46-54, 2001.

3. **Cuppoletti J, Tewari KP, Sherry AM, and Malinowska DH.** Activation of human ClC-2 $Cl^-$ channels: implications for cystic fibrosis. Clin Exp Pharmacol Physiol 27: 896-900, 2000.

4. **Cuppoletti J, Baker AM, and Malinowska DH.** Acidophillic $Cl^-$ and $K^+$ channels of the gastric parietal cell: a new model of regulated acid secretion. In: NATO ASI Cell Biology Series H89: Molecular and Cellular Mechanisms of H+ Transport, edited by Hirst BH. Springer-Verlag, 1994, p. 383-390. (Proc. NATO Advanced Research Workshop, 1994)

5. **Cuppoletti J, Baker AM, and Malinowska DH.** $Cl^-$ channels of the gastric parietal cell that are active at low pH. Am J Physiol Cell Physiol 264: C1609-C1618, 1993.

6. **Cuppoletti J and Sachs G.** Regulation of gastric acid secretion via modulation of a chloride conductance. J Biol Chem 259: 14952-14959, 1984.

7. **Grahammer F, Herling AW, Lang HJ, Schmitt-Grass A, Wittekindt OH, Nitschke R, Bleich M, Barhanin J, and Warth R.** The cardiac $K^+$ channel KCNQ1 is essential for gastric acid secretion. Gastroenterology 120: 1363-1371, 2001.

8. **Grahammer F, Herling AW, Lang HJ, Schmitt-Grass A, Wittekindt OH, Nitschke R, Bleich M, Barhanin J, and Warth R.** Identification of potaassium channels associated with $H^+/K^+$-ATPase in the parietal cell. Proceedings of the 9th International Proton Transport Conference "Mechanisms and Consequences of Proton Transport", p. 92-93, 2001.

9. **Horio Y, Mouri T, Fujita A, Inanobe A, Tanemoto M, and Kurachi Y.** Inward rectifying $K^+$ channel Kir4.1 participates in gastric acid secretion. Physiologist 42: A-18, Abstract 10.14, 1999.

10. **Im WB, Blakeman DP, and Davis JP.** Studies on $K^+$ permeability of rat gastric microsomes. J Biol Chem 260: 9452-9460, 1985.

11. **Malinowska DH, Kupert EY, Sherry AM, and Cuppoletti J.** A candidate $K^+$ channel involved in gastric HCl secretion. Experimental Biology Meeting, San Diego, CA (April 2000). FASEB J 14: A338, Abstract #275.4, 2000.

12. **Malinowska DH, Kupert EY, Bahinski A, Sherry AM, and Cuppoletti J.** Cloning, functional expression, and characterization of a PKA-activated gastric $Cl^-$ channel. Am J Physiol Cell Physiol 268: C191-C200, 1995.

13. **Reenstra WW and Forte JG.** Characterization of $K^+$ and $Cl^-$ conductances in apical membrane vesicles from stimulated rabbit oxyntic cells. Am J Physiol Gastrointest Liver Physiol 259: G850-G858, 1990.

14. **Saccomani G, Psarras CG, Smith PR, Kirk KL, and Shoemaker RL.** Histamine-induced chloride channels in apical membrane of isolated rabbit parietal cells. Am J Physiol Cell Physiol 260: C1000-1011, 1991.

15. **Sherry AM, Malinowska DH, Morris RE, Ciraolo GM, and Cuppoletti J.** Localization of ClC-2 $Cl^-$ channels in rabbit gastric mucosa. Am J Physiol Cell Physiol 280:C1599-1606, 2001.

16. **Tewari KP, Malinowska DH, Sherry AM, and Cuppoletti J.** PKA and arachidonic acid activation of human recombinant ClC-2 chloride channels. Am J Physiol Cell Physiol 279: C40-50, 2001.

17. **Wolosin JM and Forte JG.** $K^+$ and $Cl^-$ conductances in the apical membrane from secreting oxyntic cells are concurrently inhibited by divalent cations. J Membr Biol 83: 261-272, 1985.

18. **Wolosin JM and Forte JG.** Stimulation of oxyntic cell triggers $K^+$ and $Cl^-$ conductances in apical $H^+$-$K^+$-ATPase membrane. Am J Physiol Cell Physiol 246: C537-C545, 1984.

19. **Wulfsen I, Hauber HP, Schiemann D, Bauer CK, and Schwarz JR.** Expression of mRNA for voltage-dependent and inward rectifying K channels in GH3/B6 cells and rat pituitary. J Neuroendocrinol 12: 263-272, 2000.

Chapter 30

# Kinetics of Activation of $H^+$ Secretion by Secretagogues in Relation to Changes in $[Ca^{2+}]_i$ in Isolated Gastric Glands

FABIÁN MICHELANGELI, JOSÉ FRANCISCO PÉREZ
and MARIE CHRISTINE RUIZ

*Centro de Biofísica y Bioquímica*
*Instituto Venezolano de Investigaciones Científicas (IVIC)*
*Caracas, Venezuela.*

## INTRODUCTION

Acid secretion by the mammalian gastric parietal cell is regulated by neural, endocrine and paracrine systems. Interaction of released secretagogues with cell surface receptors triggers changes in the concentration of intracellular second messengers, mainly cAMP and calcium. These in turn activate the $H^+/K+$ ATPase by an unknown sequence of processes (18). It has been well established that binding of ACh, released from enteric nerve terminals close to parietal cells, to a muscarinic M3 receptor induces the release of intracellular $Ca^{2+}$ by activation of phospholipase C (6). On the other hand, endogenous histamine can be released from neighboring ECL cells in the gland by gastrin and cholinergic stimuli. The role of the histaminergic component in these responses is controversial and may depend on the species studied (11, 15, 16). Released histamine binds to an $H_2$-receptor coupled to an adenyl cyclase, resulting in an increase in cAMP (7). In addition, $H_2$-receptor activation can also increase $[Ca^{2+}]_i$ by yet undefined mechanisms (5, 10). The increase in the two messengers following stimulation by histamine may result in a potentiated secretory response. The physiological response is likely to be the result of synergistic interactions.

Secretion of HCl by the parietal cell is a very complex regulated transport process. The end result of the signal transduction cascade is the activation $K^+$ and $Cl^-$ channels and the translocation of $H^+/K^+$ pumps to the apical membrane by fusion. In this mechanism, KCl leaks from the parietal cell into the lumen and $K^+$ is recycled by the pump, leaving HCl as the final product. (18). The time course of this chain of events that leads to $H^+$ secretion is yet unknown. While fluorescence methods reveal fast changes in $[Ca^{2+}]_i$ as a second messenger in the order of seconds, activation of acid secretion in isolated gastric glands or cells could only be detected in the order of minutes with the amynopyrine accumulation ratio (2). This index of acid secretion lacks the spatio-temporal resolution needed to address questions about the kinetics of the activation process. The method we have recently developed using Lysosensor yellow-blue fluorescence (LYB) (14), takes advantage of the accumulation of the dye exclusively in acid spaces and of its property of changing fluorescence with changes in pH, which can be followed in specific areas of the gland in real time. In this paper we characterize the cholinergic and histaminergic responses and its interactions, using dual-dye fluorescence measurements of $H^+$ secretion and $[Ca^{2+}]_i$.

## Methods

Glands were prepared by collagenase digestion as described (2, 10). Isolated glands were loaded with 2 μM LYB and 5μM Fluo3-AM at room temperature for 40 minutes.

Fluorescence imaging and photometry of stained glands was made using a fluorescence inverted microscope (Nikon TE-300) equipped with a computer controlled cooled CCD camera system (Newcastle Photometric Systems), a filter wheel for excitation wavelength changes, shutters and computer controlled image and data acquisition. Glands were adhered to polylysine coated coverslips and mounted in a flow-through perfusion microscope chamber. Fluorescence was measured from up to 16 regions of interest at 16 bit resolution, selected from an initial fluorescence image, including background. Images can be recorded at given times at 8-bit resolution. Simultaneous fluorescence measurements were performed at alternating excitations of 380 and 500 nm, and emission at 530 nm, enabling the measurement of acid changes together with $Ca^{2+}$ using LYB and Fluo3, respectively. All measurements were made at room temperature. Fluorescence data were normalized according to the equation: $\Delta F/Fo = (F - Fb) / (Fb - Background)$, where Fb is the fluorescence just before secretagogue addition.

## RESULTS

**The response of double stained gastric glands to cholinergic stimulation.**

Images of double stained glands show a differential pattern of staining, with the dyes accumulating in the predicted structures within the gland. Fluo3 stained predominantly parietal cells, whereas LYB stained the intracelular canaliculi of parietal cells and the gland lumen (Fig. 1C). Stimulation with 10μM CCh produced a transient increase in fluo3 fluorescence that peaked and descended to a plateau that was higher than resting fluorescence. Peak stimulation was reached about 10 seconds after initiation of signal rise. The plateau was maintained as long as the stimulant was present. The trace shown in fig 1 A corresponds to a single parietal cell (enclosed). Similar $Ca^{2+}$ responses were obtained from all other parietal cells in the gland. These fluorescence changes correspond to an initial $Ca^{2+}$ release from intracellular stores and a subsequent $Ca^{2+}$ influx through the basolateral membrane, as widely shown in parietal cells and many other cell types (3, 5, 10, 12).

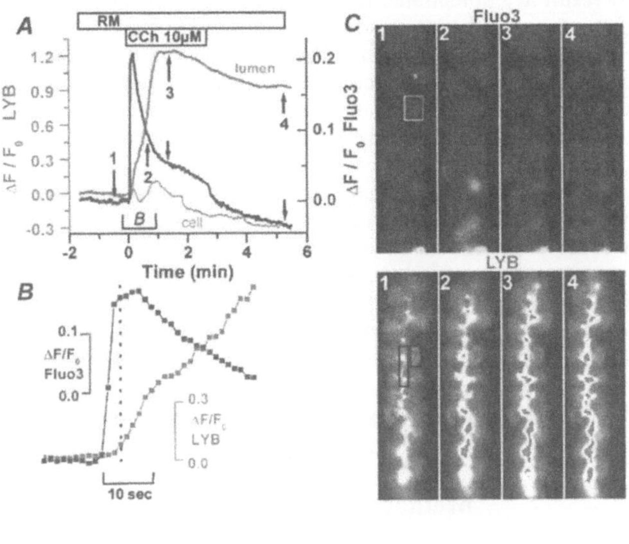


**Fig. 1. Carbachol stimulation rapidly increases $[Ca^{2+}]_i$ and $H^+$ secretion in single gastric glands.** Glands were double stained with LYB and Fluo3 *A*, normalized Fluo3 and LYB fluorescence of a single representative parietal cell and LYB fluorescence in the adjacent luminal region during carbachol stimulation, as remarked in images. *B*, amplification of time scale during the rise phase indicated in *A*. Acid in the lumen was detected 3 seconds after the initiation of the $[Ca^{2+}]_i$ rise recorded in adjacent parietal cell. *C* Fluo3 (top) and LYB (bottom) images taken at time points indicated by numbers on the trace.

Acid secretion increased very rapidly after stimulation, mostly in the gland lumen, with a lag of only 3 seconds after the initiation of the $Ca^{2+}$ signal rise in an adjacent parietal cell. Almost no acid secretion signal was recorded in parietal cells. The change in fluorescence appeared to be synchronous in all gland luminal regions after stimulation. Lumen appeared to increase not only in fluorescence intensity but also in volume. Moreover, fluorescence increase was not homogeneous along the entire length of the gastric lumen which sometimes appeared to be collapsed at some points. Fluorescence of LYB rapidly attained a plateau suggesting that carbachol could be inducing a short transient pulse of acid, concomitant with the $Ca^{2+}$ release phase of the spike. The transitory nature of the carbachol response was further supported by the results presented in figure 2. Repetitive and short (20 sec) pulses of CCh produced repetitive responses in Fluo3 and LYB fluorescence. This type of $[Ca^{2+}]_i$ response has been associated to the release and recharge of $Ca^{2+}$ from stores during addition and removal of the stimulus (12). Acid secretion increased stepwise with each stimulation, every time attaining a new, higher and accumulative level. The derivative function of LYB fluorescence plotted together with the $Ca^{2+}$ response clearly shows that carbachol induces the secretion of a transitory pulse of acid, associated in amplitude and time with the release phase of the spike.

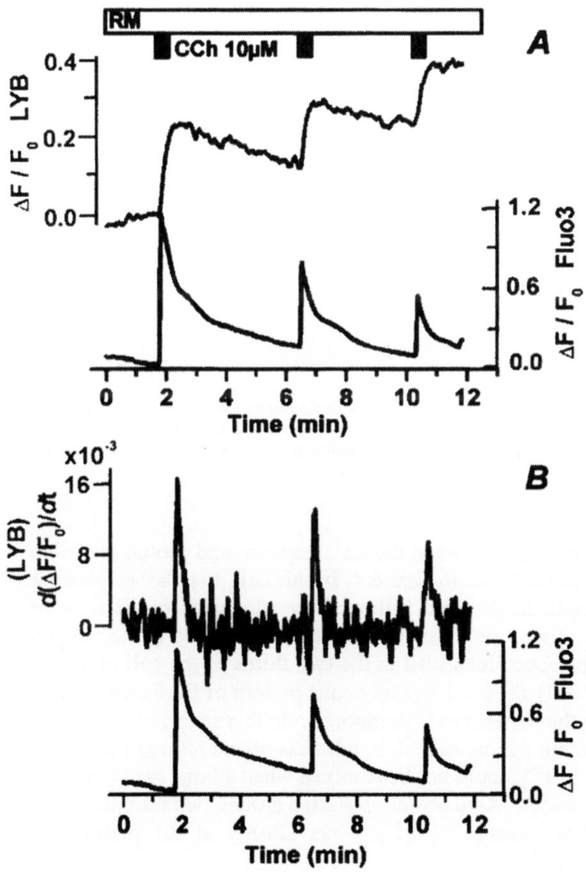


**Fig. 2. Repetitive stimulations with carbachol elicit repetitive responses in $[Ca^{2+}]_i$ and $H^+$ secretion.** Gastric glands were stimulated at indicated times with 10 μM carbachol. Representative traces of Fluo3 fluorescence recorded in a parietal cell and LYB fluorescence in an adjacent luminal region (top). First derivative of the LYB signal as a function of time (bottom).

**Activation of the gastric secretory process by cAMP dependent secretagogues**

Histamine has been shown to increase acid secretion by increasing cAMP levels and intracellular $[Ca^{2+}]_i$ (5, 7, 10). The response to histamine was very different from that obtained with carbachol stimulation. Histamine induced a $Ca^{2+}$ response characterized by oscillations with irregular spikes above an elevated basal level, as earlier shown (Fig. 3) (9). Along the gland, parietal cells responded with a varied and a non-synchronous pattern of oscillatory response. Removal of the stimulant from the medium, brought about a decrease of $[Ca^{2+}]_i$ to the resting level with attenuated oscillations. At difference with carbachol, images and traces of LYB fluorescence show a clear increase of acid secretion both in intracellular canaliculi of parietal cells and the gland lumen. Onset of acid secretion was much slower than in the case of carbachol stimulation, lagging 30-300 seconds behind the initiation of the $[Ca^{2+}]_i$ signal, and it was not synchronically activated in all cells along the gland.

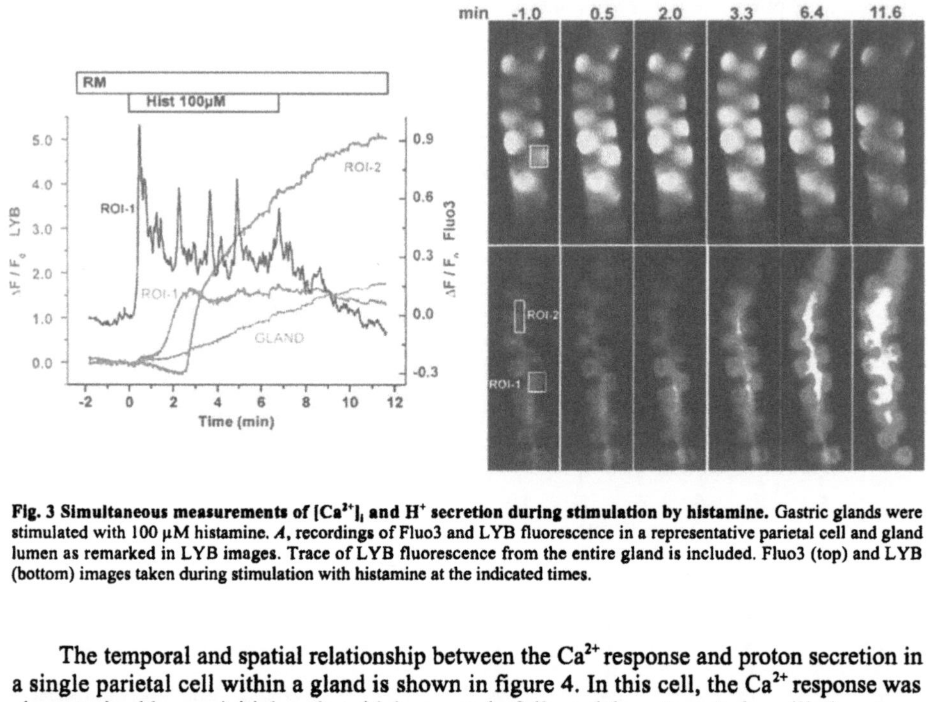


**Fig. 3 Simultaneous measurements of $[Ca^{2+}]_i$ and $H^+$ secretion during stimulation by histamine.** Gastric glands were stimulated with 100 μM histamine. *A*, recordings of Fluo3 and LYB fluorescence in a representative parietal cell and gland lumen as remarked in LYB images. Trace of LYB fluorescence from the entire gland is included. Fluo3 (top) and LYB (bottom) images taken during stimulation with histamine at the indicated times.

The temporal and spatial relationship between the $Ca^{2+}$ response and proton secretion in a single parietal cell within a gland is shown in figure 4. In this cell, the $Ca^{2+}$ response was characterized by an initial and rapid large peak, followed by attenuated oscillations on a plateau above the resting level. The Fluo 3 image taken at the peak (n° 2) shows a homogenous distribution of fluorescence localized in the two thirds of the cell in the basal region. The following images (n° 3-7) show a heterogeneous pattern of fluorescence within the cell (spatial) that varied along the experiments (temporal). In this sense, we can observe regions of the cells whose fluorescence remained high, whereas others returned to the basal level. These localized changes in $[Ca^{2+}]_i$ could not be resolved when a large area (entire cell) was measured by photometry (see trace). Acid secretion started around two minutes after the initiation of the $Ca^{2+}$ signal, when average $[Ca^{2+}]_i$ was already at the plateau level. Acidification was firstly observed in the deeper part of the intracellular canaliculi closest to basolateral side of the cell. A wave of activation of acid secretion seemed to follow from the basal to the apical pole, to dump acid into the gland lumen. Although there was no apparent temporal relationship between the $[Ca^{2+}]_i$ oscillation spikes and increases in the rate of acid secretion (Fig. 3), it appears from this figure that high $Ca^{2+}$ fluorescence regions were juxtaposed to the secretory canaliculus area where secretion was initiated.

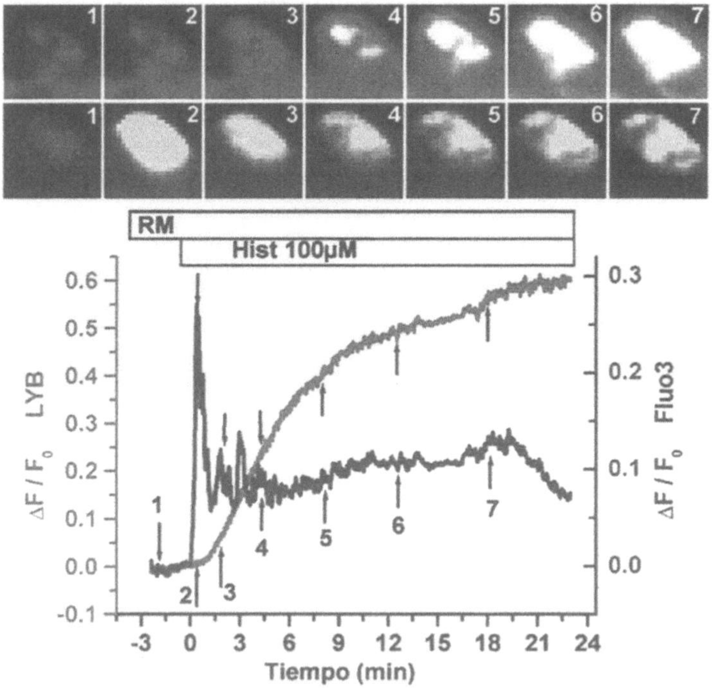


**Fig. 4 Temporal and spatial relationship between the $Ca^{2+}$ response and proton secretion in a single parietal cell within a gland..** LYB (top) and Fluo 3 images (bottom) taken at indicated times.

The pattern of activation of the $[Ca^{2+}]_i$ response and acid secretion when the gland was stimulated by forskolin was very similar to that observed with histamine. Since forskolin activates the adenyl cyclase directly, this suggests that $Ca^{2+}$ oscillations were related to increases in cAMP rather than to occupation of the $H_2$-receptor by histamine. In this particular experiment we can observe that after reaching a high level, LYB signal decreased below basal, probably due to the opening of the gland and leakage of the marker.

**Sequential stimulation of the glands by carbachol and cAMP-dependent secretagogues**

When glands were sequentially stimulated by carbachol and histamine the two patterns of response in $Ca^{2+}$ and secretion shown above were clearly discernible. In figure 6 (left panel) a carbachol pulse was given, immediately followed by sustained stimulation with histamine. Carbachol elicited the typical $Ca^{2+}$ spike, accompanied by a small increase in acid secretion observable in the gland lumen but not in parietal cells. Replacing carbachol by histamine as a stimulant produced a new elevation of $[Ca^{2+}]_i$ in the form of high frequency and low amplitude oscillations, followed by a new increase in acid secretion. This was observed in the cells as well as in the gland lumen. Joint addition of both stimulants induced an identical type of response (fig. 6, right panel). Similar results were obtained when forskolin was used after carbachol (Fig. 7). These results suggest that different mechanisms may be operating in the activation of the pump by these two types of secretagogues.

The addition of carbachol to glands previously stimulated by histamine provoked a small $Ca^{2+}$ spike and an immediate rise in the rate of secretion, that was observed in the gland lumen but not in the cell (Fig.8). This increase was large, despite the small increase in $[Ca^{2+}]_i$ elicited by carbachol in this case. This may indicate that the release of calcium potentiates an already established level of secretion induced by cAMP.

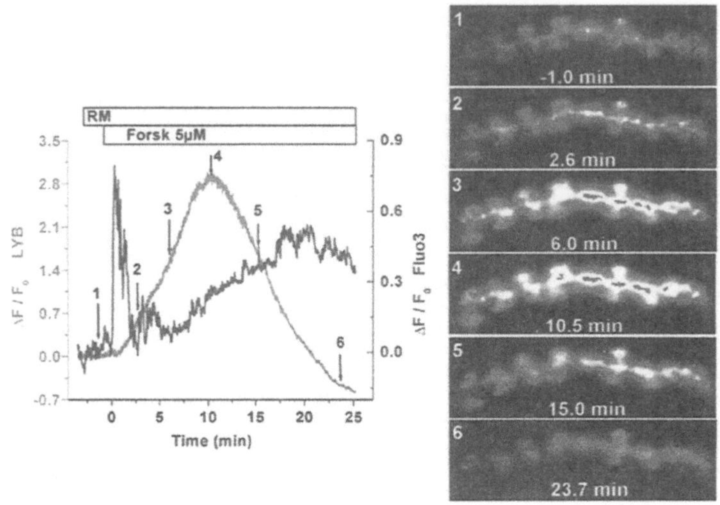


**Fig. 5 Simultaneous measurements of $[Ca^{2+}]_i$ and $H^+$ secretion during stimulation by forskolin.** *Left*, normalized Fluo3 and LYB fluorescence of a single representative parietal cell and LYB fluorescence in the adjacent luminal region during forskolin stimulation. *Right*, LYB fluorescence images taken at indicated times during the experiment.

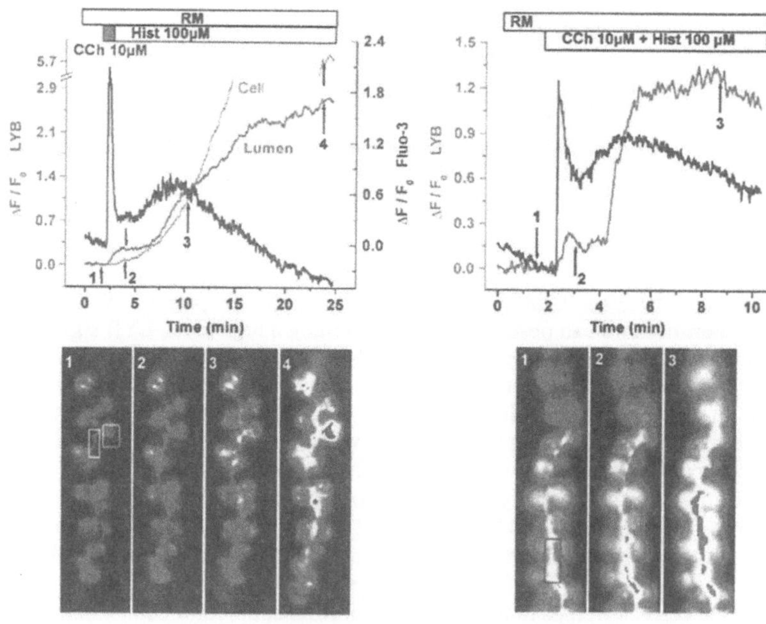


**Fig. 6 Sequential stimulation of the glands by carbachol and histamine.**

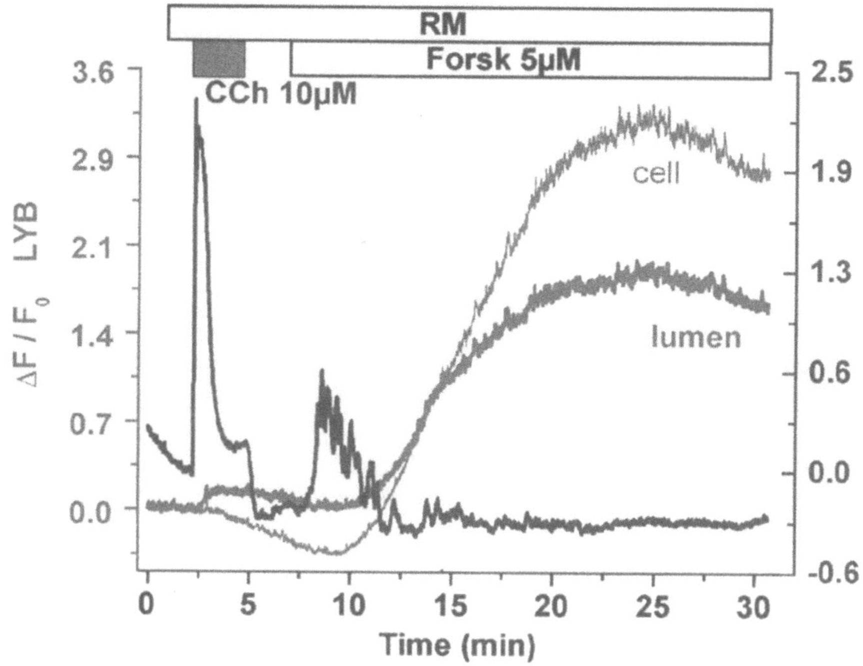


**Fig. 7 Sequential stimulation of the glands by carbachol and forskolin.**

# DISCUSSION

The experiments presented in this paper reveal for the first time the kinetics of activation of $H^+$ secretion by the parietal cell, in relation to changes in intracellular calcium stimulated by carbachol and histamine. Cholinergic stimulation induced a short transient pulse of acid that was related in time and magnitude to the $Ca^{2+}$ release component of the spike. This supports the contention that acid secretion is triggered by $Ca^{2+}$ release during carbachol stimulation. This is in accordance with experiments in populations of isolated gastric glands using aminopyrine as an index of secretion, where the $[Ca^{2+}]_i$ and the secretory responses were abolished by intracellular $Ca^{2+}$ buffering with Bapta (10, 13). It is worth pointing out that acid secretion under carbachol stimulation was clearly observed in the lumen, whereas a barely detectable increase was measured in the cell.

On the other hand, histamine or forskolin stimulation induced $[Ca^{2+}]_i$ oscillations, as earlier reported (9). These may be coupled to changes in cAMP and PKA phosphorylation of ion channels in the plasma membrane or the ER (4, 8). At difference with carbachol, $H^+$ secretion stimulated by histamine and forskolin was slower, larger and sustained, and with no temporal relationship to the $Ca^{2+}$ oscillations. In many experiments a large $[Ca^{2+}]_i$ transient was observed with histamine or forskolin. However, this was not accompanied by a fast secretory response as observed with carbachol. These two distinct patterns of proton secretion activation elicited by carbachol and histamine may be also observed during sequential or joint addition of the secretagogues.

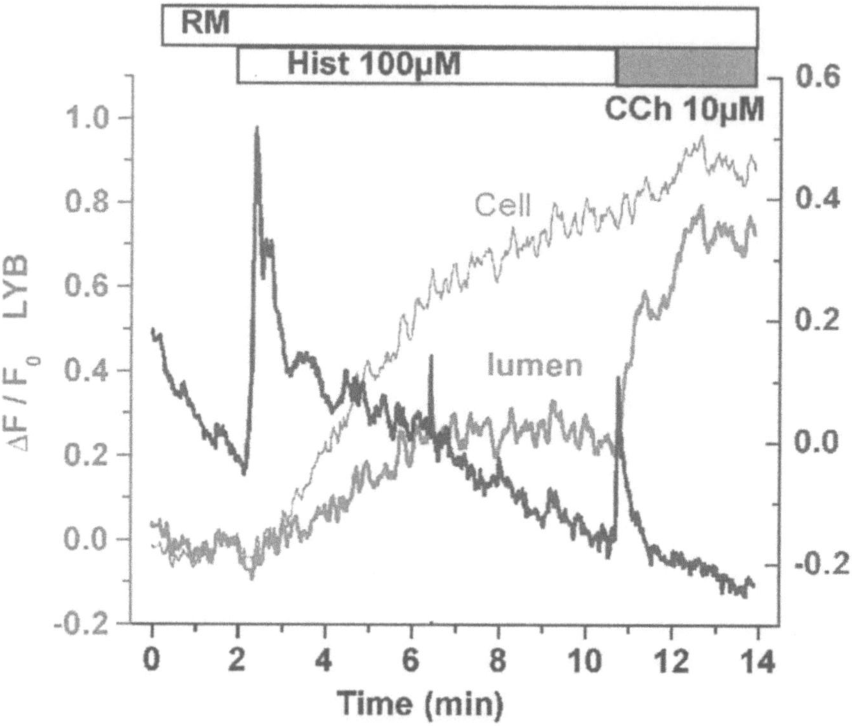


**Fig. 8 Stimulation of the gland by the glands by sequential addition of histamine and carbachol**

The $[Ca^{2+}]_i$ changes induced by carbachol and histamine may have a different spatial distribution and regulation in relation to the localization of $H^+/K^+$ pumps. Recently, it was shown that the $Ca^{2+}$ signal during histamine stimulation was initially observed at the basolateral region of parietal cells, followed by a faint wave spreading to the apical pole (1). However, we observed a spatiotemporal heterogeneous distribution of $Ca^{2+}$ within the parietal cells following histamine stimulation. On the other hand, cholinergic stimulation of pancreatic acinar cells released calcium mostly in the apical region (17, 19). This might also occur in the parietal cell.

Based on the difference in the kinetics and the localization of activation of $H^+$ secretion by the two classes of secretagogues, we propose two distinct mechanisms for stimulation. Cholinergic agonists would increase apical $[Ca^{2+}]_i$, open calcium dependent K+ and/or Cl- channels, thereby activating pumps that are already present in the apical membrane close to the gland lumen. In the case of histamine, secretion would be activated by a slower mechanism involving the cAMP-dependent translocation of pumps to the intracellular canaliculus and activation of PKA phosphorylation dependent ion channels.

Research supported by CONCIT grant 95000519.

## REFERENCES


1. **Athmann, C., N. Zeng, D. R. Scott, and G. Sachs.** Regulation of parietal cell calcium signaling in gastric glands. *Am J Physiol* 279: G1048-58., 2000.
2. **Berglindh, T.** Gastric glands and cells: preparation and in vitro methods. *Methods in Enzymology* 192: 93-107, 1990.
3. **Berridge, M. J.** Elementary and global aspects of calcium signalling. *J Physiol* 499: 291-306., 1997.
4. **Bugrim, A. E.** Regulation of $Ca^{2+}$ release by cAMP-dependent protein kinase. A mechanism for agonist-specific calcium signaling? *Cell Calcium* 25: 219-26., 1999.
5. **Chew, C. S.** Cholecystokinin, carbachol, gastrin, histamine, and forskolin increase $[Ca^{2+}]$i in gastric glands. *Am J Physiol* 250: G814-23, 1986.
6. **Chew, C. S., and M. R. Brown.** Release of intracellular $Ca^{2+}$ and elevation of inositol trisphosphate by secretagogues in parietal and chief cells isolated from rabbit gastric mucosa. *Biochim Biophys Acta* 888: 116-25., 1986.
7. **Chew, C. S., S. J. Hersey, G. Sachs, and T. Berglindh.** Histamine responsiveness of isolated gastric glands. *Am J Physiol* 238: G312-20, 1980.
8. **Joseph, S. K., and S. V. Ryan.** Phosphorylation of the inositol trisphosphate receptor in isolated rat hepatocytes. *J Biol Chem* 268: 23059-65., 1993.
9. **Ljungstrom, M., and C. S. Chew.** Calcium oscillations and morphological transformations in single cultured gastric parietal cells. *Am J Physiol* 260: C67-78., 1991.
10. **Michelangeli, F., M. C. Ruiz, E. Fernandez, and A. Ciarrocchi.** Role of $Ca^{2+}$ in $H^+$ transport by rabbit gastric glands studied with A23187 and BAPTA, an incorporated $Ca^{2+}$ chelator. *Biochim Biophys Acta* 983: 82-90, 1989.
11. **Michelangeli, F., D. M. Sulcas, and M. C. Ruiz.** Ultrastructural studies of endocrine-like cells in the fundic gastric mucosa of the bullfrog, Rana catesbeiana. *Cell and Tissue Res* 250: 413-9, 1987.
12. **Negulescu, P. A., and T. E. Machen.** Release and reloading of intracellular Ca stores after cholinergic stimulation of the parietal cell. *Am J Physiol* 254: C498-504, 1988.
13. **Negulescu, P. A., W. W. Reenstra, and T. E. Machen.** Intracellular Ca requirements for stimulus-secretion coupling in parietal cell. *Am J Physiol* 256: C241-51, 1989.
14. **Perez, J. F., M. C. Ruiz, and F. Michelangeli.** Simultaneous measurement and imaging of intracellular $Ca^{2+}$ and $H^+$ transport in isolated rabbit gastric glands. *J Physiol-London* 537:735-745, 2001.
15. **Prinz, C., M. Kajimura, D. R. Scott, F. Mercier, H. F. Helander, and G. Sachs.** Histamine secretion from rat enterochromaffinlike cells. *Gastroenterology* 105: 449-61., 1993.
16. **Sachs, G., N. Zeng, and C. Prinz.** Physiology of isolated gastric endocrine cells. *Annu Rev Physiol* 59: 243-56, 1997.
17. **Thorn, P., A. M. Lawrie, P. M. Smith, D. V. Gallacher, and O. H. Petersen.** Local and global cytosolic $Ca^{2+}$ oscillations in exocrine cells evoked by agonists and inositol trisphosphate. *Cell* 74: 661-8., 1993.
18. **Urushidani, T., and J. G. Forte.** Signal transduction and activation of acid secretion in the parietal cell. *J Memb Biol* 159: 99-111, 1997.
19. **Xu, X., W. Zeng, J. Diaz, and S. Muallem.** Spacial compartmentalization of $Ca^{2+}$ signaling complexes in pancreatic acini. *J Biol Chem* 271: 24684-90., 1996.

Chapter 31

# Measurement of Intracellular Calcium and Gastric Acid Secretion in Real Time Using Fluorescent Probes

J.F. PÉREZ, M.C. RUIZ and F. MICHELANGELI.
*Centro de Biofísica y Bioquímica, Instituto Venezolano de Investigaciones Científicas (IVIC) Caracas, Venezuela.*

## INTRODUCTION

Studies of HCl secretion by the gastric parietal cell, have been based on isolated oxyntic cells or glands in vitro (3, 5). Accumulation of amynopyrine has been used to measure secretory activity in these preparations (2). This weak base can be protonated at acid pH, accumulating in acid spaces as a function of the pH gradient. Stimulation of acid secretion leads to an enhanced accumulation ratio of the radioactively labeled amynopyrine (AP ratio). This index has been invaluable in our understanding of the mechanisms of activation of the parietal cell by secretagogues (1). However, for kinetic studies, the AP ratio method lacks spatio-temporal resolution. We have used Lysosensor Yellow-Blue (LYB), a new acidotropic fluorescence pH indicator (4), to monitor $H^+$ secretion in isolated gastric glands in real time at one second resolution. We characterize the fluorescence properties of this dye in the $H^+$ concentrating spaces of isolated gastric glands. Using this dye in conjunction with Fluo3 we have made simultaneous measurements of intracellular calcium and $H^+$ secretion during stimulation by carbachol.

## METHODS

Isolated gastric glands were prepared by collagenase digestion (1). For experiments of simultaneous measurements of $[Ca^{2+}]_i$ and $H^+$ secretion, gland suspensions were incubated for 30 min at room temperature with 2μM LYB and 5μM Fluo-3/AM. Emission and excitation spectra of LYB in solution at different pH and LYB fluorescence of loaded gland suspensions were obtained using a spectrofluorimeter. Fluorescence imaging and photometry of stained glands was made using a fluorescence microscope equipped with a cooled CCD camera, shutters and a filter wheel for excitation wavelength changes. Stained glands were adhered to polylysine coated coverslips and mounted in a flow-through perfusion microscope chamber. Fluorescence was measured from up to 16 regions of interest at 16 bit resolution. All measurements were made at room temperature. Fluorescence data were normalized according to: $\Delta F/F_0 = (F - Fb) / (Fb - \text{background})$, where Fb is the fluorescence just before secretagogue addition.

## RESULTS

Lysosensor Yellow-Blue is a pyridyl oxazole with an alkyl amino sidechain (Fig. 1A). The only protonable groups with a predicted pKa between 0 and 14 were the pyrydyl and the alkyl amino group, with values of 3.2 and 8.8, respectively (predictions were obtained using the ACD/I-Lab service, http: www.acdlabs.com). The protonation of either group would lead to an accumulation of the dye in glandular acid spaces in the presence of a pH gradient. Accumulation of a weak base (wb) such as LYB in acidic spaces of the gastric gland is a function of the pH gradient and the pKa of their protonable groups. Assuming a pH value within the resting isolated gland of pHig = 4.4 and pHo = 7.4 for the extracellular medium, we plotted theoretical values of wb accumulation vs pKa (Fig. 1B). Accumulation occurs at pKa values higher than 6; thus, involving only protonation of the amino group. With a pKa value for this group of around 8.8, the accumulation of LYB in the acidic spaces of the gland is close to 1000 fold.

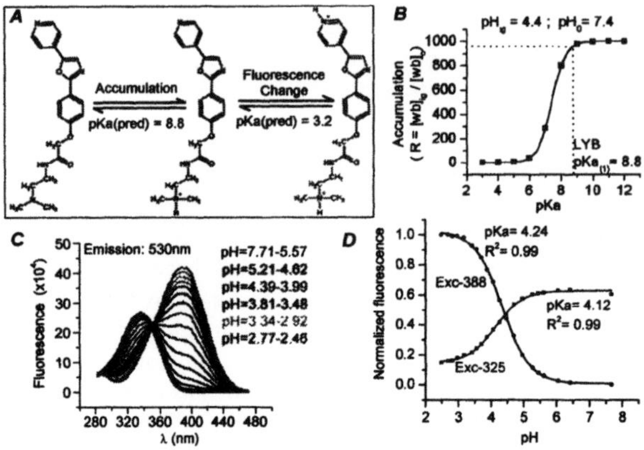


$$\frac{[wb]_{ig}}{[wb]_o} = \frac{10^{(pKa-pHig)}+1}{10^{(pKa-pHo)}+1}$$

Fig.1. **Characterization of LYB: structure, accumulation and spectra as a function of pH.** ***A***: Structure of LYB ***B***: Accumulation of a weak base in acidic compartments as a function of pKa. ***C***: Excitation spectra of LYB at different pH. Fluorescence spectra of 1 μM LYB were obtained in 50 mM citric acid/citrate buffer. ***D***: Normalized fluorescence values at excitation 325 and 388 nm were plotted against pH. Boltzmann fits obtained from the excitation spectra gave a single pKa of around 4.2.

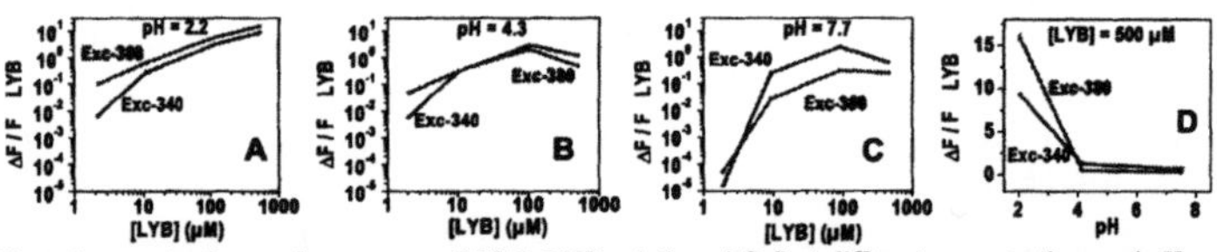


Fig. 2. **Effect of concentration on fluorescence of LYB.** LYB solutions (10μl) at different concentrations and pH were placed on a coverslip and the fluorescence measured at 530 nm using excitation at 340 and 380 nm. (A, B and C) Fluorescence of LYB as a function of dye concentration at pH of 2.2, 4.3 and 7.7, respectively. (D) Plot of data for 500 μM LYB as a function of pH .

The fluorescence ratios 388/325 or 430/530 could be used to calculate the pH of acid spaces, provided the spectral characteristics did not change upon accumulation. Therefore, we studied the effects of concentration on the fluorescence of LYB (Fig. 2). Fluorescence increased as a function of concentration (Fig. 2*A-C*). At pH 2.2, this relationship was linear. However, at pH 7.7 and 4.3, it showed some quenching at high concentration (0.5 mM) for both excitation wavelengths (340 and 380 nm). Plotting the 500 μM LYB data as a function of pH reveals that in the range 4.3-2.2, there is a strong increase in signal for both wavelengths (Fig. 2*D*). The increase at 340 did not follow the behavior predicted by the spectra at low concentration. In the range of 4.3 to 7.7 the dye becomes insensitive to pH. These results suggest dequenching of the dye due to protonation of the nitrogen in the pyridyl moiety. Fig 3 shows the kinetics of LYB accumulation by gastric glands and response to forskolin stimulation. Addition of LYB to the cuvet resulted in a step increase at 325 nm corresponding to the fluorescence of LYB in the extracellular medium. The subsequent decrease in signal at 325 nm and increase at 388 nm correspond to dye entering acidic compartments within the gland. Forskolin (2.5μM) increased fluorescence at 388 nm but not much at 325 nm. As predicted from Fig.2D this corresponds to further acidification of the compartments where the dye is concentrated. Dissipation of the pH gradient with the H+/K+ exchanger nigericin (5μM) decreased fluorescence at 388 nm with a concomitant increase at 325 nm, due to unprotonated dye release.

Measuring fluorescence of LYB at 380/530 permits the simultaneous measurement of other dyes such as fluo3, using 380-500/530 nm(6). Fig. 4 shows one such experiment, stimulating double-dye loaded glands with carbachol. Calcium and pH are readily discernible in images and photometric traces with no superposition of signals.

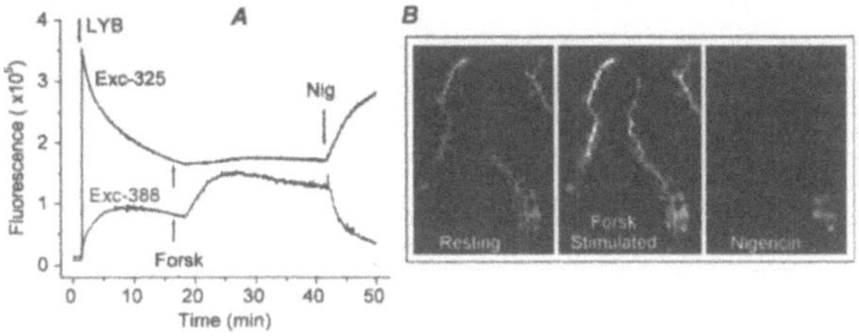


**Fig. 3. Kinetics of LYB accumulation by gastric glands and response to forskolin.** ***A***: time course of fluorescence change by addition of 1 μM LYB to a gland suspension measured in a spectrofluorimeter. Excitation at 325 and 388 nm; emission 530 nm. ***B***: Fluorescence images of a similar experiment performed in the microscope.

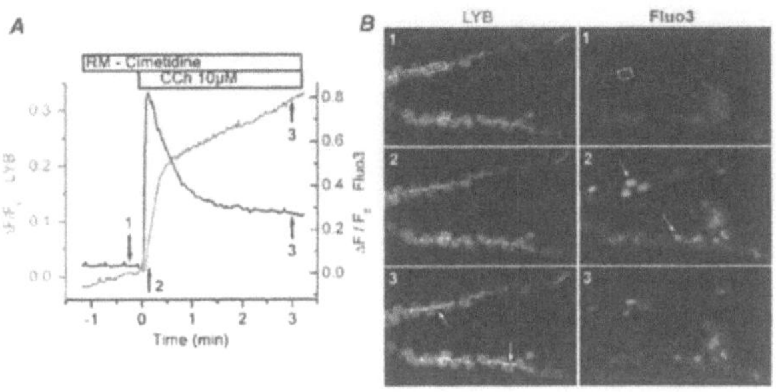


**Fig. 4. Simultaneous measurement of $[Ca^{2+}]_i$ and gastric acid secretion stimulated by carbachol.** Carbachol (10μM CCh) stimulation increased fluorescence of LYB and Fluo3 ***B***: LYB and Fluo3 images taken at time points indicated by numbers on the trace. Fluo3 stained predominantly parietal cells; LYB stained the intracellular canaliculi of parietal cells and the gland lumen.

## DISCUSSION

In this work we describe a new method to measure gastric acid secretion with spatio temporal resolution using the fluorescent probe LYB. We characterized the responses of this dye that were used to sense pH changes in the intracellular canaliculi and lumen of gastric glands. LYB accumulates exclusively in acid spaces, with good retention and no binding to cellular structures. Once accumulated, the dye can be washed out of the extracellular space and its pH dependent fluorescence followed in specific areas of the gland in real time. LYB might be used with other fluorescein-like indicators to study the coupling of $H^+$ transport with other intracellular events as shown here with Fluo3. This technique allows us for the first time to study the kinetics of activation of $H^+$ transport stimulated with secretagogues in relation with the increases in $[Ca^{2+}]_i$.

Research supported by CONICIT, Venezuela (Grant 95-000519).

## REFERENCES

1. **Berglindh, T.** Gastric glands and cells: preparation and in vitro methods. *Meth Enzymol* 192: 93-107, 1990.

2. **Berglindh, T., D. R. Dibona, S. Ito, and G. Sachs.** Probes of parietal cell function. *AmJ Physiol* 238: G165-176, 1980.

3. **Berglindh, T., and K. J. Obrink.** A method for preparing isolated glands from the rabbit gastric mucosa. *Acta Physiol Scand* 96: 150-159, 1976.

4. **Diwu, Z., C. S. Chen, C. Zhang, D. H. Klaubert, and R. P. Haugland.** A novel acidotropic pH indicator and its potential application in labeling acidic organelles of live cells. *Chemistry Biology* 6: 411-418, 1999.

5. **Michelangeli, F.** Acid secretion and intracellular pH in isolated oxyntic cells. *J Memb Biol* 38: 31-50, 1978.

6. **Perez, J. F., M. C. Ruiz, and F. Michelangeli.** Simultaneous measurement and imaging of intracellular $Ca^{2+}$ and $H^{+}$ transport in isolated rabbit gastric glands. *J Physiol* 537: 735-745, 2001.

Chapter 32

# Tissue and Cell Distribution of Parchorin, a Chloride Intracellular Channel-Related Protein

YUMIKO MIZUKAWA[1], TOMOHIRO NISHIZAWA[1], TAKU NAGAO[1], KEN KITAMURA[2], AND TETSURO URUSHIDANI[1]

*[1]Laboratory of Pharmacology and Toxicology, Graduate School of Pharmaceutical Sciences, The University of Tokyo, Tokyo 113-0033, Japan, and [2]Department of Otolaryngology, Tokyo Medical and Dental University, School of Medicine, Tokyo 113-8519, Japan.*

## INTRODUCTION

We previously reported a 120 kDa phosphoprotein that translocated from cytosol to the apical membrane of gastric parietal cells in association with stimulation of HCl secretion (8). Recently, we cloned this protein from rabbit choroid plexus and named it "parchorin" based on its high enrichment in parietal cells and choroid plexus (6). Parchorin is a novel protein with 64 kDa that has significant homology to the family of chloride intracellular channels (CLIC), especially the chloride channel from bovine kidney, p64, in its C-terminal 235 amino acids (Fig. 1). In contrast to other CLIC family members, most of which are considered to be $Cl^-$ channels in intracellular vesicles (7), parchorin is a soluble cytosolic phosphoprotein which translocates to the plasma membrane under stimulation. When acid-secreting parietal cell is stimulated, tubulovesicles containing H,K-ATPase fuse with the apical membrane, and both $K^+$ and $Cl^-$ permeability in the apical membrane is increased. During this activation process, it was observed that parchorin translocated from cytosol to the apical membrane (6, 8). It was also observed that parchorin, transfected to LLC-PK1 cell, translocated to the plasma membrane and accelerated the $Cl^-$ efflux rate when $Cl^-$ efflux was caused by elimination of $[Cl]_o$ (6). Using Northern and Western blotting, it has been found that parchorin preferentially distributes to tissues related to water movement, i.e., gastric mucosa, choroid plexus, salivary gland, and kidney (6,8). Parchorin was named by its characteristic distribution, i.e., the highest expression in parietal cell and choroid plexus. These observations strongly suggest that parchorin plays an important role in the regulated movement of body fluid via $Cl^-$ transport. In the present study, we examined the cellular localization of parchorin in various tissues using immunohistochemistry under light microscopic level. We also observed development-related changes of parchorin expression in some exocrine glands.

## IDENTIFICATION OF THE EPITOPE OF ANTI-PARCHORIN ANTIBODY

In order to assure the specificity of the antibody for recognizing parchorin among CLIC family members, we first identified the epitope of the presently used monoclonal anti-parchorin antibody. As shown in Fig. 1, we created various deletion mutants of parchorin as

GFP-fusion protein, transiently expressed them in COS-7 cells and then partially purified. The monoclonal anti-parchorin antibody (8) recognized shaded fragments in the figure, i.e., any fragments having the amino acid sequence 159-262. This indicated that the epitope for the antibody exists in the sequence which contains the characteristic GGSVDA repeat which is specific for parchorin (6). We conclude here that the antibody could specifically distinguish parchorin from other CLIC family members reported to be ubiquitously distributed in the organs.

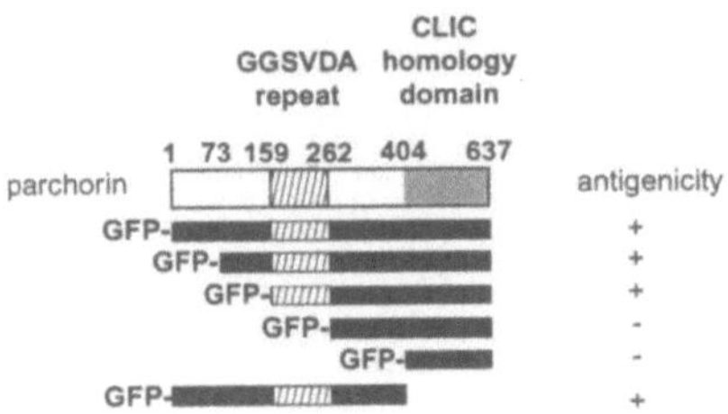


***Fig. 1.*** *Monoclonal antibody to parchorin recognizes the characteristic GGSVDA repeat. The deletion mutants with various length of parchorin were produced as GFP-fusion proteins and subjected to Western blotting using anti-parchorin monoclonal antibody. Posibive (+) and negative (-) responses are indicated on the right. Note that the epitope is within the GGSVDA repeat from 159 to 262, which is specific for parchorin among CLIC family members.*

## TISSUE AND CELL DISTRIBUTION OF PARCHORIN

We fixed various tissues from rabbit and then stained with anti-parchorin monoclonal antibody using FITC- or Cy3-anti-mouse IgG (for confocal laser scanning system) or biotin-conjugated anti-mouse IgG with horseradish peroxidase-conjugated biotin-avidin complex (visualized with diaminobenzidine). The results are summarized in Table 1.

***Table 1.*** *Tissue distribution of parchorin in rabbit. As for the cells where parchorin exists, physiological functions related to water movement is described.*

| | tissue | parchorin expression | the water-moving events that involvement of parchorin is suggested |
|---|---|---|---|
| circulating system | heart | - | |
| | vessel | - | |
| digestive system | gastric parietal cell | + | secretion of gastric acid |
| | salivary duct | + | secretion of saliva |
| | pancreatic duct | + | secretion of pancreatic juice |
| | gallbladder epithelium | + | re-absorption of bile fluid |
| | rectum | - | |
| respiratory system | airway epithelia | + | airway fluid re-absorption |
| | type II alveolar epithelial cell | + | alveolar fluid re-absorption |
| urinary system | kidney tubule (mainly distal) | + | urinary re-absorption |
| | urinary bladder | - | |
| genital organs | glandular epithelia of prostate | + | secretion of prostatic fluid |
| | epithelium of rete testis | + | secretion of semen |
| | mammary gland | + | secretion of milk |
| | placenta | - | |
| sensory organs | ciliary body epithelium, retinal pigment epithelium | + | secretion and transport of aqueous humor |
| | lacrimal duct | + | secretion of tear |
| | cochlea and hair cell of semicircular canal | + | homeostatic maintenance of endolymph and perilymph |
| nervous system | brain (except choroid plexus) | - | |
| | choroid plexus | + | secretion of cerebrospinal fluid |

Parchorin-immunoreactivity was completely absent in the aorta and heart. Cardiac tissue is reportedly enriched in other CLIC family members, so parchorin is exceptional in that regard. In the digestive tract, only parietal cell showed highly positive staining to parchorin antibody within the fundic mucosa. In the pyloric mucosa where parietal cell is absent, no reactivity was noticed. In the intestine via duodenum to rectum, no positive staining was obtained.

Among various tissues, parchorin preferentially resides in the epithelium of ducts of lacrymal, parotid, submandibular, and mammary glands (of pregnant or lactating female), and that of pancreas, prostate, and testis. Fig 2 shows the stereo image of parchorin staining of isolated submandibular glands. It appears that parchorin exists on the apical membrane of epithelial cells of the intercalated duct of the gland. In pancreatic ductal cells, parchorin existed mainly in the epithelial cells of the middle sized duct, less in the large sized one, and was practically absent in the intercalated duct. It is of interest to investigate the correlation between the differences in the physiological function and the contents of parchorin in these ductal cells.

In the gallbladder, the epithelia that elicit water re-absorption from the bile were strongly positive, whereas the smooth muscle layer was negative. Within the epithelial cells, parchorin appeared to be present in the apical membrane.

In the trachea and lung, parchorin was found to be in the airway epithelia and type II alveolar cells. In kidney, parchorin distributed mainly from the thick ascending limb to distal convoluted tubule, and somewhat in the part of the inner medullary collecting duct. In eye, both pigment and nonpigment epithelia of ciliary body were positive, whereas only the pigment epithelia was positive in the retina. In the ear, parchorin was enriched in the cochlea and semicircular canal.

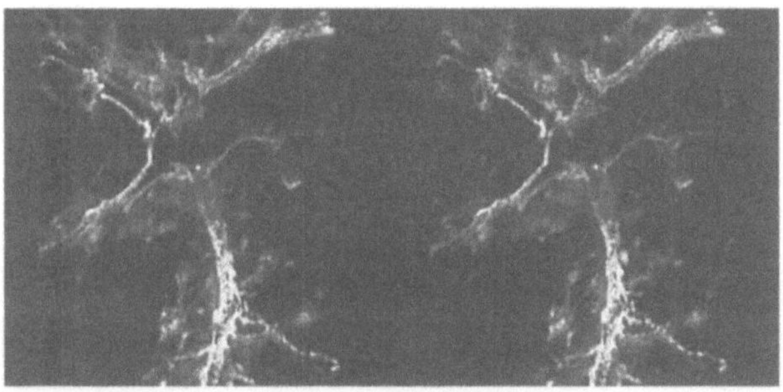

***Fig. 2.*** *Immunostaining of parchorin in rabbit submundibular glands isolated by collagenase digestion (stereo image). Although parchorin is present in the aciar cells, it is markedly enriched in the apical side of the ductal cells which secrete water within the salivary gland.*

## DEVELOPMENTAL CHANGES IN THE EXPRESSION OF PARCHORIN

In the previous section, it was clearly shown that parchorin distributes preferentially to the cells that play an important role in water movement. If so, the contents of parchorin in a tissue is expected to be proportional to its function. It is well known that gastric acid secretion drastically increases after weaning (2) but salivary secretion does not (5). Gastric mucosa and submandibular glands were isolated from suckling and adult rabbits. The amount of parchorin per mg protein in the gastric mucosa was several fold higher in adult rabbits as compared with suckling rabbits, whereas that in submandibular glands was unchanged. It is also well known that the secretory capacity of milk increases during pregnancy and lactation. Although parchorin was undetectable by Western blotting in the mammary gland of normal female rabbit, it was clearly found in the glands of pregnant rabbit, and it increased several fold in lactating rabbit. Therefore, it is obvious that the expression of parchorin increases with functional development.

## CONCLUSION

The cellular distribution of parchorin and its change in expression with development indicate that parchorin plays an important role in cells involved in water movement, possibly via chloride transport. It has been claimed that two chloride channels, CFTR and ClC2, play important roles in the physiological control of water movement. The importance of the former is established by the fact that a genetic defect of the protein function is the cause of cystic fibrosis (4). For the latter, ClC2 deficient mouse was recently reported (1). Although a diverse phenotype had been expected from the wide distribution of this channel, tissue-specific diseases emerged: male infertility and blindness - because of loss of cells maintaining the blood-testis barrier and the blood-retina barrier respectively. These results suggest the pathophysiological significance of $Cl^-$ transport in Sertoli cells and retinal pigment epithelia. Considering the present observation that parchorin existed both in epithelia of testis and renital pigment epithelia, we feel male genital organ and eye would also be quite interesting future targets. The enrichment of parchorin in cochlea and semicircular canal of the ear is also quite suggestive considering the observation that $Na^+$-$K^+$-$2Cl^-$ co-transporter 1 deficient mice showed diseases arising from impediments in the inner ear (3). Together with the observation that expression of parchorin was proportional to physiological function, it is strongly suggested that parchorin plays an indispensable role in water movement. The molecular mechanism of exactly how parchorin participates in ion transport, possibly $Cl^-$ channel activity, awaits future study. It is also necessary to design experiments to probe the physiological status of organs expressing parchorin, including the creation of a gene-targeting animal.

This study was supported in part by Japanese Ministry of Education, Science, Sports, and Culture Grants 13470511 and 13557220.

## REFERENCES

1. **Bösl MR, Stein V, Hübner C, Zdebik AA, Jordt S-E, Mukhopadhyay AK, Davidoff, MS, Holstein A-F, and Jentsch TJ.** Male germ cells and photoreceptors, both dependent on close cell-cell interactions. degenerate upon ClC-2 $Cl^-$ channel disruption. *EMBO J* **20**: 1289-1299, 2001
2. **Crothers JM, Reenstra WW, and Forte JG.** Ontogeny of gastric H(+)-K(+)-ATPase in suckling rabbits. *Am J Physiol Gastrointest Liver Physiol* **259**: G913-G921, 1990.
3. **Delpire E, Lu J, England R, Dull C, and Thorne T.** Deafness and imbalance associated with inactivation of the secretory Na-K-2Cl co-transporter. *Nature Genet* **22**: 192-195, 1999.
4. **Grubb BR, and Boucher RC.** Pathophysiology of gene-targeted mouse models for cystic fibrosis. *Physiol Rev* **79**: S193-S214, 1999.
5. **Leeson CR, and Forman DE.** Postnatal development and differentiation of the secretory elements of the rabbit parotid and submandibular glands. *Anat Anz* **149**: 210-225, 1981.
6. **Nishizawa T, Nagao T, Iwatsubo T, Forte JG, and Urushidani T.** Molecular cloning and characterization of a novel chloride intracellular channel-related protein, parchorin, expressed in water-secreting cells. *J Biol Chem* **275**: 11164-11173, 2000.
7. **Tulk BM, Schlesingers PH, Kapadia SA, and Edwards JC.** CLIC-1 functions as a chloride channel when expressed and purified from bacteria. *J Biol Chem* **275**: 26986-26993, 2000.
8. **Urushidani T, Chow D, and Forte JG.** Redistribution of a 120 kDa Phosphoprotein in the Parietal Cell Associated with Stimulation. *J Memb Biol* **168**: 209-220, 1999.

Chapter 33

# Scanning EM Studies of Gastric Oxyntic Cells with Special Reference to the Translocation of Tubulovesicular System towards the Intracellular Canaliculus

TAKURO OGATA

*Department of Surgery and Physiology, Kochi Medical School, Nankoku, Kochi, Japan*

## ABSTRACT

The membrane systems of the rat and frog oxyntic cells were examined by ultra-high-resolution SEM. Using a recently developed favorable fixation technique for parietal cells by Sugai et al. (20), rat gastric mucosae were microwave fixed in a cacodylate buffer, 334 milliosmoles/kg$H_2O$ (mOsm), to which 1.0% glutaraldehyde and 0.5% formaldehyde were added. Specimens examined by TEM of thin sections showed the cytoplasm packed with tubular membranes similar to images found after rapid-freeze/freeze-substitution fixation, which is generally considered to cause minimal structural alterations. To render the cytoplasmic membranes visible by SEM, fixed mucosae were frozen, fractured and macerated by the A-ODO procedure. With much of the cell matrix and filaments removed, SEM revealed numerous 30-60 nm tubules formed a three-dimensional network with small cisternae about 100 nm in diameter having a small fenestration at their center. This network was designated the "tubulocisternal network (TCN)". The TCN occasionally connected with the rough-ER and the Golgi apparatus. Vesicles or isolated tubules were not found in appropriately macerated parietal cells. The intracellular canaliculus (IC) was lined with numerous microvilli. In favorable sites connections of the TCN to the IC were clearly visible. In the frog oxyntic cells, the TCN were more frequently connected with the luminal membrane than in the rat parietal cells. Connections between these two membrane compartments suggest the probability of rapid membrane transposition.

## INTRODUCTION

Mammalian parietal (oxyntic) cells have two distinctive membrane systems. One is the intracellular canaliculus (IC) narrow canals lined with numerous microvilli and the other is the tubular and vesicular membrane system usually called tubulovesicles (TV). There are two unsolved questions on the membrane systems of the parietal cell. The first question: What is the true morphology of TV system. In the previous studies, the TV membrane system has been variously described as vacuoles, vesicles, tubules, or mixtures of these forms (4). The variations described by different investigators appear to be due to tissue preparation methods. As a compromise this membrane system has been designated tubulovesicles, a term which is not entirely accurate or appropriate. The second question: In the resting state, there are numerous TV in the cytoplasm and the area of the IC is small. But after the stimulation to secrete acid the area of TV decreased with the concomitant increase of the IC. However, there is no convincing evidence the connections between the TV and the IC. How the TV membrane transforms into the IC membrane without connections after the stimulation is the enigma for more than 40 years. To solve these questions, the rat and frog oxyntic cells were treated with the aldehyde–osmium-DMSO-osmium (A-ODO) method (21), which eliminates cell matrix and expose intracellular membrane systems, and were observed with an ultra-high resolution scanning electron microscope (SEM).

Recently, Sugai et al. (20) evaluated numerous fixation solutions for electron microscopy of mammalian parietal cells. Their aim was to duplicate the extensive preservation of tubules obtained by rapid-freezing followed by freeze-substitution fixation, which is believed to cause minimal structural changes (19). They found that the osmolality of the buffer alone rather than that of the complete fixative was found to be crucial in preserving the parietal cell cytoplasmic membrane predominately as tubules rather than as vesicles and some tubules. Only with fixative solutions containing buffer osmolality of 290 milliosmoles/kg$H_2O$ (mOsm) or higher resulted in retaining of the predominant tubular configuration. When lower osmolality buffers were used vesicles and vacuoles predominated. Variations in the concentration of glutaraldehyde between 1.0 and 3.0% in the fixatives did not alter tubule preservation.

For the present study, we used a buffer concentration of 334 mOsm with added aldehydes and consistently obtained rat parietal cells with extensive meshwork of tubules and cisternae, designated as the tubulocisternal network (TCN). At certain points the membranous continuity between the TCN and the IC was clearly visible in our preparations. In the frog oxyntic cells, the TCN connected more frequently with the luminal membrane than in the rat parietal cells. These continuities provide further insight into the probability that membranes between these two compartments are rapidly transposed when the cell is stimulated to secrete acid and that the luminal surface membrane rapidly increases while the TCN membrane concomitantly decreases.

## MATERIALS and METHODS

Male Wistar rats weighing 200-250 g and Japanese meadow frogs weighing 50-100 g were used for this study. The precise tissue preparation methods were described in the following papers (12-14). In brief, small pieces of fundic mucosa were microwave fixed in 0.5% formaldehyde and 1.0% glutaldehyde in 0.16 M cacodylate buffer pH 7.4 for 20 seconds. After irradiation the mucosae were fixed for 15 min at 4°C in the same solution.

For TEM specimens, the fixed mucosa were post-fixed in 1.0 % osmium solution in 0.16 M cacodylate buffer (pH 7.4) for 1 hr. ethanol dehydrated, and embedded in Epon. Thin sections were stained uranyl acetate and lead hydroxide and examined in a JEOL 1200 TEM.

For immunogold electron microscopy, specimens, small pieces of fundic mucosa were fixed with 2.0 % formaldehyde and 0.4 % glutaraldehyde in PBS for 1 hr at 4°C.

The fixed tissues were embedded in Lowicryl K4M. Thin sections were treated for 30 min with a monoclonal antibody: 1H9 specific for α-$H^+K^+$ATPase. The sections were then labeled with 10 nm colloidal gold for 1 hr. Sections were stained with uranyl acetate and lead citrate, and observed in a TEM.

For SEM specimens, the aldehyde-osmium-DMSO-osmium method (21) was used. Fixed pieces of mucosa were immersed in 50 % dimethy sulfoxide, frozen in cooled liquid propane and freeze-cracked. Tissues were thawed and post-fixed with 1.0 % $OsO_4$ in 0.16 M cacodylate buffer (pH 7.4) for 1 hr then transferred to 0.1 % osmium solution in 67 mM cacodylate buffer. The tissues were macerated at 20°C for 96 hr to remove the cytoplasmic matrix. Then tissue immersed for 2 hr in 1.0 % tannic solution in the same buffer and post-fixed for 1 hr in 1.0 % $OsO_4$. The specimens were dehydrated in ethanol and critical point dried. The dried specimens were impregnated with osmium by osmium-hydrazine method (7). Observations were made with a Hitachi S-100 ultra-high resolution SEM.

## RESULTS

***TEM observations on the resting parietal cell***

Thin sections of resting rat parietal cells had cytoplasm with its major components, including closely packed tubular membranes, which were randomly packed and thus sectioned in various planes (Fig. 1). The IC was lined with characteristic microvilli often occupying much of the lumen.

***Immunocytochemical staining of $H^+K^+$ATPase***

Gold particles representing the α-subunit of $H^+K^+$ATPase were seen on the membranes of IC and cytoplasm tubular membranes (Fig. 2).

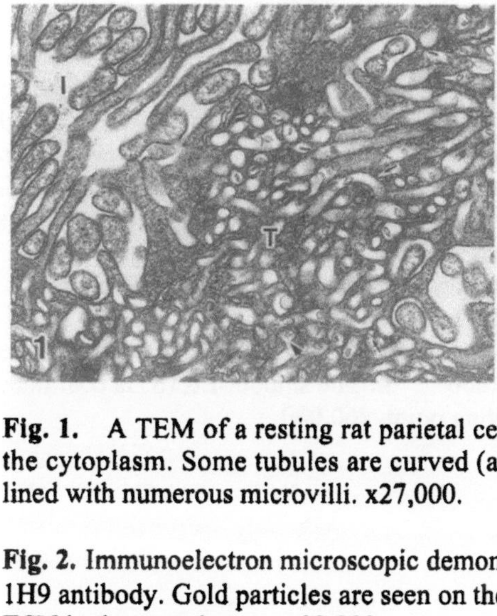


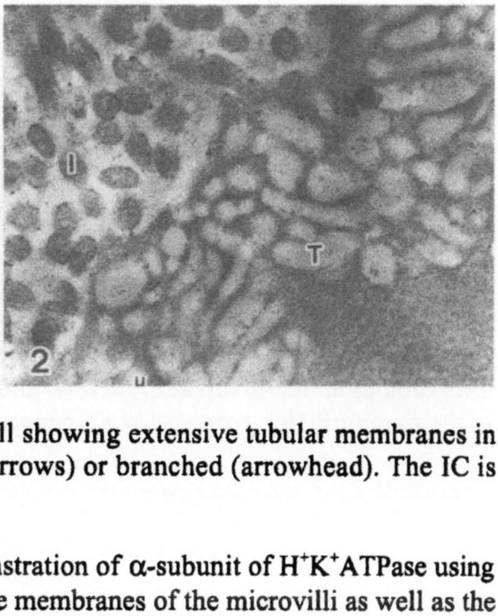


**Fig. 1.** A TEM of a resting rat parietal cell showing extensive tubular membranes in the cytoplasm. Some tubules are curved (arrows) or branched (arrowhead). The IC is lined with numerous microvilli. x27,000.

**Fig. 2.** Immunoelectron microscopic demonstration of α-subunit of $H^+K^+$ATPase using 1H9 antibody. Gold particles are seen on the membranes of the microvilli as well as the TCN in the cytoplasm. x32,000

***SEM observations the resting parietal cells***

In the parietal cell treated with A-ODO method, the cytoplasmic matrix and cytoskeletons were removed and the cytoplasmic membranes could be observed clearly (Figs. 3-5). The low magnification SEM of a fractured parietal cell showed the IC were lined with numerous microvilli (Fig. 3a). In addition, the TV membranes appeared as a reticulated meshwork of tubules. The high magnification SEM revealed the 30-60 nm tubules connected with cisternae of 100-600 nm wide forming the three-dimensional network, the "tubulocisternal network" (TCN) (Fig. 3). Near the center of the some small cisternae, there was a small 10 nm fenestration (Figs. 3b,c). The TCN occasionally connected with the rough ER (Fig. 3d,e) and the Golgi apparatus (Fig. 3e). Polyribosomes were not found on the surface of the tubules or the IC. The Golgi apparatus was recognized by its closely stacked smooth cisternae (Fig. 3e).

When the IC was cross-fractured and the lumen was exposed, some irregularly arrayed microvilli were visible (Fig. 5a,b). On the cytoplasmic surface of the IC, round openings representing the macerated bases of the microvilli about 100 nm in diameter were characteristic of IC (Fig. 5c,d). In favorable areas where the IC could be visualized in association with the TCN, continuities or connections between these two compartments were observed (Fig. 5).

---

**Fig. 3.** SEM images of resting parietal cells. **a.** Lower magnification. The IC is lined with numerous microvilli and numerous mitochondria are accumulated in the cytoplasm. Slender tubules and small cisternae form three-dimensional TCN in the cytoplasm. x7,000. **b.** Slender 30-60 nm tubules in diameter branch and anastomose or join to cisternae forming TCN. A tiny fenestration (arrow) is seen at the center of small cisternae. X60,000. **c.** Higher magnification of TCN. A tiny round fenestration is seen at the center of small cisterna. The TCN connect to the large cisterna (L). x120,000 **d.** The TCN connects to the large cisterna with a polyribosome (P) by slender tubules (arrows). x85,000. **e.** The TCN connects to the Golgi apparatus (G) at the point of arrows. Several layers of Golgi stacks in close array are seen. P: polysome x80,000

**Fig. 4.** The smooth-ER of the rat liver cell. **a.** Cisternae about 100 nm are connected with tubules about 40 nm at the arrow. x80,000 **b.** The smooth-ER (S) is continuos to the rough ER (R). Arrow shows a connecting point. x60,000

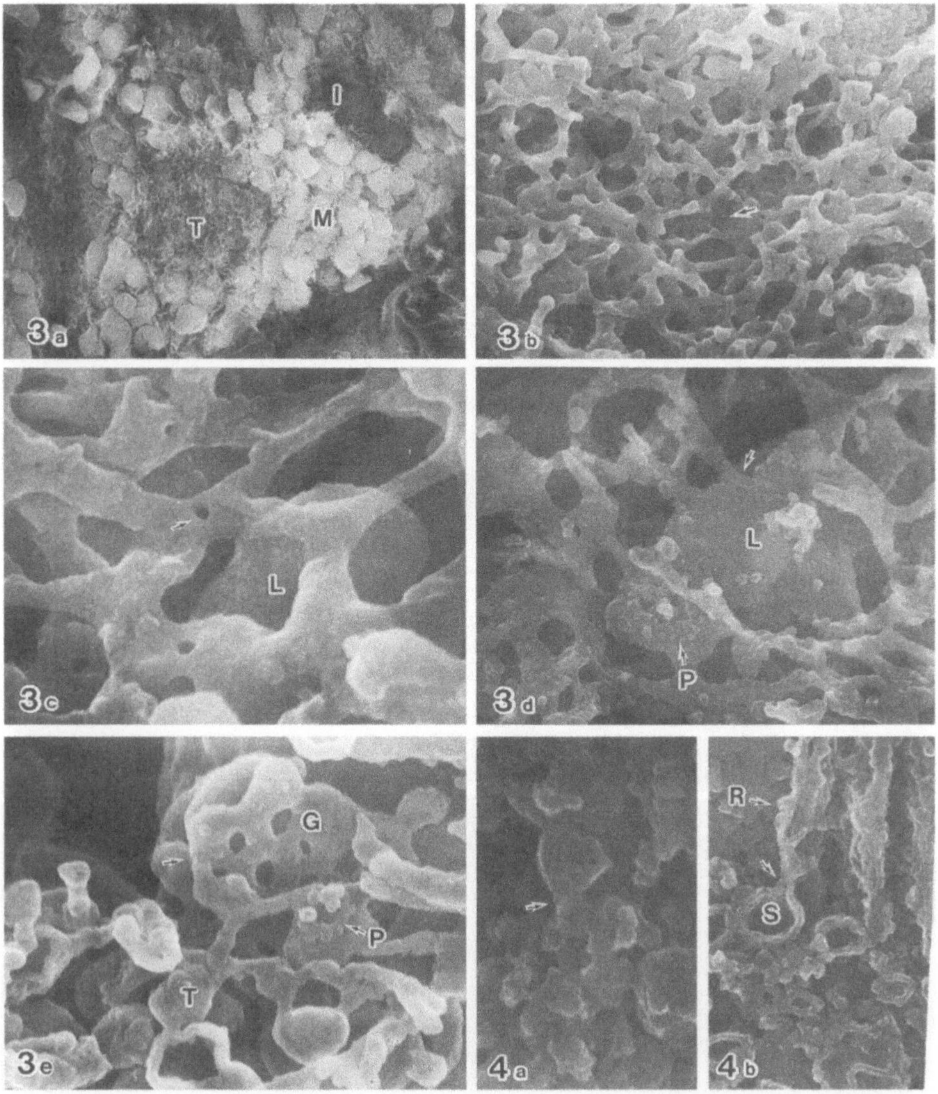

3a
T
M
I
3b
3c
L
3d
L
P
3e
G
P
T
4a
4b
R
S

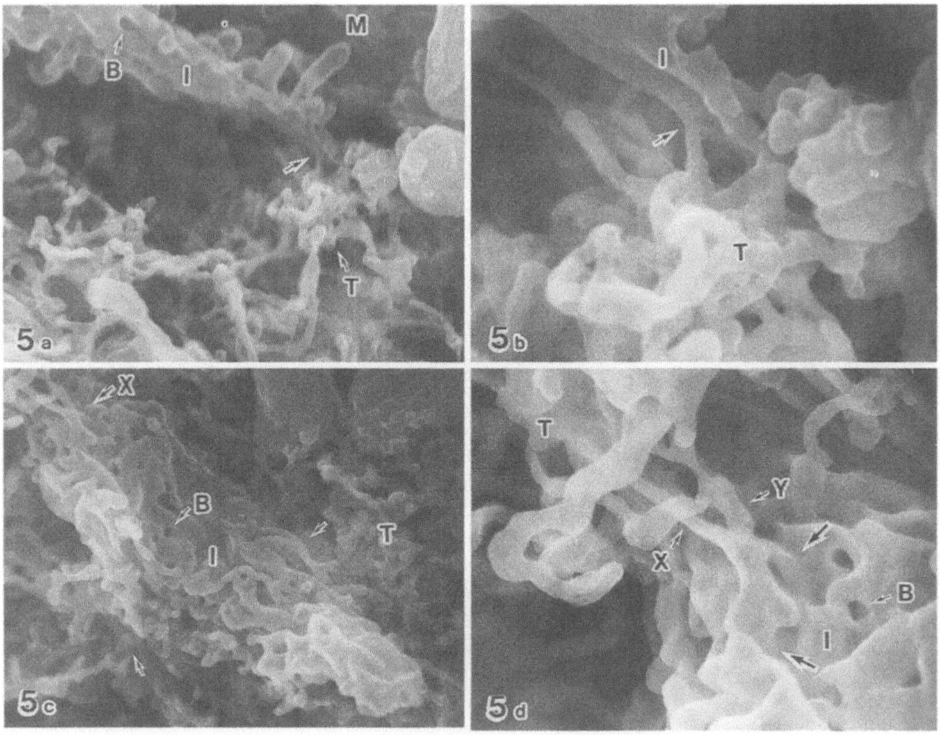


**Fig. 5.** SEM images of resting parietal cells. **a.** A fractured IC lined with microvilli. The round opening (B) is to the empty macerated base of a microvillus. The TCN is shown connected to the IC at the arrow. x43,000. **b.** A higher magnification of an area indicated with arrow in Fig. 5a. The TCN connection to the IC is shown at the arrow. x106,000. **c.** Cytoplasmic side view of the IC. Round holes (B) indicate the macerated bases of microvilli. TCN connect the IC at the several points indicated with arrows. x33,000 **d.** Higher magnification of the area indicated with arrow X in Fig. 5c. A straight tubule (X) and undulating tubule (Y) are connected to the IC at the arrows. x74,000.

***Parietal cell s after tetragastrin stimulation***

After gastrin stimulation, the TCN decreased while the IC increased. However, the basic structure of TCN is similar as that of the resting parietal cells. In addition, the connection part of TCN to the IC remained (Fig. 6a). The parietal cell 15 minutes after gastrin stimulation, a cisterna with a tiny fenestration connected to the IC (Fig. 6b) and similar size fenestration was seen on the surface of the IC suggesting that a small cisterna was just transformed into the IC membrane. On the cytoplasmic side, the holes in the IC at the base of the microvilli were smaller and more widely spaced near the connection site between the TCN and the IC than those more distant from the connection (Fig. 6c). On the luminal side, the microvilli were less numerous and smaller in size around the connection sites and gradually increased their number and size apart from them (Fig. 6d).

***Frog oxyntic cells***

SEM pictures of the frog oxyntic cell revealed that the luminal membrane formed a rather flat surface with a few microvilli (Fig. 7a). In the cytoplasm, round cisterna are connected with slender tubules forming the TCN. The continuities between the luminal membrane and TCN were more frequently seen in the frog than in the rat. The TEM picture also showed the TCN was continuos with the luminal membrane (Fig. 7b).

## DISCUSSION

In the gastric parietal cell there has been a long-standing controversy on the form and configuration of the cytoplasmic or the tubulovesicular membrane system. Initially parietal cells cytoplasm was described as being replete with vacuoles and vesicles and numerous large mitochondria along the IC. Ito (3) showed that the addition of 0.25M sucrose to the osmium fixative resulted in the preservation of tubules and cisternae in place of vesicles and vacuoles reported by others. In spite of these observations reports of vesicles and vacuoles persisted. Vial et al. (1985) reported that sucrose in concentrations higher than 0.2M in the fixative preserved the tubular configuration of this membrane system consistently as tightly packed tubules.

The present study was carried out after Sugai et al. (20) reported that the osmolality of the buffer was critical for parietal cell tubulovesicles fixation. In our previous SEM studies (8-11), we used a 0.067M cacodylate buffer, which was 135 mOsm, and the cytoplasmic membranes were predominantly vesicular with some tubular elements. In TEM preparations fixed with a fixative containing a buffer of 334 mOsm, the cytoplasm of the parietal cell had closely packed tubular membranes (Fig. 1). Immunocytochemical staining of $H^+K^+$ATPase showed that gold particles were seen on the membranes of IC as well as of TCN membranes (Fig. 2). In our macerated SEM preparations, three-dimensional networks of tubules and cisternae were seen in the parietal cell cytoplasm. The absence of vesicles or vacuoles and the occurrence of interconnected cisternae it seems reasonable to designate these membranes as "tubulocisternal". However, this may be due to the maceration procedure, which removed isolated, unattached vesicles as well as many tubules. If all the tubules as revealed by TEM were retained, the network would not be visible. Therefore, it is probable that some vesicles and isolated or shorter tubules were present. However, the great preponderance of tubules in favorably preserved parietal cells indicated that this might more truly represent the form of the tubulocisternal membrane system in the living cell (13,14).

In parietal cells prepared by a modified rapid-freeze/freeze-substitution method, Pettitt et al. (15) showed the TV membranes comprised of helical coils of tubules. Using a similar technique Sugai et al. (19) reported predominantly tubular arrays with some curved, branched, or circular arrays of tubules. Helical coils were not observed in our SEM specimens

The rough ER has been found as occasional cisternae in the parietal cells (4, 16). In the present study, we have observed the TCN connected to the rough ER with a few polyribosomes. In other cell types, the continuity between the rough ER and smooth ER has been observed by SEM (22). In addition, parietal cell TCN resembles the SEM images of the SER of the ductus deferens epithelial cells (22) and those of rat liver cells (Fig. 4). Early studies (3, 17) have described the tubular membrane system in the oxyntic cell cytoplasm as the agranular (smooth). Based on these and our current findings, the TCN may be considered to be a specialized type SER in the hydrochloric acid secreting parietal cell.

A considerable number of studies (1, 5) have shown that there is a rapid transformation of parietal cell morphology when stimulated to secrete acid. The abundant TV membranes appear to be transferred or translocated to form the IC membrane. Quantitative studies have indicated that the reduction in cytoplasmic membrane is equal to the increase in microvillar membrane (2). There is further evidence that these two membrane compartments share similar properties. Immunocytochemical localization of $H^+/K^+$-ATPase (6, 18) indicated that this enzyme is located on both the cytoplasmic and plasma membranes.

---

**Fig. 6.** Parietal cells after gastrin stimulation. **a.** The TCN connects to the IC at the arrow. x110,000 **b.** The TCN connection to the IC (large arrows). A tiny fenestration (small arrow) is seen on the surface of the IC near the connection. Note similar size fenestration (small arrow) is seen at the center of the small cisterna (C). B: macerated base of microvillus. x126,000. **c.** The round holes corresponding to the base of the microvilli are less numerous and smaller in size at the right side of the figure (small *), where the TCN is connected (arrow) than at the left portion (large *). X20,000 **d.** On the lower side of the picture (small *), the microvilli are scarce and small where the TCN are connected (small arrows), gradually increasing in size (large arrows) and in number toward farther area at the upper side (large *). x30,000

**Fig. 7.** Frog oxyntic cells. **a.** A SEM micrograph. Luminal membrane is tightly ruffled with a few microvilli. Connections between the luminal membrane and TCN are seen (arrows). x38,000 **b.** A TEM micrograph. The TCN (*) is continuous with the luminal membrane (L) at the point of arrow. x50,000

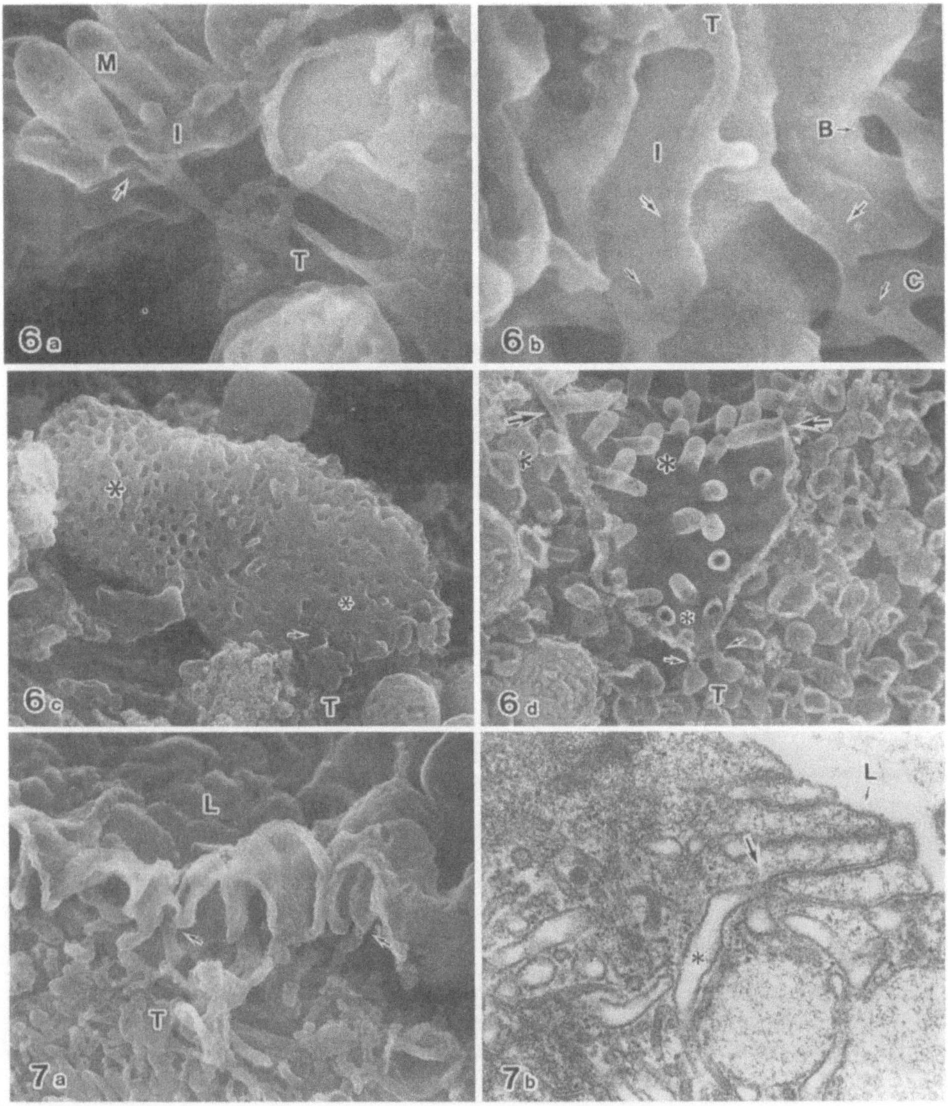
M
I
T
6a
T
I
B
C
6b
T
6c
T
6d
L
T
7a
L
7b

Heretofore, there has been no convincing demonstration of the continuity between the TV membrane and the IC membrane in resting gastric parietal cells. Our demonstration of connections between the TCN and the IC in rat parietal cells (8, 13, 14) as well as between the TCN and the luminal membrane in frog oxyntic cells (12) strongly supports the hypothesis of direct transfer of these membranes from the cytoplasmic compartment to the luminal surface. Moreover, after gastrin stimulation, the microvilli on the luminal surface of the IC at the sites of connections between the TCN and the IC were relatively scare and small, gradually increasing in number and size as the distance from them. On the cytoplasmic side the round holes of the IC indicating the base of the microvilli were also fewer and smaller near the connection site. These results provide the assumption that the TCN membrane is transformed into the IC membrane and the excess membrane is stored as the membrane of the microvilli.

The author's interpretation of the three dimensional structure of the TCN and the IC of the rat parietal cell is schematically represented in Figure 8.

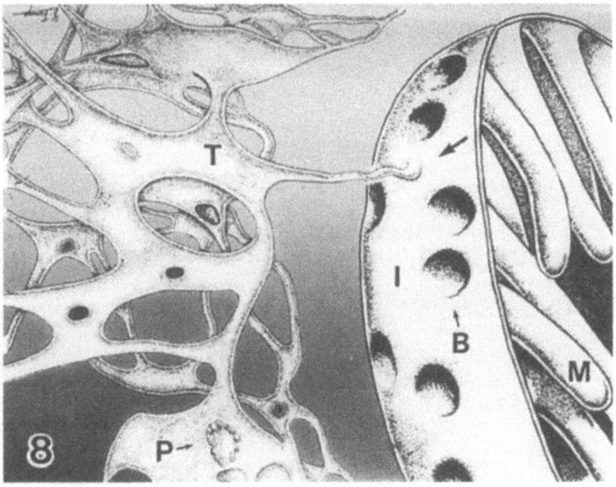


**Fig. 8.** A schematic drawing of a typical rat gastric parietal cell cytoplasmic membrane system based on our SEM observations. B: basal part of the microvillus, I: IC, M: microvilli, P: polyribosome, T: TCN, Arrow: a connection site of the TCN to the IC.

*Figures 2-5, 6a and 8 were reproduced from Anat. Rec. 258:15-24, 2000, Figure 6c,d from Anat. Rec. 237:208-219, 1993 and Figure 7 from Anat. Rec. 245:559-567, 1996 with permissions. Abbreviations used in figures as follows: B: macerated base of microvillus, I: intracellular canaliculus, M: mitochondria, T: tubulocisternal network.*

## REFERENCES


1. **Forte JG, Black JA, Forte TM, Machen TE, and Wolosin JM.** Ultrastructural changes related to functional activity in gastric oxyntic cells. *Am J Physiol* 241: G349-G358, 1981.
2. **Helander HF and Hirschowitz BI.** Quantitative ultrastructural studies on inhibited and on partly stimulated gastric parietal cells. *Gastroenterology* 67: 447-452, 1974.
3. **Ito S.** The endoplasmic reticulum of gastric parietal cells. *J Biophys Biochem Cytol* 11: 333-347, 1961.
4. **Ito S.** Functional gastric morphology. In: Johnson LR, editor. *Physiology of the Gastrointestinal Tract.* 2nd ed. New York: Raven Press. 1987, p 817-851.
5. **Ito S and Schofield GC.** Studies on the depletion and accumulation of microvilli and changes in the tubulovesicular compartment of mouse parietal cells in relation to gastric acid secretion. *J Cell Biol* 63: 364-382, 1974.
6. **Kobasyashi A, Araki K, Ando T, and Ogata T.** An immuno- and enzyme cytochemical study of the $H^+$-$K^+$ATPase in human parietal cells after administration of tetragastrin and omeplazole. *Arch Histol Cytol* 60: 205-214, 1997.
7. **Kubotsu A and Ueda M.** A new conductive treatment of the specimen for scanning electron microscopy. *J Electron Microsc* 29: 45-53, 1980.
8. **Ogata T.** Electron microscopic studies of the normal and pathological stomach. Part I. Ultra-high resolution SEM study of the rat parietal cell. *J Clin Electron Microsc* 25: 323-324, 1992.
9. **Ogata T.** Gastric oxyntic cell structure as related to secretory activity. A review. *Histol Histopathol* 12: 739-754, 1997.
10. **Ogata T, Araki K, and Yamasaki Y.** Ultra-high-resolution scanning electron microscopic studies on the membrane system of the rat parietal cells after tetragastrin stimulation. *It J Anat Embryol* 100: 393-401, 1995.
11. **Ogata T and Yamasaki Y.** Ultra-high-resolution scanning electron microscopic studies on the membrane system of the parietal cells of the rat in the resting state and shortly after stimulation. *Anat Rec* 237: 208-219, 1993.
12. **Ogata T and Yamasaki Y.** Ultra-high-resolution scanning electron microscopy of the continuity of cytoplasmic and luminal membranes in frog oxyntic cells. *Anat Rec* 245: 559-567, 1996.
13. **Ogata T and Yamasaki Y.** Scanning EM of resting gastric parietal cells reveals a network of cytoplasmic tubules and cisternae connected to the intracellular canaliculus. *Anat Rec* 258:15-24, 2000a.
14. **Ogata T and Yamasaki Y.** Morphological studies on the translocation of tubulovesicular system toward the intracellular canaliculus during stimulation of the gastric parietal cell. *Micro Res Tech* 48:282-292, 2000b.
15. **Pettitt JM, Humphris DC, Barrett SP, Toh B-H, van Driel IR, and Gleeson PA.** Fast freeze-fixation/freeze-substitution reveals the secretory membranes of the gastric parietal cell as a network of helically coiled tubule. *J Cell Sci* 108: 1127-1141, 1995.
16. **Rubin W, Ross LL, Sleisenger MH, and Jeffries GH.** The normal human gastric epithelia. A fine structural study. *Lab Invest* 19: 598-626, 1968.

17. **Sedar AW.** 1961. Electron microscopy of the oxyntic cell in the gastric glands of the bullfrog (*Rana catesbiana*). II. The acid-secreting gastric mucosa. *J Biophys Biochem Cytol* 10: 47-57.
18. **Smolka A, Helander HF, and Sachs G.** Monoclonal antibodies against gastric $H^+/K^+$ ATPase. *Am J Physiol* 245: G589-G596, 1983.
19. **Sugai N, Ito S, Ichikawa A, and Ichikawa M.** The fine structure of the tubulovesicular system in mouse gastric parietal cell processed by cryofixation method. *J Electron Microsc* 34: 113-122, 1985.
20. **Sugai N, Okamura H, Tsunoda R, Terashima H, and Suzuki K.** Morphological changes of tubulovesicles of rat parietal cells according to the different osmolalities of fixatives. (in Japanese). *Acta Anat Nipponica* 70: S79, 1995.
21. **Tanaka K and Mitsushima A.** A preparation method for observing intracellular structures by scanning electron microscopy. *J Microsc* 133: 213-222, 1984.
22. **Tanaka K and Naguro T.** High resolution scanning electron microscopy of cell organelles by a new specimen preparation method. *Biomed Res* 2 (Suppl): 63-70, 1981.

Chapter 34

# 3-Dimensional Modeling of the Gastric Parietal Cell

JOHN G. FORTE[1], JOSEPH G. DUMAN[1], NIMESH J. PATHAK[1], AND KENT L. MCDONALD[2]

[1]*Department of Molecular and Cell Biology and* [2]*Electron Microscope Lab, University of California, Berkeley, CA 94720*

## ABSTRACT

This article reviews recent studies to evaluate the morphology of intracellular membrane systems in the resting parietal cell using high quality fixation/preservation, serial sectioning and modeling techniques. The findings are as follows: (i) The apical canaliculus is a microvilli-decorated, branching membrane network that extends into and throughout the parietal cell. This agrees with many earlier studies. (ii) The abundant parietal cell mitochondria form an extensive reticular network throughout the cytoplasm. This interconnectivity has not been previously demonstrated in the parietal cell and may have functional significance for these metabolically active secretory cells. (iii) The compartment of H,K-ATPase-rich membranes is chiefly comprised of small stacks of cisternae, thus the tubulocisternae designation given by Ogata & Yamasaki (Anat. Rec. 258:15, 2000) seems appropriate. However, contrary to their conclusions, in the resting cell there are no continuities between the apical canaliculus and the tubulocisternae or among tubulocisternae. These data support the recruitment-recycling model of parietal cell stimulation.

## BACKGROUND

The secretion of gastric hydrochloric acid secretion involves enormous cellular metabolic demands, highly specialized membrane systems, and elaborate membrane rearrangements that accompany the activation process. There is general agreement on two basic sets of facts: 1) the H, K-ATPase, discovered in 1973, is the primary gastric proton pump (10, 6); and 2) receptor-activated secretion leads to morphological transformations whereby the apical surface of the acid-secreting cell becomes greatly elaborated in the form of long microvillar projections (12, 14, 15). On the other hand, there has been much controversy regarding the mechanism by which the apical surface transformation occurs. The major disagreement concerns whether H, K-ATPase-rich cytoplasmic membranes reside in resting cells as distinct tubular or vesicular structures (called tubulovesicles) or whether they are continuous with the apical plasma membrane. These alternatives lead to very different models to account for stimulus dependent morphological transformation. The membrane recycling hypothesis proposes that expansion of apical surface occurs as a result of recruitment of H, K-ATPase-rich membranes to the apical plasma membrane (7, 8). The alternate osmotic swelling theory proposes that the H, K-ATPase-rich membranes are really in physical contiguity with the apical plasma membrane (2), thus no fusion is required, rather an expansion of the highly involuted membrane system using osmotic work or cytoskeletal forces.

Acid-secreting cells are easily distinguished from other gastric glandular cells by numerous large mitochondria and by the profusion of vesicular and tubular membrane profiles that contain H, K-ATPase. In the mammal, these so-called parietal cells have an additional degree of morphological specialization in the form of secretory canaliculi, an elaborate system of invaginations or canals extending into the cell from the apical surface. In non-secreting parietal cells, the apical surface contains many short (0.2-0.5 µm) microvilli, each with a system of underlying microfilaments. After stimulation, the apical surface becomes greatly expanded as the microvilli become elongated (8, 12, 15). A measured 5- to 10-fold increase in apical surface area is quantitatively accounted for by the decrease in area of H, K-ATPase-rich tubulovesicles (12) The membrane recycling hypothesis proposes that tubulovesicular membranes are recruited to the apical membrane by some sort of trafficking/fusion machinery, thus relocating the H, K-ATPase to the membrane where active secretion occurs. Functional studies show that the apical surface of the stimulated parietal cell also has channels for $K^+$ and $Cl^-$ flux, which are needed along with $H^+/K^+$ exchange driven by the H, K-ATPase to produce net HCl secretion (9, 28).

The osmotic expansion hypothesis, historically less favored, has been rejuvenated by several recent morphological approaches. Pettitt et al. (23, 24) have contested earlier electron microscopic evidence, noting that glutaraldehyde fixatives can fragment intracellular organelles. From their own results with rapid-freeze fixation they suggested that the tubulovesicular compartment is arranged as a series of tightly coiled membranes. They further proposed that small sized openings provided "connections" to the surface as well as among the membrane coils. An alternative interpretation of tubulovesicular morphology based on high-resolution scanning electron microscopy was offered by Ogata and his colleagues (19, 20, 21). They proposed that the compartment is actually composed of multiple flattened membrane sacs, or cisternae, and that these cisternae intercommunicate with each other by means of very narrow tubular connections that are present even in the resting cell. Thus, it is clear that important structural details must be elucidated for the function of gastric acid secreting cells to be fully appreciated.

In this paper we review recent studies from our own laboratory using high quality fixation/preservation, and serial sectioning and modeling techniques to re-evaluate the morphology of resting parietal cells. Much of the data have been presented in two recent publications that should be consulted for greater experimental detail (5, 22). Here we show reconstructed 3-dimensional models of the parietal cell interior from which the morphology of the tubulovesicular compartment can be appreciated in the context of other membranes within the cell. The experimental data fully support the claim that tubulovesicles are distinct from the canalicular membrane, and they provide evidence consistent with the tubulocisternal morphology promoted by Ogata. However, there is no evidence of permanent continuities between the individual elements of the tubulocisternal system or with the apical canalicular surface of the resting parietal cell.

## MATERIALS AND METHODS

*Sample Preparation.* Rabbit gastric glands were isolated as described in previous studies (25). The glands were incubated at 37°C for 20 min with 100 μM cimetidine to assure that they were in a resting, non-secreting, condition. The glands were gently sedimented and transferred to Type A high-pressure freezing planchettes, and frozen in a Bal Tec HPM 010 High Pressure Freezer. Samples were post-fixed by freeze-substitution on dry ice in acetone using 2% osmium tetroxide plus 0.1% uranyl acetate in acetone for optimal morphology in the modeling studies, or 0.2% glutaraldehyde plus 0.1% uranyl acetate in acetone for immunolocalization studies. After 3 days, the samples were warmed to room temperature, rinsed with pure acetone, and embedded in Epon-Araldite resin for morphological examination or LR White resin for immunolocalization (18).

*Image collection.* Ultra-thin (60-70 nm thick) serial sections were post-stained with lead citrate and uranyl acetate. Sections were visualized in a JEOL 100CX transmission electron microscope operating at 80 kV. Micrographs were taken for the same parietal cell in a series of thin sections and the images were digitized.

*Model Construction.* Models were constructed on Silicon Graphics computers running MIDAS and IMOD software (17). Because the electron beam mildly distorts ultra-thin sections, the images were aligned with respect to specific structures of interest using MIDAS. IMOD was used to stack the aligned images and to draw contours (outlines) on specific cellular structures. Contours of the same structure from different serial sections were arranged into objects using IMOD. The IMODmesh feature of IMOD was used to join the contours of each object forming a 3-D models of structures of interest. Movies of images similar to those shown in this study can be viewed on the web at http://jcs.biologists.org/supplemental/.

## RESULTS

The technique of high pressure, rapid freeze fixation offers significant advantages over conventional methods. The contents of the sample are immobilized in milliseconds, leading to

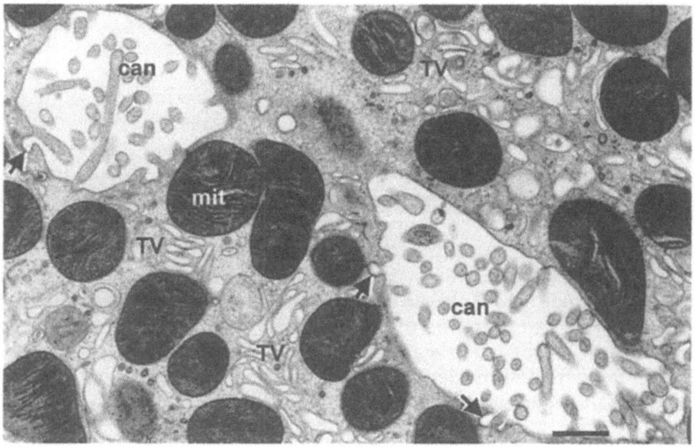


**Figure 1.** Electron micrograph of parietal cell prepared by high-pressure freezing. Cell interior is shown featuring profiles of two intracelluar canaliculi (can) with many microvilli projecting into the canalicular lumen. Prominent endocytotic figures are apparent (arrows). Mitochondria (mit) abound in the cell, as do the abundant cross-sectional profiles of the tubulovesicular membrane system (TV). Bar marker is 0.5 μm. Reprinted from (5) with permission.

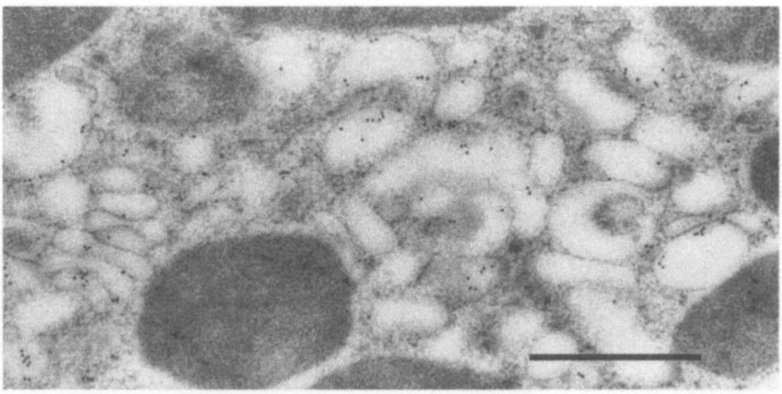

**Figure 2.** Immunogold labeling of H,K-ATPase in the parietal cell. Ultrathin sections were first incubated with the 2G11 antibody against the β-subunit of H,K-ATPase and then with secondary antibody labeled with colloidal gold. H,K-ATPase is clearly localized to the tubular and vesicular membrane structures throughout the cytoplasm. Bar marker is 0.5 μm.

a much higher level of tissue integrity than standard fixation, which may take up to an hour. Also, since high pressure freezing allows the post-fixation to occur at extremely cold temperatures, cell components are not lost or rearranged during the course of sample preparation.

Figure **1** is a micrograph of a well-preserved parietal cell, post-fixed with osmium for the best structural detail. In this section there are two cross-sectional profiles of intracellular canaliculi complete with many short microvillli projected into the lumen. The canalicular membrane also displays a many coated vesicles marking it as an endocytically active site, even in the resting parietal cell, as described by Okamoto et al. (22). Also note, critical to the present study, the extensive network of cytoplasmic membranes collectively known as tubulovesicles. These profiles of tubular and vesicular structures predominate the cytoplasmic space between the prominent mitochondria.

Figure **2** shows parietal cells prepared by high pressure freezing using the procedure to preserve immunogenicity and probed with an antibody against the β-subunit of H,K-ATPase followed by immunogold staining (see methods). Colloidal gold particles are heavily represented over the tubular and vesicular membrane profiles that abound in the cytoplasm, but virtually absent over mitochondria and nuclei. Thus, as demonstrated in previous studies the tubulovesicles are H,K-ATPase rich-membranes.

To appreciate the organization and arrangements of parietal cell organelles three-dimensional models were reconstructed from serial ultrathin sections. Ribbons of sections were cut and electron micrographs taken from the same regions on successive sections. Figure **3** provides a schematic representation of the image stack. Outlines (contours) were drawn on selected individual structures, then assembled into objects either as a stack of contours or meshed to create the finished model. Using this technique, we modeled three of the major membrane systems in the parietal cell: intracellular canaliculi, tubulovesicles and mitochondria.

Figure **4** shows a stack of canalicular contours (Fig. 4A) and the same meshed canalicular structures (Fig. 4B) extending through 29 sections accounting for a depth of ~2 μm. Because of morphological complexity it was difficult to trace the outline of the canalicular surface including each microvillar extension, so our initial models included only the surface that outlined the canaliculus at the base of the microvilli. Segments of canaliculi can be visualized traversing the cell, and there is evidence of a branching network that can be traced throughout the parietal cell if one includes sufficient numbers of sections. Thus the modeling system reinforces at the EM level previous data obtained by conventional light microscopy and as evidenced by the early camera lucida renderings of Golgi (11).

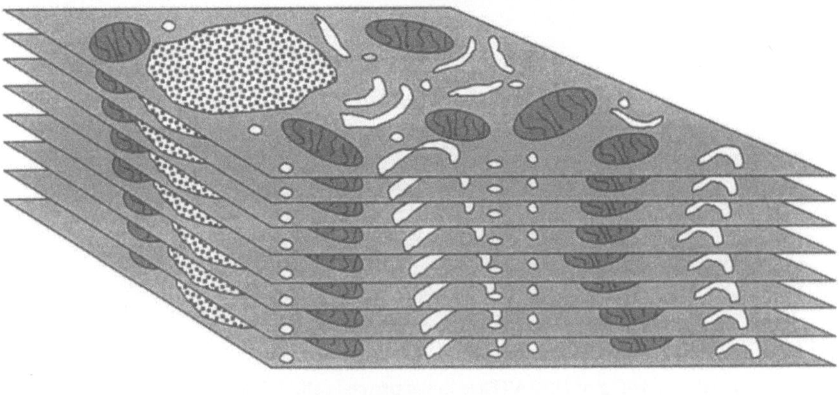

**Figure 3.** Schematic representation of a stack of ultrathin sections. Digitally rendered images of individual sectionare aligned and contours drawn around selected organelles udentified in successive members of the stack. These data are reconstructed into the 3-dimensional models that follow.

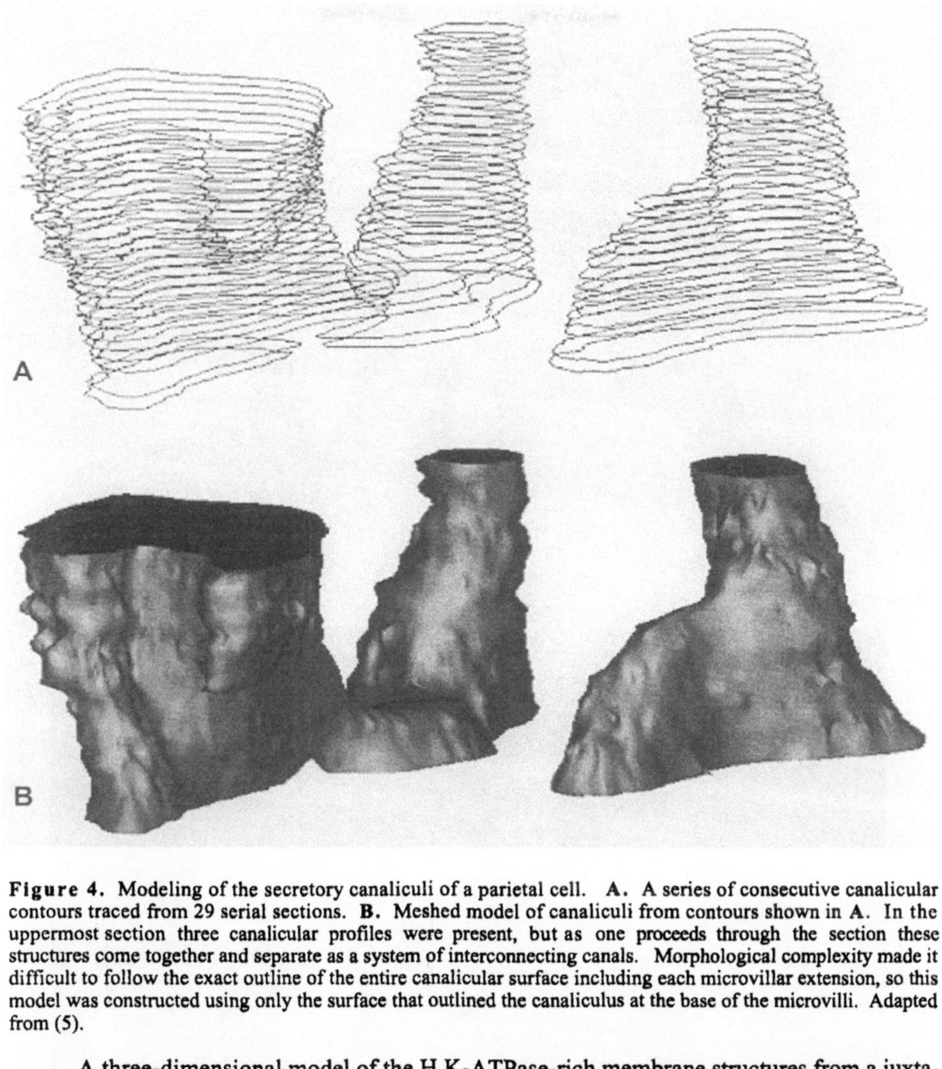


**Figure 4.** Modeling of the secretory canaliculi of a parietal cell. **A.** A series of consecutive canalicular contours traced from 29 serial sections. **B.** Meshed model of canaliculi from contours shown in **A**. In the uppermost section three canalicular profiles were present, but as one proceeds through the section these structures come together and separate as a system of interconnecting canals. Morphological complexity made it difficult to follow the exact outline of the entire canalicular surface including each microvillar extension, so this model was constructed using only the surface that outlined the canaliculus at the base of the microvilli. Adapted from (5).

A three-dimensional model of the H,K-ATPase-rich membrane structures from a juxta-canalicular region is shown in Figure **5a**. It is clear from this reconstruction that the H,K-ATPase-rich membranes exist primarily in the form of small cisternal stacks, each stack including from three to six cisternae. While there are often small tubular or vesicular structures near or within these stacks, the cisternal form is the most prominent. Review of many sections revealed that each tubulocisternal element appeared to be a discrete membranous unit with no connections between or within the stack; furthermore, there was no preferred orientation of the stacked membranes. Figure **5b** shows a more expansive region of tubulocisternal stacks surrounding an apical canalicular membrane. While the membrane stacks are apposed quite closely to one another and frequently are subadjacent to the canaliculi, it is clear that there are

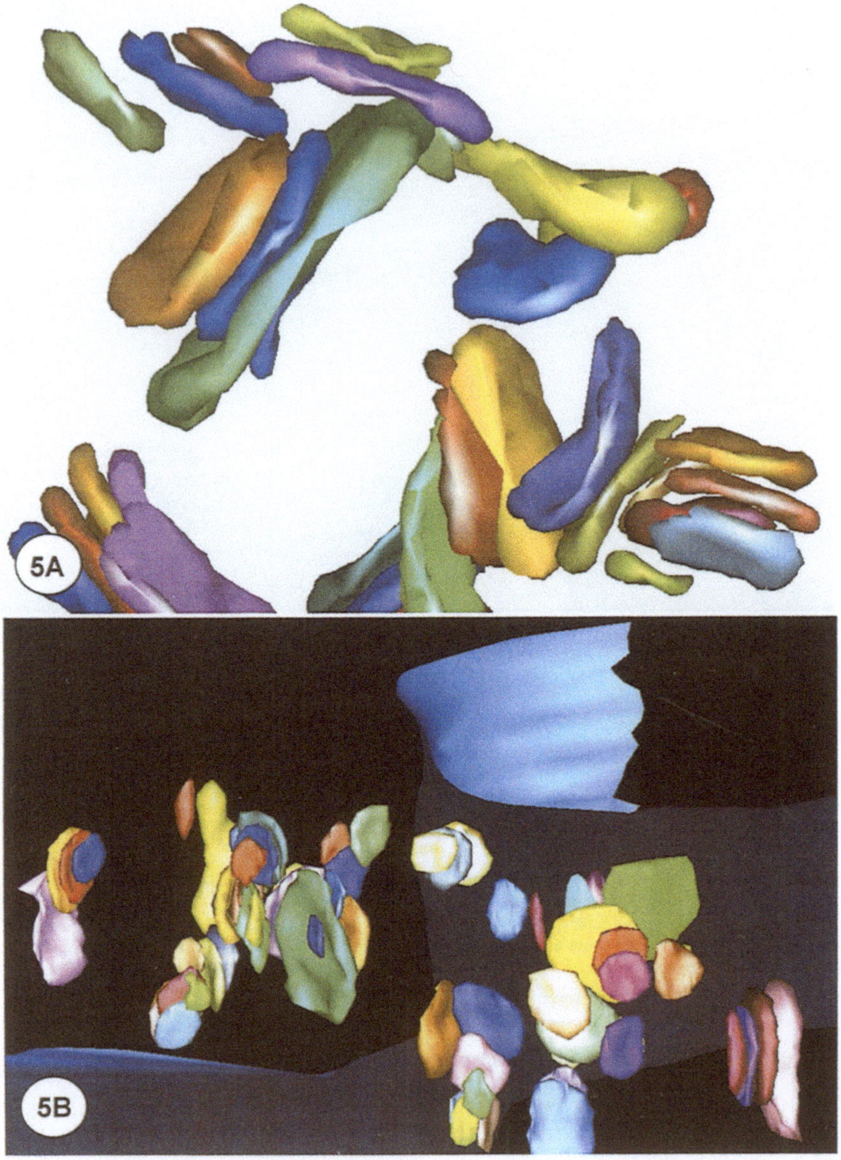


**Figure 5.** Modeling of H,K-ATPase-rich membrane system in parietal cells. **A.** Each membrane unit was defined as an object and assigned a separate color. 3-D models were reconstructed from contours drawn around each membrane unit. While some membranes appear as tubules, the majority are flattened cisternal discs, often arranged in small stacks. **B.** Reconstruction of tubulocisternae from a larger area surrounding a canalicular lumen (large blue structure). Adapted from (5).

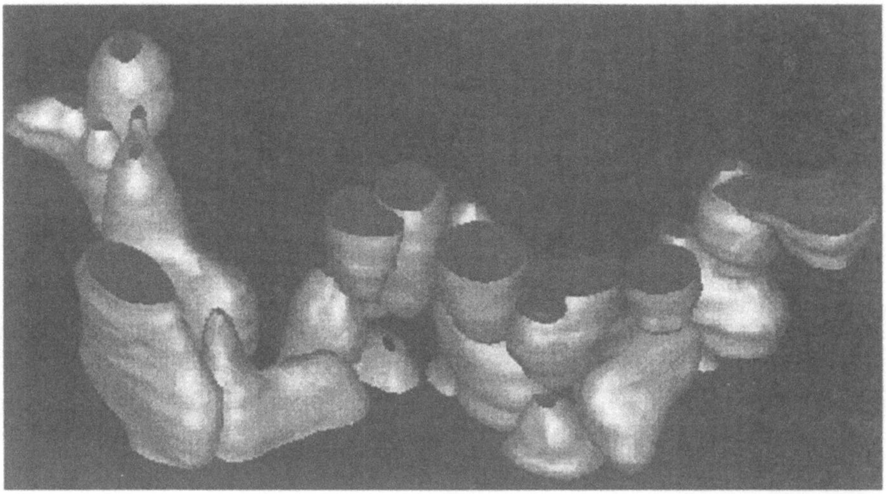

**Figure 6.** Meshed model of parietal cell mitochondria reconstructed from a parietal cell cytoplasmic space of about 7μm x 4μm x 2 μm. Note that virtually every mitochondrion appears to be interconnected with an adjacent mitochondrion, either by major branching or via more diminished tight annular interconnections. Adapted from (5).

no connections among tubulocisternae or between tubulocisternae and the canaliculus.

Another characteristic feature of parietal cells is the abundance of mitochondria as can be seen from Figures 1 and 2. After excluding nuclei and canalicular compartments, we found that mitochondria occupy 40.3 ±1.8% (sem, n = 5) of resting rabbit parietal cell volume. These data are consistent with several previous morphometric studies indicating that mitochondria occupy between 20% and 42% of the parietal cell volume (26, 32, 3). However, there are some new features to be observed from the three dimensional modeling of mitochondrial profiles as shown in Figure **6**. The model, reconstructed from a region of ~50 $\mu m^3$ of parietal cell cytoplasm, clearly demonstrates that the mitochondria do not resolve into discrete units. Rather the mitochondria appear to form an extensive reticular network coursing throughout the cytoplasm. The tight radii of some of the interconnections suggest the possibility of transient connections that may open and close in time.

## DISCUSSION

The system of extensive intracellular parietal cell membranes has been of great interest, especially with respect to its possible role in the HCl secretory process. Various investigators have called these H,K-ATPase-rich membranes vesicles, tubules, bulbotubules, tubulovesicles, coiled tubules and, more recently, tubulocisternae. For simplicity, most authors tend to refer to the system generically as "tubulovesicles", though there still remains controversy on the specific morphological form. Often the descriptive differences are related to the methods of tissue preservation and in some instances to the animal species examined. The discovery of the proton pump as the predominant protein in tubulovesicles (10) fit well into the earlier membrane recycling hypothesis of HCl secretion (16) for which H,K-ATPase is now identified as the cargo for stimulus-dependent recruitment of pumps into the apical membrane, and provided a mechanism for regulating pump activity. However, alternative views of parietal cell structure and secretion related changes in membrane morphology have persisted (23, 24, 19-21).

The reconstructed cellular models presented earlier by our laboratory (5) and reviewed here provide a definitive position on parietal cell ultrastructure. The high-pressure freezing preserved the cells in a pristine morphological state, offering opportunity to construct 3-dimensional views that addresse the long standing controversy regarding the form of the H,K-ATPase-rich tubulovesicle compartment. We conclude that the H,K-ATPase-rich intracellular membranes in rabbit parietal cells predominantly take the form of flattened cisternae, although tubules and vesicles can also be found. More importantly, no permanent connections among the tubulocisternal components, or between the tubulocisternae and the apical plasma membrane were found. Thus any communication that occurs after stimulation must be the result of some fusion based reaction that greatly expands the functional (secretory) surface area. Immunocytochemistry has also confirmed the work of several others that these membranes are in fact the loci of H,K-ATPase.

This general structural form of the tubulocisternal compartment partially agrees with the SEM of Ogata & Yamasaki (21), but we do not agree on the issue of connectivities between individual units. Furthermore, our 3-D reconstructions do not support the claim of Pettitt et al. (23, 24) that the membrane system was composed of densely packed helical tubular coils, although the close packing of some cisternae could possibly have misdirected their thinking. In addition, these authors predicted that the coiled tubules were interconnected with the apical canalicular membrane, and that the secretagogue-mediated resting/stimulated transition was the result of unwinding of the coils. We found no such interconnections between the cistrernal compartment and the apical surface, although there was evidence of considerable endocytic recycling at the apical surface, even in resting parietal cells. Thus, while our data clearly show membrane interconnections between canalicular segments, and among mitochondria, they argue against connections between or among the tubulovesicular system and other membrane compartments in the resting parietal cell. The functional implication of our model is that it clearly adds support to the membrane recruitment/recycling hypothesis.

In addition to the tubulocisternal system, the modeling data provided an interesting perspective on parietal cell mitochondria. Not only are they abundant, but there is clearly a great deal of branching and interconnection among the mitochondria, supporting the idea of a large dynamic mitochondrial network within cells. This perspective of mitochondrial connectivity has not previously been raised for the parietal cell, although the proposal of a mitochondrial reticular network has been made for other systems (1, 4, 13, 27). These studies have variously suggested that mitochondrial networks are dynamic (1, 4) and may contribute to power transmission in the form of a transmembrane electrical potential difference over long distances within, or even between, cells (27). In the case of the parietal cell, mitochondrial connectivity may have special relevance due to the activity and enormous energy requirements of the proton pump. There are large distances, up to 20 μm, between basal and apical cellular surfaces, separating the respective loci of energy input and output. A mechanism for power transmission along mitochondrial networks could reduce diffusional limitations and provide efficient use of the entire mitochondrial compartment in the face of the localized energy sink of the proton pump.

## REFERENCES

1. **Amchenkova, A.A., L.E. Bakeeva, Y.S. Chentsov, V.P. Skulachev, and D.B. Zorov.** Coupling membranes as energy-transmitting cables. I. Filamentous mitochondria in fibroblasts and mitochondrial clusters in cardiomyocytes. *J. Cell Biol.*, 107:481-495, 1988.
2. **Berglindh, T., D.R. Dibona, S. Ito and G. Sachs.** Probes of parietal cell function. *Am. J. Physiol.*, 238:G165-G176, 1980.
3. **Black, J.A., T.M. Forte, and J.G. Forte.** Inhibition of HCl secretion and the effects of ultrastructure and electrical resistance in isolated piglet gastric mucosa.

*Gastroenterology* 81:509-519, 1981.
4. **De Giorgi, F., L. Lartigue and F. Ichas.** Electrical coupling and plasticity of the mitochondrial network. *Cell Calcium*, 28:365-70 2000.
5. **Duman, J.G., N. Pathak, M.S. Ladinsky, K. L. McDonald and J. G. Forte.** Three-Dimensional Reconstruction of Cytoplasmic Membrane Networks in Parietal Cells. *J. Cell Science*, 2002, in press
6. **Forte, J. G., A.L. Ganser, R. Beesley, and T.M. Forte.** Unique enzymes of purified microsomes from pig fundic mucosa. *Gastroenterology* 69, 175-189, 1975.
7. **Forte, J. G., and H.C. Lee.** Gastric adenosinetriphosphatases: a review of their possible role in HCl secretion. *Gastroenterology 73*, 921-926, 1977.
8. **Forte, T. M., T. E. Machen, and J. G. Forte.** Ultrastructural changes in oxyntic cells associated with secretory function: a membrane recycling hypothesis. *Gastroenterology* 73: 941-955, 1977.
9. **Forte, J. G., J.A. Black, T.M. Forte, T.E. Machen and J.M. Wolosin.** Ultrastructural changes related to functional activity in gastric oxyntic cells. *Am. J. Physiol.* 241, G349-G358, 1981.
10. **Ganser, A.L., and J.G. Forte.** $K^+$-stimulated ATPase in purified microsomes of bullfrog oxyntic cells. *Biochim. Biophys. Acta* 307:169-180, 1973.
11. **Golgi, C.** Sur la fine organisation des glandes peptiques des mammiferes. Arch. Italiennes de Biologie 19:448-453, 1893.
12. **Helander, H. F., and B.I. Hirschowitz.** Quantitative ultrastructural studies on inhibited and on partly stimulated gastric parietal cells. *Gastroenterology 67*, 447-452, 1974.
13. **Hoffmann, H.P. and C.J. Avers.** Mitochondrion of yeast: ultrastructural evidence for one giant, branched organelle per cell. *Science*, 181:749-751, 1973.
14. **Ito, S.** Functional gastric morphology. In: *Physiology of the Gastrointestinal Tract*, ed. L.R. Johnson, New York: Raven Press, Vol. 1, 1987, p. 817-851.
15. **Ito, S., and G.C. Schofield.** Studies on the depletion and accumulation of microvilli and changes in the tubulovesicular compartment of mice parietal cells in relation to gastric acid secretion. *J. Cell Biol. 63*, 364-382, 1974.
16. **Kasbekar, D.K., G.M. Forte and J.G. Forte.** Phospholipid turnover and ultrastructural changes in resting and secreting bullfrog gastric mucosa. *Biochim Biophys. Acta*163:1-13, 1968
17. **Kremer, J.R., Mastronarde, D.N. and McIntosh, J.R.** Computer Visualization of Three-Dimensional Image Data Using IMOD. J. Struct. Biol. 116: 71-76, 1996.
18. **McDonald, K.** High pressure freezing for presentation of high resolution fine structure and antigenicity for immunolabeling. Meth. Molec. Biol. 117: 77-97, 1999.
19. **Ogata, T.** Gastric oxyntic cell structure as related to secretory activity. *Histol Histopath*, 12:739-54, 1997.
20. **Ogata, T. and Y. Yamasaki, Y.** Morphological studies on the translocation of tubulovesicular system toward the intracellular canaliculus during stimulation of the gastric parietal cell. *Micr. Res. Tech.* 48:282-92, 2000.
21. **Ogata, T. and Y. Yamasaki.** Scanning EM of resting gastric parietal cells reveals a network of cytoplasmic tubules and cisternae connected to the intracellular canaliculus. Anat. Rec. 258:15-24, 2000.
22. **Okamoto, C.T., J.G. Duman, K. Tyagarajan, K.L. McDonald, Y.Y. Jeng,. J. McKinney, T.M. Forte and J.G. Forte.** Clathrin in gastric acid secretory (parietal) cells: biochemical characterization and subcellular localization. *Am. J. Physiol., Cell Physiol*, 2000, 279:C833-C851, 2000.
23. **Pettitt, J.M., D.C. Humphris, S.P. Barrett, B.H. Toh, I.R. van Driel and P.A. Gleeson.** Fast freeze-fixation/freeze-substitution reveals the secretory membranes of the gastric parietal cell as a network of helically coiled tubule. A new model for parietal cell transformation. *J. Cell Sci.* 108:1127-1141, 1995.
24. **Pettitt, J.M., I.R. van Driel, B.H. Toh, and P.A. Gleeson.** From coiled

tubules to a secretory canaliculus: a new model for membrane transformation and acid secretion by gastric parietal cells. *Trend Cell Biol.* 6:49-53, 1996.
25. **Reenstra, W.W. and Forte, J.G.** Isolation of H,K-ATPase-containing membranes from the gastric oxyntic cell. Meth. Enz - Biomemb, Part W. 192:151-164, 1990
26. **Schofield, G.C., S. Ito and R.P. Bolender.** Changes in membrane surface areas in mouse parietal cells in relation to high levels of acid secretion. *J. Anat.* 128:669-92, 1979.
27. **Skulachev, V.P.** Power transmission along biological membranes. *J Memb Biol*, 114:97-112, 1990.
28. **Wolosin, J.M., and J.G. Forte.** Stimulation of oxyntic cell triggers $K^+$ and $Cl^-$ conductances in apical $(H^++K^+)$-ATPase membrane. *Am. J. Physiol.* 246:C537-C545, 1984.

Chapter 35

# Pharmacological Control of Gastric Acid Secretion via the Apical Membrane of Parietal Cells in Dogs

SUSUMU OKABE, KENICHI SHIMOSAKO, KIKUKO AMAGASE, AND TERUAKI FUJISHITA

*Department of Applied Pharmacology,*
*Kyoto Pharmaceutical University,*
*Misasagi, Yamashina, Kyoto 607-8414, Japan*

## SUMMARY

This report aims to highlight drugs that are able to inhibit parietal cells from the luminal side, resulting in suppressed acid secretion. ME3407 (a myosin light chain kinase inhibitor) was locally applied to denervated gastric pouches in dogs for 5 to 30 min. The drug, administrated 1 hr after histamine infusion was commenced, significantly inhibited stimulated-gastric acid secretion in a time-dependent manner. ME3407 also significantly inhibited histamine-stimulated acid secretion upon application to the pouches for 30 min prior to histamine infusion. The antisecretory effect persisted for more than 5 hr when ME3407 was applied at a dosage of 3 mg/pouch for 30 min. The drugs, however, had no antisecretory effect when intravenously administered after histamine infusion, or orally administered before histamine infusion. In addition, acid-stable local anesthetic drugs (sulcain and oxethazaine) had no antisecretory effect after local application. Such results strongly suggest that the apical membrane of parietal cells possesses a ME 3407 sensitive portion similar to the basolateral membrane that usually mediates gastric acid secretion. The apical membrane represents an intriguing target for developing new antisecretory drugs, as well as for elucidating the functional features of parietal cells.

## INTRODUCTION

Most antisecretory drugs, regardless of the route of administration, inhibit gastric acid secretion by blocking various receptors (histamine $H_2$ receptors, muscarinic cholinergic $M_3$ or $M_1$-receptors) or enzymes ($H^+$,$K^+$-ATPase) after absorption (8, 9, 16, 26). It is of interest that several drugs have been reported to have an antisecretory effect when locally applied to the stomach (2, 4, 6, 11, 12, 18, 24). Nonetheless, attention has not been directed towards such drugs as either clinically useful antisecretory drugs or investigational tools for the elucidation of parietal cell function. We examined whether or not certain drugs could inhibit histamine-stimulated gastric acid secretion when locally applied to denervated (Heidenhain) gastric pouches in dogs. It was found that ME3407 (a myosin light chain kinase inhibitor and a functional analogue of wortmannin (22)(Figure 1), omeprazole and leminoprazole (acid pump inhibitors) (12, 21), and FPL52694 (a mast cell stabilizer)(4, 6) significantly inhibited histamine- or gastrin-stimulated acid secretion. The current review, mainly based on our unpublished results, describes the effects of locally applied ME3407 and sulcain and oxethazaine (acid stable local anesthetics) and the underlying mechanisms of action, emphasizing the important role of parietal cell apical membranes for acid secretion.

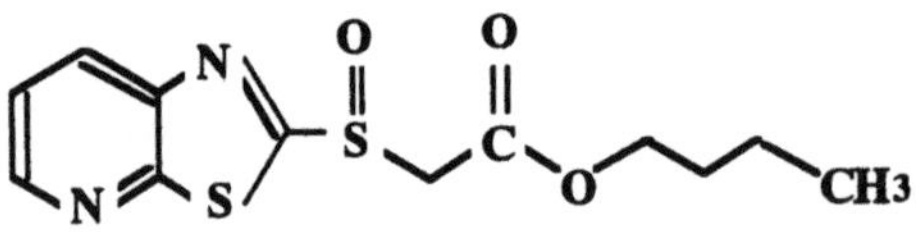

**ME 3407**

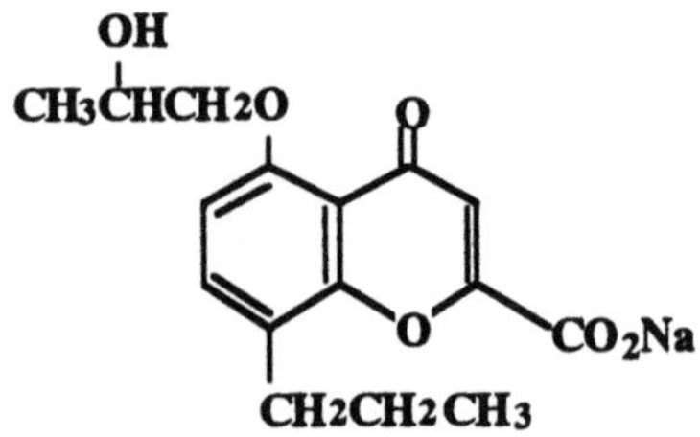

**FPL 52694**

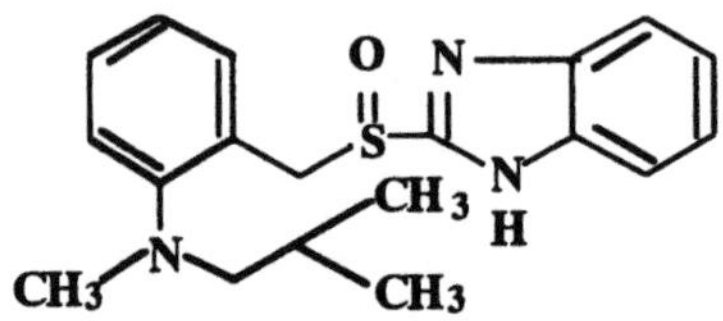

**Leminoprazole**

***Figure 1.*** *Chemical structures of the compounds that applied in this study.*

## MATERIALS AND METHODS

### Animals

Ten male and female beagles (10-13 kg), each with a cannulated Heidenhain pouch, were used in the study no earlier than 2 months after their operation. The interval between the experiments was at least 7 days. Food was withheld for 18 hr before each experiment, but water was freely provided.

### Measurement of Gastric Acid Secretion

The pouch of each animal was washed with 15 ml of warm saline several times until the washings became clear. Gastric juice samples were collected throughout the experiments every 15 min by gravity drainage (Figure 2). Following collection of the basal secretion of the initial 30 min, acid secretion was stimulated by continuous intravenous infusion of histamine 2HCl (Nacalai Tesque, Kyoto, Japan) at a dose of 160 μg/kg/hr and a rate of 10 mL/hr via a catheter inserted into a leg vein. Test drugs were generally applied to the pouches 1 hr after commencement of histamine infusion, yet in certain cases drugs were either intravenously or orally administered. In certain cases, ME 3407 was applied to the pouches for 30 min prior to the commencement of histamine infusion. During the application of the drugs, histamine infusion was maintained at the same rate. After local application, test drugs were removed from the pouches and the pouches were washed with saline three times. Thereafter, gastric juice samples were continuously collected every 15 min for 1.5 to 5 hr and analyzed for volume and acidity. Total acidity was determined by titration of the gastric juice against 0.1N NaOH to pH 7.0, using an automatic titrator (Radiometer Copenhagen, Denmark); acid output was expressed as mEq/15min.

**Drugs**

ME3407 (Meiji Seika Kaisha, Yokohama, Japan) sulcain (Nippon Shinyaku, Kyoto, Japan) and oxethazaine (Wako, Osaka, Japan) were suspended in 0.5% carboxymethylcellulose (CMC, Nacalai Tesque, Osaka, Japan). Drugs were prepared directly before administration.

**Statistics**

Data are presented as means ± 1S.E. Statistical differences were evaluated using the Student's t-test or the Dunnett's multiple comparison test, with a P value of <0.05 regarded as significant.

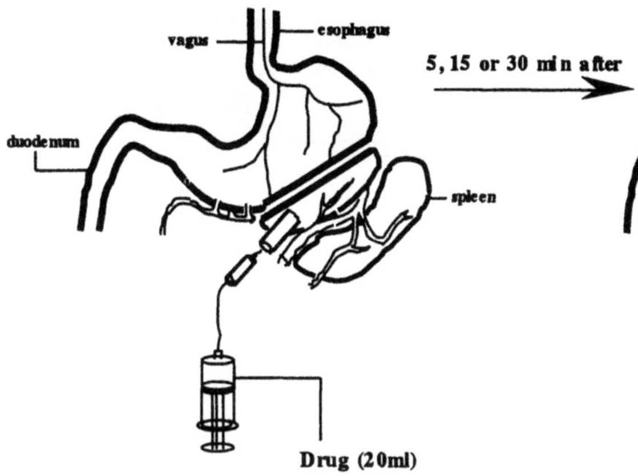


***Figure 2.*** *General design for the experimental procedure.*

## RESULTS

Continuous infusion of histamine 2HCl (160 μg/kg/hr) inevitably induced maximal stimulation of gastric acid secretion 1 hr later; the plateau level was maintained for more than 3 hr. During histamine infusion, the local application of 0.5% CMC alone for 30 min essentially exerted no effect on gastric acid secretion. Acid output was transiently reduced for 15 min after removal of the solution, but returned to the stimulated level thereafter.

**Effects of ME3407 on Gastric Acid Secretion**

Local application of ME3407 at doses of 1 and 3 mg/pouch for 30 min after histamine infusion significantly inhibited gastric acid secretion in a dose-related manner; the rate of inhibition was greater than 90% (Figure 3). Even in the case of local application prior to histamine infusion, ME3407, at a dose of 3 mg/pouch, significantly inhibited histamine-stimulated acid secretion (Figure 4). In contrast, ME3407 intravenously administered at a rate of 1 mg/kg had little or no effect on gastric acid secretion. In addition, oral administration of the drug 2 hr before histamine infusion at a rate of 1 mg/kg exerted no effect on gastric acid secretion. In regards to application time, local application of ME3407 at a dose of 3 mg/pouch for 15 min also significantly inhibited gastric acid secretion for more than 1.5 hr (Figure 5). The degree of inhibition approximated the level observed following a 30 min

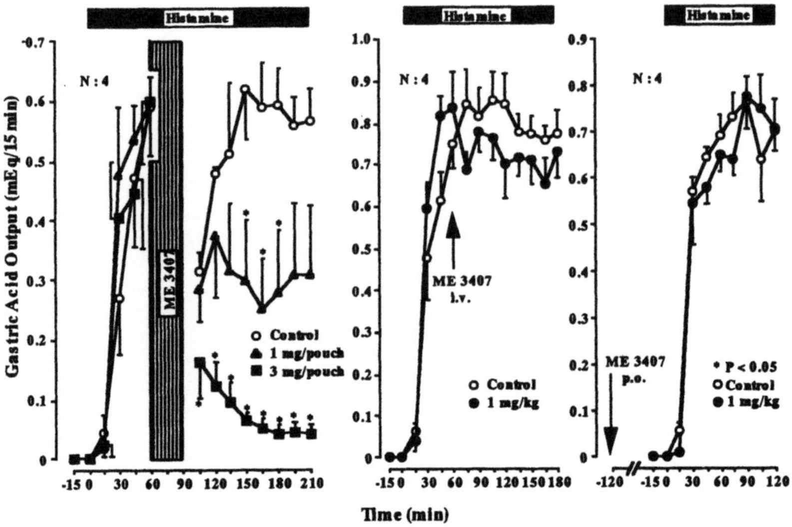


***Figure 3.*** *Effects of ME3407 on histamine-stimulated gastric acid secretion in Heidenhain pouch dogs. ME3407 was administered directly into the pouch for 30 min, intravenously 1 hr after intravenous histamine infusion, or orally 2 hr before histamine infusion. Note that ME3407 significantly suppressed gastric acid secretion only when administered into the pouch. Data are expressed as means ± 1S.E. for 4 dogs. * indicates a statistically significant difference from the corresponding control, with p < 0.05.*

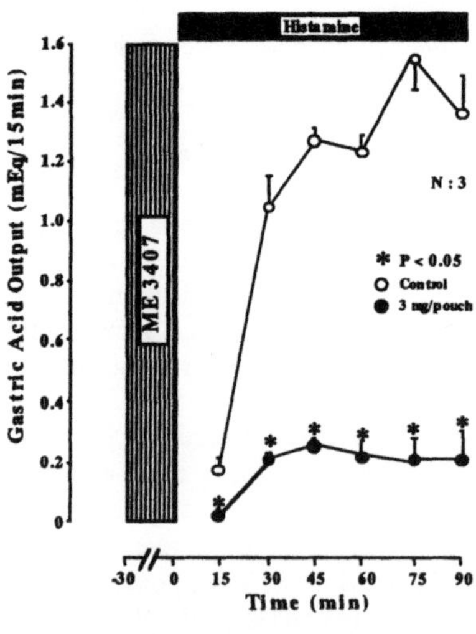


***Figure 4.*** *Effects of ME3407 on histamine-stimulated gastric acid secretion in Heidenhain pouch dogs. ME3407 was administered directly into the pouch for 30 min immediately before the commencement of histamine infusion. Data are expressed as means ± 1S.E. for 3 dogs. *indicates a statistically significant difference from the corresponding control, with p < 0.05.*

application. Although a 5 min application of the drug revealed a tendency to inhibit gastric acid secretion, the response was not significant. It should be noted that the antisecretory effect of ME3407, administered at a dose of 3 mg/pouch, persisted for more than 5 hr (Figure 6). Nonetheless, such an effect on gastric acid secretion was not demonstrated upon examination 9 to 10 hr after local application.

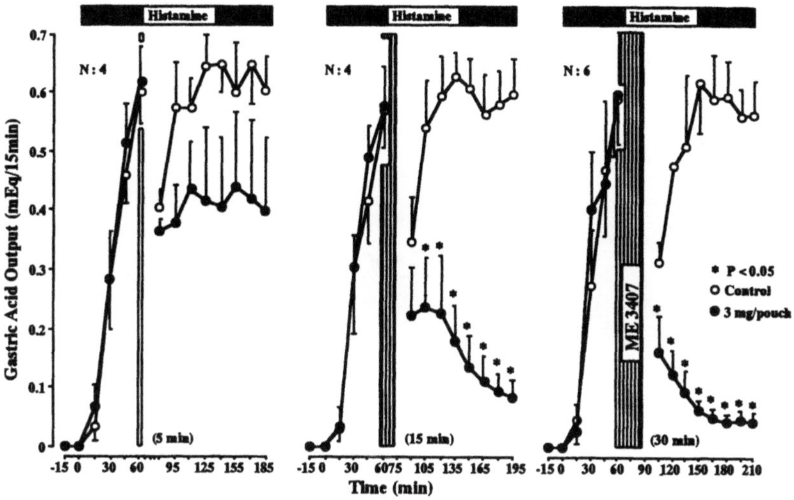


***Figure 5.*** *Effects of ME3407, locally applied in Heidenhain pouch dogs for 5, 15, or 30 min, on histamine-stimulated gastric acid secretion. Note that local antisecretory effects were observed even after only the 15 min application. Data are expressed as means ± 1S.E. for 4-6 dogs. * indicates a statistically significant difference from the corresponding control, with p<0.05.*

**Effects of Local Anesthetics on Gastric Acid Secretion**

Topical application of sulcain (200 mg/pouch) and oxethazaine (15 mg/pouch) for 30 minutes exerted no effect on histamine-stimulated gastric acid secretion (Figure 7). The dosages of the drugs were based on the oral dose used for patients suffering from gastritis and ulcers.

## DISCUSSION

The above results clearly indicate that there exist drugs that are able to directly inhibit gastric acid secretion from the luminal side (Figure 8). In particular, it was found that ME 3407 possesses a strong inhibitory effect on acid secretion even at doses of 1 to 3 mg/pouch and even after only a 15 min application to the pouches. It should be noted that the antisecretory effect of ME3407 against parietal cells was observed to a similar degree for both non-stimulated and stimulated conditions. Such a result indicates that ME 3407 is able to inhibit parietal cell function without resulting in expansion of secretory canaliculi. It is even more interesting that the antisecretory effect persisted for more than 5 hr, a duration that was slightly longer than that observed following a single intravenous injection of histamine $H_2$-

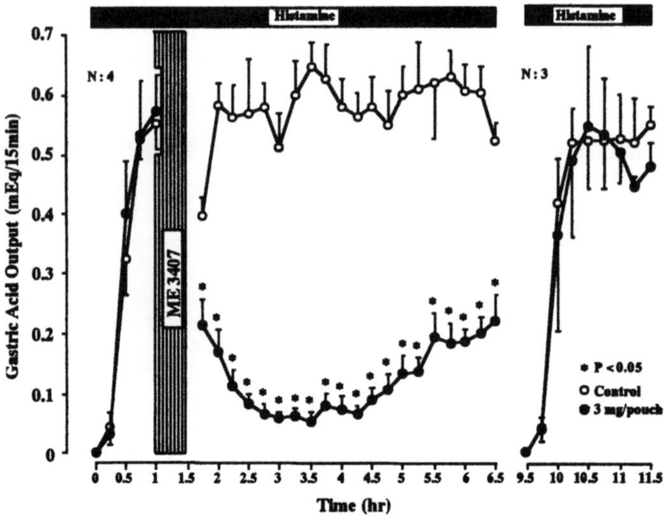


*Figure 6. Duration of antisecretory effects of ME 3407, locally applied in Heidenhain pouch dogs for 30 min, on histamine-stimulated gastric acid secretion. Note that the inhibition persisted for more than 5 hr. Data are expressed as means ± 1S.E. for 3-4 dogs. * indicates a statistically significant difference from the corresponding control, with p<0.05.*

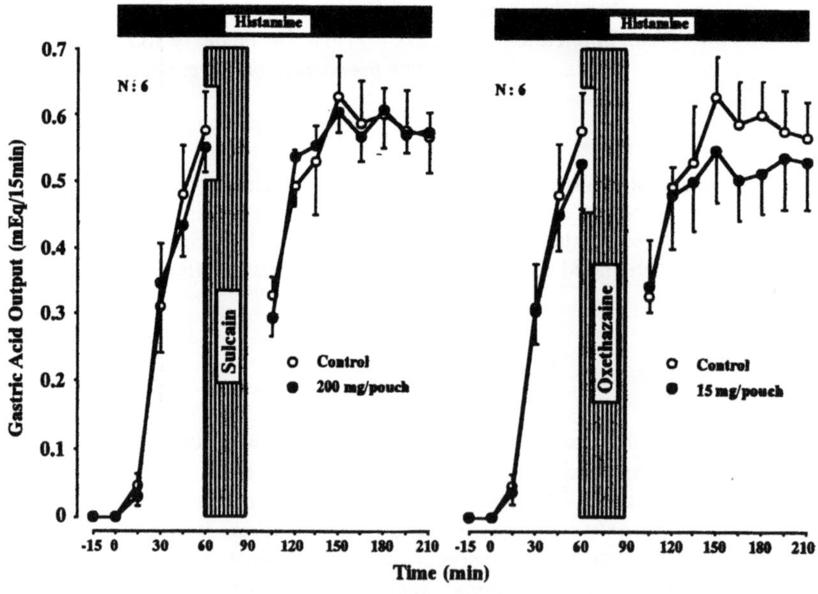


*Figure 7. Effects of locally applied local anesthetics in Heidenhain pouch dogs for 30 min. Note that the drugs had little or no antisecretory effect. Data are shown as means ± 1S.E. for 6 dogs.*

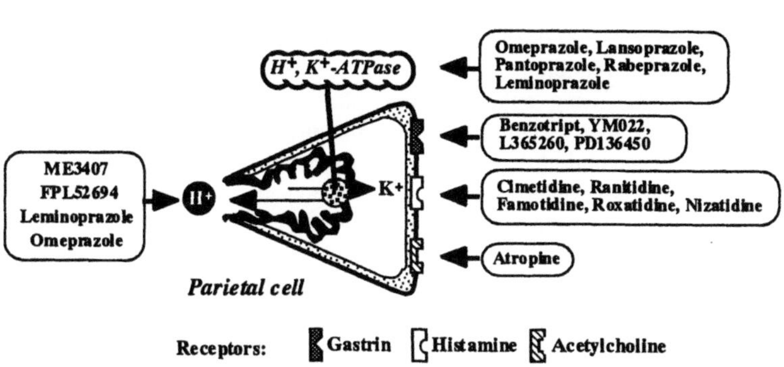


***Figure 8.*** *Schematic drawing for the effective site of drugs that inhibit the secretory function of parietal cells via the basolateral membrane (receptors), the acid pump, and the apical plasma membrane.*

receptor antagonists ($H_2$-blodckers), such as cimetidine, ranitidine, and famotidine. The underlying mechanism remains unknown, but ME3407, similar to FPL52694 (4, 6), omeprazole, and leminoprazole (12, 21), appears to exert its effect on parietal cells from the mucosal side.

Originally it appeared possible that ME 3407 might be absorbed from the gastric mucosal surface into the blood and exert its effect on parietal cells from the basolateral plasma membrane, as is the case for $H_2$-blocers and gastric acid pump inhibitors. In the case of ME3407, however, no inhibitory effect was observed following intravenous and oral administration, excluding the possibility of inhibition of receptors and $H^+,K^+$-ATPase. Furthermore, the dose used for local application was 1 or 3 mg/pouch, while the intravenous dose administered was 1 mg/kg (approximately 10 to 13 mg/dog). Accordingly, even if the locally applied drug was all absorbed from the mucosa into the circulation, the dose was less than one-third of the intravenous dose. Such results strongly suggest that the antisecretory effect of the drug is solely explained by a local effect, not a systemic effect.

Urushidani et al. (31) reported that ME3407 significantly enhanced the healing of acetic acid-induced gastric ulcers in rats, most likely by exerting a potent antisecretory effect that did not inhibit the acid pump. The study found that the inhibitory effect of ME3407 on gastric acid secretion was more potent when orally administered than when subcutaneously or intraperitoneally administered. Citing our preliminary report (22), Urushidani et al. also concluded that ME3407 acts directly on the mucosa to inhibit acid secretion. In addition, using isolated rabbit gastric glands, Urushidani demonstrated that ME3407 inhibited acid secretion by not only inhibiting the stimulation-associated redistribution of $H^+,K^+$-ATPase from microsomes to the apical plasma membrane but also by delocalizing ezrin from the apical membrane. Should their findings be applicable to our study, it would appear that ME3407 exerts its local antisecretory effect through both inhibition of redistribution of the acid pump and delocalization of ezrin after penetrating into the parietal cells from the mucosal side. Since ME3407 resulted in no inhibitory effect by intravenous administration, penetration of the drug into the parietal cells to exert its effect must occur from the apical membrane side, quite possibly from the membranes of the secretory canaliculi. Certainly, there also exists the possibility that ME3407 directly acts on secretory canaliculi itself without penetrating into the cytoplasm of parietal cells. Although only a speculation, ME3407 might delocalize ezrin from the plasma membrane and prevent the fusion of tubulovesicles to secretory canaliculi. Since the inhibitory effect disappeared approximately 6 to 7 hr following local application, the reaction of the apical membrane, in particular secretory canaliculi, to ME3407 appears to be reversible.

Agnew et al. (1), using cultured rabbit parietal cells, also provided evidence that ME3407 is able to prevent histamine (+isobutylmethylxanthine)-stimulated acid secretion. Agnew contends that the underlying mechanism follows from prevention of translocation of $H^+,K^+$-ATPase rich tubulovesicles to apical membrane vacuoles and dissociation of ezrin from the putative cytoskeletal-membrane functional site. Similar to Urushidani et al. (31), however, Agnew et al. did not describe whether or not ME3407 either penetrated into the parietal cells to exert its effect or acted on the secretory canaliculi membrane itself leading to prevention of the secretion-association cascade. Clarification of the action of ME3407 on secretory canaliculi will represent an important step towards elucidation of the mechanism of acid secretion from the apical membrane.

Local anesthetics are reportedly known to sedate the pain of gastritis and ulcers and inhibit the release of gastrin and secretin from the gastrointestinal tract (3, 25, 34, 35). It is of interest that Venables (32) demonstrated that oxethazaine applied to denervated and innervated gastric pouches in dogs for 20 min significantly inhibited pentagastrin-stimulated gastric acid secretion. Tani et al. (30) reported that oxethazaine inhibited carbachol-stimulated $^{14}C$-aminopyrine accumulation in parietal cells isolated from guinea pigs. Based upon the results, Tani suggested that oxethazaine inhibits acid secretion by mechanisms other than inhibition of muscarinic receptors. In contrast to Venables's findings, we discovered that locally applied sulcain and oxethazaine exerted no inhibitory effect on histamine-stimulated gastric acid secretion in dogs. The difference in the results might stem from the different stimulants used (pentagastrin vs. histamine). Such data suggest that the mechanism of the antisecretory effect of ME3407 is different from that observed with local anesthetics.

In recent years, most of the genes for the expression of the above receptors or related enzymes for acid secretion have been cloned, offering the opportunity to generate gene deficient, i.e. knockout (KO) animals. To date, histamine $H_2R$-KO mice (19, 23), muscarinic cholinergic $M_3R$-KO mice (15, 36), $CCK_2R$-KO mice (13, 17), histidine decarboxylase KO mice (29), gastrin KO mice and gastrin transgenic mice (7, 33), and $H^+,K^+$-ATPase ($\alpha$, $\beta$) KO mice (27, 28) have been generated. It is also of note that parietal cell deficient mice were generated and found to live normal lifespans without any indication of excessive problems (5, 14), suggesting that parietal cells are not requisite cells for stomach development or function. Based upon physiological and pharmacological analyses of such KO mice, comprehension of the regulatory mechanisms of gastric acid secretion is rapidly progressing.

Elucidation of the antisecretory effect of locally active drugs, such as ME3407, FPL52694, and leminoprazole on $H^+,K^+$-ATPase KO mice and receptor KO mice will afford new and important insights requisite for understanding the functional mechanisms underlying parietal cells.

Incidentally, the origin of parietal cells, which inordinately secrete high concentrations of inorganic acid in the stomach, has yet to be thoroughly investigated. The key question that remains is why parietal cells with P-type $H^+,K^+$-ATPase differentiated from stem cells in the oxyntic gland mucosa? One of the present authors (SO) hypothesizes that parietal cells might have evolved from transfer of the $H^+,K^+$-ATPase gene from a microbe with a P-type ATPase during evolution (19, 20). The hypothesis is based upon the following observations: 1) microbes secrete $H^+$ to intake nutrients from the extracellular environment; and 2) the high conservation of amino acid sequence between $H^+,K^+$-ATPase in parietal cells and $H^+$-ATPase present in yeast. Further investigation of the origin of parietal cells is required for a better understanding of not only the stomach, but also the evolution of human beings.

In summary, the above results suggest that ME3407 directly suppresses the function of parietal cells on the luminal surface, resulting in a reduction of acid secretion. Regardless of the precise mechanism for a local antisecretory effect, the above drugs represent a new armamentarium that targets the apical membrane of parietal cells to control gastric acid secretion in the treatment of acid-related diseases. Moreover, such drugs will no doubt prove useful for the study of the function of parietal cells, particularly in elucidating the role of the apical plasma membrane.

## ACKNOWLEDGEMENTS

We wish to thank CJ Hurt (John Hopkins University School of Medicine, USA) for a critical reading of the manuscript.

## REFERENCES

1. **Agnew BJ, Duman JG, Watson CL, Coling DE, Forte JG.** Cytological transformations associated with parietal cell stimulation: critical steps in the activation cascade. J Cell Sci 112:2639-46, 1999
2. **Bond AM, Hunt JN.** The effect of sodium fluoride on the output of some electrolytes from the gastric mucosa of cats. J. Physiol 133:317-329, 1956
3. **Bower RJ, Myerson RM.** Double-blind, controlled study of an anesthetic-antacid agent in the treatment of peptic ulcer, esophagitis and gastritis. Am. J. Gastroenterol 42:515-520, 1964
4. **Canfield SP, Curwain BP.** Inhibition of gastric acid secretion in the conscious dog by the mast-cell stabilizing agent, FPL-52694. Br. J. Pharmacol 80:27-32, 1983.
5. **Canfield V, West AB, Goldenring JR, Levenson R.** Genetic ablation of parietal cells in transgenic mice: a new model for analyzing cell lineage relationships in the gastric mucosa. Proc Natl Acad Sci USA. 93:2431-5, 1996
6. **Curwain BP, Canfield SP.** Inhibition of gastric acid secretion in conscious dog with fistula following intragastric administration of FPL 52694. Int J Tiss React 5:245-247, 1983
7. **Hinkle KL, Samuelson LC.** Lessons from genetically engineered animal models. III. Lessons learned from gastrin gene deletion in mice. Am J Physiol 277:G500-G505, 1999
8. **Hirschowitz BL, Keeling D, Lewin M, Okabe S, Parsons M, Sewing K, Wallmark B, Sachs G**: Pharmacological aspects of acid secretion. Dig Dis Sci 40:3S-23S, 1995
9. **Huang JQ, Hunt RH**. Pharmacological and pharmacodynamic essentials of $H_2$ -receptor antagonists and proton pump inhibitors for the practicing physician. Bailleres Best Pract Res Clin Gastroenterol 15:355-370, 2001
10. **Kobayashi T, Tonai S, Ishihara Y, Koga R, Okabe S, Watanabe T.** Functional and morphological regulation of gastric mucosa by histamine; studies of histamine $H_2$-receptor deficient mice. J Clin Invest 105:355-370, 2000
11. **Konturek SJ, Cieszkowski M, Kwiecien N, Harrison C.** Gastric acid response to topical or intravenous histamine and topical $H_2$-receptor blockade in dogs. Agents Actions 11:437-441, 1981
12. **Konturek SJ, Cieszkowski M, Kwiecien N, Konturek J, Tasler J, Bilski J.** Effects of omeprazole, a substituted benzimidazole, on gastrointestinal secretion, serum gastrin, and gastric mucosal blood flow in dogs. Gastroenterol 96:71-77, 1984
13. **Langhans N, Rindi G, Chiu M, Rehfeld JF, Ardman B, Beinborn M, Kopin AS.**

Abnormal gastric histology and decreased acid production in cholecystokinin-B/gastrin receptor-deficient mice. Gastroenterol 112:280-286, 1997
14. **Li QT, Karam SM, Gordon JI.** Diphteria toxin-mediated ablation of parietal cells in the stomach of the transgenic mice. J Biol Chem 271: 3671-3676, 1996
15. **Matsui M, Motomura D, Karasawa H, Fujikawa T, Jian J, Komiya Y, Takahashi S, Taketo MM.** Multiple functional defects in peripheral autonomic organs in mice lacking muscarinic acetylcholine receptor gene for the $M_3$ subtype. Proc Natl Acad Sci USA 97:9579-9584, 2000
16. **Modlin I, Sachs G.** Acid related diseases. Biology and treatment. Gmbh D-Konstanz, Schnetztor-Verlag, 1998
17. **Nagata A, Ito M, Iwata N, Kuno J, Takano H, Minawa O, Chihara K, Matsui T, Noda T.** G protein-coupled cholecystokinin-B,gastrin receptors are responsible for physiological cell growth of the stomach mucosa in vivo. Proc Natl Acad Sci USA 93:11825-11830, 1996
18. **Nicol AK, Thomas M, Wilson J.** Inhibition of gastric acid secretion by sodium cromoglycate and FPL52964. J. Pharm. Pharmacol 33:554-556, 1981
19. **Okabe S.** Hypothesis — Origin of the parietal cell: microorganism? J Clin Gastroenterol 25(Suppl. 1): S141-S148, 1997.
20. **Okabe S**: Hypothesis —Origin of the parietal cells: transfer of the $H^+,K^+$-ATPase gene from parasitic microorganisms to cnidaria? Chinese J Physiol 42:121-128, 1999
21. **Okabe S, Shimosako K, Harada H.** Antisecretory effect of leminoprazole on histamine-stimulated gastric acid secretion in dogs: potent local effect. Jpn. J Pharmacol 69:91-100, 1995.
22. **Okabe S, Shimosako K, Yuda Y.** Effect of topical application of ME 3407 on histamine-stimulated gastric acid secretion in Heidenhain pouch dogs. Gastroenterol 104:A164, 1993.
23. **Okabe S, Tonai S, Furutani K, Maeda K, Kobayashi T, Watanabe T.** Menetrier's disease-like changes in the gastric mucosa of histamine $H_2$ receptor deficient mice. Role of gastrin. Gastroenterol 120:A19, 2001.
24. **Reed JD, Smy JR.** The effects of sodium fluoride on gastric acid and electrolyte output in the anaesthetized cat. J. Physiol 301:39-48, 1980.
25. **Rheault M, Sem LS, Harkins HN, Nyhus LM.** Effect of oxethazaine and procaine on the gastrin mechanism in the isolated canine antrum. Am J Dig Dis 10:128-134, 1965.
26. **Sachs G, Prinz CK, Hersey SJ.** Acid related disorders. Palm Beach, Sashu Publishing, 1995.
27. **Scarff KL, Judd LM, Toh BH, Gleeson PA, Van Driel IR.** Gastric $H^+,K^+$-adenosine triphosphatase $\alpha$ subunit is required for noraml function, development, and membrane structure of mouse parietal cells. Gastroenterol. 117:605-608. 1999
28. **Spicer, Z, Miller ML, Andringa, A, Riddle TM, Duffy JJ, Doetschman T, Shull GE.** Stomachs of mice lacking the gastric H,K-ATPase $\alpha$–subunit have achlorhydria, abnormal parietal cells, and ciliated metaplasia J Biol Chem 275: 21555-21565, 2000
29. **Tanaka S, Ichikawa A, Ohtsu H, Watanabe T, Okabe S.** Enhancement of gastric acid secretion by histamine in histidine decarboxylase-deficient mice. Jpn. J Pharmacol. 38 (Suppl. 1) 31P.
30. **Tani N, Watanabe Y, Suzuki T, Muramatsu S, Miyazawa M, Kimura N, Miwa T.** Effect of procaine and oxethazaine on muscarinic receptors of parietal cells. Tokai J Exp Clin Med 22: 103-109, 1997.
31. **Urushidani T, Muto Y, Nagao T, Yao X, Forte JG.** ME3407, a new antiulcer agent, inhibits acid secretion by interfering with redistribution of $H^+,K^+$-ATPase. Am J Physiol

272:G1122-G1134, 1997
32. **Venables CW.** The effects of oxethazaine upon acid and pepsin secretion in the dog. Brit J Surg 58:297-298, 1971
33. **Wang TC, Dockray GJ.** Lessons from genetically engineered animal models. I. Physiological studies with gastrin in transgenic mice. Am J Physiol 277:G6-G11, 1999
34. **Woodward ER, Schapiro H.** Inhibition of acid-induced peptic ulcer pain by local anesthetics. J Appl Physiol 8:357-360, 1956
35. **Woodward ER, Schapiro H.** Effect of local anesthetics on the isolated antrum of the stomach in dogs. Am J Physiol 192:479-481, 1958
36. **Yamada M, Miyakawa T, Duttaroy A, Yamanaka A, Moriguchi T, Makita R, Ogawa M, Chou CJ, Xia B, Crawley JN, Felder CC, Deng CX, Wess J.** Mice lacking the $M_3$ muscarinic acetylcholine receptor are hypophagic and lean. Nature 410:207-212, 2001

Chapter 36

# The Cytoskeleton As a Modulator of Gastric Secretion

DAVID A AMMAR AND JOHN G FORTE
*Dept. of Molecular & Cell Biology,*
*University of California, Berkeley, CA 94720*

## ABSTRACT

Within the gastric parietal cell, actin and the membrane-cytoskeletal linker protein ezrin are localized to the apical canalicular membrane. Drugs that inhibit the translocation of $H^+$,$K^+$-ATPase (HK) containing tubulovesicles can also have the effect to disrupt either the actin cytoskeleton or membrane localization of ezrin.

In order to investigate actin turnover in the stimulated parietal cell, we have investigated the effects of Latrunculin B (Lat B), which binds to and sequesters monomeric actin [3,6]. In isolated gastric glands, relatively high concentrations of Lat B were required to inhibit acid accumulation ($ED_{50}$ ~70 μM). In resting cultured parietal cells, there were only minor effects on morphology and F-actin staining up to 50 μM Lat B. When stimulated in the presence of low Lat B (0.1-1 μM), parietal cells have reduced lamellipodia formation and some aberrant punctate phalloidin-stained structures, but translocation of HK and vacuolar swelling appeared unaffected at these concentrations. Higher Lat B (10-50 μM) resulted in punctate phalloidin-stained structures throughout the cell and nucleus as well as reduced translocation of HK and vacuolar swelling. When resting cells were treated with up to 50 μM Lat B, but washed immediately before stimulation, they exhibited a normal stimulated morphology. These data suggest distinct pools of parietal cell actin: a pool highly susceptible to Lat B primarily involved in motile function of cultured cells; and a Lat B-resistant pool, most likely microvillar filaments, that is essential for secretion. Furthermore, the stimulation process appears to accentuate the effects of Lat B, most likely through Lat B binding to monomer actin liberated by the turnover of the motile actin filament pool.

We have also developed a permeabilized gland system in order to study the mechanisms involved in tubulovesicle translocation. Pores formed by streptolysin-O (SLO) allow molecules of up to 500 kDa to pass through the cell membrane [5]. Glands are permeabilized to the extent that exogenously added fluorescently-labeled actin incorporates into the cytoskeleton in a pattern that mimics endogenous F-actin. SLO-permeabilized glands are stimulated to secrete acid by the addition of cAMP and ATP and are inhibited by proton pump inhibitors. Direct visualization using Lysosensor as a fluorescent pH probe shows accumulation of acid in the glandular lumen and in parietal cell canaliculi.

The SLO-permeabilized gland model presented us with a new way to study the effects of the kinase inhibitors ME-3407 and wortmannin on both ezrin and acid secretion. After SLO glands were stimulated for 20 min at 37°C, very little of the cellular actin can be detected in the extracellular medium, and staining of SLO glands with fluorescent phalloidin indicates a typical cytoskeletal structure. In contrast, stimulation of SLO glands occurs even while most of the ezrin is lost to the extracellular medium. While immunofluorescence analysis indicates a reduction in ezrin staining in SLO glands as compared to intact preparations, the remaining ezrin is still localized to the apical canalicular surface. ME-3407, an anti-ulcer drug whose inhibitory action has been implicated to involve ezrin, inhibited AP uptake in intact and SLO-glands, and effectively released ezrin from both. By contrast, wortmannin, an effective inhibitor of secretion in intact glands, had minimal effects on either ezrin or AP accumulation in SLO-glands. Taken together, this suggests that if ME-3407 and wortmannin inhibit acid secretion by actin as kinase inhibitors, they have different targets in the protein kinase-signaling pathway of the parietal cell. This raises the possibility that either the residual ezrin in control- and wortmannin-treated parietal cells is adequate for the activation process, or that ezrin is not required for activation and ME-3407 therefore affects ezrin as well as some other critical kinase activity.

## INTRODUCTION

Acid secretion by the gastric parietal cell is thought to be regulated by the recruitment of vesicles containing the primary proton pump, the H,K-ATPase, from a broadly distributed compartment of cytoplasmic vesicles to the apical plasma membrane [7, 10]. Although the actual mechanism for the movement and fusion of these vesicular structures is not completely understood, there is evidence to suggest the involvement of the actin cytoskeleton. Cytochalasin D was also found to inhibit acid secretion and increase the proportion of G-actin in isolated gastric glands [8]. Thus it is not certain whether, or how, the actin cytoskeleton changes as the structural and functional activities of the cell are transformed by stimulation.

The present studies were undertaken to re-examine the role of the cytoskeleton in gastric HCl secretion. The principal approach was to determine the structural and functional effects of Latrunculin B (Lat B) on isolated gastric glands and cultured parietal cells. Latrunculins (A and B) bind only the monomeric form of actin, forming a complex that is incompetent in both nucleation and polymerization reactions [3, 6]. Latrunculins sequester the actin monomer away from sites of actin polymerization, but not influencing the rate of actin turnover. Since the steady state of parietal cell actin is known to exist in the predominant F-actin form [8], we thought tested the structural and functional effects of Lat B, which sequesters cellular actin away from the F-actin form.

The development of permeabilized cell preparations has provided a valuable experimental tool linking *in vitro* studies that use subcellular fractions with direct assays of cellular function. We have utilized the bacterial pore forming toxin isolated from β-hemolytic group A streptococci, streptolysin-O (SLO). The pores formed by SLO are large enough to allow passage of proteins of up to 200-500 kDa [5]. Here we test the effects of several kinase inhibitors on acid secretion by SLO-permeabilized glands, as well as there effects on actin and the cytoskeletal linker protein ezrin.

## METHODS

*Isolation of gastric glands.* Intact gastric glands were isolated from New Zealand White rabbits as described in [4]. Briefly, the stomach was perfused under high pressure with phosphate-buffered saline (PBS) containing 1 mM each $CaCl_2$ and $MgSO_4$. The gastric mucosa was scraped from the smooth muscle layer, finely minced, and then washed three times in PBS and twice with minimal essential medium buffered with 20 mM HEPES, pH 7.4 (HEPES-MEM). The minced mucosa was digested at 37°C for approximately 30 min in a minimal amount (~20 mL) of HEPES-MEM containing 15 mg of collagenase (Sigma Blend) and 20 mg each of bovine serum albumin (BSA: Fraction V, Sigma) and HEPES acid. The digestion was terminated by diluting the solution into 200 mL of HEPES-MEM, followed by filtration through a nylon mesh to remove large pieces of undigested mucosa. Intact gastric glands were isolated by settling the digestion mixture for 10-15 min, and then washed by settling three times in HEPES-MEM. In all subsequent gland experiments (intact and permeabilized), glands were resuspended at 5% cytocrit (vol/vol) in the appropriate buffer for final assay. Cultures enriched in parietal cells were obtained from the above cell suspension as described in [1]. Briefly, debris was removed by straining the cell suspension through a 40 μm mesh. Intact cells were recovered by three repetitions of centrifugation at 200 x g for 5 min followed by resuspension in fresh HEPES-MEM.

*Streptolysin-O permeabilization of gastric glands.* Streptolysin-O (SLO) was obtained from S. Bhakdi (University of Mainz, Germany) and from H-P Moore (University of California, CA). Intact glands in HEPES-MEM were washed two times by settling at 4°C in ice-cold K-buffer (10 mM TRIS base, 20 mM HEPES acid, 100 mM KCl, 20 mM NaCl, 1.2 mM $MgSO_4$, 1 mM $NaH_2PO_4$, 40 mM Mannitol, pH 7.4). SLO was added to a final concentration of 1 μg/ml (750 lytic units per μg), the glands (at 5% cytocrit) were mixed by inversion and then incubated on ice until the glands had completely settled (10 to 15 min). The K-buffer supernatant containing excess SLO was removed and the glands were washed two more times by settling at 4°C in ice-cold K-buffer. Finally, glands were resuspended in K-buffer solution containing 1 mM pyruvate and 10 mM succinate, and then incubated at 37°C for 3 min to allow the SLO pores to form.

*$^{14}$C-Aminopyrine uptake assays.* Stimulation of gastric glands or parietal cells was quantified using the aminopyrine (AP) uptake assay. The performance of and calculations for the aminopyrine uptake assay in glands were essentially as described by [4] with modifications described by [15]. For both intact and permeabilized preparations, glands were kept in the resting state with cimetidine (100 μM). Intact preparations were stimulated with 3-isobutyl-1-methylxanthine (IBMX: 30 μM) and histamine (100 μM). Permeabilized preparations were stimulated with cAMP (100 μM) plus ATP (1 mM). Other drugs were also added where designated; the actin binding drug latrunculin B (Lat B; Calbiochem), the H,K-ATPase specific proton pump inhibitor SCH28080 (10 μM) [[14]], and the kinase inhibitors ME-3407 (300 μM) [13] and wortmannin (10 μM). Preparations were incubated for 20 min at 37°C with gentle shaking then centrifuged briefly to pellet the glands. The pellets were dried and weighed, and aliquots of both the supernatant and pellet were counted in a Beckman liquid scintillation counter. These data were used to calculate the AP accumulation ratio according to the method of Berglindh [4]. In order to normalize AP uptake values among the various preparations, the data are expressed as a fraction of the stimulated control in some experiments.

*Immunofluorescence and confocal microscopy.* Cultured parietal cells were treated as described above. Cells were fixed by treatment with 3.7% formaldehyde in PBS for 20 min, followed by permeabilization in 0.5% Triton X-100 in PBS for 15 min and blocking in 2% BSA in PBS for 15 min. All subsequent antibody dilutions were made in PBS containing 2% BSA. Ezrin was detected with a mouse monoclonal antibody 4A5 [11]. H,K-ATPase was detected with a mouse monoclonal antibody, 2G11, against its β-subunit (Affinity Bioreagents). Primary antibodies were detected by FITC-labeled goat anti-mouse IgG (Jackson Labs). Actin was detected by coincident incubation with 80 nM rhodamine-labeled phalloidin (Molecular Probes). Cells were visualized by conventional fluorescence microscopy with a Nikon Microphot FX-2 using Inovision software to collect images. Confocal microscopy was performed with a Zeiss Axioplan microscope and a Biorad MRC 1024 laser system, using COMOS software to collect images.

## RESULTS

*Lat B inhibits acid accumulation and increases soluble actin.* As described in detail in Ammar *et al* [2], Lat B inhibited acid accumulation in glands only at the highest concentrations tested. Treatment with 50 μM Lat B reduced secretion to 64%±10% of stimulated controls, and 100 μM reduced secretion to 31%±6% controls. There was an increase in the detergent-soluble actin (and corresponding decrease in the cytoskeletal actin) in both resting and stimulated gland preparations as the concentration of Lat B was increased. The extractable actin in stimulated glands increased by three-fold at 10 μM Lat B and by six-fold at 100 μM [2]. When these experiments were repeated in cultured parietal cells, Lat B inhibited secretion in a dose dependent manner, but at lower doses. From these limited data the $ED_{50}$ for Lat B was considerably higher for intact glands (~70 μM) compared to the cells in culture (~8 μM), thus the isolated, cultured cells appeared more sensitive.

*Lat B affects actin structures in stimulated parietal cells at concentrations that do not affect acid accumulation.* To determine the morphological effects of Lat B on actin, and to ascertain whether Lat B interferes with the translocation of the H,K-ATPase, cultured parietal cells were prepared for immunofluorescence. Cells stimulated in the presence of 0.1 μM Lat B had the same morphological characteristics as control secreting cells; that is, vacuoles were swollen and there was reasonably good overlap in the probes for F-actin and H,K-ATPase (Fig. 1C,C'). However, even at these low concentrations, Lat B had major morphological effects on lamellipodial formation (Fig. 2B); the lamellipodia were very much condensed, or absent altogether. Where present, the leading edge of remnant lamellipodia stained brightly with phalloidin, almost as if the rest of the structure had been dragged back toward the cell. Further perturbations of the cytoskeleton were seen in stimulated parietal cells treated with 1 μM Lat B. At this concentration, stimulated cells had intense punctate phalloidin-stained structures that were especially apparent in the plane of the retracted lamellipodia (Fig. 2C), and the staining of the surrounding plasma membrane was often less intense and discontinuous. However, in most other respects the morphological response to stimulation in the presence of 1 μM Lat B appeared normal, with large swollen vacuoles (Fig. 1D,D').

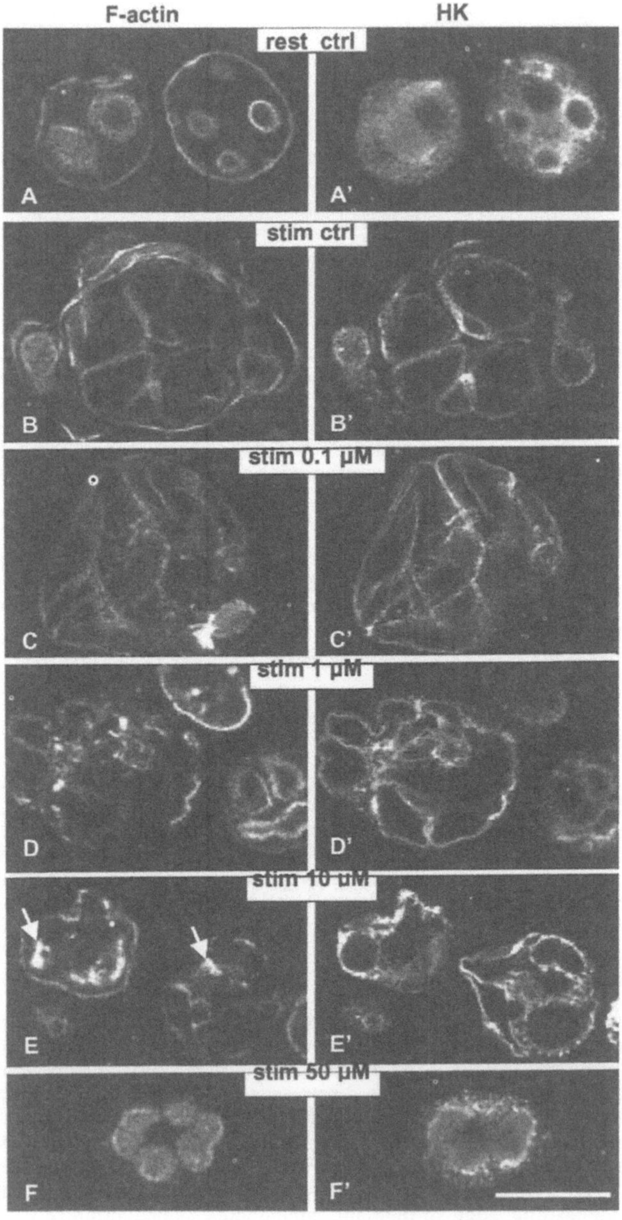


**Figure 1.**
The morphological effects of Lat B on cultured parietal cells. Cultured parietal cells were treated as described in METHODS. At 0.1 μM Lat B (C, C') cells were nicely swollen with no consistent differences noted from control stimulated cells. At 1 μM Lat B (D, D') cells also looked well stimulated; staining pattern was similar to control cells except for the appearance of some more punctate F-actin staining. By 10 μM Lat B (E, E') the morphology was distinctly different from control although much of the HK was still localized to the partially swollen vacuoles; F-actin staining tended to be non-uniform and punctate; frequently, phalloidin stained rod-like structures accumulated in the nucleus (E, arrows). At 50 μM Lat B (F,F') the cells showed little characteristic of the stimulated morphology except for some accumulation of HK on the non-swollen vacuoles. Bar marker is 20 μm.

*Lat B causes gross changes in actin at 10 to 50 μM.* At 10 μM Lat B, the number of swollen parietal cells and the magnitude of swelling were much reduced, compared with lower Lat B treatments (Fig 1E). Moreover, those cells with a stimulated morphology did not appear normal. While the vacuoles of the stimulated cells were somewhat expanded, they did not appear to have many large vacuolar spaces. Often intense punctate H,K-ATPase staining co-localized with the punctate actin staining. Furthermore, the H,K-staining frequently appeared heavier on the outer surfaces of vacuolar membranes facing the plasma membrane, and lighter on the vacuolar membranes facing toward the center of the cell (Fig 1E'). The lamellipodia are greatly reduced (Fig. 2C). All of these characteristics worsen when Lat B was increased to 50 μM (Fig. 1F). Few cells on the stimulated coverslips showed translocation of the H,K-ATPase. Cells that did have some vacuolar accumulation of H,K-ATPase were rarely swollen (Fig. 1F'), looking like stimulated cells treated with a pump inhibitor.

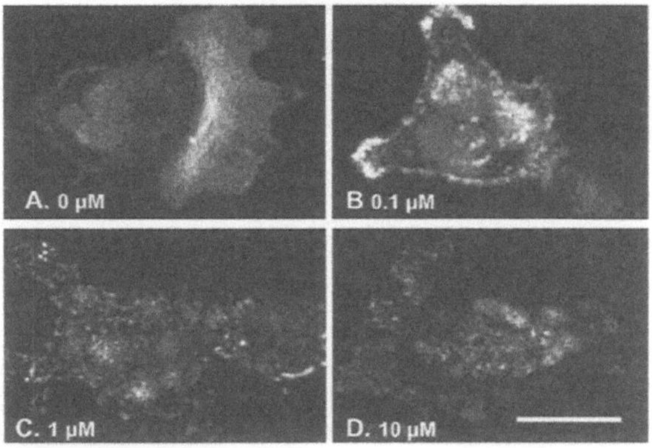


**Figure 2.** Effects of Lat B on cultured parietal cells stimulated in the presence of a proton pump inhibitor. All cell cultures treated as in Figure 1 in the presence of 10 μM SCH28080 as a proton pump inhibitor. When control parietal cells were stimulated in the presence of a pump inhibitor (A) large F-actin-rich lamellipodial extensions form. Lat B at 0.1 (B) and 1 μM (C) causes a major disruption of the lamellipodial F-actin as well as discontinuities of F-actin staining at the surrounding plasma membrane. At 10 μM Lat B, F-actin staining appears to be more spotty and punctate, and lamellipodial F-actin is completely disrupted (D). Bar marker is 20 μm.

*Pretreatment of resting cells with Lat B has minimal effects on subsequent response to stimulation.* In contrast to what was seen for stimulated parietal cells, treatment of cultures that were maintained in the resting condition (cimetidine) with Lat B for 30 min had relatively minimal effects on F-actin staining. When 50 μM Lat B was applied to resting cells for 30 min the vacuolar F-actin staining appeared normal but staining at the plasma membrane tended to become spotty and we often saw the pattern of nuclear rod-like phalloidin staining (Fig. 3A). Because of the striking differences in the morphological effects of Lat B on resting and stimulated cells, we tested the responses of cells pretreated for 30 min in the resting state with Lat B, rinsed briefly, and then stimulated (± SCH) in the absence of the drug. Stimulation of cells, after pretreated with up to 50 μM Lat B, produced morphological secretory responses that were indistinguishable from control stimulated cells (Fig. 3B). The vacuoles appeared swollen, and there was good co-localization of H,K-ATPase and F-actin. The cells were entirely lacking of the punctate phalloidin staining that was so prevalent in cells stimulated in the presence of 1 μM or greater Lat B (cf. Figs. 3 & 4). Consistent with the washout results from stimulated cells, the SCH cells pretreated with up to 50 μM Lat B showed good translocation of the H,K-ATPase from the cytoplasm to vacuolar membranes. Thus this aspect of stimulation-associated membrane recruitment was not apparently altered by the pretreatment of resting cells (Fig. 3C). In fact SCH cells pretreated with 1 μM Lat B displayed excellent lamellipodial formation and appeared similar to control cells in all respects. However, lamellipodial formation in the SCH cells became diminished when the pre-incubation was 10 μM Lat B, and was eliminated at 50 μM Lat B (Fig. 3C), even though all other morphological characteristics of these cells appeared quite normal.

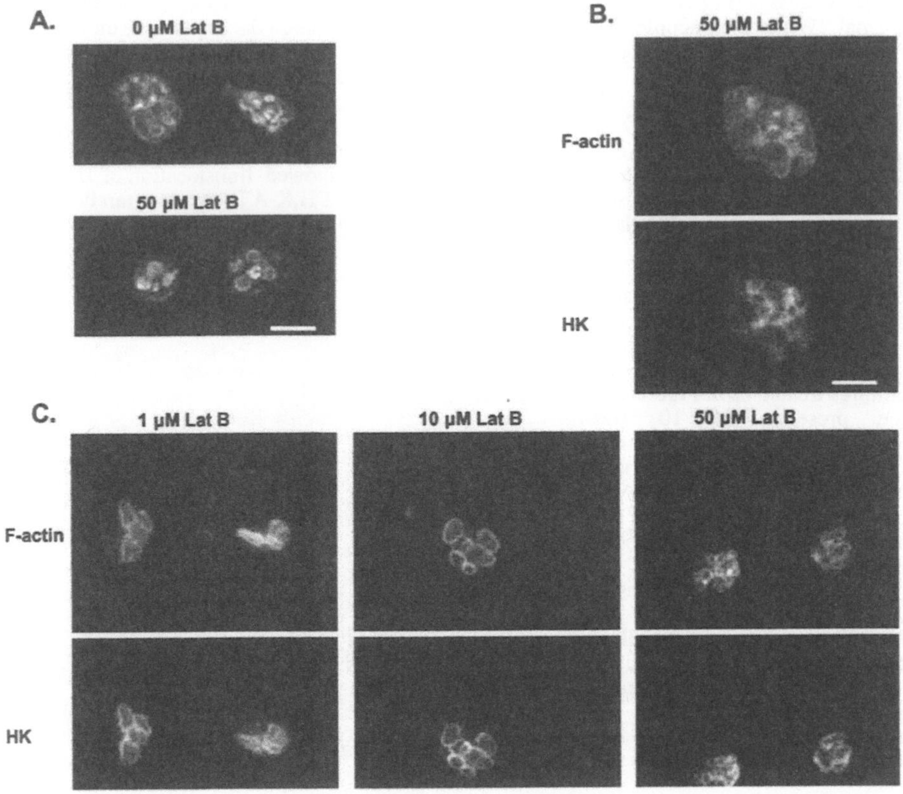


**Figure 3.** Effects of Lat B on resting parietal cells and on parietal cells that are stimulated after a pre-incubation and wash-out of Lat B. All incubations were at 37°C, after which the cells were fixed, permeabilized, and stained for F-actin and H,K-ATPase (HK) as described in METHODS. **A.** At 50 μM Lat B, the vacuolar staining of F-actin was comparable to controls with some diminution of F-actin staining at the surrounding plasma membrane. **B.** Resting parietal cells were pre-incubated in the indicated concentration of Lat B for 30 min, the reagent was then washed free and the cells were stimulated for 30 min. The morphological responses to stimulation were normal even up to 50 μM Lat B. **C.** Resting parietal cells were pre-incubated in the indicated concentration of Lat B for 30 min, the reagent was then washed free and the cells were stimulated for 30 min in the presence of SCH28080. For all concentrations of Lat B the recruitment of HK to the secretory vacuoles, and co-localization with F-actin, was normal. Lamellipodial formation at 1 μM Lat B was also normal (in contrast to Fig. 4), although these extensions were reduced with 10 μM and eliminated with the 50 μM Lat B pre-incubation. Bar marker is 20 μM.

*LO-permeabilized glands secete acid.* The effects of several inhibitors of secretory function were assayed in order to compare the characteristics SLO-permeabilized glands (nucleotide-dependent stimulation) to intact glands (histamine/IBMX stimulation). SCH28080 inhibits AP accumulation by directly inhibiting the proton pump, the H,K-ATPase [12]. Wortmannin and the anti-ulcer drug ME-3407 are both inhibitors of acid secretion that do not act directly on the proton pump, but rather through some intermediary signaling step such as phosphorylation or the action of ezrin [13]. As shown in Figure 4, addition of 10 μM SCH28080 to either intact or SLO glands completely abolished AP accumulation, and actually reduced the AP uptakes to levels below that seen in non-stimulated glands. Addition of 0.3 mM ME-3407 also reduced AP uptakes for both intact and SLO glands, although the degree of inhibition tended to be somewhat more effective for the intact glands. Surprisingly, while 10 μM wortmannin effectively inhibited intact glands (15 ±6% of stimulated controls), AP uptake by the SLO glands was only slightly reduced by wortmannin (86 ±11% of stimulated controls; $p > 0.05$).

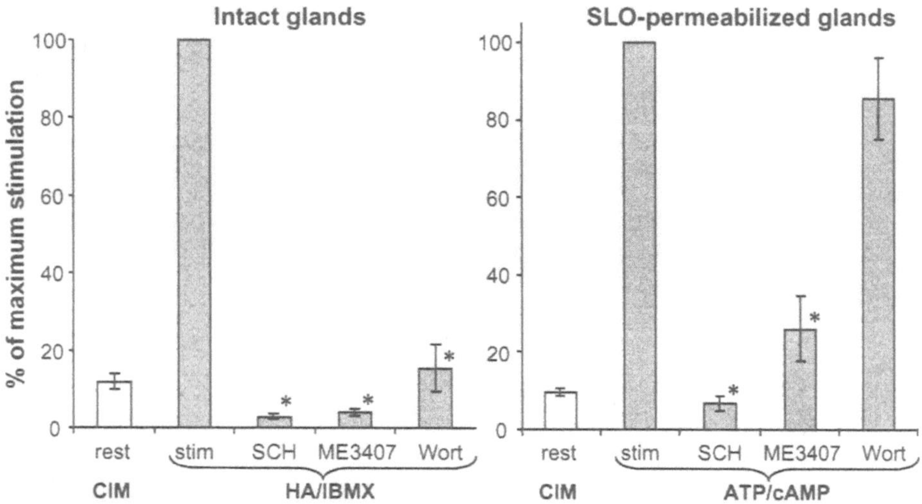


**Figure 4.** Inhibitors of AP uptake in gastric glands. AP uptake was measured for intact and SLO-permeabilized gastric glands in the resting (open bars) and stimulated states (shaded bars). In addition the stimulated glands were treated with several inhibitors as indicated, including the H,K-ATPase-specific inhibitor SCH28080 (SCH, 10 μM), and kinase inhibitors wortmannin (Wort, 10 μM) and ME-3407 (300 μM). Error bars are the SEM of at least three experiments; * indicates a significant difference with respect to stimulated control ($P < 0.05$).

*Ezrin in SLO-permeabilized glands.* To further explore this difference in action between wortmannin and ME-3407, we looked at the effects of both drugs on the cytoskeletal-membrane linking protein ezrin. ME-3407 had previously been shown to cause a delocalization of ezrin away from the plasma membrane and into the cytoplasm [13]. Here, we compared intact and SLO-glands that had been prepared for immunofluorescence using an anti-ezrin monoclonal antibody. TRITC-conjugated phalloidin was used as a second probe to localize the F-actin in the glandular cells. In general, the ezrin staining was weaker in SLO-glands as compared to intact preparations [not shown], most likely due to some ezrin exiting the cell through SLO-pores. However, the remaining ezrin signal was clearly localized to the canalicular membranes of parietal cells in both intact and SLO-glands [Fig. 5], and in both the resting (not shown) and stimulated states. ME-3407 caused a remarkable loss of ezrin staining in the intact and SLO glands [Fig. 5, ME-3407]. Furthermore, the remaining ezrin signal was diffusely distributed throughout the parietal cell cytoplasm, and entirely lacking in association with membrane structures. Wortmannin treatment may also have reduced ezrin staining in intact glands, but not to the same extent as did ME-3407 [not shown]. After wortmannin treatment, canalicular structures were still visible in many, but not all parietal cells and the diffuse ezrin signal in the cytoplasm was increased. In contrast, wortmannin had little effect on the ezrin staining pattern of SLO-glands, i.e., canalicular structure remained distinct and clearly defined [Fig. 5, Wort].

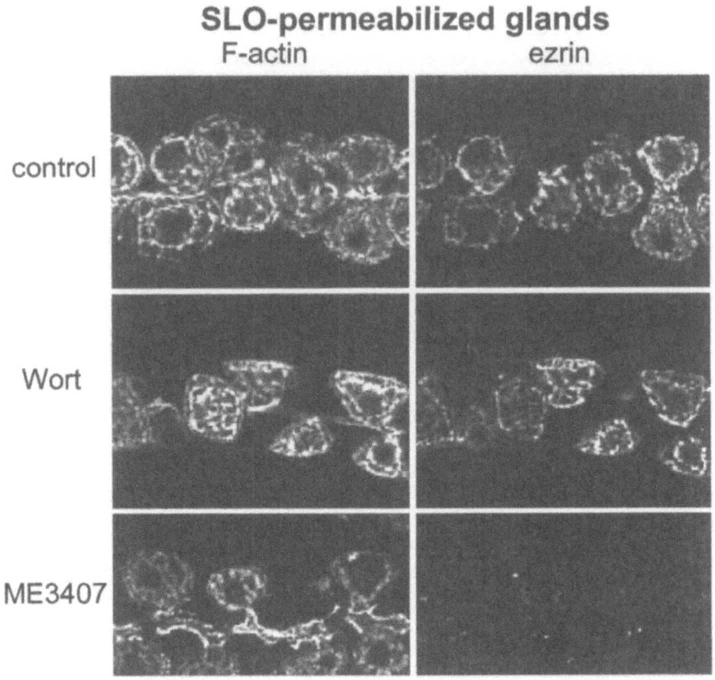


**Figure 5.** Ezrin and actin staining SLO-permeabilized glands respond differentially to kinase inhibitors. SLO-permeabilized glands were prepared as described in METHODS. In all cases the pattern of F-actin staining was not significantly altered, either by SLO-permeabilization or by the inhibitors, with the prominent actin cytoskeleton localization along the entire gland lumen and the apical/canalicular and basolateral membranes of the parietal cell. For intact control glands ezrin was localized as previously reported [11] to the subset of actin associated with the apical/canalicular membrane and to a lesser extent the basolateral membrane of parietal cells. This same general ezrin staining pattern was observed for SLO-permeabilized control glands, except with a somewhat reduced intensity than for intact, presumably due to ezrin leakage from permeabilized cells. ME3407 produced a total loss of parietal cell ezrin in SLO as well as intact glands (not shown), whereas F-actin was not altered. Wortmannin caused some reduction of ezrin staining in the intact glands but no differences were detected for the SLO-permeabilized glands.

## DISCUSSION

The diminution of acid secretion by latrunculin B (Lat B), coupled with morphological assessments that show inhibition of secretagogue-dependent translocation of H,K-ATPase, support the hypothesis that actin plays a role in acid secretion . However, the mechanism of Lat B action allows further dissection of this actin-dependent process not previously characterized [rev in 9]. Phalloidin staining reveals that parietal cell F-actin is abundant at or near the plasma membranes, especially within the microvillar filaments lining the tortuous secretory canaliculi that constitute the apical plasma membrane [10]. Relatively high concentrations of Lat B are required to alter the histamine/IBMX secretory responses of parietal cells in freshly isolated glands. The resistance of glandular F-actin to Lat B was validated by the pattern of phalloidin staining, which was fairly normal with Lat B concentrations up to 50 μM. Since Lat B depletes the G-actin pool as F-actin turns over, it affects all actin filaments uniformly.

Parietal cells maintained in culture were more sensitive than intact glands to Lat B, yet inhibition of functional responses, such as acid accumulation and secretagogue-dependent morphological rearrangement, required much higher levels of Lat B than were needed for disruption of more rudimentary cytoskeletal structures. Doses of 1 μM Lat B or greater were required to block the morphological responses of cultured parietal cells to stimulation by secretagogues, including the stimulation-associated recruitment of H,K-ATPase. Notwithstanding the relative refractoriness of the acid secretory functions, there were obvious cytoskeletal effects of Lat B on parietal cells and other cell types in culture at much lower concentrations of the drug. The most obvious effects of low dose Lat B on parietal cells were observed on the lamellipodial formations that accompany secretagogue stimulation in the cultured system. These flattened, F-actin-rich, highly mobile protrusions were significantly altered at sub-micromolar concentrations of Lat B, and we frequently observed accumulations of punctate foci of phalloidin staining at these low drug levels. While Lat B destabilizes some actin structures through sequestation of actin monomer, there also appears to be some relatively resistant F-actin structures in the parietal cell, and it seems likely that the secretory response is more closely tied to the more stable pool of actin filaments.

We propose that the resistance of acid secretory function of parietal cells to Lat B is the result of very stable actin filament turnover pathways that minimize the accessibility of the actin monomer to Lat B. High doses of Lat B and/or long incubation times are required to disrupt the parietal cell cytoskeleton. We propose that when parietal cells are placed into culture, a more primitive cytoplasmic motile activity becomes manifest that is more sensitive to Lat B. If a high F/G turnover in this pool were easily accessible to Lat B this could represent a significant 'drain' on the total cellular actin when the lamellipodial activity was stimulated by histamine. Such an hypothesis would be consistent with (i) the high sensitivity of lamellipodia, (ii) the resistance of F-actin structure in presumed microvillar loci, and (iii) the refractory nature of the cytoskeleton to Lat B in cultured parietal cells at rest. For the latter case where there is no activation/turnover of a lamellipodial pool, there is little stress on the flow of actin to the Lat B-sensitive state. Accordingly, we did not see any notable morphological effects of Lat B on resting cells, and more importantly, the functional responses of cultured parietal cells to histamine were totally unabated by pretreatment with Lat B while in the resting state.

In order to further examine those signaling events downstream of histamine receptor activatioin, we have developed a permeabilized gastric gland model. While the cAMP-mediated stimulation of SLO-glands is in many ways similar to histamine/IMBX stimulation of intact glands, two known kinase inhibitors have different effects on these two systems. Previous experiments had shown that both ME-3407 and wortmannin inhibit acid secretion in intact glands, and it was suggested that they had a common target in the enzyme myosin light chain kinase [13]. ME-3407 was also shown to have the further effect of delocalizing ezrin from its usual membrane bound location in parietal cells. In SLO-glands, we found that ME-3407 both inhibited AP uptake and caused the delocalization of ezrin. Wortmannin, on the other hand, had rather minimal effect on ezrin staining in either intact or SLO-glands. Furthermore, it had radically different effects on acid secretion in the intact and SLO-permeabilized preparations. Consistent with earlier work [13] wortmannin inhibited AP uptake by intact glands but, unpredictably, had no significant effect on SLO-glands. This suggests that ME-3407 and wortmannin have different targets, unlike previously proposed. The wortmannin-sensitive factor may be released into the extracellular medium by permeabilization; whereas, the ME-3407-sensitive factor appears to be retained by the permeabilized gland.

Based on these findings, it is not unreasonable to suggest that the inhibitory effect of ME-3407 on acid secretion may be due to some ezrin-related effect. While total ezrin levels were reduced after SLO-permeabilization, the remnant ezrin staining pattern is reminiscent of staining seen in intact glands. Higher laser power was required to get the same level of intensity, but there remained distinct canalicular staining in parietal cells. The finding of reduced ezrin levels is at odds with the hypothesis that ezrin is an integral component of the stimulation process [see 14]. However, it may be that the remnant ezrin is sufficient for the process of transforming the parietal cell from the resting to the stimulated state. Or, it is possible that membrane-associated ezrin is not required for the immediate events leading to acid secretion, although ezrin may still be required as a membrane organizing protein.

## ACKNOWLEDGEMENTS

This work was supported in part by NIH Grants DK10141 and DK38972 (to J.G.F.) and NIH NRSA Award DK09984-01 (to D.A.A.). We would like to thank Jenny Liang, Jenny Chen, Sharon Kim, Yu-Tsun Chen, and Joseph Eisenberg for their assistance in experimental procedures.

## REFERENCES

1. **Agnew BJ, Duman JG, Watson CJ, Coling DJ, and Forte JG.** Cytological transformations associated with parietal cell stimulation: critical steps in the activation cascade. *J Cell Sci* 112:2639-4266, 1999.
2. **Ammar DA, Nguyen PNB, and Forte JG.** Functionally distinct pools of actin in secretory cells. *Am J Physiol Gastrointest Cell Physiol* 281:C407-C417, 2001.
3. **Ayscough KR, Stryker J, Pokala N, Sanders M, Crews P, and Drubin DG.** High rates of actin filament turnover in budding yeast and roles for actin in establishment and maintenance of cell polarity revealed using the actin inhibitor latrunculin-A. *J Cell Biol* 137:399-416, 1997.
4. **Berglindh T.** Gastric glands and cells: preparation and in vitro methods. *Methods Enzymol* 192:93-107, 1990.
5. **Buckingham L and Duncan JL.** Approximate dimensions of membrane lesions produced by Streptolysin S and Streptolysin O. *Biochim Biophys Acta* 729:115-122, 1993.
6. **Coue M, Brenner SL, Spector I, and Korn ED.** Inhibition of actin polymerization by latrunculin A. *FEBS Lett* 213:316-318, 1987.
7. **Forte JG, Ganser A, Beesley R, and Forte TM.** Unique enzymes of purified microsomes from pig fundic mucosa. $K^+$-stimulated adenosine triphosphatase and $K^+$-stimulated pNPPase. *Gastroenterology* 69:175-189, 1975.
8. **Forte JG, Ly B, Rong Q, Ogihara S, Ramilo M, Agnew B, and Yao X.** The state of actin in gastric parietal cells. *Am J Physiol Cell Physiol* 274:C97-C104, 1998.
9. **Forte JG and Yao X.** The membrane recruitment and recycling hypothesis of gastric HCl secretion. *Trends Cell Biol* 6:45-48, 1996.
10. **Forte TM, Machen TE, and Forte JG.** Ultrastructural changes in oxyntic cells associated with secretory function: a membrane recycling hypothesis. Gastroenterology 73:941-955, 1977.
11. **Hanzel DK, Urushidani T, Usinger WR, Smolka A, and Forte JG.** Immunological localization of an 80-kDa phosphoprotein to the apical membrane of gastric parietal cells. *Am J Physiol Gastrointest Liver Physiol* 256:G1082-G1089, 1989.
12. **Urushidani T, Hanzel DK, and Forte JG.** Characterization of an 80-kDa phosphoprotein involved in parietal cell stimulation. *Am J Physiol Gastrointest Liver Physiol* 256:G1070-G1081, 1989.
13. **Urushidani T, Muto Y, Nagao T, Yao X, and Forte JG.** ME-3407, a new antiulcer agent, inhibits acid secretion by interfering with redistribution of H+-K+-ATPase. *Am J Physiol Gastrointest Liver Physiol* 272:G1122-G1134, 1997.
14. **Wallmark B, Briving C, Fryklund K. Munson R. Jackson J. Mendelein Rabon E, and Sachs G..** Inhibition of gastric $H^+$-$K^+$-ATPase and acid secretion by SCH 28080, a substituted pyridyl(1,2a)imidazole. *J Biol Chem* 262: 2077-2084, 1987.
15. **Yao X, Thibodeau A and Forte JG.** Ezrin-calpain I interactions in gastric parietal cells. *Am J Physiol Cell Physiol* *265*: C36-C46, 1993.

Chapter 37

# Clathrin And Associated Proteins On Tubulovesicles And Apical Membranes Of Parietal Cells

CURTIS T. OKAMOTO, RUI LI, AND CATHERINE S. CHEW*
*Department of Pharmaceutical Sciences, University of Southern California, Los Angeles, CA, 90089-9121 and *Institute of Molecular Medicine and Genetics, Medical College of Georgia, Augusta, GA, 30912-2630*

## INTRODUCTION

In the unstimulated gastric parietal cell, a major proportion of the gastric H,K-ATPase resides in an intracellular tubulocisternal (tubulovesicular) membrane compartment. Upon the stimulation of the parietal cell, the H,K-ATPase is delivered to the canalicular (apical) membrane as a result of fusion of tubulocisternal membranes with the apical membrane. Upon cessation of stimulation, the H,K-ATPase is retrieved from the canalicular membrane, and the tubulocisternal compartment is re-established. This membrane recycling model of the H,K-ATPase necessitates that protein sorting machinery is present in relative abundance to regulate the trafficking of the H,K-ATPase. Thus, the identification and functional characterization of the sorting machinery should be facilitated. To a significant extent, key components of the sorting machinery in parietal cells have in fact been characterized (reviewed in ref. 21). This report summarizes the identification and characterization of clathrin and clathrin-associated proteins in the parietal cell.

## IDENTIFICATION OF CLATHRIN IN THE PARIETAL CELL

During the differentiation of the parietal cell, tubulocisternae appear to arise from a dramatic elaboration of the trans-Golgi network (TGN) (10). Thus, the sorting machinery associated with tubulocisternae in fully differentiated parietal cells might be related to that which regulates membrane traffic from the TGN. Indeed, a well-known family of sorting machinery associated with budding vesicles from the TGN, clathrin and clathrin adaptors, has been identified on tubulocisternae.

Clathrin and clathrin adaptors are the best-characterized family of so-called vesicular coat protein complexes of intracellular protein sorting machinery (13,15,25). Vesicular coat proteins are cytosolic protein complexes that cycle on and off membranes. They regulate the production of transport vesicles from a donor membrane by providing a mechanism both to concentrate cargo into the nascent vesicle and to deform the planar lipid bilayer into a vesicle by a controlled assembly (polymerization) process. Conventional clathrin is a heterodimeric complex, comprised of a 160 kDa heavy chain and a 32 kDa light chain, either light chain A or light chain B. The clathrin coat is linked to vesicle membrane proteins via heterotetrameric clathrin adaptors which contain binding sites for both clathrin and the cargo membrane protein. Clathrin adaptors belong to a growing family of proteins that appear to regulate protein sorting along the exocytotic and endocytotic pathways (13,25).

On Western blots of density gradient-purified gastric microsomes derived from tubulocisternae, clathrin heavy chain was readily detected with antibodies against conventional clathrin heavy chain (18). The positive identification of gastric microsomal clathrin heavy chain as the conventional isoform resulted from mass spectrometry (20).

This identification was important, since recently another clathrin heavy chain homologue has been cloned that is expressed ubiquitously in developing organisms but is predominantly expressed in muscle cells in adult organisms (14). Western blots of gastric microsomes probed with isoform-specific anti-light chain antibodies also demonstrated that light chain A is the predominant light chain isoform associated with gastric microsomal clathrin (20).

The clathrin adaptor associated with tubulocisternae is the AP-1 clathrin adaptor (18), the isoform normally associated with clathrin-coated vesicles in the TGN. However, a unique feature of the tubulocisternal AP-1 adaptor is that one of its subunits, the β-adaptin subunit, is immunologically distinct from that found in AP-1 adaptors of the TGN (19). The nature of the immunological difference has not been characterized but could represent either a consequence of post-translational modification, e.g., phosphorylation, of the β-adaptin or a distinct β-adaptin isoform. Furthermore, the functional consequences of the presence of the immunologically distinct β-adaptin is not known.

Clathrin and AP-1 clathrin adaptors associated with gastric microsomes can be stripped from the membranes and re-polymerized in vitro into ultrastructurally identifiable clathrin baskets (20). However, unlike the distinctive clathrin-coated vesicles formed at the plasma membrane and the TGN, clathrin-coated vesicles could not be identified ultrastructurally on tubulocisternae until recently, when clathrin was localized to tubulocisternae by immunoelectron microscopy of unstimulated parietal cells that were fixed by a high pressure freezing and freeze-substitution protocol (20). Thus, despite the similarity of clathrin baskets polymerized in vitro from gastric microsomal clathrin to those polymerized in vitro from clathrin isolated from other sources, it appears that, in vivo, clathrin on tubulocisternae may possess particular properties, e.g., alternative assembled complexes, that allow it to evade detection when standard electron microscopic protocols are used.

The identification of clathrin on tubulocisternae by immunoelectron microscopy suggests that vesicles are actively budding from this compartment in resting parietal cells. Presumably, AP-1 adaptors are also present within these clathrin-coated profiles, and it would be important to confirm the presence of AP-1 adaptors on tubulocisternae by immunoelectron microscopy. AP-1 adaptors in tubulocisternae could bind directly to the cytoplasmic domains of a number of integral membrane proteins found in this membrane compartment, which then would regulate their incorporation into clathrin-coated vesicles. Such cargo could be the H,K-ATPase, vesicle-associated membrane protein 2 (VAMP 2) (3), secretory carrier associated membrane protein (SCAMP) (3), or syntaxin 3 (3,22). Indeed, published work suggests that the AP-1 adaptor appears to reside in a complex with the H,K-ATPase (18). The β-subunit of the H,K-ATPase is an attractive candidate for the binding of clathrin adaptors. Its cytoplasmic domain contains a tyrosine-containing motif 8, 26) and an acidic di-leucine type motif (Okamoto, unpublished), the location of which are shown in Figure 1. Both of these motifs are well-conserved among all H,K-ATPase β-subunits cloned to date. These two motifs have been documented to regulate the sorting of other membrane proteins by a clathrin- and clathrin adaptor-dependent mechanism (1,2,13,32). The interaction of the β-subunit of the H,K-ATPase with clathrin adaptors may be characterized in the future by recent popular approaches to detect protein-protein interactions, such as the yeast two-hybrid screen and "pull-down" assays with glutathione S-transferase fusion proteins. In addition, the identification of the other cargo molecule(s) of the tubulocisternal clathrin-coated buds should provide significant insight into the role of clathrin and AP-1 adaptors on tubulocisternae.

In addition to its localization to tubulocisternae, clathrin has been localized to the canalicular membrane of unstimulated parietal cells by immunofluorescence (20). Clearly identifiable clathrin-coated pits at the canalicular membrane have recently been documented by electron microscopy (20). This unexpected finding suggests that the canalicular membrane of the resting parietal cell is endocytotically active. The clathrin

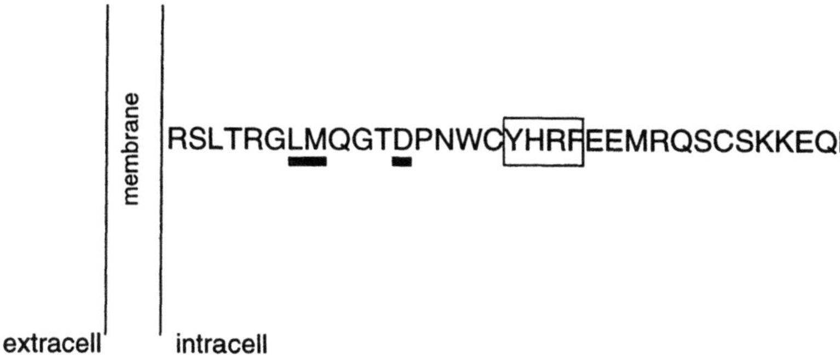


**Figure 1.** The cytoplasmic domain of the β-subunit of the H,K-ATPase possesses a tyrosine-based motif (boxed) and an acidic di-leucine type (Met-Leu) motif (bold underline). These motifs have been shown to regulate clathrin adaptor-mediated sorting of other membrane proteins.

AP-2 adaptor also has been localized to the canalicular membrane by immunofluorescence (20). As in the case of identification of the cargo for the tubulocisternal AP-1 adaptor, the identification of the cargo for the AP-2 adaptor at the canalicular membrane should provide insight into the role of clathrin and AP-2 adaptors in the regulation of apical membrane protein sorting in the parietal cell.

## PUTATIVE ROLES OF CLATHRIN IN PARIETAL CELLS

Clathrin and clathrin adaptors thus constitute at least a significant part, if not the majority, of the vesicular coat protein complexes necessary to effect membrane and protein kinesis occurring during the secretory cycle in parietal cells. In one model of the secretory cycle, the first step of the massive internalization of the H,K-ATPase from the canalicular membrane upon cessation of stimulation may be dependent upon clathrin-mediated endocytosis. A transgenic mouse has been created in which the putative internalization motif in the cytoplasmic domain of the β-subunit of the H,K-ATPase was mutated by substitution to alanine (8). These mice develop histological pathologies consistent with a hypersecretory phenotype, and it was suggested that the H,K-ATPase is unable to be internalized from the canalicular membrane, thereby resulting in chronic gastric HCl secretion. Alternatively, however, this hypersecretory phenotype is also consistent with an inability to retain the H,K-ATPase in tubulocisternae. In this case, the intracellular tubulocisternal retention of the H,K-ATPase in resting cells would be dependent upon the tyrosine-based "internalization" motif, and if absent would result in the steady-state accumulation of H,K-ATPase at the canalicular membrane.

The presence of clathrin-coated membrane domains on tubulocisternae and canaliculi suggests that even unstimulated parietal cells are active with respect to vesicular traffic and that an exchange of membrane and membrane proteins is occurring between the canalicular membrane and tubulocisternae. The recycling of membrane between the canaliculus and tubulocisternae may represent the constitutive recycling of yet unidentified "housekeeping" membrane proteins. Alternatively, this recycling could represent a quality control mechanism to return membrane protein cargo, misplaced during the secretory cycle, to their proper locations during interdigestive periods. For example, when the stimulus for secretion is withdrawn, large portions of apical membrane (and therefore a major proportion of the H,K-ATPase) may be internalized in a manner analogous to phagocytosis, with little regard to cargo selectivity. Once the mass uptake is complete, clathrin-coated vesicles forming at the canaliculus may return any H,K-ATPase remaining there to tubulocisternae, while vesicles budding from tubulocisternae may return a subset of SNARES to the plasma membrane. Both of these models are summarized in Figure 2.

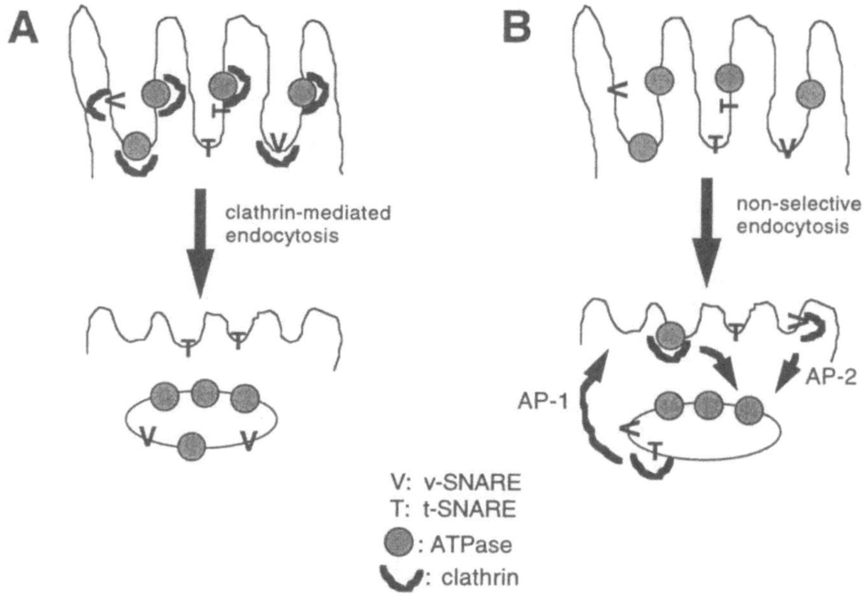


**Figure 2.** Two models of clathrin function in the secretory cycle of the parietal cell. **A.** Upon withdrawal of the stimulus for secretion, clathrin and the AP-2 clathrin adaptor mediate the endocytosis of H,K-ATPase (and v-SNARES) from the apical membrane . **B.** Upon withdrawal of the stimulus for secretion, there is a bulk, non-selective internalization of apical membrane, followed by a prolonged period of clathrin-mediated sorting. Any H,K-ATPase and v-SNAREs remaining at the apical membrane are retrieved via clathrin and AP-2 clathrin adaptors, while inadvertently internalized t-SNARES (and other apical membrane proteins) are recycled from intracellular membranes to the apical membrane via a clathrin- and AP-1 clathrin adaptor-dependent mechanism.

The parietal cell may provide unique insights into the association of clathrin and clathrin adaptors with intracellular organelles and their roles in the regulation of recycling of apical membrane proteins in epithelial cells. Indeed, tubulocisternae can be considered a novel clathrin- and AP-1 clathrin adaptor-coated organelle that may regulate apical membrane recycling in many other epithelial cell types (31). In addition, the parietal cell may represent a good model system in which to characterize the regulation of clathrin-mediated protein sorting and vesicular traffic by signal transduction pathways.

## CLATHRIN-ASSOCIATED PROTEINS--DYNAMIN

Dynamins are a family of large (~100 kDa) GTPases with a multidomain structure (Fig. 3) (5,27,28). These domains participate in protein-protein interactions that regulate dynamin's function and/or localization. The most well-characterized function of dynamin is its regulation of the fission of membrane vesicles from a donor membrane organelle, particularly in clathrin-coated vesicle budding from the plasma membrane. In the unstimulated parietal cell, dynamin has been localized by immunfluorescence to the two

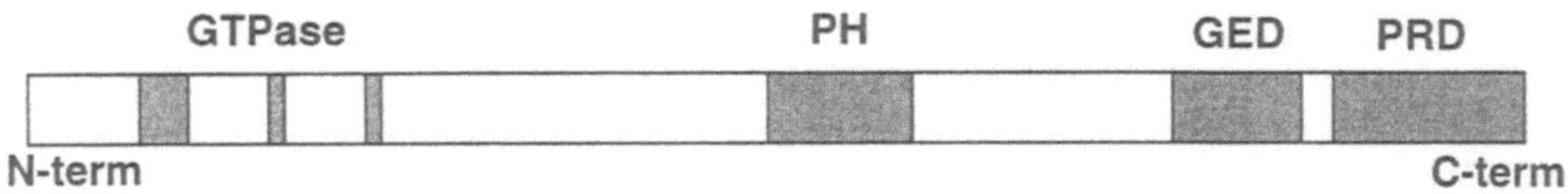


**Figure 3.** The multidomain structure of dynamin. GTPase: N-terminal, tripartite GTP binding domain; PH: pleckstrin homology domain thought to interact with phosphatidylinositol lipids; GED: GTPase effector domain which appears to behave as an intramolecular GTPase activating protein (GAP); PRD: C-terminal, proline-arginine-rich domain which interacts with src homology 3 (SH3) domain-containing proteins.

major sites of membrane-associated clathrin, the canaliculus (3,20) and tubulocisternae (20), where it would be predicted to regulate clathrin-coated vesicle fission at these sites.

Dynamins interact via their carboxy-terminal proline-arginine rich domain (PRD) with a select group of proteins that contain a src homology 3 (SH3) domain. This PRD-SH3 interaction regulates dynamin GTPase activity in vitro (11) and regulates endocytosis in vivo (12,29) and in permeabilized cell models (30). A recent exciting development is that several of these SH3 domain-containing proteins have also been demonstrated to regulate various aspects of the actin cytoskeleton, particularly with respect to cortical membrane dynamics and cell shape (12,16,17,23,24). Of significant interest would be to identify such SH3 domain-containing, dynamin-interacting proteins in the parietal cell, that might regulate dynamin function in vivo.

One such candidate protein in the parietal cell is lasp-1 (for LIM- and SH3-domain containing protein) (6,7). It is a multidomain-containing, 40 kDa protein that is phosphorylated upon stimulation of cAMP-dependent signaling pathways. In addition to consensus phosphorylation sites, it contains an N-terminal LIM domain, nebulin repeat domains, and a C-terminal SH3 domain. In the unstimulated parietal cell, lasp-1 is localized to certain regions along the canaliculus and to the basolateral cortical membrane. Upon stimulation, a significant fraction of lasp-1 appears to translocate from the basolateral cortical membrane to the canaliculus (7). The localization of a significant fraction of lasp-1 to the canaliculus would then allow for its interaction with not only dynamin but also the actin cytoskeleton at the canalicular membrane. This interaction would be predicted to occur by binding of the lasp-1 SH3 domain with dynamin's PRD domain; this interaction, in turn, may be regulated by phosphorylation of lasp-1.

There are several criteria that need to be fulfilled to identify an SH3 domain-containing protein such as lasp-1 as a binding partner for dynamin. One basic requirement is that of co-localization in cells. As mentioned above, dynamin and lasp-1 are both localized to the canalicular membrane, with the inferred co-localization being stronger in stimulated parietal cells compared to resting parietal cells. However, of co-localization of lasp-1 and dynamin in parietal cells has not yet been formally demonstrated. Alternatively, one could express lasp-1 in a heterologous expression system and assess the degree to which dynamin and lasp-1 are co-localized.

A second criterion would be to demonstrate direct interactions by the use of a number of currently popular approaches to detect protein-protein interactions. For example, co-immunoprecipitation of lasp-1 and dynamin from parietal cells would demonstrate the presence of a complex containing lasp-1 and dynamin, although it would not necessarily indicate a direct interaction between lasp-1 and dynamin. In vitro assays, such as "pull-down" assays with glutathione S-transferase fusion proteins, yeast two-hybrid screens, and protein overlay ("far Western") blots offer greater opportunities to screen for direct interactions, since these approaches often utilize specified domains of a protein, rather than the entire protein.

The third criterion is to demonstrate the functional interaction of dynamin and the SH3 domain-containing protein by heterologous expression of the entire protein or its SH3 domain in fibroblasts and assaying its effects on internalization of transferrin, a clathrin- and dynamin-mediated process. Of the SH3 domains tested thus far, all appear to be inhibitory (12,23,24,29,30).

The demonstration of a lasp-1-dynamin interaction would open new avenues in the investigation into the links between signal transduction, the actin cytoskeleton, and vesicular trafficking machinery in parietal cells. The characterization of the functional consequences of such interactions would also shed light on the mechanism of apical membrane remodeling during the secretory cycle of the parietal cell.

## REFERENCES

1. **Blagoveshchenskaya, A.D., Hewitt, E.W., and Cutler, D.F.**. Di-leucine signals mediate targeting of tyrosinase and synaptotagmin to synaptic-like microvesicles with PC12 cells. *Mol. Biol. Cell* 10: 3979-3990, 1999.

2. **Bonifacino, J.S. and Dell'Angelica, E.C.** Molecular bases for the recognition of tyrosine-based sorting signals. *J. Cell Biol.* 145: 923-926, 1999.

3. **Calhoun, B.C. and Goldenring, J.R.** Two rab proteins, VAMP-2, and SCAMPS are present on immunoisolated parietal cell tubulovesicles. *Biochem. J.* 325: 559-564, 1997

4. **Calhoun, B.C., Lapierre, L.A., Chew, C.S., and Goldenring, J.R.** Rab11a redistributes to apical secretory canaliculus during stimulation of gastric parietal cells. *Am. J. Physiol. Cell Physiol.* 275: C163-C170, 1998.

5. **Cao, H., Garcia, F., and McNiven, M.A.** Differential distribution of dynamin isoforms in mammalian cells. *Mol. Biol. Cell* 9: 2595-2609, 1998.

6. **Chew, C.S., Parente, J.A. Jr., Zhou, C.-J., Baranco, E., and Chen, X.** Lasp-1 is a regulated phosphoprotein within the cAMP signaling pathway in the gastric parietal cell. *Am. J. Physiol. Cell Physiol.* 275: C56-C67, 1998.

7. **Chew, C.S., Parente, J.A. Jr., Chen, X., Chaponnier, C., and Cameron, R.S.** The LIM and SH3 domain-containing protein, lasp-1, may link the cAMP signaling pathway with dynamic membrane restructuring activities in ion transporting epithelia. *J. Cell Sci.* 113: 2035-2045, 2000.

8. **Courtois-Coutry, N., Roush, D., Rajendran, V., McCarthy, J.B., Geibel, J., Kashgarian, M., and Caplan, M.J.** A tyrosine-based signal targets H/K-ATPase to a regulated compartment and is required for the cessation of gastric acid secretion. *Cell* 90: 501-510, 1997.

9. **Dunbar, L.A. and Caplan, M.J.** Ion pumps in polarized cells: sorting and regulation of the $Na^+$,$K^+$- and $H^+$,$K^+$-ATPases. *J. Biol. Chem.* 276: 29617-29620.

10. **Forte, G.M., Limlomwongse, L., and Forte, J.G.** The development of intracellular membranes concomitant with the appearance of HCl secretion in oxyntic cells of the metamorphosing bullfrog tadpole. *J. Cell Sci.* 4: 709-727, 1969.

11. **Gout, I., Dhand, R., Hiles, I.D., Fry, M.J., Panayotou, G., Das, P., Truong, O., Totty, N.F., Hsuan, J., Booker, G.W., Campbell, I.D., and Waterfield, M.D.** The GTPase dynamin binds to and is activated by a subset of SH3 domains. *Cell* 75: 25-36, 1993.

12. **Kessels, M.M., Engqvist-Goldstein, A.E.Y., Drubin, D.G., and Qualmann, B.** Mammalian Abp1, a signal-responsive F-actin-binding protein, links the actin cytoskeleton to endocytosis via the GTPase dynamin. *J. Cell. Biol.* 153: 351-366, 2001.

13. **Kirchhausen, T.** Adaptors for clathrin-mediated traffic. *Annu. Rev. Cell Dev. Biol.* 15: 705-732, 1999.

14. **Long, K.R. Trofatter, J.A., Ramesh, V., McCormick, M.K., Buckler, A.J.** Cloning and characterization of a novel human clathrin heavy chain gene (CLTCL). *Genomics* 35: 466-472, 1996.

15. **Marsh, M. and McMahon, H.T.** The structural era of endocytosis. *Science* 285: 215-219.

16. **McNiven, M.A., Kim, L., Krueger, E.W., Orth, J.D., Cao, H., and Wong, T.W.** Regulated interactions between dynamin and the actin-binding protein cortactin modulate cell shape. *J. Cell Biol.* 151: 187-198, 2000.

17. **Ochoa, G.-C., Slepnev, V.I., Neff, L., Ringstad, N., Takei, K., Daniell, L., Kim, W., Cao, H., McNiven, M., Baron, R., and De Camilli, P.** A functional link between dynamin and the actin cytoskeleton at podosomes *J. Cell Biol.* 150: 377-389, 2000.

18. **Okamoto, C.T., Karam, S.M., Jeng, Y.Y., Forte, J.G., and Goldenring, J.R.** Identification of clathrin and clathrin adaptors on tubulovesicles of gastric acid secretory (oxyntic) cells. *Am. J. Physiol. Cell Physiol* 274: C1017-C1029, 1998.

19. **Okamoto, C.T. and Jeng, Y.Y.** An immunologically distinct β-adaptin on tubulovesicles of gastric oxyntic cells. *Am. J. Physiol. Cell Physiol.* 275: C1323-C1329, 1998.

20. **Okamoto, C.T., Duman, J.G., Tyagarajan, K., McDonald, K.L., Jeng, Y.Y., McKinney, J., Forte, T.M., and Forte, J.G.** Clathrin in gastric acid secretory (parietal) cells: biochemical characterization and subcellular localization. Am. J. Physiol. Cell Physiol. 279: C833-C851, 2000

21. **Okamoto, C.T. and Forte, J.G.** Vesicular trafficking machinery, the actin cytoskeleton, and $H^+$-$K^+$-ATPase recycling in the gastric parietal cell. *J. Physiol.* 532: 287-296, 2001.

22. **Peng, X.-R., Yao, X., Chow, D.-C., Forte, J.G., and Bennett, M.K.** Association of syntaxin 3 and vesicle-associated membrane protein (VAMP) with $H^+/K^+$-ATPase-containing tubulovesicles in gastric parietal cells. *Mol. Biol. Cell* 8: 399-407, 1997.

23. **Qualmann, B. and Kelly, R.B.** Syndapin isoforms participate in receptor-mediated endocytosis and actin organization. *J. Cell Biol.* 148: 1047-1061, 2000.

24. **Qualmann, B., Kessels, M.M., and Kelly, R.B.** Molecular links between endocytosis and the actin cytoskeleton. *J. Cell Biol.* 150: F111-F116, 2000.

25. **Robinson, M.S.** Coats and vesicle budding. *Trends Cell Biol.* 7: 99-102, 1997.

26. **Roush, D.L., Gottardi, C.J., Naim, H.Y., Roth, M.G., and Caplan, M.J..** Tyrosine-based membrane protein sorting signals are differentially interpreted by polarized Madin-Darby canine kidney and LLC-$PK_1$ epithelial cells. *J. Biol. Chem.* 273: 26862-26869, 1998.

27. **Schmid, S.L., McNiven, M.A., and DeCamilli, P.** Dynamin and its partners: a progress report. *Curr. Op. Cell Biol.* 10: 504-512, 1998.

28. **Sever, S., Damke, H., and Schmid, S.** Garrotes, springs, ratchets, and whips: putting dynamin models to the test. *Traffic* 1: 385-392, 2000.

29. **Shupliakov, O., Löw, P., Grabs, D., Gad, H., Chen, H., David, C., Takei, K., De Camilli, P., and Brodin, L.** Synaptic vesicle endocytosis impaired by disruption of dynamin-SH3 domain interactions. *Science* 276: 259-263, 1997.

30. **Simpson, F., Hussain, N.K., Qualmann, B., Kelly, R.B., Kay, B.K., McPherson, P.S., and Schmid, S.L.** SH3-domain-containing proteins function at distinct steps in clathrin-coated vesicle formation. *Nature Cell Biol.* 1: 119-124, 1999.

31. **van IJzendoorn, S.C.D. and Hoekstra, D.** The subapical compartment: a novel sorting centre? *Trends Cell Biol.* 9: 144-149, 1999.

32. **Vowels, J.J. and Payne, G.S.** A dileucine-like sorting signal directs transport into an AP-3-dependent, clathrin-independent pathway to the yeast vacuole. *EMBO J.* 17: 2482-2493, 1997.

Chapter 38

# Rab11 Interacting Proteins as Regulators of Parietal Cell Apical Recycling: Lessons from the Master

JAMES R. GOLDENRING, CHADWICK M. HALES AND LYNNE A. LAPIERRE

*Institute of Molecular Medicine and Genetics, Medical College of Georgia and theAugusta Department of Veterans Affairs Medical Center, Augusta, GA 30912*

The gastric parietal cell in mammals is a highly differentiated cell with a specialized function: the secretion of HCl into the gastric lumen. To accomplish this highly regulated function, the parietal cell sequesters the H/K-ATPase responsible for pumping protons within a intracellular tubulovesicular membrane network (5). The tubulovesicular network lies deep to an elaborate infolding of the apical plasma membrane designated as the intracellular canaliculus. The intracellular canaliculus represents a target membrane surface for regulated fusion with tubulovesicles (38). In response to either cAMP-dependent (histamine) or calcium-dependent (muscarinic cholinergic) agonists, tubulovesicle membrane fuse with the apical canalicular target membranes to deliver the H/K-ATPase to the luminal surface of activated parietal cells. Following the cessation of agonist stimulation, the H/K-ATPase is internalized and reassembled into the tubulovesicular system (12).

Previously, many authors have treated the parietal cell as a very specialized example of epithelial cell differentiation. However, recent investigations have demonstrated that, rather than a specialized case, the parietal cell trafficking of the H/K-ATPase represents an amplification of a function common to many epithelial cells: apical membrane recycling. Indeed, Forte coined the term "apical recycling" years before its general investigation in more traditional epithelial cell systems (12). The apical recycling system in epithelial cells was predominantly characterized as a tubulovesicular compartment in cultured epithelial cells such as MDCK and Caco-2 cells (20, 22, 29, 42). Both cargoes recycling to the apical membrane following apical endocytosis as well as cargoes transiting through basolateral to apical transcytosis passed through this tubulovesicular system. Thus, in the paradigm of MDCK cells stably transfected with the polymeric IgA receptor, IgA recycling following apical internalization and IgA transcytosing to the apical membrane following basolateral internalization both passed through the common apical recycling system (2;26;27;42). Recent investigations, detailed below, have demonstrated that the vesicles responsible for apical recycling in epithelial cells and parietal cell tubulovesicles are regulated by a similar cadre of proteins. These results establish the parietal cell as the paradigm for apical recycling in epithelial cells.

## PARIETAL CELL TUBULOVESICLES CONTAIN CHARACTERISTIC SNARE PROTEINS AND THE SMALL GTPASE, RAB11A

Rabbit parietal cells are an important model for the study of gastric acid secretion because they manifest a profile for agonist dependent activation which is similar to human (10). In addition, a number of investigators have developed enriched preparations of tubulovesicles from several species (19). We further adapted these density gradient procedures with immunoisolation using monoclonal antibodies specific for the alpha subunit of the H/K-ATPase (37). Immunoisolation of tubulovesicles allowed detailed examination of the components of these tubulovesicles in a highly purified membrane preparation. Numerous investigations over the past several years have demonstrated the presence of characteristic SNARE proteins which are involved in docking vesicles to their proper target membrane (17;35). Our group as well as Forte and colleagues both demonstrated that immunoisolated tubulovesicles contained both a v-SNARE, VAMP-2, and a t-SNARE, syntaxin 3 (6;30). These results strongly supported the model of vesicle fusion for regulated apical trafficking of H/K-ATPase.

## RAB PROTEINS ARE ASSOCIATED WITH PARIETAL CELL TUBULOVESICLES

Rab proteins are a diverse family of over fifty small GTPases. Rab proteins associate with particular vesicle populations and are hypothesized to mediate specificity in vesicle trafficking (28). GTP overlay studies of gastric parietal cell proteins demonstrated a set of one major and two minor small GTP-binding protein species in enriched populations of parietal cell tubulovesicles (4). The major GTP-binding protein species was originally identified as Rab2 using a polyclonal antiserum (39). However, this antiserum was developed against whole recombinant Rab2 at a time when only five Rab proteins were identified. Later studies with monoclonal antibodies specific for Rab2 demonstrated that this Rab generally associated with the endoplasmic reticulum-Golgi intermediate compartment (40)was not present on tubulovesicles (6). Rather, the major species associated with the tubulovesicles represented a cross reactivity with Rab11a.

The association of Rab11a with parietal cells was first recognized when we utilized degenerate oligonucleotide primers for the major WDAGQE sequence of known Rab GTPases for a 3'-Rapid amplification of cDNA ends (RACE) screen for Rab sequences. The majority of clones from yielded sequences for Rab11a (14;16). In addition, we also identified a novel sequence, Rab25, with greater than 65% identity with Rab11a (14). Interestingly, the wild-type Rab25 demonstrated a novel sequence in its GTP-binding sequence of WDTAGLE. The presence of a leucine instead of a glutamine had previously been associated with inactivation of GTPase activity in small GTP-binding proteins leading to a dominant active protein phenotype. However, subsequent studies have demonstrated that wild type Rab25 as well as Rab11aQ70L are both active GTPases (8).

Rab11a was highly enriched in parietal cells and was also present in subapical vesicles in a number of epithelial cells including kidney collecting duct, colonic crypt cells, gastric surface, hepatocytes as well as pancreatic and biliary duct cells (13;15). Immunoisolation studies demonstrated that Rab11a and Rab25 were enriched on parietal cell tubulovesicles immunoisolated with H/K-ATPase antibodies (6;7). In addition, in correlation with the original observations for GTP-binding proteins (4), Rab11a redistributed to the secretory canaliculus along with H/K-ATPase upon stimulation of parietal cells with histamine (7). All these studies suggested that Rab11a was a critical Rab protein associated with H/K-ATPase containing tubulovesicles and that Rab11a remained associated with the membranes when they fused with the secretory canaliculus. More recently, Forte and colleagues (11) have demonstrated that overexpression of a dominant negative Rab11a (Rab11aN124I) in gastric glands leads to an inhibition in aminopyrine accumulation. These data support the hypothesis that Rab11a is a major regulator of H/K-ATPase apical recycling.

## MYOSIN VB IS A DOWNSTREAM TARGET OF RAB11 FAMILY MEMBERS

While data was accumulating to suggest that Rab11a was associated with the regulation of vesicle trafficking in parietal cells, the mechanisms involved remained elusive. Initial investigations by both Sabatini and Takai had identified a Rab11 interacting protein (designated Rab11-BP or rabphillin-11) (24;45). However, antibodies to Rab11-BP/Rabphillin-11 (a gift of Dr. David Sabatini) showed only cytosolic localization in parietal cells (Goldenring, unpublished data). Thus, the proximate effectors for Rab11a on tubulovesicles remained unidentified. To approach this question, we utilized yeast 2-hybrid screening of an oligo-dT-primed rabbit parietal cell cDNA library with Rab11aS20VDC, a construct for a dominant active Rab11a with truncation of the carboxyl terminal cysteines utilized for isoprenylation and membrane insertion. The screen identified two identical independent clones coding for 60 kDa of carboxyl-terminal sequence for the rabbit homolog of myr6, also known as myosin Vb (23). The myosin V class of unconventional myosin motors are double-headed dimers which have been associated with vesicle trafficking (34). Mutations of myosin Va have been associated with changes in murine coat color (dilute) (25;33) and more recently and association of myosin Va with Rab27a on melanosomes has been suggested (3;21;44). While myosin Va has been studied extensively since its first purification by Mooseker and colleagues (9), little is known of myosin Vb (46).

Utilizing yeast two-hybrid screening, we studied the interaction of myosin Vb with members of the Rab11 family. Myosin Vb interacted preferentially with GTP bound Rab11a (Rab11aS20V), but did not interact with either GDP-bound Rab11a (Rab11aS25N) or Rab11a constructs with site-directed mutations in the effector domain binding site (Rab11aT43A or Rab11aI44N). Myosin Vb also interacted with wild type constructs of both Rab11b and Rab25. In contrast, myosin Vb did not interact with Rab2 or Rab3a. Thus, while we have only screened a limited number of Rab proteins, it appears that myosin Vb interacts specifically with only the Rab11 family of small GTPases. Using yeast 2-hybrid screening, we also were also able to map the regions of myosin Vb that were necessary for Rab11a binding. These studies demonstrated that the binding region was confined to the globular region of the carboxyl terminus. Thus, dimerization of the myosin Vb was not necessary, since the globular tail does not contain the requisite coiled-coil regions. Nevertheless, the Rab11a binding region was complex. Thus, sequences separated almost 400 amino acids were needed to maintain binding to the myosin Vb globular tail.

To investigate whether myosin Vb might be associated with general apical plasma membrane recycling systems, we studied its association with the apical recycling system in MDCK cells. We observed significant overlap of myosin Vb immunostaining with Rab11a as assessed in MDCK cells stably transfected with GFP-Rab11a (Figure 1). In addition, we observed that both GFP-Rab11a and myosin Vb staining were dispersed following disruption of microtubules with nocodazole (23). In addition, both GFP-Rab11a and myosin Vb staining relocated to the subapical perijunctional region following treatment with the microtubule stabilizing agent taxol. Interestingly, colocalization of myosin Vb and Rab11a was far less apparent in nocodazole treated cells, suggesting that myosin Vb might associate with vesicles after their transport into the recycling system on microtubules. While these studies supported the association of myosin Vb with Rab11a and the recycling system, we were never able to immunoprecipitate the complexed proteins. This pattern is similar to findings with other myosin V species (36). It appears that many of the associations of the myosin Vb tail with other proteins are highly detergent sensitive. This has left the question of

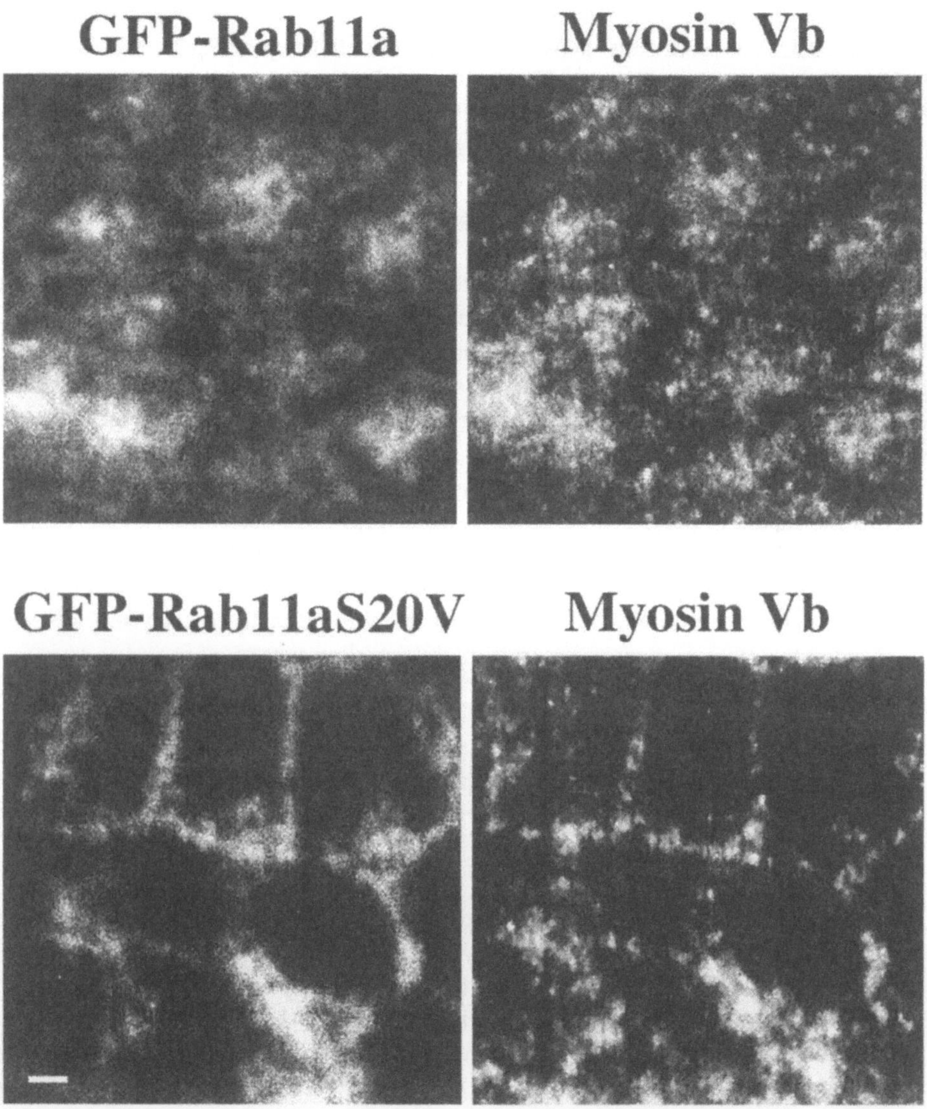


**Figure 1.** Distribution of myosin Vb in cell lines transfected with GFP-Rab11a and GFP-Rab11aS20V. MDCK cells stably transfected with GFP-Rab11a or GFP-Rab11aS20V were grown to confluence on permeable filters and then fixed with 4% paraformaldehyde. Cells were stained with a rabbit polyclonal anti-myosin Vb and Cy3-anti-rabbit IgG secondary antibody mand examined with confocal fluorescence microscopy. Note the redistribution of myosin Vb to the cell periphery in cells transfected with the dominant active Rab11aS20V. Bar=4μm.

whether Rab11a anchors myosin Vb to the membrane. To address this, we have investigated the distribution of myosin Vb in cells stably transfected with GFP-Rab11aS20V, the dominant active Rab11a. We have previously reported that Rab11aS20V is distributed more peripherally to the subapical and junctional regions in transfected cells (43). In GFP-Rab11aS20V cells, myosin Vb was also relocated to a more peripheral perijunctional region. These results indicate that Rab11a likely anchors myosin Vb to vesicles.

An indication of the functional role of myosin Vb was identified by examining the effects of a GFP-chimera of the myosin Vb tail transfected into MDCK cells stably-transfected with the polymeric IgA receptor (pIgAR). This tail construct lacked the motor domain but still retained the requirements of vesicle binding in the globular tail. The GFP-myosin Vb tail caused compaction of Rab11a-containing vesicles within a pericentrosomal region, but had no effect on the Golgi apparatus or on early and late endosomal markers. In addition to affecting the distribution of Rab11a, myosin Vb also caused accumulation of pIgAR, a known cargo for traffic through the apical recycling system (27). Most importantly, when fluorescently-labeled IgA was incubated with the basolateral media of polarized cells, transcytosing IgA was accumulated and retained within the GFP-myosin Vb tail containing vesicles. Fluorescent IgA was retained even after extensive chase periods of up to 12 hours. Since the cargoes could enter the recycling system, but could not leave, these results indicated that myosin Vb was involved in the process of cargo export from the recycling system.

## IDENTIFICATION OF A FAMILY OF RAB11 INTERACTING PROTEINS

In addition to the yeast 2-hybrid screens that led to the identification of myosin Vb, we also performed screens with Rab11aS20V and Rab25S21V without a carboxyl terminal truncations (18). These screens led to the identification of a novel protein which we designated Rab11-Family Interacting Protein 1 (Rab11-FIP1) because it interacted with all three of the Rab11 family members. This protein contained an amphipathic alpha-helical domain in its carboxyl terminus which was required for interaction with Rab11a. The rest of the protein was unremarkable except for high proline content in the amino terminal region. A search of Genbank and dbEST entries led to the identification of three other human proteins that contained homologous alpha helices all near the carboxyl termini. Rab11-FIP2 contained an identical helix sequence along with an amino terminal C2 domain with highest homology to C2 domains from synaptotagmin. Rab11-FIP3 contained a homologous helix along with a central ERM motif and a paired EF hand motif near the amino terminus (31). Finally, we identified a protein, pp75, which was originally characterized by Chan and coworkers as a binding partner for the Ro autoantigen (41). Subsequently, this protein was also identified as a Rab11a binding protein by protein association assays and was designated Rip11 (32). pp75/Rip11 has a homologous alpha helical Rab11binding motif and also contains a protein kinase C-like C2 domain. Our investigations demonstrated that deletion of the amphipathic alpha helix of Rab11-FIP proteins eliminated their ability to bind to bind Ra11a. All four of the Rab11-FIP members interacted with dominant active Rab11a. However, interestingly Rab11aS25N, a dominant negative Rab1a construct, also interacted with Rab11a-FIP2. No other Rab proteins tested interacted with the Rab11-FIP proteins, suggesting that the binding helix is specific for the Rab11 family. Also of interest for future investigation, we observed that Rab11-FIP2 also bound to the tail of myosin Vb. Preliminary investigations suggest that Rab11-FIP2 binds to a region of the myosin Vb tail separable from the Rab11a biding site.

We raised antibodies to Rab11-FIP1 and Rab11-FIP2 and demonstrated using immunocytochemistry that they were present on Rab11a-containing recycling system

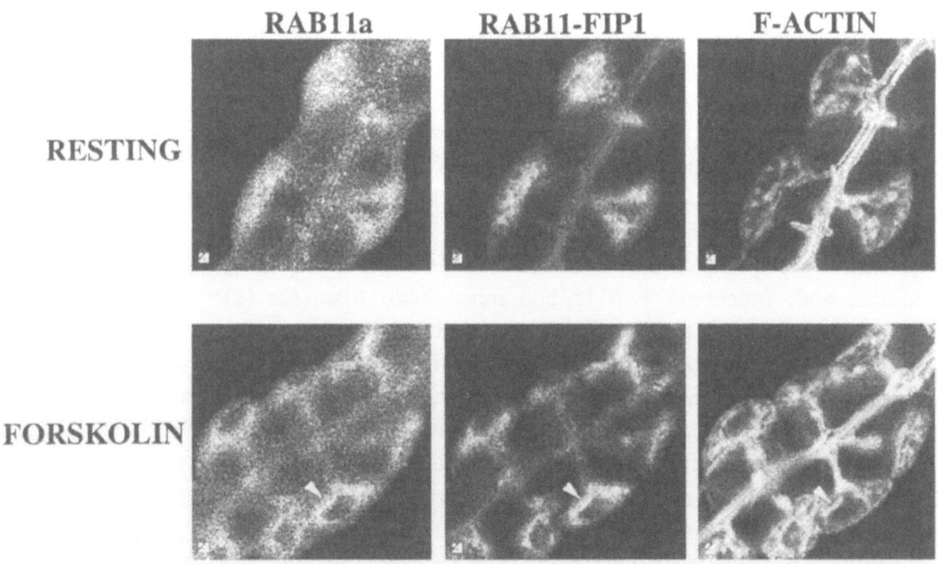


**Figure 2.** Rab11-FIP1 in isolated rabbit gastric glands. Isolated gastric glands were incubated for 30 min in either 100 μM ranitidine (resting) or 10 μM forskolin followed by fixation in 4% paraformaldehyde. The fixed glands were stained with monoclonal antibodies raised against Rab11-FIP1 and rabbit polyclonal anti-Rab11a (Zymed). Confocal fluorescence images demonstrated that Rab11-FIP1 was enriched in parietal cells and showed a diffuse distribution in resting cells, similar to Rab11a. Stimulation of glands with forskolin led to colocalization (arrowheads) of Rab11a and Rab11FIP1 staining with F-actin stain (FITC-phallacidin, Molecular Probes). Bars=2 μm.

vesicles in MDCK cells (18). In addition, we isolated enriched fraction of rabbit parietal cell tubulovesicular membranes and demonstrated that immunostaining for H/K-ATPase, Rab11a, Rab11-FIP1, Rab11-FIP2 and pp75/Rip11 all co enriched throughout the vesicle preparations. Using immunocytochemistry, we observed colocalization of Rab11-FIP1 with Rab11a in resting glands (Figure 2). Interestingly, upon stimulation with 100 μM histamine, we observed relocation of the both Rab11a and Rab11-FIP1 to the secretory canaliculus as exemplified by colocalization of immunostaining with F-actin staining. Similar results were obtained for both Rab11-FIP1 and Rab11-FIP2 in cultured parietal cells (18). These results suggested that Rab11a and its effectors were all present on tubulovesicles..

## MODELS FOR THE REGULATION OF H/K-ATPASE APICAL RECYCLING

These investigations of Rab11a began with a search for the regulators of apical recycling in parietal cells, but have now led to insights that bear on the regulation of epithelial cell recycling systems in general. The success of these is studies, largely initiated yeast 2-hybrid screens, likely accrues from the use of a library prepared from parietal cells that have massively amplified their apical recycling system. These studies suggest that the association of Rab11a with recycling system vesicles leads to the anchoring of a key motor protein, myosin Vb, which is responsible for transit of vesicles out of the recyclingsystem, likely on actin filaments. This generalized hypothesis fits well with prevailing knowledge of parietal cells apical recycling which requires both functional Rab11a and intact actin filaments (1). Nevertheless, the existence of four proteins which all can bind Rab11a through the same amphipathic alpha helical motif suggests that some level of complexity must underlie Rab11a-deendent function. Several possibilities may explain this observation. First, the tubulovesicular cisternae that mediate apical recycling in epithelial cells in general, and in an amplified form in parietal cells, may contain regional diversity. Such regional heterogeneity may account for the regulated trafficking of multiple cargoes through the recycling system. Alternatively, regionalization may account for "maturation" of recycling cargoes as they pass through the system. In that case, certain Rab11-FIP proteins might hand off to others during progression through the recycling system. Finally, it is possible that Rab11a molecules serve as a scaffolding agent for the assembly of a multiprotein complex. In support of this hypothesis, we have observed that Rab11a does have the capacity to form oligomers in solution (Goldenring, unpublished results). Thus, oligomerization could allow multiple Rb11-FIP proteins to become part of a larger signaling complex also associating with the myosin Vb motor protein. The latter hypothesis appears to fit the data from parietal cells better since Rab11a, Rab11-FIP1 and Rab11-FIP2 all maintain their association with the membrane during fusion with the apical canaliculus. This pattern would support the assembly of a functional signaling complex amplified in the parietal cells. Further studies will be required to understand the molecular composition of Rab11a-based signaling complexes.

## ACKNOWLEDGEMENTS

This work was supported by grants to JRG from awards from NIH NIDDK (DK48370 and DK43405) and a Department of Veterans Affairs Merit Review award.

## REFERENCES

1. **Ammar, D. A., P. N. Nguyen, and J. G. Forte.** Functionally distinct pools of actin in secretory cells. *Am.J.Physiol.* 281: C407-C417, 2001.

2. **Apodaca, G., L. A. Katz, and K. E. Mostov.** Receptor-mediated transcytosis of IgA in MDCK cells is via apical recycling endosomes. *J.Cell Biol.* 125: 67-86, 1994.

3. **Bahadoran, P., E. Aberdam, F. Mantoux, R. Busca, K. Bille, N. Yalman, G. de Sain-Basile, R. Cararoli-Marano, J. P. Ortonne, and R. Ballotti.** Rab27a: A key to melanosome transport in human melanocytes. *J.Cell Biol.* 152: 843-850, 2001.

4. **Basson, M. D., J. R. Goldenring, L. H. Tang, J. J. Lewis, P. Padfield, J. D. Jamieson, and I. M. Modlin.** Redistribution of 23 kDa tubulovesicle-associated GTP-binding proteins during parietal cell stimulation. *Biochem.J.* 279: 43-48, 1991.

5. **Black, J. A., T. M. Forte, and J. G. Forte.** Structure of oxyntic cell membranes during conditions of rest and secretion of HCl as revealed by freeze-fracture. *Anat. Rec.* 196: 163-172, 1980.

6. **Calhoun, B. C. and J. R. Goldenring.** Two Rab proteins, VAMP-2, and SCAMPs are present on immunoisolated gastric tubulovesicles. *Biochem.J.* 325: 559-564, 1997.

7. **Calhoun, B. C., L. A. Lapierre, C. S. Chew, and J. R. Goldenring.** Rab11a redistributes to apical secretory canaliculus during stimulation of gastric parietal

cells. *Am.J.Physiol.* 275: C163-C170, 1998.

8. **Casanova, J. E., X. Wang, R. Kumar, S. G. Bhartur, J. Navarre, J. E. Woodrum, G. S. Ray, and J. R. Goldenring.** Rab11a and Rab25 association with the apical recycling system of polarized MDCK cells. *Mol.Biol.Cell.* 10: 47-61, 1999.

9. **Cheney, R. E., M. K. O'Shea, J. E. Heuser, M. V. Coelho, J. S. Wolenski, E. M. Espreafico, P. Forscher, R. E. Larson, and M. S. Mooseker.** Brain myosin-V is a two-headed unconventional myosin with motor activity. *Cell* 75: 13-23, 1993.

10. **Chew, C. S.** Parietal cell culture: new models and directions. *Annu.Rev.Physiol.* 56: 445-461, 1994.

11. **Duman, J. G., K. Tyagarajan, M. S. Kolsi, H. H. Moore, and J. G. Forte.** Expression of rab11a N124I in gastric parietal cells inhibits stimulatory recruitment of the $H^+$-$K^+$-ATPase. *Am.J.Physiol.* 277: C361-C372, 1999.

12. **Forte, T. M., T. E. Machen, and J. G. Forte.** Ultrastructural changes in oxyntic cells associated with secretory function: a membrane recycling hypothesis. *Gastroenterology.* 73: 941-955, 1977.

13. **Goldenring, J. R., C. J. Soroka, W. Rodriguez, L. H. Tang, K. R. Shen, H. D. Vaughan, and I. M. Modlin.** Rab11 is highly expressed in gastric parietal cells. *Gastroenterology* 104: A826, 1993.

14. **Goldenring, J. R., K. R. Shen, H. D. Vaughan, and I. M. Modlin.** Identification of a small GTP-binding protein, Rab25, expressed in the gastrointestinal mucosa, kidney and lung. *J.Biol.Chem.* 268: 18419-18422, 1993.

15. **Goldenring, J. R., J. Smith, H. D. Vaughan, P. Cameron, W. Hawkins, and J. Navarre.** Rab11 is an apically located small GTP-binding protein in epithelial tissues. *Am.J.Physiol.* 270(33): G515-G525, 1996.

16. **Goldenring, J. R., C. J. Soroka, K. R. Shen, L. H. Tang, W. Rodriguez, H. D. Vaughan, S. A. Stoch, and I. M. Modlin.** Enrichment of rab11, a small GTP-binding protein, in gastric parietal cells. *Am.J.Physiol.* 267: G187-G194, 1994.

17. **Gonzales, Jr. L. and R. H. Scheller.** Regulation of membrane trafficing: structural insights from a rab/effector complex. *Cell.* 96: 755-758, 1999.

18. **Hales, C. M., R. Griner, K. C. Hobdy-Henderson, M. C. Dorn, D. Hardy, R. Kumar, J. Navarre, E. K. L. Chan, L. A. Lapierre, and J. R. Goldenring.**

Identification and characterization of a family of Rab11-interacting proteins. *J.Biol.Chem.* 276: 39067-39075, 2001

19. **Hirst, B. H. and J. G. Forte.** Redistribution and characterization of ($H^+$,$K^+$)-ATPase membranes from resting and stimulated parietal cells. *Biochem.J.* 231: 641-649, 1985.

20. **Hughson, E. J. and C. Hopkins.** Endocytic pathways in polarized caco-2 cells: identification of an endosomal compartment accessible from both apical and basolateral surfaces. *J.Cell Biol.* 110: 337-348, 1990.

21. **Hume, A. N., L. M. Collinson, A. Rapak, A. Q. Gomes, C. R. Hopkins, and M. C. Seabra.** Rab27a regulates the peripheral distribution of melanosomes in melanocytes. *J.Cell Biol.* 152: 795-808, 2001.

22. **Knight, A., E. Hughson, C. R. Hopkins, and D. F. Cutler.** Membrane protein trafficking through the common apical endosome compartment of polarized Caco-2 cells. *Mol.Biol.Cell.* 6: 597-610, 1995.

23. **Lapierre, L. A., R. Kumar, C. M. Hales, J. Navarre, S. G. Bhartur, J. O. Burnette, J. A. Mercer, M. Bahler, and J. R. Goldenring.** Myosin Vb is associated with and regulates plasma membrane recycling systems. *Mol.Biol.Cell* 12: 1843-1857, 2001.

24. **Mammoto, A., T. Ohtsuka, I. Hotta, T. Sasaki, and Y. Takai.** Rab11BP/Rabphillin- 11, a downstream target of Rab11 small G protein implicated in vesicle recycling. *J.Biol.Chem.* 274: 25517-25524, 1999.

25. **Mercer, J. A., P. K. Seperack, M. C. Strobel, N. G. Copeland, and N. A. Jenkins.** Novel myosin heavy chain encoded by murine dilute coat colour locus. *Nature* 349, Feb. 21: 709-712, 1991.

26. **Mostov, K., G. Apodaca, B. Areoti, and C. Okamoto.** Plasma membrane protein sorting in polarized epithelial cells . *J Cell Biol* 116: 577-583, 1992.

27. **Mostov, K. E. and D. L. Deitcher.** Polymeric immunoglobulin receptor expressed in MDCK cells transcytoses IgA. *Cell* 46: 613-621, 1986.

28. **Nuoffer, C. and W. E. Balch.** GTPases: Multifunctional molecular switches regulating vesicular traffic. *Ann.Rev.Biochem.* 63: 949-990, 1994.

29. **Odorizzi, G., A. Pearse, D. Domingo, I. D. Trowbridge, and C. R. Hopkins.**

Apical and basolateral endosomes of MDCK cells are interconnected and contain a polarized sorting mechanism. *J Cell Biol* 135: 139-152, 1996.

30. **Peng, X.-P., X. Yao, D.-C. Chow, J. G. Forte, and M. K. Bennett.** Association of syntaxin 3 and vesicle associated membrane protein (VAMP) with $H^+/K^+$-ATPase-containing tubulovesicles in gastric parietal cells. *Mol.Biol.Cell.* 8: 399-407, 1997.

31. **Prekeris, R., Davies, J. M., and Scheller, R. H.** Identificaition of a novel Rab11/25 binding domain in eferin and Rip proteins. *J.Biol.Chem.* 276:38966-38970, 2001

32. **Prekeris, R., J. Klumperman, and R. H. Scheller.** A Rabll/Rip11 complex regulates apical membrane trafficking via recycling endosomes. *Molec.Cell* 6: 1437-1448, 2000.

33. **Provance, D. W. Jr., M. Wei, V. Ipe, and J. A. Mercer.** Cultured melanocytes from dilute mice exhibit dendritic morphology and altered melanosome distribution. *Proc.Natl.Acad.Sci., USA* 93: 14554-14558, 1996.

34. **Reck-Peterson, S. L., D. W. Provance, M. S. Mooseker, and J. A. Mercer.** Class V myosins. *Biochem.Biophys.Acta* 1496: 36-51, 2000.

35. **Rothman, J. E. and G. Warren.** Implications of the SNARE hypothesis for intracellular membrane topology and dynamics. *Curr.Biol.* 4: 220-233, 1994.

36. **Schott, D., J. Ho, D. Pruyne, and A. Bretscher.** The COOH-terminal domain of Myo2p, a yeast myosin V, has a direct role in secretory vesicle targeting. *J.Cell Biol.* 147: 791-807, 1999.

37. **Smolka, A., H. F. Helander, and G. Sachs.** Monoclonal antibodies against gastric H/K-ATPase. *Am.J.Physiol.* 245: G589-G596, 1983.

38. **Soroka, C. J., C. S. Chew, I. M. Modlin, D. Hanzel, A. Smolka, and J. R. Goldenring.** Characterization of membrane and cytoskeletal compartments in cultured parietal cells using immunofluorescence and confocal microscopy. *Eur.J.Cell Biol.* 60: 76-87, 1993.

39. **Tang, L. H., S. A. Stoch, I. M. Modlin, and J. R. Goldenring.** Identification of rab2 as a tubulovesicle-membrane associated protein in rabbit gastric parietal cells. *Biochem. J.* 285: 715-719, 1992.

40. **Tisdale, E. J. and W. E. Balch.** Rab2 is essential for the maturation of pre-Golgi

intermediates. *J.Biol.Chem.* 271: 29372-29379, 1996.

41. **Wang, D., J. P. Buyon, W. Zhu, and E. K. Chan.** Defining a novel 75 kDa phosphoprotein assocaited with SS-A/Ro and identification of distinct human autoantibodies. *J.Clin.Invest.* 104: 1265-1275, 1999.

42. **Wang, E., P. S. Brown, B. Aroeti, S. J. Chapin, K. E. Mostov, and K. W. Dunn.** Apical and basolateral endocytic pathways of MDCK cells meet in acidic common endosomes distinct from a nearly neutral apical recycling endosome. *Traffic.* 1: 480-493, 2000.

43. **Wang, X., R. Kumar, J. Navarre, J. E. Casanova, and J. R. Goldenring.** Regulation of vesicle trafficking in Madin-Darby Canine Kidney cells by Rab11a and Rab25. *J.Biol.Chem.* 275: 29138-29146, 2000.

44. **Wu, X., K. Rao, M. B. Bowers, N. G. Copeland, N. A. Jenkins, and J. A. Hammer.** Rab27a enables myosin Va-dependent melanosome capture by recruiting the myosin to the organelle. *J.Cell Sci.* 114: 1091-1100, 2001.

45. **Zeng, X., M. Ren, D. Gravotta, C. De Lemos-Chiarandini, M. Lui, H. Erdjument-Bromage, P. Tempst, G. Xu, T. H. Shen, T. Morimoto, M. Adesnick, and D. D. Sabatini.** Identification of a putative effector protein for rab11 that participates in transferrin recycling. *Proc.Natl.Acad.Sci., USA* 96: 2840-2845, 1999.

46. **Zhao, L. P., J. S. Koslovsky, J. Reinhard, M. Bahler, A. E. Witt, D. W. Jr. Provance, and J. A. Mercer.** Cloning and characterization of myr6, an unconventional myosin of the dilute/myosin-V family. *Proc.Natl.Acad.Sci., USA* 93: 10826-10831, 1996.

Chapter 39

# Regulation of Acid Secretion and Mucosal Permeability by F-actin in the Bullfrog Gastric Mucosa

SUSAN J. HAGEN and TARIK A. ABDUL-GHAFFAR AL-SHAIBANI

*Department of Surgery*
*Beth Israel Deaconess Medical Center, Dana 805*
*330 Brookline Avenue, Boston, MA 02215*

## INTRODUCTION

Actin accounts for nearly 5% of the total protein in parietal cells, of which 90% is filamentous, or F-actin, and 10% is globular, or G-actin (9). At the apical surface of mammalian parietal and frog oxyntic cells, F-actin is associated with microvilli or surface folds, respectively, and with the tight junction (13, 24, 30, 31). To determine that F-actin is involved in gastric acid secretion, many studies have inhibited the polymerization of G- to F-actin in both intact tissues (4, 11) and in isolated gastric glands (9, 23) with cytochalasin (C), B, D, or E. In each study, the disruption of actin filaments and resulting inhibition of acid secretion was attributed to a direct effect of the disrupted actin filaments on some aspect of stimulated acid secretion. The role of tight junctions in regulation of paracellular permeability, essential for trapping $H^+$ in the lumen so it can be measured, was not considered.

In epithelial cells, tight junctions are structurally associated with a perijunctional actomyosin ring that contracts and acts to regulate paracellular permeability of the mucosa (28). In the intact ileum (20, 21), jejunum (19), and gallbladder (3), as well as in cultured Caco-2 (18), T84 (12) and MDCK (15, 17, 26, 27) cells, a significant reduction in transepithelial resistance (TER) and an increase in paracellular permeability accompanies the disruption of actin filaments with CB or CD. In addition, permeability in the presence of CD is size-selective, where mannitol (3.6 Å) but not inulin (11-15 Å) can cross the disrupted junction (20). Chen, et al (5) found that CD also decreased TER and increased mannitol, but not inulin, flux in monolayer cultures of canine gastric epithelial cells. These results suggest that TER and paracellular permeability are also regulated by F-actin (at the tight junction) in gastric epithelial cells.

Thus, the purpose of this study was to determine whether apical F-actin regulates gastric acid secretion or influences mucosal permeability such that hydrogen ion back-diffusion (mucosal to serosal) and bicarbonate ion flux (serosal to mucosal) titrate secreted $H^+$ when actin filaments are disrupted with CD. We found that despite a significant increase in paracellular permeability in the presence of CD, the back-diffusion of hydrogen ion and flux of bicarbonate are minimal and cannot account for the significant reduction in acid secretion. Our results demonstrate that actin filaments specifically influence some aspect of stimulated acid secretion in the bullfrog gastric mucosa.

## MATERIALS AND METHODS

*Preparation of the bullfrog gastric mucosa for the Ussing chamber studies.* Animals used for this study were maintained in accordance with the guidelines of the Committee on Animals at the Beth Israel Deaconess Medical Center and those prepared by the committee on Care and Use of Laboratory animals by the National Research Council. Bullfrogs (*Rana catesbeiana*), caught in the wild, were purchased from West Jersey (Wenonah, N.J.) and kept at room temperature in large water tanks until use. Stomachs were removed from pithed frogs, and the fundic mucosa was stripped from underlying external muscle layers and submucosa to bare the muscularis mucosa as described previously (13). Stripped mucosae were divided into two halves (one experimental and one control); each was mounted between two Lucite halves of an Ussing-type chamber with an exposed mucosal area of 0.636 $cm^2$. Mucosal surfaces (from here onward called "luminal") were bathed with a solution containing (in mM) 102.4 $Na^+$, 4.0 $K^+$, 0.8 $Mg^{2+}$, 1.8 $Ca^{2+}$, 91.4 $Cl^-$, 10.1 $SO_4^{2-}$, and 19.3 mannitol and continuously gassed with 100% $O_2$. Luminal solutions were unbuffered and kept at pH 4.7 with a pH-stat device (Radiometer America, Cleveland OH). Serosal surfaces (from here onward called "nutrient") were bathed with (in mM) 102.4 $Na^+$, 4.0 $K^+$, 0.8 $Mg^{2+}$, 1.8 $Ca^{2+}$, 91.4 $Cl^-$, 0.8 $SO_4^{2-}$, 0.8 $H_2PO_4^{2-}$, 10 glucose, 17.8 $HCO_3^-$ (pH 7.2) and continuously gassed with 95% $O_2$ / 5% $CO_2$. Acid secretion was measured using a Radiometer pH-stat device and calculated from the volume of 10 mM NaOH needed to titrate the nutrient solution to a constant pH of 4.7. Transepithelial resistance (TER) was calculated from Ohm's law using measurements of potential difference that were monitored continuously by KCl-saturated agar bridges connected via two calomel electrodes to a voltmeter.

*Mannitol flux studies with CD in stimulated and inhibited tissues.* To stimulate acid secretion, 0.1 mM histamine and 1 mM carbachol (histamine/carbachol) were added to the nutrient solution. To inhibit acid secretion, both luminal and nutrient compartments were washed with buffer and then the tissues were incubated with 1 mM aqueous cimetidine until acid secretion reached 0 μEq $hr^{-1}cm^{-2}$. Once the tissues were stimulated or inhibited, 0.1% DMSO or 20 μM cytochalasin D (CD) in 0.1% DMSO was added to the nutrient solution and $^3$H-mannitol (50 μCi, 15-30 Ci/mmol, NEN Life Science Products, Boston, MA) was added to the luminal solution. For mannitol flux studies, duplicate aliquots (0.25 ml each) were taken every 30 min from the nutrient solution, replacing this aliquot with an equal volume of unlabeled nutrient buffer containing histamine/carbachol or cimetidine and DMSO or CD/DMSO, respectively. Samples from both stimulated and inhibited tissues were diluted with 3 ml of scintillation fluid (Atomlight, NEN Life Science Products, Boston, MA) and the amount of $^3$H in each sample was determined by liquid scintillation (Packard Instrument Company, Meriden, CT). Mucosal-to-serosal flux was calculated by standard techniques.

*Electron microscopy.* Frog tissues from the Ussing-chamber were fixed overnight at 4°C with 2 % glutaraldehyde in 0.1 M cacodylate buffer (pH 7.4), postfixed for 1 hr at 4°C with 1 % osmium tetroxide in 0.1 M cacodylate buffer (pH 7.4), and stained overnight at 4°C with 2 % aqueous uranyl acetate. Tissues were dehydrated in graded alcohols and propylene oxide and embedded in LX112 resin. Thin sections, cut parallel to the long axis of gastric glands, were placed on Formvar- and carbon-coated grids and examined with a JEOL 100CX electron microscope.

*Measurement of bicarbonate transport (serosal to mucosal) in inhibited tissues.* Tissues were stripped and mounted in Ussing chambers, as described above, except that nutrient and luminal solutions were maintained at pH 7.4. To inhibit tissues, 0.3 mM omeprazole (Astra Hassle AB, Sweden) was added to the nutrient solution until the rate of acid secretion reached 0 μEq $hr^{-1}cm^{-2}$ and TER was at least 400 Ohm $cm^2$. Bicarbonate transport was measured using a pH-stat device and was calculated from the volume of 5 mM HCl needed to titrate the luminal solution to a constant pH of 7.4. Anoxia was applied to differentiate between passive and active bicarbonate transport. For this, 100% $N_2$ was applied to the luminal solution and 5% $CO_2$/ 95% $N_2$ to the nutrient solution for 30 min.

*$^3$H-mannitol / $^{22}$Na flux studies.* Tissues were stripped, mounted in Ussing chambers (nutrient solution pH 7.2 and luminal solution pH 4.7), and inhibited with 0.3 mM omeprazole until the rate of acid secretion reached 0 μEq $hr^{-1}cm^{-2}$ and TER was at least 400 Ohm $cm^2$. Buffer solutions were replaced (on both sides) with nutrient buffer (pH 7.4) and DMSO or CD/DMSO were added for 90 min. This is the time required to maximally increase paracellular permeability. Tissues were then switched from open- to closed circuit conditions and allowed to equilibrate for 30 min in the presence of luminal $^3$H-mannitol (50 μCi) and $^{22}$Na (10 μCi, 287.3 mCi/mg, NEN Life Science Products, Boston, MA). Four duplicate aliquots (0.25 ml) were then taken every 30 min from the nutrient solution, replacing this aliquot with an equal volume of unlabeled nutrient buffer containing omeprazole and DMSO or CD/DMSO, respectively. Samples were diluted with 3 ml of scintillation fluid (Atomlight) and the DPM of $^3$H/$^{22}$Na in each sample was determined by liquid scintillation. Quench curves for $^3$H and $^{22}$Na were done prior to counting. Mucosal-to-serosal flux was calculated by standard techniques.

*Statistical analysis.* Combined data were expressed as mean ± SEM. Statistical analyses of data were done with SigmaStat software (Jandel Scientific Software, San Rafael, CA) using the unpaired t-test for analysis of two groups or one-way analysis of variance for many groups. Best-fit regression lines were calculated in SigmaPlot for dual $^3$H/$^{22}$Na experiments. Differences were regarded as statistically significant at $P < 0.05$.

## RESULTS

*Cytochalasin D (CD) inhibits stimulated acid secretion and decreases TER in inhibited tissues.* Acid secretion in the frog fundic mucosa attained a maximal rate of 6.68 ± 0.17 μEq $hr^{-1}cm^{-2}$ after stimulation in the presence of 0.1% DMSO (Fig. 1*A*). In contrast, tissues that were maximally stimulated and then incubated from the serosal surface with 20 μM CD in 0.1% DMSO (CD/DMSO) showed a significant reduction in acid secretion to 1.69 ± 0.54 μEq $hr^{-1}cm^{-2}$ (Fig. 1*A*). In stimulated tissues treated with DMSO, transepithelial resistance (TER) was 81.6 ± 5.79 Ohm $cm^2$ and was 78.4 ± 16.39 Ohm $cm^2$ in tissues treated with CD/DMSO (Fig. 1*B*). Tissues that were inhibited with cimetidine had an acid secretion rate of 0.00 μEq $hr^{-1}cm^{-2}$ and a TER of approximately 500 Ohm $cm^2$ (data not shown). Incubation of inhibited tissues with DMSO resulted in a modest decline in TER to 305.3 ± 17.8 Ohm $cm^2$ (Fig. 1*C*). In contrast, incubation of inhibited tissues with CD/DMSO caused a significant decrease in TER to 106.8 ± 12.1 Ohm $cm^2$ (Fig. 1*C*).

*CD causes an increase in paracellular permeability.* The significant decrease in TER in inhibited tissues treated with CD/DMSO suggested that treatment with CD results in an increase in paracellular permeability. Thus, unidirectional (mucosal to serosal) mannitol fluxes were studied in both inhibited and stimulated tissues that were treated with DMSO or with CD/DMSO.

In inhibited tissues treated with DMSO (30-240 min), or CD/DMSO for 30-60 min, the rate of mannitol flux was comparable (Table 1). Stimulated tissues treated with DMSO had a slower initial rate of mannitol flux, which increased by 90-240 min (Table 1). Stimulated tissues incubated with CD/DMSO for 30-60 min had a similar rate of mannitol flux to those of DMSO-treated tissues (Table 1). In contrast, there was a significant increase in mannitol flux in both inhibited and stimulated tissues by 90-240 min after incubation with CD/DMSO. In inhibited tissues, the rate of mannitol flux increased by 282% and, in stimulated tissues, by 239% (Table 1). In all cases, the rate of mannitol flux was greater in inhibited, when compared to stimulated, tissues.

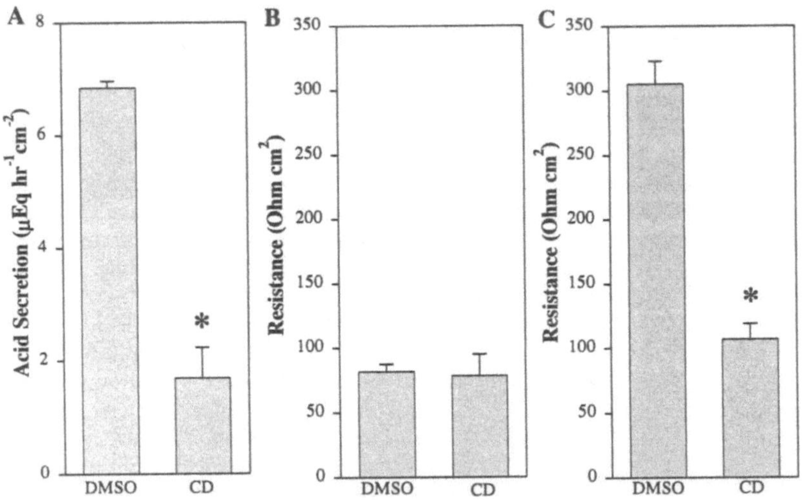


**Figure 1.** Effect of 0.1% DMSO or 20 µM cytochalasin D in 0.1% DMSO (CD/DMSO) on acid secretion and transepithelial resistance (TER) in the stimulated and inhibited bullfrog gastric mucosa. *A, B*: in tissues stimulated with histamine plus carbachol for 60 minutes, addition of CD/DMSO reduced acid secretion within 3 hr (*A*) without causing a change in TER (*B*). *C*: in contrast to stimulated tissues, TER dropped significantly in inhibited tissues that were treated with CD/DMSO. Data were obtained from 12 experiments and are expressed as means ± SE. (Adapted from Al-Shaibani and Hagen, 2001).

**Table 1.** *Rates of mannitol flux from tissues stimulated with histamine carbachol or inhibited with cimetidine and then treated with DMSO or with cytochalasin D (CD). (Adapted from Al-Shaibani and Hagen, 2001)*

| | STIMULATED Mannitol Flux (µmol $hr^{-1}$ $cm^{-2}$) | INHIBITED Mannitol Flux (µmol $hr^{-1}$ $cm^{-2}$) |
|---|---|---|
| **DMSO** | | |
| 30-60 min | 4.17 ± 0.32 (x $10^{-2}$) | 11.68 ± 1.20 (x $10^{-2}$) |
| 90-240 min | 7.15 ± 0.91 (x $10^{-2}$) | 12.3 ± 1.22 (x $10^{-2}$) |
| | | |
| **CD** | | |
| 30-60 min | 7.24 ± 0.11 (x $10^{-2}$) | 13.26 ± 0.76 (x $10^{-2}$) |
| 90-240 min | 20.92 ± 1.77 (x $10^{-2}$) | 32.32 ± 3.75 (x $10^{-2}$) |

*CD does not disrupt the structure of oxynticopeptic cells as determined by electron microscopy.* To determine if CD/DMSO damages the mucosa to result in an increase in paracellular permeability, mucosal structure was evaluated in inhibited tissues by electron microscopy. Low magnification micrographs showed that changes in mucosal structure in inhibited tissues treated for 4 hr with CD/DMSO were limited to dilation of the gland lumen and a reduction in the number of surface folds at the apical surface of oxynticopeptic cells. Tight junctions in oxynticopeptic cells treated with CD/DMSO were identical to those from control tissues and showed no increase in membrane dilatations or condensed perijunctional actin. High and low magnification images can be found in the original article (1).

*$HCO_3^-$ flux contributes little to luminal acid neutralization during acid secretion.* To test the hypothesis that an increase in paracellular permeability would increase the serosal to mucosal $HCO_3^-$ flux and serve to neutralize luminal acid, we measured the rate of $HCO_3^-$ flux in inhibited tissues treated with DMSO or with CD/DMSO (Fig. 2). In tissues treated with DMSO, the serosal to mucosal $HCO_3^-$ flux was 0.14 ± 0.03 μEq $hr^{-1}cm^{-2}$ after 5 hr. The serosal to mucosal $HCO_3^-$ flux in tissues treated with CD/DMSO was 0.34 ± 0.03 μEq $hr^{-1}cm^{-2}$ after 5 hr, which was a significant (3-fold) increase when compared to control tissues (Fig. 2).

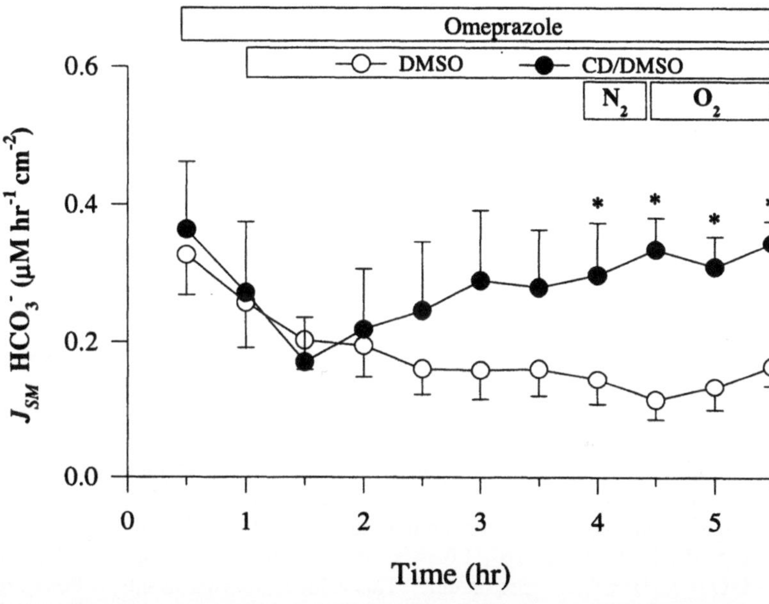


**Figure 2.** Effect of 0.1% DMSO or 20 μM cytochalasin D in 0.1% DMSO (CD/DMSO) on the serosal to mucosal flux ($J_{SM}$) of bicarbonate ($HCO_3^-$) in the bullfrog gastric mucosa. $HCO_3^-$ flux increased significantly within 5 hr in tissues treated with CD/DMSO. Anoxia ($N_2$) for 30 min followed by re-oxygenation for 60 min did not cause a significant change in $HCO_3^-$ flux in either group. These results demonstrate that the increase in $HCO_3^-$ flux in tissues treated with CD/DMSO is by passive paracellular diffusion. Data are from 8 experiments and represent means ± SE. * Denotes significantly different ($P < 0.01$) when compared to control tissues treated with DMSO. (With permission, Al-Shaibani and Hagen, 2001).

To determine whether the mean increase in $HCO_3^-$ flux in CD/DMSO-treated tissues was due to active $HCO_3^-$ transport or by passive paracellular diffusion, anoxia ($N_2$) was applied to the tissues for 30 minutes (Fig. 2). The rate of $HCO_3^-$ flux in tissues treated with CD/DMSO during anoxia was not significantly different from that of tissues in normal oxygen conditions (Fig. 2). In addition, re-oxygenation of tissues treated with CD/DMSO resulted in no significant difference in $HCO_3^-$ transport (Fig. 2). Anoxia followed by re-oxygenation also had no significant effect on control tissues treated with DMSO. These results show that bicarbonate transport occurred by passive paracellular diffusion in inhibited tissues treated with DMSO or with CD/DMSO.

*Back-diffusion of $H^+$ contributes little to the reduction in acid secretion that occurs with CD.* With increased paracellular permeability, $H^+$ may be lost from the lumen by movement across the tight junction into the serosal solution. Since it is difficult to study the flux of $H^+$, *per se*, $H^+$ movement was predicted from measurements of $Na^+$ and mannitol flux, under short circuit conditions, across the tight junction. Rationale for this study was based on the validated assumption that the movement of mannitol, $Na^+$ and $H^+$ through the paracellular space would be related to the movement of these species in free solution (8). We can determine the rate of $Na^+$ and mannitol flux across the paracellular space under our experimental conditions. Then, the rate of $H^+$ flux across the paracellular space can be predicted, under the same experimental conditions, from its diffusion coefficient in free solution. When dual $Na^+$ and mannitol fluxes were measured in the inhibited stomach, in the presence of DMSO or CD/DMSO, the ratio of $Na^+$ to mannitol flux was 10.63 (Fig. 3*A*). $Na^+$ and mannitol fluxes had a strong correlation ($r^2 = 0.91$) and the intercept was different from 0 (Fig. 3*A*). The ratio of the free-solution diffusion coefficients of each species corrected for the concentration in the luminal solution, $D^{Na}$ ($1.334 \times 10^{-5}$ cm²/sec) x 102.4 mM $Na^+$ / $D^{man}$ ($0.677 \times 10^{-5}$ cm²/sec) x 19.3 mM mannitol, is 10.46. Since the calculated ratio of 10.63 is nearly identical to the ratio due to diffusion alone (10.46), these results verify that the increase in ion permeability, in the presence of CD/DMSO, is by passive diffusion through the paracellular space.

Using data from the known rate of mannitol flux in Fig. 3*A*, it is possible to calculate the rate of $Na^+$ flux that would be predicted to occur by diffusion alone. This would be done by plotting the data for mannitol flux (from Fig. 3*A*) with a slope of 10.46 (ratio of the free-diffusion coefficients as calculated above) and extrapolating the predicted $Na^+$ concentration on the Y-axis. Since there were 6 experiments done with DMSO- and 7 experiments with CD-treated tissues, the data were extrapolated and then averaged for each group. Using extrapolated data, the mean rate of $Na^+$ flux with DMSO is $0.519 \pm 0.097$ µEq $hr^{-1}cm^{-2}$. This rate is not significantly different from the measured rate of $Na^+$ flux with DMSO, which is $0.669 \pm 0.075$ µEq $hr^{-1}cm^{-2}$ (Fig. 3*B*). Likewise, the predicted rate of $Na^+$ flux in the presence of CD/DMSO is $2.918 \pm 0.214$ µEq $hr^{-1}cm^{-2}$. This value is not significantly different from the measured rate of $Na^+$ flux with CD/DMSO, which is $3.08 \pm 0.327$ µEq $hr^{-1}cm^{-2}$ (Fig. 3*B*). Thus, equivalent results can be obtained concerning the movement of $Na^+$ across the tight junction either experimentally (by measuring ion flux directly) or by predicting the movement of $Na^+$ from its diffusion coefficient in free solution.

We used the above strategy to predict the rate of $H^+$ diffusion through the paracellular space (Fig. 4). The ratio of $H^+$ to mannitol was calculated, using the $H^+$ concentration in the luminal solution which, when titrated to pH 4.7 during acid secretion, is $10^{-4.7}$ M (or 1.99 x

$10^{-5}$ M). The ratio, $D^H$ (9.311 x $10^{-5}$ $cm^2$/sec) x 0.0199 mM $H^+$ / $D^{man}$ (0.677 x $10^{-5}$ $cm^2$/sec) x 19.3 mM mannitol, is 0.0142. We plotted the data for mannitol flux in Fig. 3*A* with a slope of 0.0142 and then extrapolated values for $H^+$ flux from the Y-axis in Fig. 4*A*. Since there were 6 experiments done with DMSO- and 7 experiments with CD-treated tissues, the data were extrapolated and then averaged for each group. The predicted mean rate of $H^+$-flux was 12.5 ± 4.9 (x$10^{-4}$) μEq $hr^{-1}cm^{-2}$ in the presence of DMSO and 39.6 ± 2.91 (x$10^{-4}$) μEq $hr^{-1}cm^{-2}$ in the presence of CD/DMSO (Fig 4*B*). It should be noted that mannitol flux in inhibited tissues is significantly greater than in stimulated tissues (Table 1). Thus, our predicted rates of $H^+$-flux through the paracellular space represent an over-estimate of the true rate of $H^+$-flux in stimulated tissues.

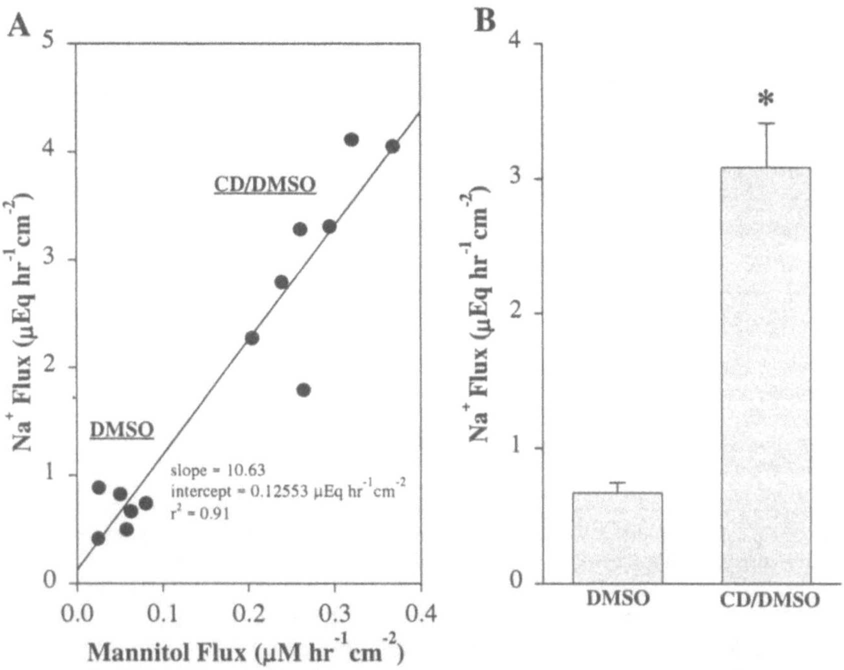


**Figure 3.** $Na^+$ back-diffusion calculated from the rate of mucosal-serosal $Na^+$ and mannitol flux across the paracellular space. *A*: frog fundic mucosa was inhibited with omeprazole and then switched to closed-circuit conditions in the presence of 0.1% DMSO (DMSO) or 20 μM CD in 0.1% DMSO (CD/DMSO). $Na^+$ and mannitol fluxes were determined simultaneously with dual tracer techniques. Data fit a line with a slope of 10.63 ($r^2$=0.91), which indicates that the increase in ion permeability during treatment with CD/DMSO is by passive diffusion across the paracellular space. *B*: the measured rates of $Na^+$ flux in 6 experiments with DMSO and 7 experiments with CD/DMSO (from Fig. 3*A*) were averaged. In the presence of CD, there is a significant increase in $Na^+$ flux across the paracellular space. The data represent means ± SEM. * Denotes a significant difference ($P < 0.005$). (With permission, Al-Shaibani and Hagen, 2001).

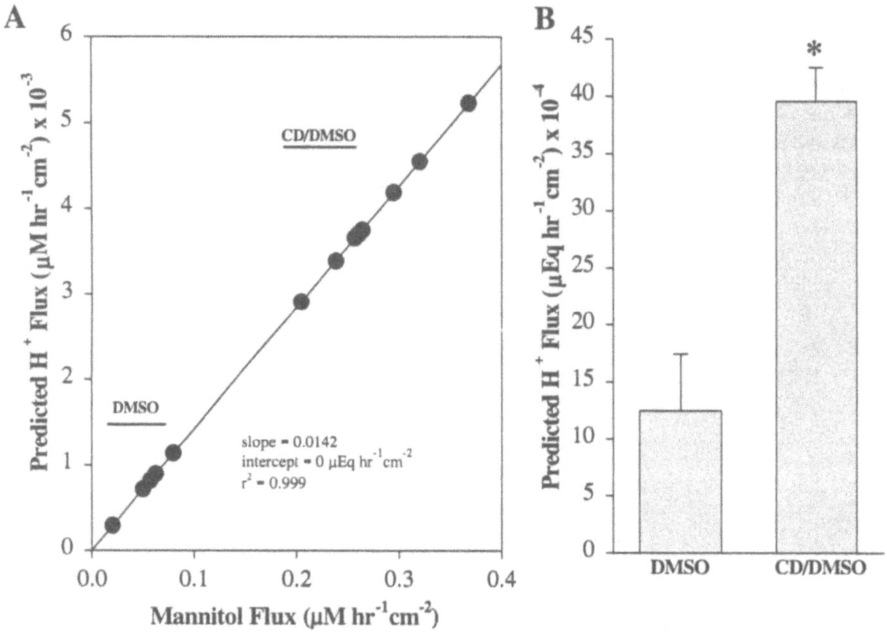


**Figure 4.** $H^+$ back-diffusion predicted from the rate of mucosal-serosal $H^+$ and mannitol flux across the paracellular space. *A*: the slope of 0.0142 corresponds to the ratio of the free-solution diffusion coefficients for $H^+$ and mannitol corrected for the concentration in the luminal solution. The data for mannitol flux (Fig. 3*A*) was plotted with a slope of 0.0142 to determine the predicted $H^+$ concentration. *B*: the predicted rates of $H^+$ flux in 6 experiments with DMSO and 7 experiments with CD/DMSO (from Fig. 4*A*) were averaged. The results show that a significant increase in $H^+$ flux across the paracellular space would be predicted to occur in the presence of CD/DMSO. The data represent means ± SEM. * Denotes a significant difference ($P < 0.005$). (With permission, Al-shaibani and Hagen, 2001).

## DISCUSSION

The present study shows that disruption of actin filaments with cytochalasin D (CD) significantly inhibits acid secretion in the frog gastric mucosa. After incubation with CD, acid secretion decreased from 6.46 ± 0.06 µEq hr⁻¹cm⁻² to 1.69 ± 0.54 µEq $hr^{-1}cm^{-2}$ and was accompanied by a significant increase in passive ion flux across the tight junction. We show here, however, that the increase in ion flux with CD would not result in significant acid depletion, either by titration with $HCO_3^-$ (serosal to mucosal flux) or by back-diffusion of $H^+$. In fact, the CD-induced increase in $HCO_3^-$ flux would reduce acid secretion (by 0.34 ± 0.03 µEq $hr^{-1}cm^{-2}$) from 6.46 to 6.12 µEq $hr^{-1}cm^{-2}$. Similarly, the CD-induced increase in $H^+$ back-diffusion would reduce acid secretion (by 0.00396 ± 0.00029 µEq $hr^{-1}cm^{-2}$) from 6.12 to 6.116 µEq $hr^{-1}cm^{-2}$. Thus, our results demonstrate that disruption of actin filaments with CD has a specific inhibitory effect on gastric acid secretion, which cannot be explained by an increase in passive ion flux across the tight junction.

Our results, using CD, agree with previously reported findings in whole tissue preparations in which CE or CB inhibited acid secretion in the rat and piglet gastric mucosa, respectively (4, 11). In addition, our results are consistent with those where CB, CE, or CD inhibited acid secretion in isolated rabbit gastric glands (7, 9). However, our results are in contrast to studies in the frog gastric mucosa, where CB had no effect on acid, or pepsinogen, secretion (14). It is not known why our results differ from other studies in the same species. Perhaps our method of stripping the submucosa from the muscularis mucosa (13) facilitates the penetration of CD in the *in vitro* preparation. Alternatively, CD may have a different effect on the frog gastric mucosa than does CB. Experiments done by us to address these issues provided the following results. First, stripping the submucosa significantly influenced the rate of stimulated acid secretion. Acid secretion was $2.14 \pm 0.28$ $\mu$Eq $hr^{-1}cm^{-2}$ in unstripped tissues, which is reduced considerably from the rate of acid secretion in stripped tissues as described in Fig. 1. Without the mucosa stimulated maximally, it is difficult to assess the effects of CD or CB on stimulated acid secretion. Second, CB reduced the rate of stimulated acid secretion more in stripped (86%) than in unstripped (55%) tissues (data not shown). These results demonstrate that both CD and CB inhibit stimulated acid secretion in the frog gastric mucosa, but that drug effectiveness is dependent on tissue preparation.

We show here that TER in the gastric mucosa is influenced by the rate of passive paracellular movement of ions, consistent with studies using CD in the intestine (20) or with chemical hypoxia in T84 cells (22). In the stimulated gastric mucosa (frog), previous studies have shown that TER is low (80-100 Ohm $cm^2$) due to a high rate of net $Cl^-$and $Na^+$ flux across the mucosa (10). In contrast, previous studies have shown that TER is high (400-600 Ohm $cm^2$) in the inhibited gastric mucosa (frog) due to a significant decrease, when compared to stimulated tissues, in net $Cl^-$ and $Na^+$ flux across the mucosa (10). In both cases, an assumption was made that the tight junction has similar permeability properties so that it is not a factor in determining TER. When the mucosa is treated with CD, however, we demonstrate a significant increase in paracellular permeability in both stimulated and inhibited tissues. In inhibited tissues, this increase in passive ion permeability fully accounts for the decline in TER, demonstrating that TER is dependent, in part, on mucosal permeability in the gastric mucosa. In stimulated tissues, treatment with CD both inhibits acid secretion (which should increase TER) and increases paracellular permeability (which should decrease TER). It is not clear, therefore, why TER remains low during treatment with CD. It is possible that the rise in TER (when acid secretion is inhibited with CD) offsets the decline in TER (when paracellular permeability increases in the presence of CD). Alternatively, $H^+$ transport may be inhibited in the presence of CD without a concomitant decrease in acid (secretion)-mediated $Cl^-$ and $Na^+$ flux. Further experiments will be required to understand, in detail, the effects of CD on TER in the stimulated gastric mucosa. In addition, studies to determine the properties of tight junctions in inhibited and stimulated tissues will be required to verify that permeability is similar in both conditions.

Even though CD influences mucosal permeability and TER in this study, we show that there are no gross alterations in tight junction structure as determined by electron microscopy. These results are consistent with studies by Unno, et al, who showed that ATP depletion induced by hypoxia increased permeability and changed the localization of actin at the tight junction, but had no effect on tight junction structure as determined by electron microscopy (29). Madara and Daharmsathaphorn (12) showed that strand counts at the tight junction, as determined by freeze fracture microscopy, are predictive of the TER and rate of passive ion flux. CD reduces the number of strands associated with the tight junction, in concert with a reduction in TER and an increase in paracellular permeability (20). Electron microscopic examination of the tight junction, however, has been beneficial to determine abnormalities in tight junction structure including contraction of the perijunctional actomyosin ring, an increase in perijunctional actin, and development of membrane dilatations (16, 21).

The results presented here show that the rate of mannitol flux is greater in inhibited than in stimulated tissues. Although the reason for this difference is not known, we propose that the "secretory flush", or the serosal to mucosal movement of water, makes it more difficult for ions to diffuse in the opposite direction (from mucosal to serosal) during acid secretion. Alternatively, there may be distinct differences in the structure of tight junctions in inhibited and stimulated tissues. Such differences could facilitate the regulation of ion permeability during acid secretion.

The mechanism by which actin filaments regulate gastric acid secretion is not known. Forte, et al suggest that actin filaments participate in vesicular trafficking and the translocation of tubulovesicular membranes to the apical surface after stimulation (9, 25). Chew et al showed that signaling proteins required for acid secretion are associated with F-actin (6). Actin filaments are thought to participate in $Ca^{2+}$ mobilization after stimulation (23) and/or may serve to organize ion exchangers after stimulation (2). Since little direct experimental evidence has been obtained to identify the way in which actin filaments regulate gastric acid secretion, this should be a priority for further investigation.

## REFERENCES

1. **Al-Shaibani, T. and S. J. Hagen.** Regulation of acid secretion and paracellular permeability by F-actin in the bullfrog, *Rana catesbieana. Am. J. Physiol. GI and Liver Physiol.* 282: G519-G526, 2002.

2. **Bachmann, O., T. Sonnentag, W.-K. Siegel, G. Lamprecht, A. Weichert, M. Gregor, and U. Seidler.** Different acid secretagogues activate different $Na^+/H^+$ exchanger isoforms in rabbit parietal cells. *Am. J. Physiol.* 275: G1085-G1093, 1998.

3. **Bentzel, C. J., B. Hainau, S. Ho, S. W. Hui, A. Edelman, T. Anagnostopoulos, and E. L. Benedetti.** Cytoplasmic regulation of tigh-junction permeability: effect of plant cytokinins. *Am. J. Physiol.* 239: C75-C89, 1980.

4. **Black, J. A., T. M. Forte, and J. G. Forte.** The effects of microfilament distrupting agents on HCl secretion and ultrastructure of piglet gastric oxyntic cells. *Gastroenterology* 83: 595-604, 1982.

5. **Chen, M. C., A. Chang, T. Buhl, M. Tanner, and A. H. Soll.** Apical acidification induces paracellular injury in canine gastric mucosal monolayers. *Am. J. Physiol.* 267: G1012-G1020, 1994.

6. **Chew, C. S., J. A. Parente, X. Chen, C. Chaponnier, and R. S. Cameron.** The LIM and SH3 domain-containing protein, lasp-1, may link the cAMP signaling pathway with

dynamic membrane restructuring activities in ion transporting epithelia. *J. Cell Sci.* 113: 2035-2045, 2000.

7. **Cuppoletti, J. and D. H. Malinowska.** Cytoskeletal determinants of control of gastric acid secretion. *Prog. Clin. Biol. Res.* 258: 23-37, 1988.

8. **Dawson, D. C.** Na and Cl transport across the isolated turtle colon: parallel pathways for transmural ion movement. *J. Membrane Biol.* 37: 213-233, 1977.

9. **Forte, J. G., B. Ly, Q. Rong, S. Ogihara, M. Ramilo, B. Agnew, and X. Yao.** State of actin in gastric parietal cells. *Am. J. Physiol.* 274: C97-C104, 1998.

10. **Forte, J. G. and W. W. Reenstra.** Chloride transport by the gastric mucosa. In: *Advances in Comparative and Environmental Physiology.* Berlin, Springer-Verlag, 1994, chapt. 13, p. 239-259.

11. **Glinsukon, T., W. Suvitayawat, S. Sahaphong, and P. Piyachaturawat.** Effects of cytochalasin E on $H^+$ and volume secretion in gastric fistula rats. *Toxicol. Lett.* 36: 95-103, 1987.

12. **Gopalakrishnan, S., N. Raman, S. J. Atkinson, and J. A. Marrs.** Rho GTPase signaling regulates tight junction assembly and protects tight junctions during ATP depletion. *Am. J. Physiol.* 275: C798-C809, 1998.

13. **Hagen, S. J., A. Yanaka, and R. Jansons.** Localization of brush border cytoskeletal proteins in grastric oxynticopeptic cells from the bullfrog *Rana catesbeiana. Cell Tissue Res* 275: 255-267, 1994.

14. **Kasbekar, D. K. and G. S. Gordon.** Effects of colchicine and vinblastine on in vitro gastric secretion. *Am. J. Physiol.* 236: E550-E555, 1979.

15. **Kovbasnjuk, O. N., U. Szmulowicz, and K. R. Spring.** Regulation of the MDCK cell tight junction. *J. Membrane Biol.* 161: 93-104, 1998.

16. **LeFevre, M. E., E. J. Gohmann, and W. S. Rhem.** An hypothesis for discovery of inhibitors of gastric acid secretion. *Am. J. Physiol.* 270: 613-618, 1964.

17. **Ma, T. Y., N. T. Hoa, D. D. Tran, V. Bui, A. Pedram, S. Mills, and M. Merryfield.** Cytochalasin B modulation of Caco-2 tight junction barrier : role of myosin light chain kinase. *Am. J. Physiol.* 279: G875-G885, 2000.

18. **Ma, T. Y., D. Hollander, L. T. Tran, D. Nguyen, N. Hoa, and D. Bhalla.** Cytoskeletal regulation of Caco-2 intestinal monolayer paracellular permeability. *J. Cell Physiol.* 164: 533-545, 1995.

19. **Madara, J. L.** Increases in guinea pig small intestinal transepithelial resistance induced by osmotic loads are accompanied by rapid alterations in absorptive-cell tight-junction structure. *J. Cell Biol.* 97: 125-136, 1983.

20. **Madara, J. L., D. Barenberg, and S. Carlson**. Effects of cytochalasin D on occluding junctions of intestinal absorptive cells: Further evidence that the cytoskeleton may influence paracellular permeability and junctional charge selectivity. *J. Cell Biol.* 102: 2125-2136, 1986.

21. **Madara, J. L., R. Moore, and S. Carlson**. Alteration of intestinal tight junction structure and permeability by cytoskeletal contraction. *Am. J. Physiol.* 253: C854-C861, 1987.

22. **Matthews, J. B., J. A. Smith, K. J. Tally, M. J. Menconi, H. Nguyen, and M. P. Fink**. Chemical hypoxia increases junctional permeability and activates electrogenic ion transport in human intestinal epithelial monolayers. *Surgery* 116: 150-158, 1994.

23. **Muto, Y., T. Nagano, M. Yamada, K. Mikoshiba, and T. Urushidani.** A proposed mechanism for the potentiation of cAMP-mediated acid secretion by carbachol. *Am. J. Physiol.* 280: C155-C165, 2001.

24. **Namikawa, T., K. Araki, and T. Ogata.** Localization of cytoskeletal filaments during membrane rearrangements in rat parietal cells stimulated with gastrin. *Arch. Histol. Cytol.* 61: 47-56, 1998.

25. **Okamoto, C. T. and J. G. Forte.** Vesicular trafficking machinery, the actin cytoskeleton, and $H^{+}$-$K^{+}$-ATPase recycling in the gastric parietal cell. *J. Physiol.* 532: 287-296, 2001.

26. **Sang, U. H., M. H. Saier, and M. H. Ellisman.** Tight junction formation in the establishment of intramemranous particle polarity in aggregating MDCK cells:effect of drug treatment. *Exp. Cell Res.* 128: 223-235, 1980.

27. **Stevenson, B. R. and D. A. Begg**. Concentration-dependent effects of cytochalasin D on tight junctions and actin filaments in MDCK epithelial cells. *J. Cell Sci.* 107: 367-375, 1994.

28. **Turner, J. R.** Putting the squeeze on the tight junction: understanding cytoskeletal regulation. *Semin. Cell Dev. Biol.* 11: 301-308, 2000.

29. **Unno, N., M. J. Menconi, A. L. Salzman, M. Smith, S. J. Hagen, Y. Ge, R. M. Ezzell, and M. P. Fink**. Hypermeability and ATP depletion induced by chronic hypoxia or glycolytic inhibition in Caco-$2_{BBe}$ monolayers. *Am. J. Physiol.* 270: G1010-G1021, 1996.

30. **Yao, X., C. Chaponnier, G. Gabbiani, and J. G. Forte**. Polarized distribution of actin isoforms in gastric parietal cells. *Mol. Biol. of the Cell* 6: 541-557, 1995.

31. **Yao, X., L. Cheng, and J. G. Forte**. Biochemical characterization of ezrin-actin interaction. *J. Biol. Chem.* 271: 7224-7229, 1996.

# Fluid Dynamics of the Excretory Flows of Zymogenic and Mucinous Contents in Rat Gastric Glands Processed by High Pressure Freezing/Freeze Substitution, with Special Reference to the Secretory Dynamics of Phospholipase $A_2$ and Phospholipase Cγ1

AKIRA SAWAGUCHI[1] and TATSUO SUGANUMA[2]

*[1]Department of Pathology and [2]Department of Anatomy, Miyazaki Medical College, Miyazaki 889-1692, Japan*

High pressure freezing (HPF) technique is capable of freezing up to 0.2 mm thick biological specimens in a vitreous or microcrystalline state suitable for high-resolution structural analysis (2, 3). The application of HPF/freeze substitution onto histochemistry has provided excellent ultrastructures and antigenicities of not only the epithelial cells of rat gastric gland but also the intraluminal exocytosed contents. Recently, we have developed the potassium permanganate oxidation-uranyl acetate-lead citrate ($KMnO_4$-UA/Pb) staining which gives a differential staining of the zymogen granules of rat gastric chief cells on Lowicryl K4M ultrathin sections prepared by HPF/freeze substitution (4). In the gastric glandular lumen, the $KMnO_4$-UA/Pb staining demonstrates the exocytosed zymogenic contents as electron opaque images and allows us to follow the secretory dynamics of the zymogenic contents in rat gastric glands. By taking these advantages, we histochemically investigated the fluid dynamics of the excretory flows of zymogenic and mucinous contents in rat gastric glands, with special reference to the secretory dynamics of phospholipase $A_2$ ($PLA_2$) and phospholipase Cγ1 (PLCγ1).

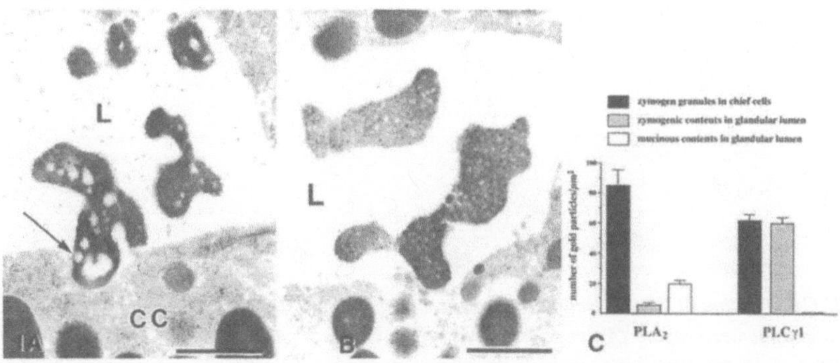


*Fig 1. The base region of rat gastric gland. Combination of $PLA_2$ immunolabeling (A), or PLCγ1 immunolabeling with $KMnO_4$-UA/Pb staining on Lowicryl K4M ultrathin sections prepared by HPF/freeze substitution. (A) The arrow indicates the zymogenic contents which have been just exocytosed. In the glandular lumen (L), the labeling for $PLA_2$ is dispersed in the electron lucent mucinous contents. (B) Note the intense PLCγ1 immunolabeling on the exocytosed zymogenic contents in the lumen. (C) Quantitative evaluations of the immunolabeling densities for $PLA_2$ and PLCγ1.* CC *chief cell. Bars = 1 μm.*

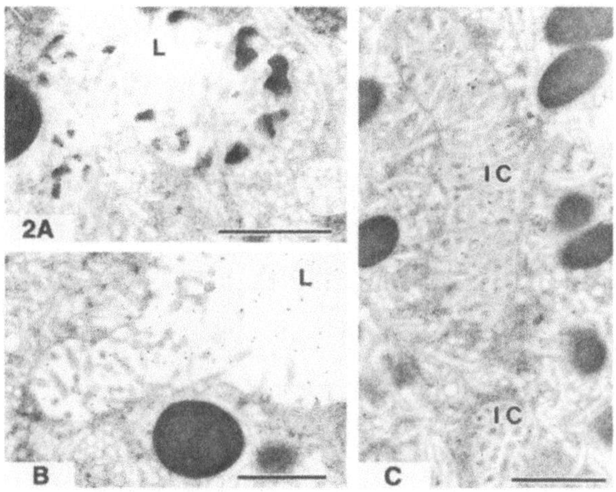


*Fig 2. Parietal cells of rat gastric gland. Combination of PLCγ1 immunolabeling (A), or $PLA_2$ immunolabeling (B, C) with $KMnO_4$-UA/Pb staining. (A) The PLCγ1 iimunolabeling is confined to the zymogenic contents among the microvilli. (B) Note the diffuse $PLA_2$ immunolabeling on the mucinous contents. Some of the gold particles are found on the microvilli. (C) The $PLA_2$ immunolabeling on the intracellular canaliculi (IC). The tubulovesicles were reactive with neither $PLA_2$ nor PLCgl immunolabeling.* L *lumen. Bars = 1 μm.*

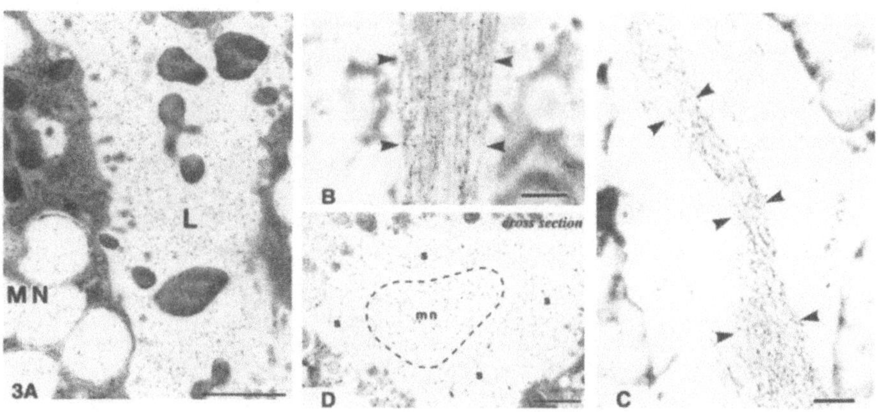


*Fig 3. Combination of GSA-II lectin labeling for mucous neck cell (MNC) mucin (1) and $KMnO_4$-UA/Pb staining at the neck region (A), and the pit region (B, C) of rat gastric gland. (A) Note the droplet-like appearance of the zymogenic contents in glandular lumen (L). (B) The appearance of exocytosed zymogenic contents changed into the several electron dense streaks within the MNC mucus channel labeled with GSA-II lectin. (C) The MNC mucus channel shows high electron density with $KMnO_4$-UA/Pb staining. (D) Dual labeling of cationic colloidal gold at pH 1.0 for sulfated mucin (5) (8 nm) and GSA-II lectin (14 nm) on a cross section of the pit region. Dashed lines indicate the boundary of MNC mucin compartment (mn).* MN *mucous neck cell,* S *sulfated mucin. Bars = 1 μm.*

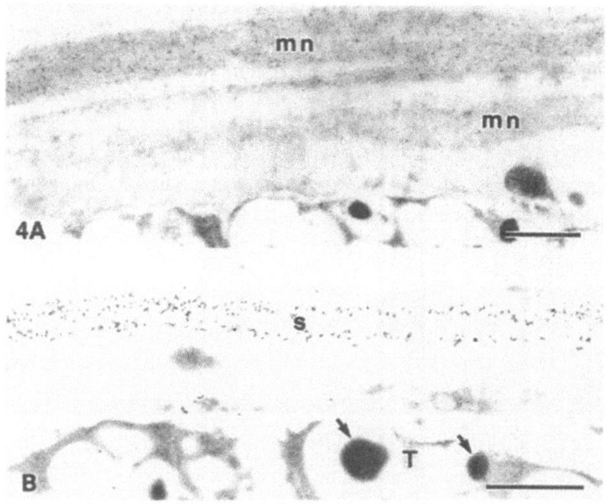


***Fig 4.*** *Laminated arrays in the surface mucous gel layer. (A) Combination of GSA-II lectin labeling and $KMnO_4$-UA/Pb staining. Note the high electron density of the MNC mucin laminae (mn). (B) Combination of cationic colloidal gold labeling at pH 1.0 and $KMnO_4$-UA/Pb staining demonstrates the sulfated mucin laminae (S). Arrows indicate the serous components of bipartite granules of top pit cell (T). Bars = 1 µm.*

## CONCLUSION

In conclusion, this study reveals the fluid dynamics of the excretory flows of zymogenic and mucinous contents in rat gastric glands. The present study first demonstrated the distinction between the $PLA_2$ and PLCγ1 in their fluid dynamics in the glandular lumen. Interestingly, some of the $PLA_2$ immunolabeling on mucinous contents was associated with the apical membranes (microvilli) of parietal cells. It could be speculated that the $PLA_2$ exerts the catalytic action on the apical membrane of parietal cells, but a further study is required to verify it. The HPF/freeze substitution of the rat gastric glands succeeded in preserving the fine structures of fluid contents, and demonstrating the intraluminal mucus channels in rat gastric glandular lumen.

## REFERENCES

1. **Ihida K, Suganuma T, Tsuyama S, Murata F.** Glycoconjugate histochemistry of the rat fundic gland using *Griffonia simplicifolia* agglutinin-II during the development. *Am J Anat* 182: 250-256, 1988.

2. **McDonald K.** High-pressure freezing for preservation of high resolution fine structure and antigenicity for immunolabeling. *Methods Mol Biol* 117: 77-97, 1999.

3. **Monaghan P, Perusinghe N, Müller M.** High-pressure freezing for immunocytochemistry. *J Microsc* 192: 248-258, 1998.

4. **Sawaguchi A, Ide S, Kawano J, Nagaike R, Oinuma T, Tojo H, Okamoto M, Suganuma T.** Reappraisal of potassium permanganate oxidation applied to Lowicryl K4M embedded tissues processed by high pressure freezing/freeze substitution, with special reference to differential staining of the zymogen granules of rat gastric chief cells. *Arch Histol Cytol* 62: 447-458, 1999.

5. **Yang DH, Kasamo H, Miyauchi M, Tsuyama S, Murata F.** Ontogeny of sulphated glycoconjugate-producing cells in the rat fundic gland. *Histochem J* 28: 33-43, 1996.

# INDEX

MIX
Papier aus verantwortungsvollen Quellen
Paper from responsible sources
FSC® C105338

If you have any concerns about our products, you can contact us on
**ProductSafety@springernature.com**

In case Publisher is established outside the EU, the EU authorized representative is:
**Springer Nature Customer Service Center GmbH**
**Europaplatz 3, 69115 Heidelberg, Germany**

Printed by Libri Plureos GmbH
in Hamburg, Germany